APPLIED GROUNDWATER STUDIES IN AFRICA

SELECTED PAPERS ON HYDROGEOLOGY

13

Series Editor: Dr. Nick S. Robins
Editor-in-Chief IAH Book Series
British Geological Survey
Wallingford, UK

INTERNATIONAL ASSOCIATION OF HYDROGEOLOGISTS

Applied groundwater studies in Africa

Editors

Segun Adelana
Department of Geology & Mineral Sciences, University of Ilorin, Nigeria

Alan MacDonald
British Geological Survey, Edinburgh, UK

Associate Editors

Tamiru Alemayehu Abiye
School of Geosciences, University of the Witwatersrand, Johannesburg, South Africa

Callist Tindimugaya
Ministry of Water and Environment, Entebbe, Uganda

CRC Press
Taylor & Francis Group
Boca Raton London New York

CRC Press is an imprint of the
Taylor & Francis Group, an **informa** business
A BALKEMA BOOK

British Geological Survey
NATURAL ENVIRONMENT RESEARCH COUNCIL

DFID Department for International Development

Published by:
CRC Press/Balkema
P.O. Box 447, 2300 AK Leiden, The Netherlands
e-mail: Pub.NL@taylorandfrancis.com
www.crcpress.com – www.taylorandfrancis.com

ISBN 13: 978-0-415-45273-1 (hbk)

Typeset by Charon Tec Ltd (A Macmillan Company), Chennai, India

Library of Congress Cataloging-in-Publication Data

Applied groundwater studies in Africa / edited by Segun Adelana, Alan MacDonald.
p. cm. — (IAH selected papers on hydrogeology ; vol. 13)
Includes bibliographical references and index.
ISBN 978-0-415-45273-1 (hbk : alk. paper)—ISBN 978-0-203-88949-7 (ebook : alk. paper) 1. Groundwater—Africa. 2. Water resources development—Africa. 3. Water-supply, Rural—Africa. I. Adelana, Segun. II. MacDonald, Alan M. III. Title. IV. Series.

GB1161.A67 2008
333.91'04096—dc22

2008025361

Table of contents

Preface

This book provides an overview of the state of knowledge of Africa's most precious natural resource: groundwater. The volume highlights the complexity and variety of issues surrounding the development and management of groundwater resources across Africa, and provides a snapshot of groundwater research and application in the early 21st century.

The volume originates from the IAH Sub-Saharan Africa group and the IAH Burdon Network (a network of professionals committed to the sustainable development of groundwater to help achieve the Millennium Development Goals). Discussions from within the groundwater community highlighted that there was a great desire to learn lessons from projects and studies across Africa, but little opportunity to do so. Also, for those outside Africa, it was difficult to get a picture of what work was being done, who the current researchers were, and what the current understanding of groundwater resources in Africa was. This volume was conceived to meet this need. People were invited to submit papers from recent projects or research in Africa – particularly from an applied perspective of developing or managing groundwater. The result is this volume of 29 papers, which range from strategic discussions of the role of groundwater in development and poverty reduction, to case studies on techniques used to develop groundwater, and numerical modelling of larger aquifers.

The editors would like to thank those who acted as referees for the papers: Osman Abdalla (Oman), Thomas Armah (Ghana), Peter Bauer-Gottwein (Denmark), Colin Booth (USA), Richard Carter (UK), Antonio Chambel (Portugal), John Chilton (UK), Jude Cobbing (Republic of South Africa), George Darling (UK), Jeff Davies (UK), Boyd Dent (Australia), AM Ebraheem (United Arab Emirates), Serigne Faye (Senegal), Wolfgang Gossel (Germany), Bi Goula (Côte d'Ivoire), Nyambe Imasiku, (Zambia), M Israil (India), GA Kazemi (Iran), Jiri Krasny (Czech Republic), David Kreamer (USA), Kwabena Kankam-Yeboah (Ghana), Bruce Misstear (Ireland), Benjamin Ngatcha (Cameroon), Daniel Nkhuwa (Zambia), Marie-Solange Oga (Côte d'Ivoire), Stephen Opoku-Duah (UK), Paul Pavelic (Australia), SN Rai (India), Kornelius Riemann (Republic of South Africa), Rebecca Schneider (USA), EM Shemang (Botswana), Semere Solomon (Sweden), Amin Shaban (Lebanon), John Sharp (USA), Mustapha Thomas (Sierra Leone), and Petr Vrbka (Germany).

We would also like to thank Stephen Foster and Andrew Skinner of the IAH Executive for their encouragement – particularly during the early stages of the project. Gill Tyson helped redraw and improve many of the illustrations, the UK Department for International Development paid for the book to be printed in colour and the British Geological Survey helped support the editing process. Many thanks to Lucinda Mileham for compiling the index.

We hope that this volume will act as a spring board for developing further collaborative groundwater research that is appropriate to Africa and African issues; and in doing so, the groundwater resources can be better understood and managed for the benefit of all people throughout Africa.

CHAPTER 1

Groundwater research issues in Africa

S.M.A. Adelana
Geology Department, University of Ilorin, Ilorin, Nigeria

A.M. MacDonald
British Geological Survey, Edinburgh, UK

ABSTRACT: Poverty reduction and economic growth drive the development of groundwater resources across Africa. Due to the ephemeral nature of surface water, groundwater abstraction is often the only realistic and affordable means of providing reliable water supply for much of Africa's needs. However, the large variability in geological and hydrological conditions have a profound influence on the availability of groundwater across the continent, and the sustainable development of the resource depends on an accurate understanding of the hydrogeology and the availability of skilled people to make informed decisions. Despite the obvious need for data, little attention has been paid to the systematic gathering of information about groundwater resources, with the result that data are patchy, knowledge is limited and investment is often poorly targeted. Given the scale of groundwater development in Africa, there is now a pressing need to take groundwater resources seriously and provide a framework and funds for applied groundwater research across Africa to underpin current and future development and management of this precious resource. This framework must recognise the issues of particular concern to Africa, such as: the requirement to develop sustainable and cost effective community water supplies across all hydrogeological environments (even challenging ones); appropriately managing and protecting groundwater resources given the rate of rapid poorly planned urbanisation and the expansion in on-site sanitation and; the imperatives of water security, from household to national levels, with unpredictability of future climate, groundwater recharge and water demand.

1 INTRODUCTION

Groundwater is Africa's most precious natural resource, providing reliable water supplies for more than 100 million people and, potentially, millions more. Groundwater has many advantages as a source of supply, particularly where populations are still largely rural and demand is dispersed over large areas. In particular, natural groundwater storage provides a buffer against climatic variability; quality is often good, and infrastructure is affordable to poor communities. Sustainable development of the resource is not a trivial task, however, and depends crucially on an understanding of groundwater availability in complex environments, and the processes through which groundwater is recharged and renewed.

Groundwater occurrence depends primarily on geology, geomorphology/weathering and rainfall (both current and historic). The interplay of these three factors gives rise to complex hydrogeological environments with countless variations in the quantity, quality, ease of access and renewability of groundwater resources. Development of the resource therefore depends on an accurate understanding of hydrogeology, and *people* with the skills to make informed decisions on how groundwater can best be developed and managed in a sustainable fashion. Despite these obvious needs, however, little attention has been paid to the systematic gathering of information about groundwater resources, with the result that data are patchy, knowledge is limited and investment is poorly targeted.

This volume highlights the complexity and variety of issues surrounding the development and management of groundwater resources across Africa, and provides a snapshot of groundwater research and application in the early 21st century. Chapters range from strategic discussions of the role of groundwater in development and poverty reduction, to case studies on techniques used to develop groundwater. Some of the issues surrounding the current status of research in Africa arising from the papers in this volume are summarized below, and a roadmap for future research is outlined.

2 AFRICAN GROUNDWATER RESOURCES

Africa has huge diversity in geology, climate and hydrology. As a result, the hydrology of Africa is probably the most variable and challenging of all populated continents (Walling, 1996). For example, annual rainfall varies from negligible over parts of the Sahara, to almost 10,000 mm in the Gulf of Guinea; continent-wide runoff (153 mm) and rainfall/runoff coefficient (0.21) are considerably lower than for any other continent; and southern Africa has the highest variation in mean annual runoff, the greatest flood variability and the highest extreme flood index of any major region (McMahon *et al.*, 1992). The great variability in rainfall, and in particular the long dry season (>5 months) over much of Africa, increases reliance on groundwater storage for water supply, providing security against dry season scarcity and longer-term drought.

Geological and hydrological variability have a profound influence on groundwater conditions. Roughly 34% of the land surface is underlain by heterogeneous Precambrian basement; 37% by consolidated sedimentary rocks; 25% by unconsolidated sediments; and 4% by volcanic rocks (MacDonald *et al.*, 2008). Some rocks form highly productive aquifers, for example the large sedimentary basins of northern Africa where porosity can exceed 20%, and permeability is sufficient to allow development of high yielding boreholes. However, many other rocks types, such as the less weathered Precambrian basement, or mudstones, are poorly yielding and groundwater may be difficult to find, or non-existent. Figure 1 shows a simplified map of the groundwater resources of Africa.

Groundwater is generally of good quality, and is largely protected from pathogenic contamination by the natural filtration of rocks and soil. However, little is known of the groundwater chemistry across much of the continent. The chemistry of the water is variable since it is largely determined by geochemical processes that take place as recharge infiltrates the ground and reacts with rock-forming minerals. In some environments, harmful concentrations of elements can occur, notably elevated concentrations of arsenic or fluoride.

Groundwater resources are often, but not always, renewable. Some of the major aquifers in the north of Africa do not receive significant present day recharge, but rely on rainfall

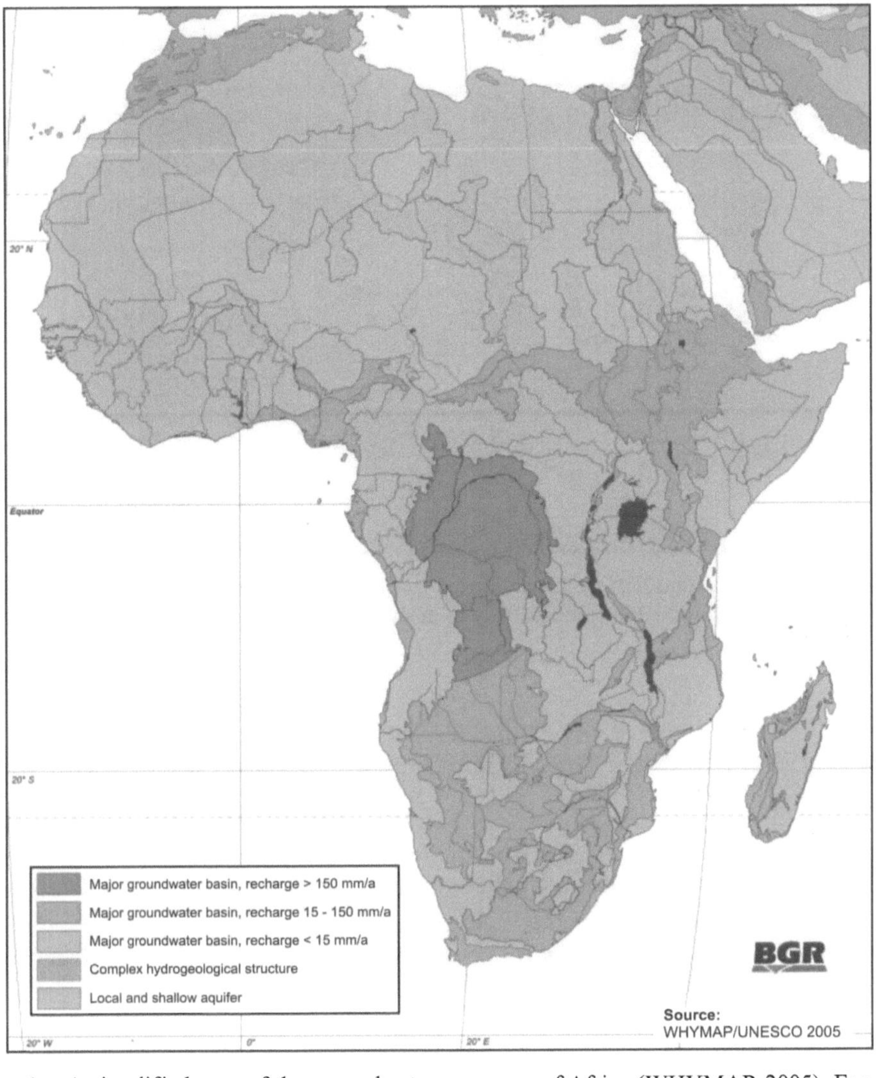

Figure 1. A simplified map of the groundwater resources of Africa (WHYMAP, 2005). For more information see Struckmier (2008).

from 5000–10,000 years ago (Edmunds, 2008). Active recharge does occur in many areas, however, and has been shown to occur in areas with annual rainfall as low as 200 mm. Recharge is a complex process and depends on many factors, including the intensity of rainfall events, and soil and aquifer conditions preceding such events. As a consequence, recharge is poorly constrained making it difficult to predict the effect of future climate scenarios.

3 GROUNDWATER DEVELOPMENT

Economic development and poverty reduction imperatives drive the development of groundwater resources across Africa. At least 320 million people in the continent still have no access

Figure 2. At least 320 million people in Africa lack access to safe water supplies. Developing groundwater resources is the only realistic way of meeting this need across Africa.

to safe water supplies, and at least 80% of these people live in rural areas (see Figure 2). In view of the ephemeral nature of many surface water bodies, and the need to develop water sources close to or within communities, groundwater development is often the only realistic and affordable means of meeting coverage targets.

How can these targets be met, and what are the main challenges in Africa? The chapters that follow (this volume) provide a number of insights, and can be summarised as follows:

1. In most areas, groundwater is the first option for water supply, and groundwater exploration and development have to take place in challenging environments – environments which many would write-off as non-aquifers. For example, 70 million people may need to rely on groundwater development in weakly permeable mudstone areas.
2. The rapid development of groundwater needed to achieve the Millennium Development Goals for water means that simple, yet fit for purpose, techniques for groundwater exploration and development must be used. Ideally, such techniques should be reliable, cost-effective and capable of being applied by local government and project staff.
3. Unchecked development of groundwater and lack of testing may mean that water quality problems remain undetected until health problems emerge. For example, quality issues associated with elevated natural fluoride concentrations have caused widespread problems in the East African rift valley.
4. In the absence of government regulation and enforcement, particularly in rural areas, communities have to devise and monitor their own rules governing groundwater use and source protection. This is likely to become increasingly problematic with the rapid increase in the use of latrines for on-site sanitation.
5. Government is unable, or unwilling, to meet the full costs of developing groundwater and maintaining infrastructure. Groundwater users – both urban and rural – now have

to meet more of the costs of service provision, so supplies need to be sustainable and affordable.

6. Climate change, and the predictions of less reliable, more extreme rainfall across parts of Africa is likely to impact groundwater resources as people seek more predictable water supplies. Demand for groundwater resources is likely to increase markedly, and in some marginal areas, recharge to groundwater may decline.

Groundwater development may also underpin urban and industrial development. However, developing and managing groundwater resources in urban Africa presents its own challenges:

1. Rapid, unplanned urbanisation can lead to the contamination of groundwater resources. There are many examples of well fields being abandoned and groundwater resources being grossly contaminated by rapid urbanisation and poor sewerage (Adelana *et al.*, 2008).
2. Increased demand for groundwater from urban centres has led to overexploitation within urban areas, and in the peri-urban or rural areas beyond them. Symptoms include rapid water-level decline, declining yields and a deterioration in water quality due to salt water intrusion.
3. Industrial sites (including those associated with oil exploration and production) can also threaten groundwater quality. Poor environmental regulation and control can lead to contamination of important aquifers, which may threaten the water supply of people in nearby communities.
4. The use of groundwater to supply rapidly growing small town supplies brings with it a new set of issues. Higher yielding, well protected boreholes are necessary, but require a greater level of expertise and exploration to site and develop, and effective methods to manage.

Groundwater is not yet used extensively for large scale irrigation in Africa, though small 'garden' irrigation is not accounted for in irrigation statistics. Foster *et al.* (2008) cite hydrogeological reasons (the crystalline basement will not support high yielding irrigation boreholes) and socio-economic factors (high capital cost, low levels of rural electrification, and lack of social tradition in irrigated crop cultivation). The challenge for Africa is to develop its irrigation potential – at a range of scales – and avoid the pitfalls of widespread groundwater degradation found in south-east Asia.

4 A WAY FORWARD FOR GROUNDWATER RESEARCH

It is clear that although groundwater supports social and economic development in Africa (particularly in rural areas), the resource is not properly understood. The papers within this volume consistently highlight the lack of systematic data and information on groundwater across Africa. This needs to be rectified. Groundwater studies have occurred on an *ad hoc* basis where resources have allowed researchers to follow an issue of interest. Data are scarce and to a large extent have not been gathered with any rigour over the last two decades. The reasons behind this change are complex (Robins *et al.*, 2006) and include unwanted impacts from decentralisation and rationalisation of government tasks, and the lack of clear demarcation of responsibilities among the various institutions involved in service delivery.

There is a pressing need for international donors and research funding bodies to take groundwater seriously, and provide a framework and funds for future research. The papers in this volume point a way forward:

- For rural water supply, identify how groundwater resources exist in difficult areas not normally taken to be aquifers and develop appropriate but effective techniques for the widespread development of groundwater.
- Groundwater quality must also be taken seriously. There is a critical need for research into the widespread contamination of groundwater resources from sanitation practices, and also more information required on the distribution of harmful substances (such as arsenic and fluoride) in groundwater.
- There must be systematic collection of information on the quality and quantity of groundwater resources to underpin groundwater management, particularly in a changing future. Groundwater models should be developed to test our understanding of the systems, where data allow.
- Management strategies must be developed that are appropriate for Africa, and which leave room for community management, while recognising the larger scale issues such as transboundary and shared aquifers.
- The current recharge and replenishment of groundwater must be known in detail across different aquifer types and climate zones. Only then can it be possible to predict how groundwater may be affected by future climates.
- Interdisciplinary research is needed that will inform understanding of the wider issues of sustainability and in particular the role that reliable groundwater sources will play in providing water security and improving livelihoods.

By learning the lessons of research in other continents, and by developing new research topics appropriate to Africa and African issues, the groundwater resources can be better understood and better managed for the benefit of all people throughout Africa.

ACKNOWLEDGMENTS

This paper is published with the permission of the Executive Director of the British Geological Survey (NERC). Thanks to Roger Calow and Nick Robins for the helpful comments on the document, and to Dr Struckmeier for permission to use the map in Figure 1.

REFERENCES

Adelana, S. M. A., Tamiru, A., Nkhuwa, D. C. W., Tindimugaya, C. & Ogam, M. S. 2008. *Urban groundwater in Sub-Saharan Africa*. In: Adelana, S. M. A. & MacDonald, A. M. Applied Groundwater Studies in Africa. IAH Selected Papers on Hydrogeology, Volume 13, CRCPress/Balkema, Leiden, The Netherlands.

Edmunds, W. M. 2008. *Groundwater in Africa – Palaeowater, climate change and modern recharge*. In: Adelana, S. M. A. & MacDonald, A. M. Applied Groundwater Studies in Africa. IAH Selected Papers on Hydrogeology, Volume 13, CRCPress/Balkema, Leiden, The Netherlands.

Foster, S. S. D., Tuinhof, A. & Garduño, H. 2008. *Groundwater in Sub-Saharan Africa – A strategic overview of developmental issues*. In: Adelana, S. M. A. & MacDonald, A. M. Applied Groundwater

Studies in Africa. IAH Selected Papers on Hydrogeology, Volume 13, CRCPress/Balkema, Leiden, The Netherlands.

MacDonald, A. M., Davies, J. & Calow, R. C. 2008. *African hydrogeology and rural water supply.* In: Adelana, S. M. A. & MacDonald, A. M. Applied Groundwater Studies in Africa. IAH Selected Papers on Hydrogeology, Volume 13, CRCPress/Balkema, Leiden, The Netherlands.

McMahon, T. A., Finlayson, B. L., Haines, A. & Stikanthan, R. 1992. Global runoff: Continental comparisons of annual flows and peak discharges. Cremlingen-Destedt, Germany.

Robins, N. S., Davies, J., Farr, J. L. & Calow, R. C. 2006. *The changing role of hydrogeology in semi-arid southern and eastern Africa.* Hydrogeology Journal 14, 8: 1483–1492.

Struckmeier, W. 2008. *Hydrogeological mapping in Africa.* In: Adelana, S. M. A. & MacDonald, A. M. Applied Groundwater Studies in Africa. IAH Selected Papers on Hydrogeology, Volume 13, CRCPress/Balkema, Leiden, The Netherlands.

Walling, D. E. 1996. *Hydrology and rivers.* In: Adams, W. M., Goudie, A. S., Orme, A. R. (eds), The physical geography of Africa, Oxford University Press, Oxford, UK, 103–121.

WHYMAP 2005. Groundwater resources of the world 1: 50,000,000. available at: http://www.whymap.org.

CHAPTER 2

Groundwater in Sub-Saharan Africa: A strategic overview of developmental issues

S. S. D. Foster*, A. Tuinhof & H. Garduño
World Bank (Groundwater Management Advisory Team), Washington DC, USA
**also serving as IAH President 2004–08*

ABSTRACT: This overview of groundwater development in Sub-Saharan Africa is based on the authors' cumulative experience, tempered by assessment of the readily-available published and grey literature, and by consultation with certain 'key players' and by dialogue at the IAH Conference on 'Groundwater for Poverty Reduction in Africa' prior to the G8 Summit in London on 29 June 2005. A systematic assessment of the status of groundwater development and use is not attempted because of the patchy nature of available information. But many countries have urgent need of substantial investment in institutional strengthening to improve management and planning of the groundwater development process. More effective use of available groundwater resources will reduce poverty and achieve enhanced livelihoods.

1 INTRODUCTION – A VITAL RESOURCE FOR HUMAN LIFE AND LIVELIHOOD

Groundwater is the critical underlying resource for human survival and economic development in extensive drought-prone areas of south-eastern, eastern and western Africa – especially where the average rainfall is less than 1000 mm/a. These areas (and more than 70% of the Sub-Saharan land area as a whole) (Figure 1) are extensively underlain by two broad aquifer classes (Foster, 1984; Wright & Burgess, 1992; Adams *et al.*, 1996; MacDonald *et al.*, 2005):

- weathered crystalline basement forming a shallow, patchy, minor aquifer system of low storage;
- consolidated sedimentary rocks which form generally deeper, but less extensive and geologically more complex, aquifers.

Traditionally throughout the Sub-Saharan Africa Region it was the accessibility of groundwater through dugwells, at springheads and in seepage areas that controlled the extent of human settlement beyond the major river valleys and riparian tracts – and this groundwater was usually developed through community initiative.

The introduction of deep drilling and pumping machinery from the 1970s enabled the area under groundwater exploitation and human settlement to be extended in response to

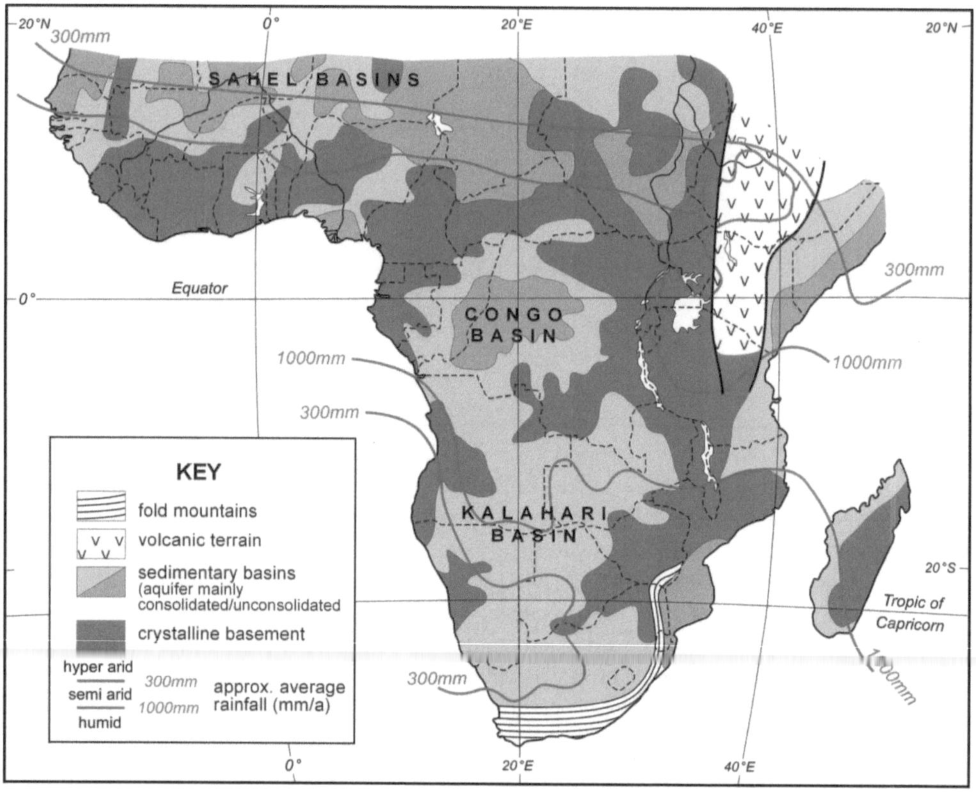

Figure 1. Simplified hydrogeological map of Sub-Saharan Africa.

increasing population and growing pressure on riparian land. Today, over very large rural land areas it is only the presence of successful boreholes equipped with reliable pumps that allow the functioning of settlements, clinics, schools, markets and livestock posts – and failure to construct and/or sustain such boreholes directly impacts, in a number of ways, on the prospects for achievement of the UN-Millennium Development Goals (MDGs).

No reliable, comprehensive, statistics on groundwater use in Sub-Saharan Africa are known to exist, but there is known to be very high dependence for domestic water-supply, rural livelihoods and livestock rearing, and increasingly for urban water supply at a range of scales. Locally there are also important examples of use for industrial and tourist development. In addition, groundwater has been a critical resource for the economical development of some major mining activity and groundwater drainage has been an issue for other mines. Although the socioeconomic effects of such mines is large, their hydrogeological impacts are relatively local.

Although the distribution of aquifers is now reasonably mapped over large areas, more quantitative information on aquifer characteristics and recharge rates, and groundwater flow regimes, abstraction rates and quality controls is very uneven and generally incomplete (UN-FAO, 2003). Moreover, regional estimates of potential groundwater resources based on analysis of climatic data alone are of little meaning. What is clear, however, is that Africa is subject to wide temporal and spatial rainfall variability, and this coupled with

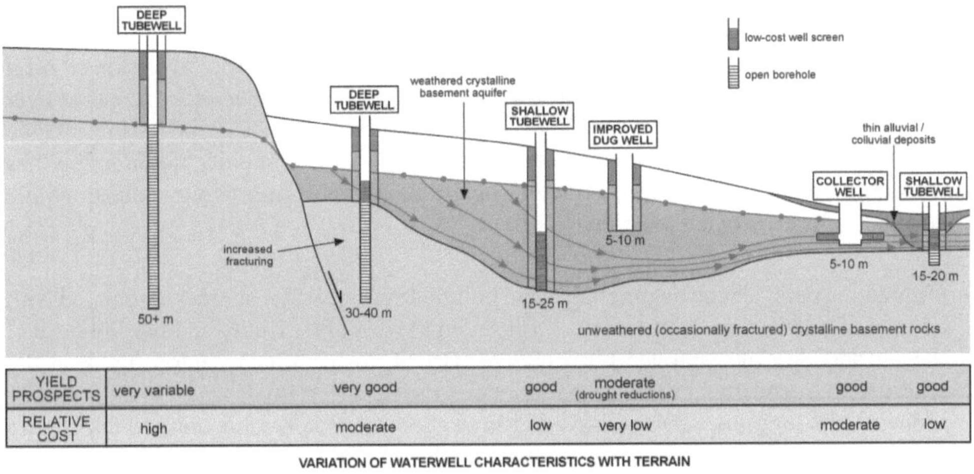

Figure 2. Harmonization of borehole and well design with local hydrogeological conditions in crystalline basement terrain.

widely-distributed superficial deposits of limited infiltration capacity tends to lead to relatively low and/or uncertain rates of groundwater replenishment.

A further generic issue which has impeded groundwater development is the high cost of borehole construction (compared for example to India and China). Although there is no such thing as a 'typical well', costs of more than US$ 120 per metre for drilling deeper (>50 m) wells are now widely reported and even shallow (<30 m) wells often cost in excess of US$ 3000. The cause of these high costs is complex, but the following factors come into play:

- lack of economy of scale and competition in borehole construction, due to absence of a large private-sector market for domestic and irrigation boreholes;
- high excise duties on imported drilling and pumping equipment, and no significant local manufacture even of spares;
- corruption in the letting and execution of borehole drilling contracts;
- inappropriate well design and excessive drilling depth for some hydrogeological conditions (Figure 2), with insufficient use of low-cost technology options.

The latter is not a new observation (Chilton & Foster, 1995) and indeed was already evident in Malawi and Zimbabwe during the UN Water Supply & Sanitation Decade – but it appears to have been overlooked in recent years due to lack of information dissemination and reduced professional capacity.

2 CRITICAL ISSUES FOR FUTURE DEVELOPMENT

An important consideration from the outset is whether a generalised overview of developmental issues is realistic, given the considerable hydrogeological variability and wide socioeconomic diversity of the region – but on consideration it was judged valid to identify 'common issues, needs and approaches', whilst pointing out some specific examples and some obvious exceptions.

2.1 *Achieving improved rural water supplies*

Groundwater (from springs, boreholes and dugwells) is the raw material of improved rural water supplies on a very widespread basis, with current level of dependency put at over 75%. This is the critical social function for groundwater resources – and its importance cannot be overstated because groundwater development for community water-supply has far-reaching benefits in terms of reducing health hazard and improving socioeconomic opportunity. Here are two examples:

- eliminating dependence on unreliable and polluted surface water sources associated with a range of waterborne diseases which cause high levels of mortality and morbidity (e.g. the introduction of improved boreholes to part of Ghana was a central plank in the eradication of endemic guinea-worm infection during the 1990s);
- reducing the long distances walked and time spent on water collection, which should enable women to engage in more productive activities and children to attend school.

It is essential for groundwater resources to be developed further if the population served by 'minimum adequate water-supply coverage' is to be rapidly expanded, since groundwater remains the only economically-viable option for improving water-supply in rural areas for many African countries. Alternatives, such as surface water from rivers and ponds or rainwater collection, are less reliable and easily contaminated, whereas aquifers and boreholes have a substantial degree of natural protection from contamination and drought.

Programmes serving the basic health and livelihood needs of rural community units of 200–500 persons usually find it difficult to support capital costs in excess of US$ 3000 per well – which implies a need to keep costs down. However, drilling unnecessarily deep wells, increased well yield failure (due to tackling more difficult terrain without adequate hydrogeological information) (Foster *et al.*, 2002–04a) and inability to meet potable water quality standards (due to toxic or troublesome soluble constituents such as F, As, Mn, Fe, $MgSO_4$ or NaCl) (Foster *et al.*, 2002–04b) are greatly increasing the cost of rural water-supply provision in some countries.

During the UN Drinking Water & Sanitation Decade (1980s) it came to be assumed that small quantities of groundwater, adequate for rural water-supply, were everywhere readily available and community considerations should be the main criteria for well siting. But whilst not questioning the important role of community management in the sustainable operation and maintenance of groundwater-based rural water-supply facilities, there has been a **breakdown of this 'decade paradigm'** with increasingly high rates of borehole construction failure (due to insufficient yield and/or inadequate quality) in the areas of more complex and/or unfavourable hydrogeology that remain to be tackled (Foster *et al.*, 2000). The hydrogeological factors may, in some circumstances, be overriding and better use of both hydrogeological expertise and data are needed to overcome this type of problem (Foster *et al.*, 2002–04a; b).

Groundwater is also usually the preferred option for the provision of vital water supplies for refugee camps endeavouring to cope with (post) conflict situations. This presents a special set of demands for technological capacity and hydrogeological information which is not the primary focus of this paper, although groundwater resources play a central role in such situations.

Whilst it is recognised that groundwater resource potential alone does not equate with improved rural livelihoods, groundwater supply availability has been a key factor in:

- water-supply for livestock rearing on which there is a high-level of economic dependence (Farr *et al.*, 1982; McIntire *et al.*, 1992), with cattle and goats still widely representing both a banking mechanism and drought-coping strategy of innumerable rural communities
- village subsistence-level cropping – with groundwater use for the cultivation of vegetable gardens and seedlings to advance the date of maize/sorghum planting being critical to the improvement of food security at local scale (Adams & Carter, 1987; Carter & Howsam, 1994)
- water-supply for community industries – such as pottery, making bricks and other construction materials.

At the larger scale, economic studies have shown the greatly increased benefits to rural livelihoods that could be achieved by operating reservoir floodwater releases to augment the recharge of riparian aquifers in West Africa (Acharya, 2004).

2.2 *Coping with small-town water-supply crisis*

In Sub-Saharan Africa the rate of population growth in small towns (and evolution of larger villages into small towns) is very high (in many cases exceeding 5% per annum) – and both community-managed facilities and service-providing utilities alike are confronted by the major challenge of having to expand the number and size of their water-supply sources. Moreover, the per capita water demand in small towns will increase steadily with growing prosperity – and there is commonly no allowance for this in groundwater demand estimation.

The urban rate of population growth (and the fact that dugwells and shallow boreholes are more vulnerable to contamination from excessively high-density and/or inadequately-designed in-situ sanitation) is widely leading to the need to locate and develop boreholes with sufficiently large individual yields to support motorised pumps and supply reticulated water distribution systems (Foster *et al.*, 2002–04a). The larger investments involved require that it is vital to put a concomitant effort into efficient borehole design, local aquifer and source protection. But as a result of poor well design, siting and/or maintenance, many boreholes developed for small-town water-supply perform considerably below their potential in terms of yield provided and energy consumed and/or experience pollution.

In all such cases an appraisal of overall groundwater recharge and pollution risk to key groundwater sources is needed to evaluate water-supply sustainability and the corresponding local resource management actions required. This type of assessment work can only be carried out effectively where a reasonable body of baseline data on the response of local groundwater to abstraction and environmental pollution is available.

2.3 *Improving the security of urban water-supply*

In a substantial number of larger towns and cities groundwater is critical to the continuity of the existing water-supply – playing a key strategic role during drought or other emergencies and an important supplementary role at other times (Foster *et al.*, 1997; Foster & Tuinhof, 2005). In a few cases the use of groundwater has evolved as part of planned urban

Table 1. Selected data on groundwater use for urban water supply in African cities (based on recent, but fragmented, partial and unverified data).

Conurbation (country)	Urban population (million)	Municipal utility water-supply (Ml/d)		Unregulated private GW use (Ml/d)	Population unserved by water-supply (%)	Estimated GW level decline (m)
		SW*†	GW*			
Nairobi (Kenya)‡	3.6	520	<20	85‡	?	40 (1970–95)
Lusaka (Zambia)	1.8	0	200	100	19	30 (1985–95)
Dar-es-Salaam (Tanzania)‡	3.2	300	50	?	39	?
Addis Ababa (Ethiopia)	3.4	220	40	70	?	some
Cape Town (South Africa)‡	3.3	260	50	?	?	?

* SW is surface water; GW is groundwater
† capacity of treatment works and/or main aqueduct but supply at this level not normally available during drought
‡ known to have been initially developed and/or used intensively in response to urban water emergency

water-supply development, but more often it has occurred in response to water shortage and/or service deficiency, and often through private initiative.

No inventory of urban groundwater dependence exists – but the above is understood to be variously the case in Lusaka, Nairobi, Dar-es-Salaam, Addis Ababa, Kampala, Cape Town, Windhoek, Gaborone, Nouakchott, Dakar, Abidjan and probably elsewhere (Table 1 provides limited provisional data for some selected cities).

Given the critical role that groundwater plays in the water-supply security of many African cities (even in those cases where most of its development has been under private rather than municipal initiative) there is an urgent need for strategic (hydrogeologic and socioeconomic) assessment of its current utilisation for water-supply provision (both for public-supply and private use) and the management actions needed to ensure future availability and greater integration with surface water-supply. In all such cases an appraisal of groundwater recharge, storage potential and pollution risk will be needed.

The main issues and needs in respect of groundwater use for small town water-supply are replicated and multiplied in the larger urban centres. In particular, wastewater infiltration (by one route or other) at larger town and major urban level is a growing concern for groundwater quality, and thus groundwater protection and improved wastewater management are of direct relevance.

Moreover, very little formalised conjunctive use of surface water and groundwater for urban water supply is practised, and groundwater resources are all too often developed anarchically (both by public utilities in response to water-supply crises and individual private users in an attempt to meet deficiencies in mains water-supply), with far from optimum use of each, even in areas which are drought prone and water stressed. A supplementary concern in coastal cities is the susceptiblity of aquifers to saline intrusion when developed without adequate control.

2.4 *Expanding irrigated agricultural*

There is scant information on which to base an estimate of current use of groundwater in irrigated agriculture, because of limitations in the returns to the UN-FAO Aquastat database. Numerous countries provide no information, and of those that do there is a lack of definition of irrigation water source, uncertainty over classification of the use of springflow and wetland seepages, and omission of traditional small-scale irrigation (UN-FAO, 2003). Additional ground survey work (Giordano, 2006) suggest that Kenya and South Africa had sound Aquastat data returns but Ghana, Mali and Zambia had major underestimates – and through data correction and grossing-up a very provisional estimate for groundwater-irrigated land of 0.85 M ha (around 1% of all arable land or 10% of all irrigated land) is obtained.

The most traditional and widespread use of groundwater in agriculture is for village garden-scale irrigation of vegetables and seedlings, which helps to improve food and nutritional security at local scale. But there are also important examples of groundwater:

- offering considerable potential to provide a supplementary source of irrigation water at small scale (plots of up to 1–2 ha) in areas with shallow water-table and intermittent surface water availability for irrigation (Burke, 1994; Carter & Howsam, 1994) – and thus offer security to farmers against the impacts of drought (e.g. in the 'fadamas' of Nigeria, Sharma *et al.*, 1996);
- being used for the commercial cultivation of high-value vegetable crops in the vicinity of some cities with developed markets and/or airports with export capacity to Europe, but to date the scale of this activity is generally small.

Certain hydrogeological and socioeconomic factors are currently interacting to deter more widespread and intensive use of groundwater for commercial irrigated agriculture:

2.4.1 *Hydrogeological factors*

- the characteristics of the weathered crystalline basement (with a high proportion of schists) and the superficial unconsolidated aquifers (often of fine grain-size) often reduce the prospects of shallow wells and boreholes (less than 30 m deep) providing yields in excess of 0.5 l/s and supporting motorised pumps (Chilton & Foster, 1995);
- the deeper sedimentary aquifers (which offer prospects of much higher well yields) are often geologically complex and with low or uncertain replenishment, and also mainly situated in areas of low rural population density.

2.4.2 *Socioeconomic factors*

- the relatively high capital cost of borehole drilling discussed previously;
- the very low levels of rural electrification, and elevated cost and intermittent supply of diesel fuel, for pumping groundwater;
- the lack of social tradition in irrigated crop cultivation, compared to rain-fed arable cropping and extensive livestock rearing (probably quite widely the consequence of a 'land abundant/labour scarce' history).

In conjunction, these factors mean that capital investment in irrigated agriculture remains a relatively high-risk venture in Sub-Saharan Africa. The question arises as how to reduce the level of this risk; at least as far as the groundwater supply dimensions are concerned. Several suggestions are given below.

Subsistence-Level Cropping. Numerous widely dispersed examples of garden-scale vegetable cropping using groundwater irrigation are believed to exist and these need to be analysed horticulturally and hydrogeologically, with the generic lessons being disseminated to guide replication.

Commercial Irrigated Agriculture. Numerous opportunities for the development of small-scale high-value irrigation in the general vicinity of city markets and export facilities are believed to exist, but are often being lost because of lack of confidence about groundwater availability and development cost. Collation, mapping and dissemination of groundwater availability data (potential borehole yields, sustainable abstraction levels and borehole development costs), together with information on agricultural soils, in areas within 30–50 km radius of potential major markets is needed to focus investment (and to reduce investment risk) in the private development of irrigated agriculture (and would also be of direct relevance to urban water-supply).

3 RELATED ENVIRONMENTAL CONCERNS

3.1 *Land degradation and recharge reduction*

Soil compaction and/or soil erosion are widely leading to reduced rates of infiltration (and increasing flash run-off), and thus the loss of critical springflows and baseflow to smaller rivers and lower discharge to vegetation in topographic lows (valley-bottom lands). There is a need to halt such processes, conserve soil cover, and find ways of enhancing groundwater recharge through land management and small-scale engineering measures.

3.2 *Groundwater-dependent ecosystems*

It is important to realise that there are a substantial number of groundwater-dependent ecosystems and some groundwater-using terrestrial ecosystems in Sub-Saharan Africa, and this aspect of groundwater service provision and its potential constraint on other uses, is only just beginning to be appreciated.

In many areas much uncertainty remains over the level of groundwater dependence of these aquatic and terrestrial ecosystems, and their susceptibility to degradation due to groundwater resource development and water-table decline. The occurrence and value of groundwater-dependent ecosystems needs to be better characterized, and the impact of groundwater use for water supply monitored to arrive at balanced approaches to their conservation.

3.3 *Climate change and drought propensity*

Extensive tracts of Sub-Saharan Africa are prone to high rainfall variability and severe drought – and drought propensity could increase in some scenarios of accelerated climate change. There is thus:

- a need to appraise the susceptibility of groundwater systems to climate-change impacts;

- an important role for groundwater storage in mitigating more frequent and extended drought episodes;
- a potential for multiple small-scale enhancement of aquifer recharge and storage use (which would be highly complementary to the construction of small dams and ponds) (Tuinhof & Foster, 2004).

There is increasing evidence of a direct correlation between drought proneness and persistent poverty in Sub-Saharan Africa – and in reality a lack of investment at all scales in drought preparedness and water storage. It is necessary to think in terms of achieving greater drought-proofing of rural livelihoods as opposed to emergency food provision to mitigate the failure of local crop production and water tankering to offset the failure of local water sources. It is also important to note that:

- groundwater sources are much less drought prone than surface water sources due to the large natural storage of aquifers;
- the impact of drought on groundwater levels (which has been termed 'groundwater drought') lags considerably behind the actual failure of rains and surface water runoff (with associated reduction in groundwater recharge).

Thus, as regards drought preparedness in terms of water-supply, it is important to invest in advance in the appraisal of drought susceptibility of aquifers (Calow *et al.*, 1997), and the drilling of new boreholes and the deepening of existing boreholes and wells with pump re-dimensioning as and where necessary.

4 WAY FORWARD ON RESOURCE GOVERNANCE

4.1 *Meeting the challenge of effective resource utilization*

The introduction of deep drilling machinery from the 1970s enabled the area under groundwater exploitation and human settlement to be extended in response to increasing population and growing pressure on land. However, regional adaptation to improving borehole technology has not, in general, been accompanied by evolution of the institutional arrangements to plan and manage groundwater resource utilisation in the broadest sense.

During the UN Water Supply & Sanitation Decade, provision of rural water supplies from groundwater with hand-pumps was closely correlated with demand levels, and the presumption that adequate groundwater resources existed was broadly valid, at least in areas with more favourable hydrogeological conditions. However, with population growth and urbanisation, many rural villages have become small towns, and many small towns have transformed into district urban centres – and reticulated water-supply systems have become an aspiration to meet population growth and increasing domestic per capita usage. As local demands on groundwater resources increase questions of sustainability and management arise (and this against a background of accelerated climate change). In addition there is a growing need for source protection against pollution associated with urbanisation.

To deal with the realities of resource use for small-scale water supply over extensive areas (Foster *et al.*, 2002–04a) there is a pressing need to introduce appropriate minimal drilling regulations, to invest in effective data collection and information processing, to develop a larger body of professional expertise, to articulate groundwater needs and constraints in development planning, to undertake critical post-case evaluation of groundwater development schemes and to empower communities to maintain and protect groundwater sources.

Furthermore the following institutional issues are of broad general concern:

- **decline of national institutions** responsible for groundwater development planning, resource administration, source protection and database management (concomitant with loss of professional personnel) since the late 1980s;
- **legislation not catering for community-based arrangements** that govern groundwater use by the rural population, and focusing only on centralised groundwater permits for larger-scale development at river- basin scale.

However, Sub-Saharan Africa is not yet experiencing the classic problems associated with major and excessive groundwater development because:

- large-scale agriculture is not the major driver for groundwater development as it is in the semi-arid areas of other continents (for certain historic and cultural reasons);
- the major aquifers tend to coincide with areas of lower population density and water demand (e.g. the Congo Basin, the Kalahari and Sahel Basins), and groundwater resources are more restricted in most other (more populated) areas.

Thus the need for conventional groundwater resource management is limited to some hotspots of more intensive groundwater use for urban and mining activities. Elsewhere (and for the most part) we are more concerned about effective planning and sustainable implementation of groundwater development (often in minor aquifers) to help meet critical social welfare targets and livelihood opportunities. In effect managed groundwater development becomes a vital component of the overall development process.

There is absolutely no doubt that institutional arrangements for groundwater will have to be given higher priority and greater investment to have any chance of achieving the UN-Millennium Development Goals in Sub-Saharan Africa. A very large proportion of Africa's population live in communities for which groundwater is likely to be the only realistic option for improved water supply. Despite the obvious need and high profile given to improved water-supply standards, funding for groundwater evaluation and development is not necessarily increasing. For example, Uganda had seen a significant decrease in such funding from 2002 till 2006, when some international donors started central budget support block funding, rather than providing finance for specific water resource and water-supply projects. Since (for the most part) there is no strong voice for groundwater in the definition of national poverty reduction strategies, such funds are usually put into other priorities – thus tendency needs to be corrected.

Moreover, money on its own is not enough, since both expertise and information about groundwater resources are also required to avoid funds being wasted on constructing unsuccessful or unreliable water points. And in this context there is serious ground for concern where hydrogeological conditions are such that exploring and developing adequate groundwater supplies will require specialist expertise and information at both the planning and implementation stages, if costs are to be kept to a minimum.

Professional expertise on how to evaluate, develop and manage groundwater appears to be decreasing in many parts of Africa (especially in government offices), due to lack of appropriate training, poor recognition of groundwater professionals, reductions in public spending, disbanding of some national offices, and with the AIDS epidemic continuing to take its toll. Existing capacity needs to be used much more effectively and significant

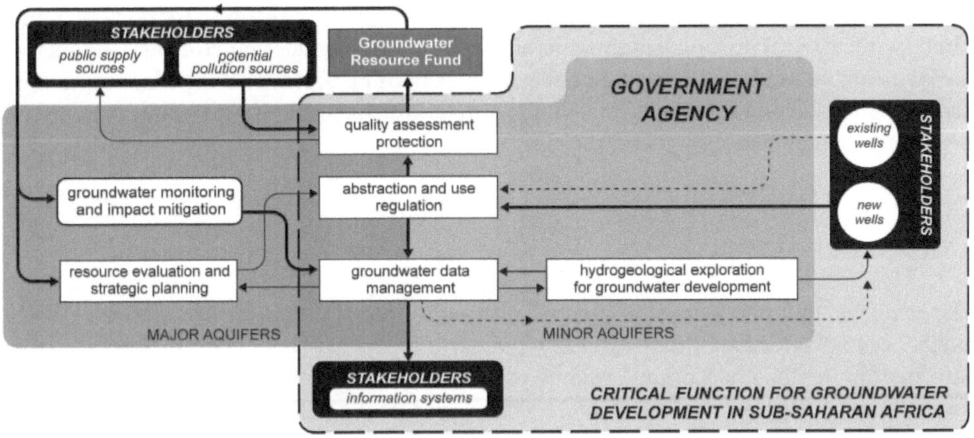

Figure 3. Essential functions for government agencies responsible for groundwater resources in Sub-Saharan Africa.

efforts are required to build new capacity through developing new training partnerships between northern and southern institutions, and the provision of training at various levels within Africa itself.

4.2 *Strengthening the essential role of government*

All of the above, results in a pressing need for strengthening appropriate government institutions to develop leadership on groundwater development planning and resource management strategy, so as to be able to interact more effectively with international donors, INGOs, local government departments, small-scale service providers, local NGOs and agricultural extension units.

The key functions for government in relation to groundwater resources are illustrated in Figure 3 – and in the case of the majority of Sub-Saharan African countries the primary need will be for emphasis on the functions on the right-hand side of this generic figure, given that most groundwater resource utilization will be at relatively small scale from so-called minor aquifers.

A related role for the responsible government institution will be to harness and apply the available information on groundwater occurrence and quality to guide better the major investments in rural water-supply provision from groundwater – this will be vital since without it the inevitable implication for many individual initiatives will be poor performance and low sustainability.

There is also a general need for scaling-up good practices in community-based small-scale rural use of groundwater for drinking water provision and rural livelihoods (such that these demands are adequately protected and not burdened with unrealistic legal requirements) and small town and village associations to promote efficient groundwater source development, maintenance and protection.

Existing information vital for efficient planning and implementation of groundwater development is often not readily accessible and in user-friendly format. The lessons learnt from successful (and unsuccessful) projects are not being collected through post-case evaluation and used for new project planning and design. As a result there remains much blind (wildcat) well drilling, resulting in high failure rates, escalating costs and ineffective use of available funds. Moreover, an effective system of Pan-African experience exchange, especially of successful best practice on the utilization and management of minor aquifers, needs to be developed.

ACKNOWLEDGEMENTS

The need for this overview was identified through interaction with David Grey (World Bank–Senior Water Resources Adviser) and various World Bank staff provided valuable information – Mukami Kariuki, Andrew Macoun, Ashok Subramanian, Rafik Hirji and Len Abrams – but the opinions expressed are those of the authors alone and do not necessarily represent those of the World Bank. The following persons willingly shared their personal vision of African groundwater issues with the authors: Jacob Burke of UN-FAO-Rome, Vanessa Tobin of UNICEF-New York, Pieter van Dongen of UNEP-Habitat-Nairobi, Alan Hall of GWP-Stockholm, Bert Diphoorn of AfDB, Alan MacDonald of IAH Burdon Commission-Edinburgh, John Chilton of BGS-Wallingford, Barbara van Koppen of IWMI-Pretoria, Gideon Tredoux of CSIR-Cape Town and Koos Groen of VU-Amsterdam. In addition the following contributors to the IAH Conference on African Groundwater (London – June 2005) provided important insights for the overview process : Callist Tindimugaya (Uganda), Lister Kongola (Tanzania), Segun Adelana (South Africa), Othniel Habiba (UNICEF), Stephen Turner (WaterAid), Nega Legasse (Oxfam), Richard Carter (Cranfield University-UK) and Roger Calow (BGS-UK).

REFERENCES

Achyara, G. 2004. *The role of economic analysis in groundwater management in semi-arid regions: the case of Nigeria.* Hydrogeology Journal, **12**, 33–39.

Adams, W. M. & Carter, R. C. 1987. *Small-scale irrigation in Sub-Saharan Africa.* Progress Physical Geography, **11**, 1–27.

Adams, W. M., Goudie, A. & Orme, A. 1996. *The physical geography of Africa.* Oxford University Press, Oxford, UK.

Burke, J. J. 1994. *Approaches to integrated water resource development and management: the Kafue Basin, Zambia.* Natural Resources Forum, **18**, 181–192.

Calow, R. C., Robins, N. S., MacDonald, A. M., Macdonald, D. M. J., Gibbs, B. R., Orpen, W. R. G., Mtembezeke, P., Andrews, A. J. & Appiah, S. O. 1997. *Groundwater management in drought-prone areas of Africa.* Water Resources Development, **13**, 241–261.

Carter, R. C. & Howsam, P. 1994. *Sustainable use of groundwater for small-scale irrigation with special reference to Sub-Saharan Africa.* Land Use Policy, **11**, 275–285.

Chilton, P. J. & Foster, S. S. D. 1995. *Hydrogeological characterisation and water-supply potential of basement aquifers in tropical Africa.* Hydrogeology Journal, **3**, 36–49.

Farr, J. L., Spray, P. R. & Foster, S. S. D. 1982. *Groundwater supply exploration in semi-arid regions for livestock extension – a technical and economic appraisal.* Water Supply & Management, **6**, 343–353.

Foster, S. S. D. 1984. *African groundwater development: the challenges for hydrologic science.* In: Challenges in African Hydrology & Water Resources, IAHS Publications **144**, 3–12.

Foster, S. S. D., Lawrence, A. R. & Morris, B. L. 1997. *Groundwater in urban development: assessing management needs and formulating policy strategies*. World Bank Technical Paper, **390**.

Foster, S. S. D., Chilton, P. J., Moench, M., Cardy, W. F. & Schiffler, M. 2000. *Groundwater in rural development: facing the challenges of supply and resource sustainability.* World Bank Technical Paper, **463**.

Foster, S. S. D., Tuinhof, A., Garduno, H., Kemper, K., Koundouri, P. & Nanni, M. 2002–04a. *Groundwater resource development in minor aquifers – management strategy for village and small-town water-supply*. World Bank GW-MATE Briefing Note, **13**.

Foster, S. S. D., Kemeper, K., Tuinhof, A., Koundouri, P., Nanni, M. & Garduno, H. 2002–04b. *Natural groundwater quality hazards – avoiding problems and formulating mitigation strategies.* World Bank GW-MATE Briefing Note, **14**.

Foster, S. S. D. & Tuinhof, A. 2005. *The role of groundwater in the water-supply of Greater Nairobi, Kenya.* World Bank GW-MATE Case Profile Collection, **13**.

Giordano, M. 2006. *Agricultural groundwater use and rural livelihoods in Sub-Saharan Africa: a first-cut assessment.* Hydrogeology Journal **14**, 310–318.

MacDonald, A. M., Davies, J., Calow, R. C. & Chilton, P. J. 2005. *Developing groundwater: a guide for rural water supply.* ITDG Publishing, Rugby, UK.

McIntire, J., Bourzat, D. & Pingali, P. 1992. *Crop-livestock interactions in Sub-Saharan Africa.* World Bank Regional & Sectoral Studies, **1**.

Sharma, N. P., Damhang, T., Gilgan-Hunt, E., Grey, D., Okam, V. & Rothberg, D. 1996. *African water resources – challenges and opportunities for sustainable development.* World Bank Technical Paper, **331**.

Tuinhof, A. & Foster, S. S. D. 2004. *Subsurface dams to augment groundwater storage in basement terrain for human subsistence – Brazilian and Kenyan experience.* World Bank GW-MATE Case Profile Collection, **5a**.

UN-FAO 2003. *Review of world water resources by country.* UN Food & Agricultural Organization Water Reports, **23**.

Wright, E. P. & Burgess, W. G. 1992. *The hydrogeology of crystalline basement aquifers of Africa.* Geological Society, London, Special Publications, **66**.

Groundwater Development and Management in Africa

CHAPTER 3

Groundwater development for poverty alleviation in Sub-Saharan Africa

R.C. Carter
Professor of International Water Development, Cranfield University, Bedford, UK

J.E. Bevan
Independent Water and Sanitation Consultant, Cambridge, UK

ABSTRACT: Nearly half of the population of Sub-Saharan Africa is in a state of severe and chronic poverty. Lack of access to safe domestic water, and indeed to significant quantities of water for other productive uses, defines and contributes to that poverty. Between one-third and one half a billion people in the region rely on unprotected and protected groundwater sources for their domestic water requirements. Further targeted development of groundwater could make a major contribution to the Millennium Development target of halving the proportion of people without access to safe and sustainable water supplies by 2015, as well as contributing significantly to incomes and livelihoods. Extending access to groundwater will be assisted by (a) significantly reducing the costs of mechanised borehole drilling and construction, (b) promoting very low cost drilling technologies in niche hydrogeological environments, preferably through the indigenous private sector, and (c) ensuring the functional sustainability of groundwater abstraction points so constructed. The development of groundwater for poverty alleviation must take account of people's requirements for domestic, agricultural and small-scale industrial water; the importance of building on user need, demand and initiative; water supply as a permanent service, not merely a short-term issue of construction and access targets; and the importance of generating far more detailed hydrogeological understanding in the region, through mapping, monitoring and groundwater data collection.

1 INTRODUCTION

Inadequate access to safe water for domestic purposes (drinking, cooking, personal and home hygiene) is an important measure of poverty. Limited access to water supply for productive uses (such as livestock watering, crop irrigation, and small scale industries) constrains households and communities in a condition of vulnerability and poverty. Extending and enhancing the development of water resources in general, and of groundwater in particular, can have significant beneficial impacts on poverty and livelihoods.

It is important to understand the constraints and opportunities posed by the poverty of the poor and vulnerable. Poverty is the multi-dimensional experience of a large number of men, women and children in Sub-Saharan Africa, especially in rural areas. It is not simply a matter of having a low or irregular income, but of lacking a wider set of assets (human, social, physical, natural as well as financial), and being vulnerable to changes which the less-poor can more readily survive. The sustainable livelihoods concept (Carney, 2003), which summarises these factors, provides a helpful integrated conceptual framework for understanding poverty. Poverty needs to be understood from the point of view of the poor, and several recent attempts to do this are now available (e.g. World Bank, 2000; CPRC, 2005).

Groundwater's potential contribution to the enhancement of livelihoods and the alleviation of poverty is great, although not without challenges and gaps in our existing understanding. One opportunity is the reduction of the costs of conventional (mechanised) drilling, and a second is the development, promotion and uptake of very low-cost well construction techniques through indigenous small-scale private enterprise.

Reducing conventional drilling costs, while demonstrably possible in many countries, requires significant changes in the practices of public and private sector institutions and donors. In this arena too, because of the inherently high costs of drilling in Sub-Saharan Africa, it is essential that a far higher proportion of new boreholes continue to function and serve their users than at present. This requires better knowledge of groundwater recharge, enhanced construction quality, and more effective practices for borehole and pump operation and maintenance.

Very low cost (manual) drilling techniques range from the traditional (Asian sludging, hand percussion) to the modern (augering, jetting), with numerous variants. The greatest possibility for directly alleviating poverty in Sub-Saharan Africa would appear to lie in the application of these technologies through small enterprises developed or strengthened to deal directly or indirectly with households, farmers, communities and institutions. It is possible that manual drilling could benefit up to 90 m people in Sub-Saharan Africa, if it were to be more widely promoted.

Sub-Saharan Africa itself poses a number of challenges. There are signs of hope, not least in the growth of indigenous entrepreneurship, and in the growing number of case studies of successful development of groundwater for poverty alleviation in the region. This paper outlines and explores some of these opportunities, and draws out conclusions for the way forward.

2 POVERTY

It is estimated that 30 to 40% of the world's poor live on the African sub-continent. The most recent Human Development Reports (UNDP, 2006, 2007) indicates that the highest regional poverty levels in the world occur in Sub-Saharan Africa, with 46.4% of the population surviving on less than US$1 a day. This is the poorest continent in the world, and despite the many interventions, projects and development initiatives that have been undertaken, the proportion has remained relatively static, and shown little sign of improvement.

Poverty is understood to be more complex and multi-dimensional than simply a lack of income, reflecting both severity (depth below the absolute poverty line of $1 a day) and duration of poverty (CPRC, 2005). The chronic poor (below the poverty line for 5 years or more) are naturally the most vulnerable in terms of security and survival – when disasters such as drought strike, they have the least protection. Long-term support that is able to lift people out of chronic poverty must reduce vulnerability and improve chances of survival. Improving livelihoods through self-help schemes that can increase food security and physical assets will help create a lasting solution to chronic poverty in Africa.

Access to water is a basic human right, but for many of Africa's poor, water is only available at a price, and supplies are often not safe. The human cost of lack of access to safe water is huge, reflected in the high child mortality rates of African countries: Africa has an average under 5 mortality rate of 170 per 1000 live births, the world average being 79 (UNICEF, 2006). The push to improve water supplies through the Millennium Development Goals is thus a great contribution in the fight to combat poverty.

3 ACCESS TO GROUNDWATER IN SUB-SAHARAN AFRICA

Availability and reliability of data on water resources and access to water has been a major problem in Sub-Saharan Africa for at least the last 40 years. Data on "informal" access to water for agricultural (livestock and crop production) and small-scale industrial uses are almost non-existent, and certainly not systematised. Data on domestic water supply coverage (distinguishing between the "served" and the "unserved") are of higher quality, having been the focus of the Joint Monitoring Programme (JMP) of WHO and UNICEF since 1991.

In Sub-Saharan Africa, approximately 326 million people, mostly (84%) in rural areas, lack adequate access to safe domestic water (JMP, 2006). These people obtain their water from a combination of unprotected and untreated surface water and groundwater sources (including unimproved hand-dug wells). It is probable that at least half of the "unserved" in Sub-Saharan Africa use unprotected groundwater.

Of those who are served by an engineered or protected water supply (approximately 409 million in total, of which about half are urban and half rural), probably half depend on motorised or hand-pumped groundwater sources. If these figures and assumptions are taken to represent the situation in the Region, then it is likely that overall at least 163 million

Table 1. Improved water supply: the served and unserved in Sub-Saharan Africa, millions (Source JMP, 2006). Urban coverage 2015 assumed to be 80%.

1990				2004				2015 projected to meet MDG[a]			
Urban		Rural		Urban		Rural		Urban		Rural	
Served	Not served	Served	Not served	Served	Not served	Served	Not served	Served	Not served	Served	Not served
119	26	134	238	212	53	197	273	321	80	286	247
Sub-total 145		Sub-total 372		Sub-total 265		Sub-total 470		Sub-total 401		Sub-total 533	
Total population 517				Total population 735				Total population 934			
Population unserved 264				Population unserved 326				Population unserved 327			
Population served 253				Population served 409				Population served 607			

[a] Note that the achievement of this target is highly unlikely. UNDP (2006) estimates that at current rates of investment the 2015 target will only be met in 2040 in SSA.

so-called "unserved" and 205 million of the "served" rely on groundwater for their domestic supplies, more than 368 million people in total.

Two major aspirations exist. The first is the subject of the Millennium Development Goal 7, Target 10, namely the increase in coverage of water supply – serving the unserved. This means more sources, more construction, more capital investment. Achieving higher coverage through groundwater development requires better hydrogeological understanding for improved siting, reduced construction costs, and increased overall investment. The second, frequently neglected, aspiration is to achieve acceptable levels of functional sustainability, in other words to maintain existing sources in a good state of repair over the long term. This goal is actually far more important than that of simply increasing coverage, since without it all capital investments are undermined. A focus on functional sustainability of groundwater development requires better understanding of the renewable resources involved (i.e. groundwater recharge), improved construction supervision practices, and efficient maintenance, repair and rehabilitation arrangements for pumps and boreholes.

4 GROUNDWATER ENVIRONMENTS

4.1 *Main environments*

There are four main hydrogeological domains in Sub-Saharan Africa (MacDonald *et al.*, 2005; MacDonald *et al.*, 2008). By far the largest of these is the ancient Pre-Cambrian Basement rocks, underlying 40% of the land area of Sub-Saharan Africa, where an estimated 220 million people live in rural areas. The Basement rocks store groundwater in the upper weathered regolith. The regolith can vary in thickness from a few metres in desert areas up to 90 m in the more humid tropics.

The second most significant hydrogeological domain is of consolidated sedimentary rocks such as sandstones and limestones (32% of the land area), containing substantial reserves of groundwater, and sustaining a rural population of approximately 110 million. Aquifers in these rock types are often productive and may be exploited to support urban populations in addition to rural crop production and livestock watering. However, in arid regions such as the Namib Desert, sedimentary aquifers may receive very little recharge, and as such constitute a non-renewable resource, whilst in coastal areas, for example along the margins of East Africa, aquifers are at risk of saline intrusion and pollution.

Unconsolidated sediments comprise about 22% of the African land area, including alluvial valleys and wind-blown deposits such as the Kalahari Sands. Water stored in these deposits is shallow (up to 50 m depth) and recharge is subject to the vagaries of climate. It is in this hydrogeological domain that many of Africa's shallow wells are found, and it is estimated that this type of groundwater supports a rural population of about 60 million (MacDonald *et al.* 2008). The aquifers can be complex, often with lenses of sand, gravel and silts, depending on the depositional environment. Flood plains of big rivers like the Zambezi, and the inland Okavango delta in Botswana are examples, and freshwater lenses can often be distinguished by subtle vegetation changes (Woodhouse, 1991), as well as by more sophisticated geophysical techniques. Concentrations of metals such as iron can be a concern in shallow unconsolidated aquifers, and faecal contamination from latrines or diffuse pollution from the intensification of agriculture is also an issue.

The fourth category of water-bearing rocks is the volcanic group, which accounts for only 6% of the land area of Sub-Saharan Africa (MacDonald *et al.*, 2008). The major volcanic area is concentrated in East Africa in a wide band across Ethiopia and Kenya. This is formed predominantly of lava flows and basalt sheets that were associated with the opening of the Rift Valley system. The volcanic rocks can be jointed and fractured, have weathered areas, and are interleaved with ash layers and palaeosoils, which are often porous and are good aquifers. Water is generally accessed from springs or from deep boreholes that can cut across several more permeable layers, but in some areas such as the Kenyan tablelands, shallow hand dug wells in volcanic terrain are often found. High concentrations of some chemical species can be a problem in these terrains, fluoride being a common natural constituent in the Rift Valley. Despite its relatively small areal coverage, water from volcanic aquifers supports an estimated 45 million rural people in the region.

The population figures cited above sum up to 435 million, the approximate year 2000 rural population of Sub-Saharan Africa. The implication is that the majority of the rural population obtain their domestic supply from groundwater.

4.2 *Niche environments*

One particular group of groundwater environments is of particular interest in the present context. In general these can be described as valley bottomlands, although they have different names and characteristics in different countries. All are low-lying, seasonally or perennially moist, and underlain by a mix of alluvial, colluvial or other superficial deposits which often contain material of sufficient permeability to yield significant quantities of groundwater. By virtue of their location they may also receive significant quantities of groundwater recharge from surface water flows or flooding. Some examples of their local names are *boliland* (Sierra Leone), *dambo* (Malawi, Zambia, Zimbabwe), *fadama* (Nigeria), *mbugua* (Tanzania), and *vlei* (South Africa). Because of their geomorphology and geo-hydrology, these features offer real potential for further poverty-focused development. However, some require significantly greater care than others in terms of possible over-exploitation and destruction of natural habitats which are important both for local economy and biodiversity.

The fadamas (seasonally flooded valley bottomlands) of northern Nigeria vary enormously in scale and usage. Those which have received the most attention, since the late 1970s, have been the major river floodplains of the north-east and north-west. In particular the Hadejia-Jama'are-Yobe river system of north-east Nigeria has been studied extensively (see for example: Carter & Alkali, 1996; Alkali, 1995; Thompson & Hollis, 1995) and developed through the construction of shallow jetted tubewells. Under natural conditions this river system experiences annual recharge. Since groundwater abstraction is exclusively by suction lift (small petrol-driven centrifugal pumps), the water table cannot be drawn down below about 6–7 m, so abstractions are self-limiting (Figure 1).

Small-scale traditional irrigation has been practised in the northern Nigerian river floodplains (fadamas) of the Hadejia, Jama'are, Yobe and Sokoto rivers, as well as many smaller areas of land with adjacent surface water or groundwater for generations. The technology used for lifting water was the Egyptian-style shaduf, a simple counter-balanced lever made from timber, mud and rope. This technology allowed the irrigation of about 0.1 ha per farmer. Dry season vegetables – tomatoes, onions, peppers – were the main crops of choice. In the late 1970s new technology was introduced: shallow jetted tubewells or washbores combined with small petrol-engined centrifugal pumps.

Figure 1. Small scale irrigation from river floodplains in Northern Nigeria using suction pumps.

These allow plots 10–20 times the size of shaduf plots to be cultivated. Investment in the technology requires only very short-term credit, since farmers can repay the full costs of the tubewell and pump with the profits from one (dry season) crop. Farmers have diversified their crops, even at times irrigating wheat, as well as the more traditional vegetables, for which ready markets exist in Nigeria. Since the early 1980s, and right up to the present, this technology has spread widely in the major northern Nigerian fadamas, initially as an activity of the World Bank-supported Agricultural Development Programmes (ADPs), and later through the private sector.

In less extensive valley bottomlands, such as many of the dambos of Malawi, judicious development based on a thorough understanding of the water balance (as well as its inter-annual variability and any long-term trends) is possible. Similar technologies to those which have been used with such success in the northern Nigerian fadamas and in Niger, described more fully in the following sections, can be mobilised. The key starting points though, before promoting technologies, is to recognise the needs of rural people and their existing water management practices, and to build on those.

5 APPROACHES FOR ACCESSING GROUNDWATER

Technologies need to be fashioned and modified to the needs of the users, and designed in such a way as to build on the attempts people have already made to solve their own problems.

Externally-driven approaches to poverty alleviation in general, and groundwater development in particular, have made significant progress in extending people's access to water. If this were not so, the water supply coverage statistics would be far worse than they are. However, technical solutions in which the users typically contribute less than 10% of the capital cost, and in which insufficient time is set aside to bring about full community participation and management, suffer major problems of lasting maintenance and functionality. The almost exclusive use of externally-driven approaches in Sub-Saharan Africa has led to an aid-mentality in which people often wait for Governments or NGOs to act for them, and are reluctant to take initiatives for themselves. In cases where there is no other option than that of deep- or hard rock drilling, then externally-initiated solutions, using mechanised drilling are necessary. However, all possible measures need to be taken to instill local ownership, responsibility (including financial responsibility) for operation and maintenance, and to ensure appropriate on-going support or back-stopping is available.

In many countries (including Uganda, Zambia, Zimbabwe, Mali, Ethiopia, Sierra Leone and Liberia) *self-initiated approaches* (also known as self-supply, see for example Sutton, 2002) are alive and well. Interventions to support such local initiatives offer significant promise, as they build on people's own attempts to solve their water problems, thereby enhancing ownership and the potential for long-term sustainability.

A third approach, described here as *enterprise-response to user-demand*, combines technology innovation, small enterprise development, and market stimulation (or demand creation), to provide locally sustainable solutions to the water problems of farmers, households and communities.

Low cost drilling technologies can also be promoted through attention to some or all of at least three aspects: *technology*, *small enterprise development*, and *market development*. In the first of these, existing technologies (hand percussion, hand augering, sludging, jetting and their variants and combinations) need to be modified and adapted to the local geological and economic environment. In particular, equipment needs to be manufactured locally from readily available parts and components (e.g. standard pipes and fittings, flexible hose, timber, rope, centrifugal pumps), and using simple fabrication technologies (e.g. cutting, hardening, welding, threading).

Small enterprises need to be introduced to the "new" technologies, trained in their use, and then provided with business training. They may need additional support in kind (e.g. equipment, transport) or in cash (seed capital or credit) in the short-term, as well as longer term assistance until they become viable business entities.

The development of the market refers to both demand creation among potential water users (who may also require short-term credit facilities to access the technology), as well as assistance with market linkages for the output of productive water uses, especially crops.

6 USING GROUNDWATER TO ALLEVIATE POVERTY

6.1 *Extending access through cost reduction*

Widening access to groundwater, whether for domestic or productive (i.e. output- or income-generating) purposes requires serious attention to cost reduction. In general, two options exist for cost-reduction in borehole construction: (i) reducing the costs of conventional (mechanised) drilling, and (ii) developing and promoting very low cost niche drilling technologies.

6.1.1 *Reducing conventional drilling costs*

Sub-Saharan Africa needs between 1 and 2 million new protected groundwater points to reach full domestic water supply coverage. This figure is based on achievement of the MDG target, reducing the number of the unserved in SSA from 326 m in 2004 to 238 m in 2015, followed by achievement of full coverage by 2025. Taking account of projected population growth (from the 2002 estimate of 683 m to 932 m by 2015 and to 1,137 m by 2025 – UN medium variant projections), 336 m people need to be served by 2015, and another 443 m by 2025. This is equivalent to 1.1 m new point sources, each serving an assumed 300 persons, needed by 2015, and a further 1.5 m by 2025. If half of these are groundwater sources, 1.3 m new boreholes will be needed. This figure takes no account of replacement of currently non-functioning sources.

If a 10% reduction could be achieved on the 'typical' US$10–15,000 cost of a borehole in much of Africa, this would result in savings of more than US$1bn, potentially giving access to safe water to an additional 30 million people who would otherwise not have been served. In Ethiopia alone, Getachew (2004) estimated that more than 80,000 new boreholes, to serve 28 million people, will be needed by 2015. A cost saving per borehole of 10% could lead to an additional 2 million people being served for the same investment – a significant achievement. Table 2, from the WSP Field Note by Carter *et al.* (2006) describing the drilling costs study in Ethiopia (Carter, 2006c), sets out 10 ways of reducing drilling costs.

Underlying the issues highlighted in Table 2 are several weaknesses which are present in Ethiopia, but also more widely in Sub-Saharan Africa. These include:

- resistance in the public sector which professionals experience when they try to change design standards or practices;
- limited expertise, and even more limited resourcing, at local Government level to permit adequate contract management and supervision;
- difficulties for the indigenous private sector of "doing business" – obtaining loans on realistic terms, importing spare parts and consumables, competing fairly in a transparent operating environment, and having some assurance of a sufficient workload;
- insufficiently detailed knowledge of groundwater conditions, introducing uncertainties into contract specifications;
- unacceptably high post-construction failure rates.

Costs of conventional drilling can be reduced, but only through committed leadership, long-term training programmes and high quality management which can allow technical and contractual solutions (such as lower-cost well designs, use of smaller rigs, and packaging of contracts) to be put in place (Figure 2).

6.1.2 *Development and promotion of manual drilling*

There has been general neglect of very low cost drilling technologies in Sub-Saharan Africa. In general terms, large scale drilling technology (capital costs in the region of US$0.5–1.0 m) translates into per-borehole costs of around US$10,000. Smaller equipment (capital costs around US$100,000) can be shown (Ball, 2004) to produce boreholes in Basement Complex for around US$3000. Very low cost technologies having a capital cost in the order

Table 2. High African drilling costs and 10 ways to reduce them (modified from Wurzel, 2001; Smith, 2003; Ball, 2004; Carter *et al.*, 2006).

Reason for high cost	Suggested means of cost reduction
Inappropriate borehole designs – for example drilling at larger diameters than are required for handpumps or modern submersibles; using steel casings rather than plastics; drilling deeper than necessary.	1. Designs which are fit for purpose: small diameters, plastic casings or no casings, drilling only to necessary depth. 2. Use of smaller, less costly rigs and support equipment to match relaxed borehole designs.
High mobilisation costs associated with long travel distances from Contractor's base.	3. Let packaged contracts for multiple boreholes in close proximity and similar geology.
Low drilling success rates resulting in a high proportion of dry holes.	4. Improve knowledge of hydrogeology, and use modern borehole siting practices.
Unnecessarily rigorous and lengthy test pumping regimes.	5. Match test pumping requirements to borehole purpose.
High rates of post-construction failure due to poor construction, inadequate assessment of groundwater resources, or unsustainable pump operation and maintenance.	6. Improve construction supervision. 7. Carry out rigorous evaluation of renewable groundwater resources, not simply test pumping. 8. Establish sustainable O&M procedures for pumped groundwater sources.
Difficulties of "doing business" in those countries most in need of boreholes. Small market leads to dis-economies of scale and limited competition.	9. Support private contractors by easing importation, facilitating local manufacture of casings and well-screens, giving tax breaks, and assuring steady and sufficient work flow.
Weaknesses in logistics compared to potential outputs from physical equipment.	10. Improve communications networks, enhance management skills of public and private sectors.

of US$500–1,000, or less in favourable circumstances, can produce boreholes at a unit cost of around US$50–200 (Figure 3).

Very low cost drilling technologies (Brush,1982; Elson & Shaw, 1995; SKAT, 2001; Carter, 2005) fit certain geological conditions, in particular unconsolidated or semi-consolidated formations in which groundwater is encountered within 20–30 m of the surface. These geological formations include alluvial and other superficial deposits (especially those found in valley bottom-lands), but under favourable circumstances they can also include the weathered mantle or regolith overlying the Crystalline Basement Complex.

These technologies are essentially low-energy drilling methods. Their ability to break indurated formations is limited and they are constrained to small diameters (generally

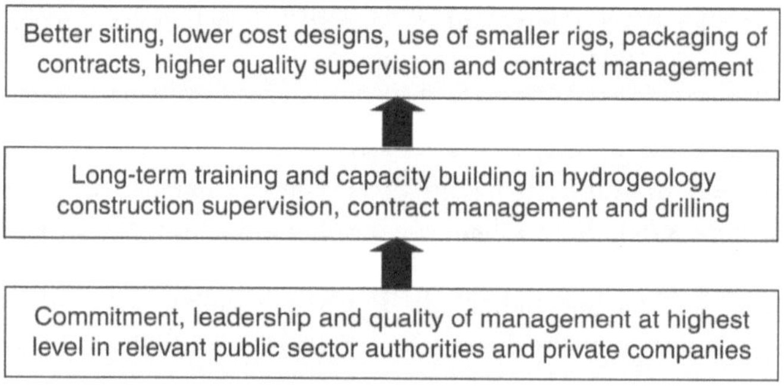

Figure 2. Technical and contractual solutions are only possible through training and resourcing, and with political will and sound management.

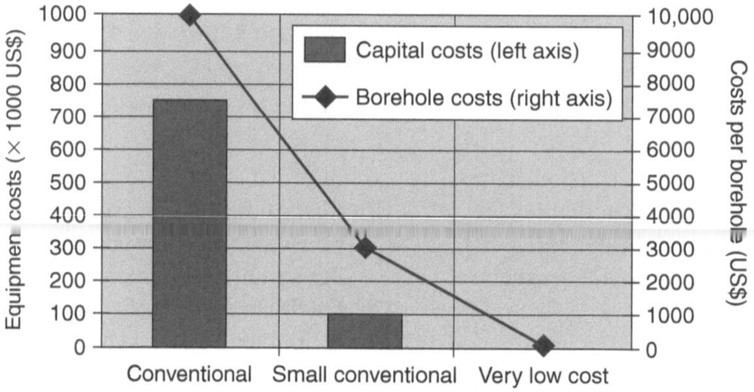

Figure 3. Cost comparisons of conventional, small conventional and very low cost drilling in SSA.

150 mm or less, although exceptionally up to 250 mm) – consequently they are restricted to niche geologies and the relatively small production rates achievable from low-lift suction pumps, medium lift handpumps, or the smallest submersibles (say 0.3–2.0 litres/sec).

The hand augering of shallow wells and the marketing of treadle pumps in Niger is a story of truly appropriate technology and sustainable development (Danert, 2006). The technique of hand augering wells to access shallow groundwater was introduced in Niger in the 1960s by the ILO*, and has been promoted over the years by the Non Governmental Organisations Peace Corps, Lutheran World Relief, Enterprise Works and more recently the World Bank. The use of a simple soil auger, which is rotated into the ground and then lifted to remove the dislodged material, can create small diameter wells up to 6–7 m deep in relatively unconsolidated sands and clays. Some examples of low cost drilling techniques are shown in Figure 4.

With continued support over the last 30 years, and particularly though the long-term championing of the approach by Jon Naugle, this technique has been successfully promoted as a small business, and there are now many well-drillers and more than 5000 such wells all over Niger. The cost of a hand-augered well is in the region of US$50 to $300, depending on

* International Labour Organisation

Figure 4. Examples of low cost drilling techniques. Top left hand percussion, top right hand augering, bottom left sludging, bottom right jetting or washboring.

the depth, a cost which compares very favourably with the more traditional hand-dug well (up to 5 times as much) and a deeper borehole, which can cost in the region of US$5000 to $20,000. This makes the technology much more affordable to subsistence farmers, who can pay off the cost fairly quickly through the sale of produce. The technique is particularly appropriate to many parts of Niger where lenses of shallow (less than 10 m) groundwater are common. Most of the wells are fitted with low cost treadle pumps, which were also developed in Niger, and are made and marketed by local enterprises, costing between US$120 and $460. Once purchased, the farmers are able to maintain and repair the pumps themselves. The wells are used for domestic drinking water and also for crop irrigation for small subsistence and market gardens, where, amongst other crops, farmers grow sugar cane, tomatoes and spices.

This project is an excellent example of how the role of external support as a catalyst for development can be a positive force, by introducing appropriate technology, understanding the local needs and constraints, and nurturing private enterprise over a long time period. Crucial to the success of the approach however, has been the productive (income-generating) aspect of water supply.

6.2 *Private water selling*

In many areas of the developing world, drinking water has become an expensive commodity. In East Africa, water vendors can charge up to US$5 per m^3 where water is scarce (Sansom *et al.*, 2004). The vendors are often youths or even children that are exploited by their employers, working long hours for little pay. The customers are those that can least afford it, and are often paying considerably more than those with formal pipeline connections. The water is in no way guaranteed to be safe.

Water vendors are generally mistrusted, but are used nonetheless, and many households will differentiate between bought water for drinking, and fetched water from an unsafe source (open pond or well) for cleaning and bathing. In areas of Uganda where water sources are scarce or far apart, there is a living to be made collecting and transporting water. Although the water might be clean enough at source, the vendor's containers and method of collection and transportation could potentially contaminate the water, making it unsafe. A recent study in Naivasha, Kenya, found that young water resellers are employed to sell 200 litre drums of water for between 10 and 40 Kenyan Shillings (US$0.14 to 0.55 or £0.07 to 0.29).

6.3 *Irrigation*

Crop irrigation is one of the most demanding uses of water, in terms of quantity, as well as being largely a consumptive use (as opposed to non-consumptive, in which "used" water is returned rapidly to the local hydrological system for potential re-use). In most of Sub-Saharan Africa growing crops require around 3–5 mm per day for transpiration (depending on the climate), and getting water to the root zone of crops usually involves significant additional "losses" due to seepage from open channels and unproductive evaporation. The inefficiencies associated with conveyance of irrigation water are usually greater in large scale public sector irrigation systems than in small-scale, farmer-managed irrigation. Even taking no account of water losses, the transpiration demand of irrigated crops translates to 30–50,000 litres per day for each hectare irrigated (equivalent to the basic domestic water needs of 2–3000 people).

The requirements for irrigated land and irrigation water which can make a real difference to impoverished farmers are not to be measured in hectares and tens of thousands of litres per day, but a few square metres and correspondingly small amounts of water. An example is the Bikita Integrated Rural Water Supply and Sanitation Project (Mathew, 2004) in Zimbabwe. In a rural district, approximately 100 km east of Masvingo, a community-managed water and sanitation programme has successfully branched out into communal and family owned productive gardens. Rains are unreliable in this area, and crops, most commonly maize and millet, may often fail if a reliable water supply cannot be secured.

The Bikita area has many boreholes of 40–60 m depth, and deep wells up to 30 m depth, all of which are equipped with the locally produced Zimbabwe Bush hand pump, and maintained within the community. The productive water point gardens use these water sources to farm communal vegetable plots with groups of up to 50 people, growing a range of crops including ground nuts, maize, cabbage, onions, tomatoes, spinach and beans.

Gardens of 0.25 hectares were allocated per household, giving the poorest and female-headed households the same amount of land as wealthier neighbours. Subsequent research in the area found that a good cross-section of the society had taken up and kept with the project, and many were selling a proportion of their produce to neighbours or at markets, to raise between US$2 and 10 each per month. The project was found to be highly sustainable, with the communities working together for a common purpose.

The even-handed allocation of the plots meant that this groundwater project has been able to be very pro-poor, in that it does not discriminate against the smaller or female-headed households. The food security and additional income allowed these individuals to help themselves out of poverty.

The 1970s was a decade of significant investment in irrigation in Sub-Saharan Africa. Most of the investment was made in medium- and large-scale Government-initiated (often donor- or lender-supported) smallholder irrigation schemes. In these schemes, Government objectives (to do with import substitution and national self-sufficiency) and foreign ideas about cropping calendars and management systems were imposed on smallholders who were reduced to little more than unpaid labourers. Not surprisingly, this top-down approach largely failed, donors withdrew and little international investment in smallholder irrigation has happened since the early 1980s in the region.

However, throughout this period and since, in probably every country of the region, small farmers have been managing and conserving water, irrigating crops on a very small-scale in niche environments (e.g. where they can divert surface water by gravity, or where groundwater is very shallow – in valley bottomlands). These activities have been largely invisible to Governments and foreign donors, and yet they represent a major potential for judicious external assistance and expansion, especially using groundwater. (For example, in Nigeria, by the end of the 1970s when about 30,000 ha of Government irrigation schemes had been developed, it was estimated that around 1 million hectares of informal, unassisted, farmer-managed or small-scale irrigation existed). As irrigation returns to the agenda of several of the larger donors and lenders, it is to be hoped that some of the lessons of the 1970s can be learned and applied.

6.4 *Groundwater for livestock*

Cattle are a status symbol as well as an investment, and a livelihood. They can be used for ploughing, to provide milk and meat, and as a source of income. They are also seen as an 'insurance policy' against lean times. In addition goats, sheep, pigs and poultry are regularly kept in large numbers. Whilst it is considered that people need at least 15–20 litres of water per day as a minimum, cattle need up to 40 litres each, and smaller animals 5 litres. Large herds can demand a significant quantity of water, and where there is scarcity, are often in direct competition with people (McDonald, 2006). It is the poorest and most vulnerable – women, children and old people – who suffer in these circumstances. Cattle can also trample the area around water points, making them muddy and unclean, and potentially contaminating the water supply.

Figure 5. Groundwater used for making bricks in Uganda.

If water points are not protected, animals can destroy pumps and platforms, rendering them unworkable, and denying access to all. Ideally there should be separate water points solely for livestock watering, or shared water points should be securely fenced, with a pipe leading to a watering trough for animals at least 10m away from the source used for human consumption (McDonald, 2006).

6.5 *Small-scale industries*

One of the most widespread uses of groundwater in rural Sub-Saharan Africa is for brick-making. In valley bottomlands the juxtaposition of suitable clays, fuelwood and groundwater makes this opportunistic enterprise possible. The environmental impacts of this activity have apparently not been documented, but clearly there are potential issues of natural resource (over-)exploitation and water pollution (Figure 5).

7 SUSTAINABILITY OF GROUNDWATER SUPPLY FOR LASTING POVERTY REDUCTION

There is little point in achieving impacts on poverty unless they are sustainable over the long term. A small impact which is sustained is of far greater significance than a large impact which fails rapidly.

7.1 *Environmental sustainability*

Environmental sustainability includes avoidance of over-exploitation, and avoidance of groundwater pollution. *Over-exploitation* results in failed wells and boreholes, and undesirable impacts on natural groundwater discharges to springs, wetlands and baseflow.

Avoidance of over-exploitation begins with a systematic analysis of groundwater resources, and especially the evaluation of groundwater recharge. Although this subject presents numerous practical difficulties – since recharge cannot be directly measured – nevertheless, methods do exist (e.g. Rushton *et al.*, 2006; Edmunds *et al.*, 2008) for recharge estimation even in the data-scarce situation of much of Sub-Saharan Africa. Whenever groundwater is exploited for poverty-reduction purposes in the region, it should be standard practice to carry out not only borehole pumping tests (which tell us nothing at all about the resource, only about the performance of the aquifer and the well), but also recharge estimations, and at least conceptual and basic quantitative modelling of inflows, throughflows and outflows from the aquifer. The latter may often need be little more than a basic water balance, but without this, it is impossible to know whether there is significant risk of over-exploitation.

Avoidance of *groundwater pollution* in groundwater-based poverty alleviation interventions is a significant challenge. Three main issues arise. First, groundwater abstraction points (wells, boreholes) may themselves act as conduits for pollution of the aquifers which they penetrate. Poor construction, with inadequate sanitary seals, is a common problem, resulting in the contamination of groundwater by polluted surface water, especially at the onset of the rains. Diarrhoeal disease outbreaks, including cholera, are common under such circumstances. Second, contamination may be introduced through open wells, for example by the use of ropes and buckets which lie on the ground while not in use. This problem can be reduced by the use of a windlass, or by covering wells and using fixed handpumps for water abstraction. Third, the use to which groundwater is put may itself contribute to groundwater pollution. For example, the use of water for irrigation commonly results in return flows to the underlying aquifer which may contain pollutants such as pesticides and fertilisers. Also poorly controlled shared use of domestic water points by humans and livestock can readily pollute the underlying aquifer.

7.2 *Functional sustainability*

Functional sustainability is concerned with whether or not groundwater abstraction points continue to work over time, thereby providing on-going services to their users and the sustained beneficial impacts which are the goal of such developments.

The *motivation* or demand of end-users is crucial. Without this, no mechanisms or systems, other than those provided free of charge by external agencies can ensure sustainability – and the latter would themselves be financially unsustainable. If households' need for their domestic water borehole, or farmers' interest in productive uses of groundwater are strong enough, then their demand will underpin the remaining pre-requisites for functional sustainability.

There is the requirement for *money*. The routine maintenance, repair and rehabilitation of groundwater abstraction points needs a steady flow of small amounts of cash, primarily from the water users, but when necessary repairs are beyond their capacity, from an outside source (Government or NGO). A recent study in Tanzania (Haysom, 2006) demonstrated that the single most influential factor in determining functionality of water points was the collection and sound management of revenues at the user level. Certainly without this resource, functionality is bound to suffer. The aspect of financial sustainability is further explored by Baumann (2006).

Third is the need for sound *maintenance structures*. Organisational hierarchies (from borehole caretaker, to community technician, to higher level backstopping agencies), appropriate skills training, provision of tools, and access to spare parts are all important. Maintenance and repair of handpumps in Sub-Saharan Africa has proved particularly problematic, not least because of the aid/public-provision culture (as opposed to more self-reliant approaches) and the dispersed rural populations and small markets.

There is the need for *on-going support*. The introduction of new technology in the form of a borehole and handpump to a community or water-user group moves people from self-reliance to inter-dependence. They now need access to technical skills and support, especially when breakdowns occur which are beyond their control or capacity to put right. The need for on-going (permanent, albeit low-level) support has only belatedly been recognised (e.g. Schouten & Moriarty, 2003; Carter & Rwamwanja, 2006), but without it functional sustainability will remain a dream.

7.3 *Groundwater regulation, monitoring and assessment*

A wide range of groundwater uses and users are to be found in the region. Many of the existing abstractions and uses of groundwater are informal, in the sense that they are invisible to the authorities, unregulated, and involving unknown quantities and qualities of abstractions and return flows. Even formally sanctioned abstractions in the form of rural domestic water points, small town water supplies and public sector or commercial irrigation or industrial boreholes can be poorly regulated and monitored.

There is an urgent need in Sub-Saharan Africa to assess and continually re-assess groundwater resources, especially as demand grows and climate change and instability take a tighter grip. So far the extent of groundwater over-exploitation and pollution in Sub-Saharan Africa has been limited to the major cities, but as populations grow generally, urbanisation continues, and demand for food and water increases correspondingly, so the urgency of mapping, quantifying, controlling and protecting groundwater resources grows.

There is significant potential for the promotion of user-based monitoring of groundwater resources, especially in the context of weak public sector institutions and the low priority which is given to data collection in general.

8 CONCLUSIONS

If groundwater development is to continue to contribute to the alleviation of poverty in Sub-Saharan Africa, a number of principles must be adopted, and actions taken, to enhance its effectiveness in that role:

- The starting point must be the needs and wants of water users, and those lacking adequate access to water for both domestic and productive uses. The ideal external interventions are those which not only respond to need and demand, but which also build on people's own initiatives. Whether interventions primarily build on self-supply initiatives, develop local entrepreneurial opportunities, or are essentially externally-driven, they must be focused on the demands of the water users.
- Where mechanised drilling is necessary (because of depth to water or hardness of rock formation), emphasis should be placed on the reduction of borehole construction costs.
- The potential for further development and deployment of very low cost drilling technologies in niche hydrogeological environments needs to be exploited much more widely in the region.

- Boreholes and other groundwater abstraction points should be seen less as water sources and more as water supplies or services. Their siting, construction, operation, maintenance and management need to be viewed in an integrated manner because of the importance of guarding investments and delivering permanent services. Increased access is important, but sustained functioning is even more so.
- Mapping, monitoring and groundwater data collection in general have been neglected for too long. Renewed emphasis needs to be placed on these keys to increasing groundwater understanding, and hence access.

REFERENCES

Alkali, A. G. 1995. *River-Aquifer Interaction in the Middle Yobe River Basin, North East Nigeria.* PhD Thesis, Silsoe College, Cranfield University.

Ball, P. 2004. *Solutions for Reducing Borehole Costs in Africa.* WSP Field Note. Water and Sanitation Programme, World Bank, Rural Water Supply Network, SKAT.

Baumann, E. 2006. *Do Operation and Maintenance Pay?* Waterlines, **25** (1), 10–12.

Brush, R. E. 1982. *Wells Construction. Hand-dug and drilled Appropriate Technologies for Development.* Manual M-9. Peace Corps, Information Collection and Exchange, 806 Connecticut Avenue, NW, Washington DC, 20525, USA.

Carney, D. 2003. *Sustainable Livelihoods Approaches: Progress and Possibilities for Change.* DFID, London.

Carter, R. C. & Alkali, A. G. 1996. *Shallow groundwater in the northeast arid zone of Nigeria.* Quarterly Journal of Engineering Geology, **29**, 341–356.

Carter, R. C. 2005 *Human Powered Drilling Technologies.* Cranfield University. Available at: http://www.rwsn.ch/documentation/skatdocumentation.2005-11-15. 5533687184

Carter, R. C. & Rwamwanja, R. 2006. *Functional sustainability in community water and sanitation: a case study from south west Uganda.* Diocese of Kigezi/Cranfield University/Tearfund.

Carter, R. C. 2006 *Drilling for Water in Ethiopia: 10 Steps to Cost-Effective Boreholes.* WSP Field Note, 2006, Water and Sanitation Program, World Bank.

Carter, R. C., Horecha, D., Berhe, E., Belete, E., Defere, E., Negussie, Y., Muluneh, N. & Danert, K. 2006. *Drilling for Water in Ethiopia: a Country Case Study by the Cost-Effective Boreholes Flagship of the Rural Water Supply Network.* Federal Democratic Republic of Ethiopia/WSP/RWSN.

CPRC 2005 *The Chronic Poverty Report.* Chronic Poverty Research Centre, Manchester, UK.

Danert, K. 2006. *A Brief History of Hand-Drilled Wells in Niger.* Waterlines, **25**(1) 4–6.

Edmunds, W. M. 2008. *Groundwater in Africa – palaeowater, climate change and modern recharge.* This volume.

Elson, R. & Shaw, R. 1995. *Technical Brief No 43: Simple Drilling Methods.* Waterlines, **13** (3) 5–18.

Getachew, H. 2004. *UNICEF Study on Groundwater and Requirements for Drilling and Other Systems Tapping Groundwater in Ethiopia.* Paper for International Groundwater Conference, Addis Ababa, 25–27 May 2004.

Haysom, A. 2006. *A study of the factors affecting sustainability of rural water supplies in Tanzania.* Unpublished MSc thesis, Cranfield University, Silsoe, UK.

JMP 2006. *Meeting the MDG drinking water and sanitation target: the urban and rural challenge of the decade.* WHO/UNICEF Joint Monitoring Programme. Available at http://www.wssinfo.org/pdf/JMP_06.pdf

Mathew, B. 2004. *Ensuring sustained beneficial outcomes for water and sanitation programmes in the developing world.* PhD thesis, Cranfield University, UK.

MacDonald, A., Davies, J., Calow, R. & Chilton, J. 2005. *Developing groundwater: a guide for rural water supply*. ITDG Publishing, Rugby, UK, 358 pp.

MacDonald A. M., Davies, J. & Calow, R. C. 2008. *African hydrogeology and rural water supply.* In: Adelana, S. M. A. & MacDonald, A. M. Applied Groundwater Studies in Africa. IAH Selected Papers on Hydrogeology, Volume 13, CRC Press/Balkema, Leiden, The Netherlands.

McDonald, J. 2006. *Understanding the problem of borehole water conflict between people and livestock in an emergency context: a case study investigation of Katakwi and Amuria districts, north eastern Uganda.* Unpublished MSc thesis, Cranfield University, UK.

Rushton, K. R., Eilers, V. H. M. & Carter, R. C. 2006. *Improved soil moisture balance methodology for recharge estimation.* Journal of Hydrology, **318**, 379–399.

Sansom, K. R., Franceys, R. W. A., Njiru, C., Kayaga, S., Coates, S. & Chary, S. 2004. *Serving All Urban Consumers: A marketing approach to water services in low and middle-income countries. Book 2 – Guidance notes for managers.* WEDC, Loughborough University, UK.

Schouten, T. & Moriarty, P. 2003. *Community water, community management: from system to service in rural areas.* ITDG Publishing, Rugby, UK.

SKAT 2001. *Manuals on Drinking Water Supply: Vol 6 Drilled Wells.* SKAT, Vadianstrasse 42, CH-9000, St Gallen, Switzerland.

Smith, C. 2003. *Rural boreholes and Wells in Africa – Economics of Construction in Hard Rock Terrain.* American Water Works Association Journal, **95** (8) 100–111.

Sutton, S. 2002. *Community-led improvements of rural drinking water supplies.* Knowledge and Research Project (KAR) R7128, Final Report, UK Department for International Development, 55 pp.

Thompson, J. R. & Hollis, G. E. 1995. *Hydrological modelling and the sustainable development of the Hadejia-Nguru wetlands, Nigeria*. Hydrological Sciences Journal, **40** (1), 97–116.

UNDP 2006. *Human Development Report. Beyond scarcity: power, poverty and the global water crisis.* United Nations Development Programme, New York, USA.

UNDP 2007. *Human Development Report 2007/2008 Fighting climate change: human solidarity in a divided world.* United Nations Development Programme, New York, USA.

UNICEF 2006. *The State of the World's Children.* United Nations Children's Fund, New York, USA.

Woodhouse, M. 1991. *Natural Indicators of shallow groundwater in Kibwezi Division, Kenya,* Journal of the East Africa Natural History Society and National Museum, **81**, 197.

World Bank 2000. Voices of the Poor. Can anybody hear us? World Bank, Washington DC, USA.

Wurzel, P. 2001. *Drilling Boreholes for Handpumps.* SKAT Working Papers on Water Supply and Sanitation, No 2, SKAT, St Gallen, Switzerland.

CHAPTER 4

A GIS-based flow model for groundwater resources management in the development areas in the eastern Sahara, Africa

W. Gossel, A.M. Sefelnasr & P. Wycisk
Geology Department, Martin-Luther University, Halle, Germany

A.M. Ebraheem
Geology Department, Assiut University, Assiut, Egypt

ABSTRACT: To simulate the response of the Nubian Sandstone Aquifer in the eastern Sahara to climate changes during the last 25,000 years, and modern pumping, a regional 3D groundwater flow model was developed and calibrated under steady state and transient conditions. Telescoping meshes were developed in this regional model to include the local details of development areas. The data for this model are held in a GIS database to allow for its implementation in different modeling systems. The simulations using finite element numerical modeling confirm that groundwater in this aquifer is likely to have originated from infiltration during the wet periods of approximately 20–25 ka BP and 5–10 ka BP. Modern recharge of groundwater due to regional groundwater flow from more humid areas to the south is highly unlikely. The model also indicates that the Nubian Aquifer System is a fossil aquifer system, which has been in an unsteady state condition since the end of the last wet period, approximately 5000 years ago. The telescoping modeling approach offers a good solution for the insufficient boundary conditions in the development areas. The simulation results of the large scale model, with telescoping meshes to account for current abstraction, demonstrates that the groundwater reserves of the Nubian Sandstone Aquifer in Egypt and Libya are being mined. By expanding the presently established well fields to their full capacity by year 2020, the water-levels will continuously decline and may fall below economic levels of abstraction.

1 INTRODUCTION

The integrated and sustainable management of the limited water resources in arid regions constitutes a significant issue today due to increased demand for water, water deficiency and ecological problems caused by overuse of available water. A classic example of the issues surrounding groundwater management in arid areas is the Nubian Aquifer System (NAS) in the eastern Sahara – the largest groundwater system in Africa, extending over more than 2 million square kilometers (Figure 1). The NAS is considered the only major fresh water resource in the eastern Sahara besides the River Nile which flows across its

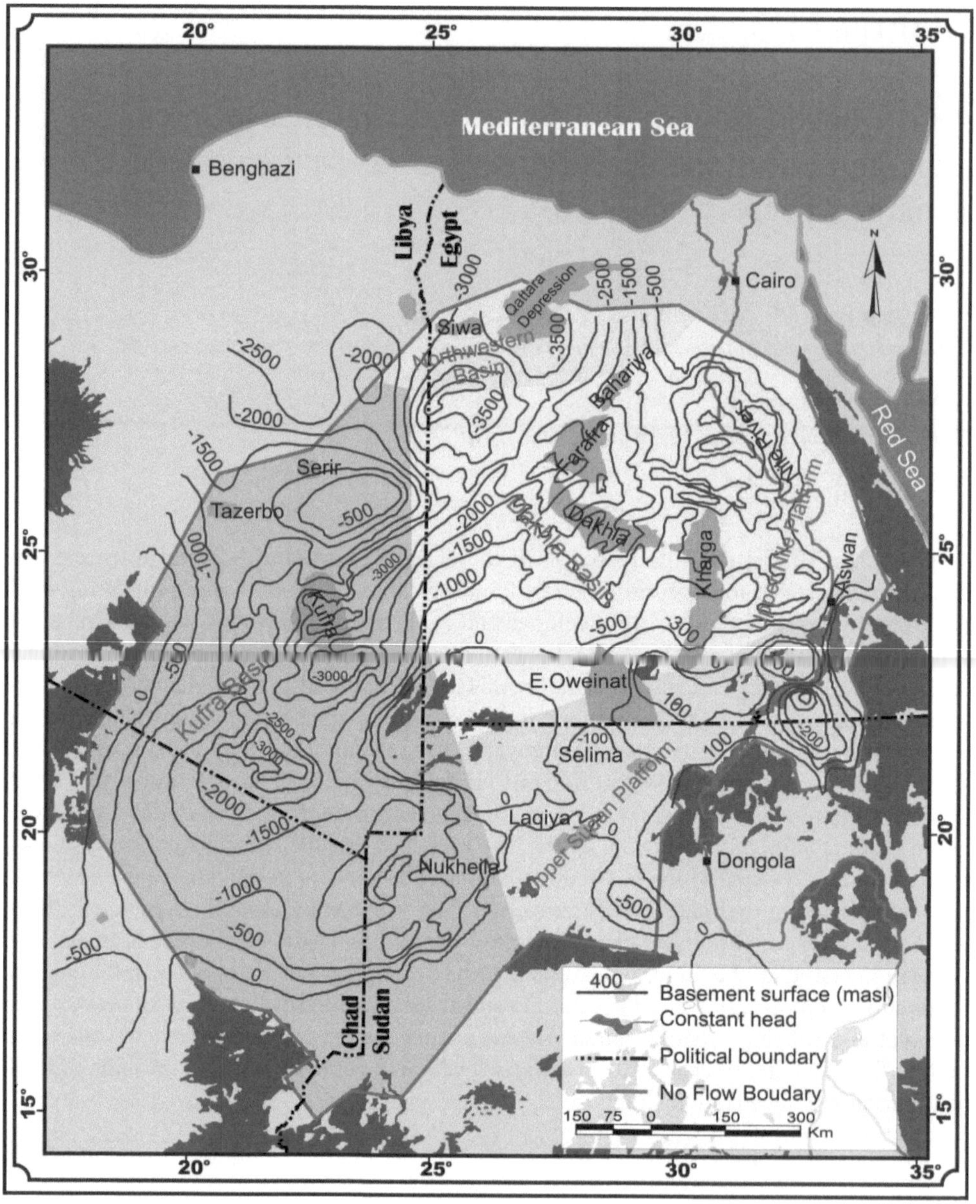

Figure 1. Extent of the Nubian Aquifer system, showing the elevation of the basement surface and the structural basins (after Klitzsch, 1983; Hissene, 1986; Hesse *et al.*, 1987). Pink areas are places of groundwater discharge.

eastern edge. Therefore, much of the population in this arid area outside the Nile valley depends totally on the NAS water for their domestic, agricultural, and industrial uses.

The NAS is formed by two major and two minor basins: (a) the Kufra Basin, which comprises the southeastern area of Libya, the northeastern area of Chad and the northwestern corner of Sudan; (b) the Dakhla Basin of Egypt; (c) the northwestern basin of

Egypt; and (d) the Sudan Platform (Wycisk, 1993). In the centre and north of the system, where a hyper-arid climate prevails, the average precipitation ranges from 0 to 5 mm/a. Consequently, there is no current groundwater recharge in most parts of the system and most of the available groundwater resources were recharged during wetter periods in Saharan history approximately 20–25 ka and 5–10 ka before present (Edmunds & Wright, 1979; Heinl & Thorweihe, 1993; Edmunds, 1994; Ebraheem *et al.*, 2002; Ebraheem *et al.*, 2004; Gossel *et al.*, 2004). Active recharge occurs close to the Nile, and from Wadi systems at the edge of the NAS (Edmunds, 1994).

The NAS has been the subject of hundreds of studies since the beginning of the nineteenth century. Most of the studies prior to the 1980s concluded that the aquifer had been under steady state conditions before development started in the 1960 (e.g. Sandford, 1935; Hellstrom, 1939; Ezzat, 1974; Amer *et al.*, 1981). The more recent studies e.g. (Heinl & Thorweihe, 1993; Churcher *et al.*, 1999; Ebraheem *et al.*, 2002; Gossel *et al.*, 2004) recognize that the aquifer system has been in transient state condition since the current arid period began about 5000 years ago.

Several groundwater modeling studies have been carried out during the last three decades and used as groundwater management tools (e.g. Nour, 1996; Heinl & Brinkmann, 1989). These models are mostly for specific small areas of the Nubian Aquifer System. In these models, the boundary conditions are poorly defined and groundwater abstraction from all other areas within the same aquifer is neglected.

Other models have been constructed covering the whole area of the Nubian Sandstone Aquifer focusing on determining the degree of non-equilibrium and time response of the aquifer to various stresses. These models ignore the local abstraction of local development areas. Only one attempt has been made by (Ebraheem *et al.*, 2002) to develop a large-scale model with refined grids on the development areas. However, the model only covered the Egyptian part of the aquifer and ignored the ongoing major groundwater extraction in Libya.

To help improve modeling and take into account both regional and local scales, Gossel *et al.* (2004) used GIS software to create a database comprising all the available hydrogeological information from the previous studies and hydrogeological data from the newly drilled water wells in Egypt and Libya up to year 2001. After constructing the GIS database, an integrated GIS-based numerical time dependent three-dimensional transient groundwater flow model for the Nubian Aquifer System in the Western Desert of Egypt and the adjacent countries was developed and used for simulating the response of the aquifer to the climatic changes that occurred in the last 25,000 years. The model calibration (discussed later) used palaeolakes that existed in this period to estimate and calibrate the groundwater recharge of this long term time horizon. Since 2004, this model has been used for:

1. Establishing a spatial and temporal prediction system for groundwater flow in whole area of NAS (the Nubian Aquifer System).
2. Developing the local scale models for the development areas by refining the grid cells of each development area in the calibrated regional model to involve the local details. In the refined grid areas, inputs from the regional model serve as boundary conditions in the refined grid. This approach allows improved analysis of pumping and the resulting drawdown (Leake & Claar, 1999) as well as taking into account the groundwater extraction from other areas in the same aquifer (which is neglected by local scale modeling studies). The resulting model is used to determine the impact of current and planned groundwater extractions.

3. Developing solute transport models in areas where sufficient hydrogeological and hydrochemical data are available.
4. Making the necessary predictions to develop a good management scheme for the whole aquifer.

2 GEOLOGICAL AND HYDROGEOLOGICAL SETTING FOR THE MODEL

The NAS is subdivided by uplifts (Figure 1 and Figure 2). The Cairo-Bahariya arch separates the northwestern Basin of Egypt from the Dakhla Basin. The Kharga uplift forms the eastern margin of the Dakhla Basin. The Oweinat-Bir Safsaf-Aswan uplift separates the Dakhla Basin from the north Sudan platform. The Howar-Oweinat uplift forms the eastern border of the Kufra Basin. However, separation of the Kufra Basin from the Dakhla Basin caused by uplift is not evident and the basins are likely connected (Wycisk, 1993).

The dominant geological units of the Nubian Sandstone System (Figure 1 and Figure 2) are the Kufra Basin and the Dakhla Basin. The formation of the Kufra Basin began in Early Paleozoic and was completed at the end of the Cretaceous. The Dakhla Basin was presumably formed at the beginning of the Cretaceous, at least its southern part. North of the Dakhla oasis latitude, Paleozoic sediments can be found (Figure 2). The Dakhla Basin is filled with continental and marine strata of the Paleozoic to early Eocene age in the northwest and of Jurassic to early Eocene in the south. The geological and lithological information was used to set the parameters of the 3D numerical groundwater model.

The top of the basement in the Kufra oasis lies at 3500 m below mean sea level where the aquifer has its maximum thickness of 4000 m. In the area of Kharga, which represents the eastern edge of the basin, the top of the basement lies at 1000 m to <500 m below mean sea level. In Dakhla oasis, the top of the basement lies at about 2000 m below mean sea level (Figure 1). Since the sediments of the Nubian Aquifer System were deposited in a predominantly continental environment, meandering rivers and deltas were the usual transport mechanism (Wycisk, 1993).

In the east, south, and west, basement outcrops bound the system of the described basins and, therefore, the Nubian Aquifer System. In the north, the fresh water aquifer is bounded by saline water that originates either from intrusion of seawater or saline groundwater that has not flushed out since the sedimentation of marine sediments. The northwestern boundary is given by a no-flow boundary according to the overall flow conditions as observed in previous investigations (e.g. Ball, 1927; Sandford, 1935). For the model, a no flow boundary was put around the boundary of the NAS, since recharge is assumed to be negligible, and only partial seepage from the Nile. Natural discharge occurs in some locations on the Nile and in the oases.

3 DEVELOPMENT OF THE GIS DATABASE

All available information was digitized and entered into the GIS database of Gossel *et al.* (2004). This included fundamental geological, hydrogeological, and climatic data from previous studies (Ambroggi, 1966; Klitzsch *et al.*, 1979; Klitzsch & Lejal-Nicol, 1984; Hesse *et al.*, 1987; Klitzsch & Squyres, 1990; Thorweihe, 1990; Meissner & Wycisk, 1993; CEDARE, 2001); information from newly drilled boreholes in Egypt and Libya up to year 2004; information from available cross-sections of the area; and the DEM (Digital

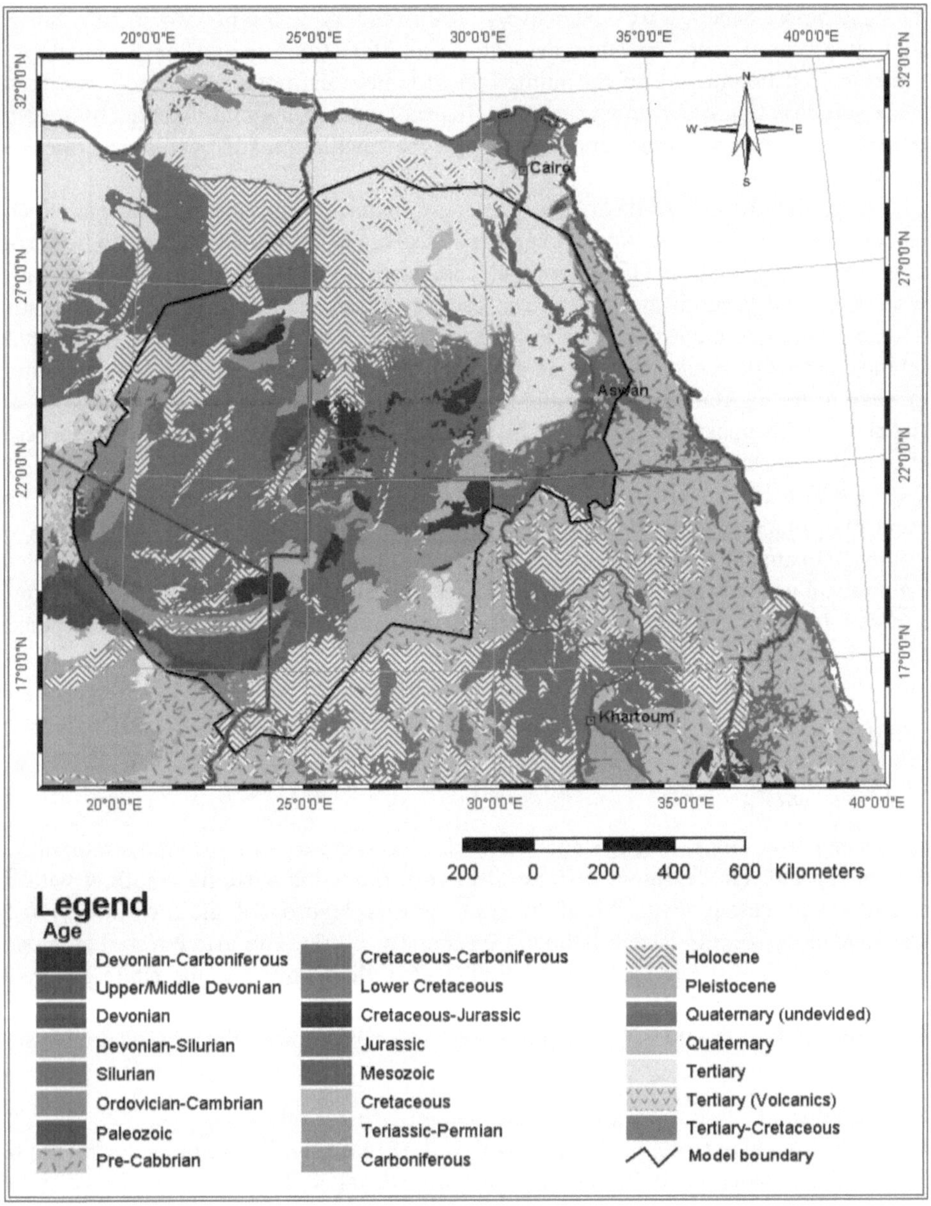

Figure 2. Map of surface geology of the Nubian Sandstone Aquifer (after Sefelnasr, 2006).

Elevation Model) distributed by SRTM-03 of Africa (NASA, 2005). Compiling the data into such coherent and logical GIS-structure helps to ensure the validity and availability of the data and provides a powerful tool for accomplishing the purposes of the study. GIS also helps in management of hydrogeological data and hydrogeological analysis, and also to provide interpreted information, such as vulnerability assessments.

The geological database was extensively used in the modeling process to calculate the model layer bottom, top, and thickness of the aquifer structure and the formulation of parameters and boundaries of the numerical groundwater model. GIS tools were used for: the interpolation of layer surfaces; the interpolation of hydraulic conductivities; the formulation of outer and inner boundaries; and the calculation and spatial description of groundwater recharge areas.

The interpolation tools in GIS are needed to get a spatial distribution of model parameters. For geological purposes statistical and geostatistical analyses of the data have to be carried out before the interpolation. Hydrogeological data with a normal distribution (e.g. elevation data or measured groundwater levels) can be interpolated with kriging and geostatistics. The map of the basement top elevation (Figure 1) represents one example of an output from the created GIS. Data sets without normal distributions (e.g. hydraulic conductivities) first have to be log-transformed to obtain a normal distribution and then analyzed and interpolated with geostatistical tools. In GIS the data can be pre- and post-processed, proved and corrected.

GIS was also conveniently used to control the surfaces of the layers of the numerical groundwater model so that there are no intersections. Although ArcGIS (ESRI, 2005) is not fully 3D capable (only 2.5D) it is sufficient to prepare the 3D structural model for the numerical groundwater models. The GIS database was also used for the calibration of the model. It is a useful tool to visualize the deviation between modeled and interpolated measured water levels as well as statistical calculations.

4 NUMERICAL GROUNDWATER FLOW MODEL

4.1 *Model setup*

Two 3D numerical modeling systems were chosen as basic tools for the simulation of the Nubian Sandstone System: A finite difference modeling system (Modflow with the Pre- and Postprocessor Visual Modflow 3.0 (Waterloo Hydrogeologic Inc., 2002)) and a finite element modeling system Feflow 5.0 (Diersch, 2003). The grid covered an area of 2.2 million km^2 and the modeled area about 1.65 million km^2. For the finite difference solution a grid of 10×10 km was used. In the finite element model the area of the triangles ranged from 10 km^2 to 100 km^2. The main reasons for the choice of a 3D modeling system is to model:

- the large distance flow from Chad to the Qattara Depression;
- the climatic change from wet to semi-arid to the present arid conditions during the last 25,000 years, which is the flow time of this large distance flow;
- the possibilities to build a transport model with implementation of slices with vertically differentiated flow and transport parameters.

The model was designed as a closed system. In this way, reliable no flow boundary could be identified at the outcrops of the basement. As the saltwater-freshwater interface seems to be highly stable, it is also implemented as a no flow boundary (Neumann condition). All groundwater flow, recharge and discharge occurred within the model. Groundwater recharge was implemented on the top layer. Grid cells of the Nile River were considered as constant head cells with spatially varying time constant heads in this long-term simulation (Dirichlet condition).

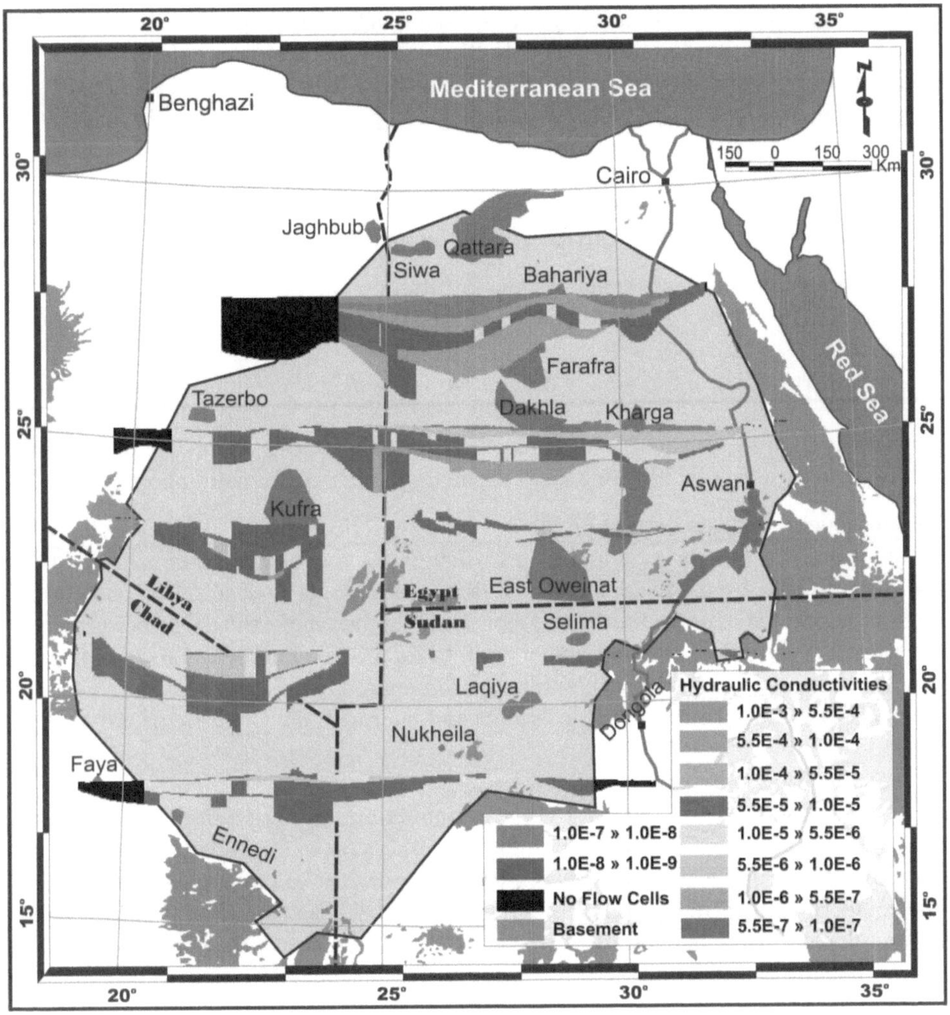

Figure 3. Cross sections of calibrated hydraulic conductivities (m/s) for the Nubian Aquifer System as they are implemented in the numerical groundwater model. Original data based on drilling information from 1960–2005 (after Sefelnasr *et al.*, 2006).

The hydraulic conductivity was evaluated for the entire area using the available drilling information up to year 2001 in the newly developed areas, e.g. Tushca (southwest of Aswan) and East Oweinat (Dahab *et al.*, 2003), as well as the results published in Thorweihe & Heinl (1999) and shown in the cross sections in Figure 3. Based on the variability of hydraulic conductivity values and the general stratigraphic setting, the Nubian Sandstone Sequence was divided into 10 layers. The layers 2 to 9 were further divided into three layers each to ensure a representative value of hydraulic conductivity and other flow parameters particularly in a local-scale solute transport model (not reported here). For the same reason, the bottom and top layer were also divided into two layers each. The confined part of the system in the north was considered as 'leaky aquifer', allowing vertical water exchange between the Nubian

Aquifer and overlying sediments. Evapotranspiration in the large Egyptian oases was made possible and implemented as discharge (Modflow) or negative recharge (Feflow). The two different numerical groundwater modeling systems have been chosen due to the problems of the finite difference system with dry cells in the upper layers.

4.2 *Simulation Runs*

The following simulations were carried out for the regional aquifer.

1. 'Steady-state' conditions simulating the infiltration on the southern highlands and evaporation in the Egyptian oases in the last 50 years neglecting the anthropogenic abstraction.
2. A long-term transient simulation of the aquifer behavior, due to climatic change followed the steady simulation.
3. A short-term time variant simulation of the period 1960–2000 using the available hydrogeological data in 1980 and 2000 in the Egyptian part for calibration.

In subsequent simulations and prediction runs (described later), the grid cells of the development areas were refined in a manner that all local details can be included. These simulations involved the development of a local-scale model for each development area within the calibrated regional model and then using them to make the necessary prediction of the impact of the different management regimes for the next hundred years. A simulation of solute transport is planned for the future. All simulations were done with both modeling systems, first with the finite difference model and later with the finite element model.

4.2.1 *Steady-State simulation*

The steady state model was run as a 50-year simulation transient model to calibrate the hydraulic conductivities. The recharge in the model was set to the hyperarid climatic conditions of the last 100 years with a few millimeters per year of recharge on the high land at the southern edge of the model area and Gilf Kebir Plateau (at Gebel Oweinat) and an average value of 30 mm/a discharge through the Egyptian oases in the northern part of the aquifer. This model achieved a 'steady state' solution for the whole aquifer area as shown in (Figure 4). The aim of the first 'steady state' model was to match the contour map of Ball (1927) and Sandford (1935) which was created before the start of pumping of groundwater in several oases in the 1960s. The 'steady state' finite difference model (Figure 4) showed at a first glance a similar pattern as the contour lines of Ball (1927). This result was improved further by using the finite element model. The root mean square error of the finite element model was only about 4 m. For an area of about 1,000,000 km^2 a difference between measured and calculated groundwater levels of less than 5 m was achieved (Gossel *et al.*, 2004).

4.2.2 *Long term simulation of the aquifer system due to climatic changes*

To clearly answer the question whether the groundwater encountered today in the Nubian Aquifer System has been formed during historical humid climate periods, or if groundwater is currently recharging from more humid areas in the south, the regional model was used to simulate the aquifer response to climatic change during the last 25,000 years (using climate data recently confirmed by isotopes and paleontological studies (Thorweihe, 1986; Pachur, 1999; and Churcher *et al.*, 1999).

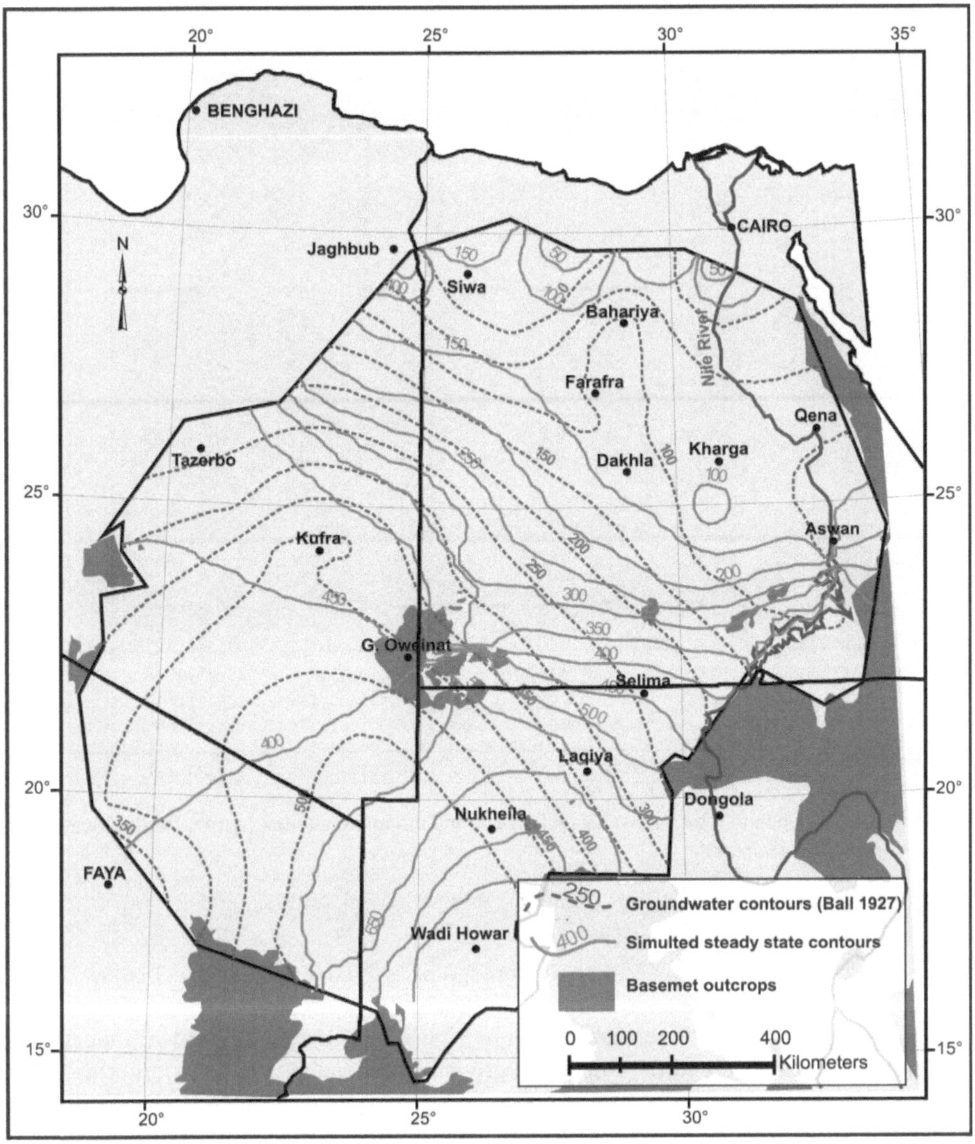

Figure 4. Simulated groundwater contours for steady state conditions with infiltration in southern highlands and discharge through the Egyptian oases (after Gossel *et al.*, 2004).

As shown in Figure 5 this long-term simulation started with a recharge period between 25–20 ka BP. In this period a 10–20 mm/a recharge rate was assumed (Pachur, 1999). Rainfall was dominated by southeast directed winds from the Mediterranean causing precipitation at the mountains in Sudan and Chad (Tibesti and Ennedi Mountains). The spatial distribution of recharge zones used in the model is shown in Figure 6. Recharge in the period between 10–5 ka BP was assumed to be up to 40 mm/a. In the periods 20–10 ka BP and 5 ka BP to present day, infiltration was set to 0 mm/a. The evaporation data for the

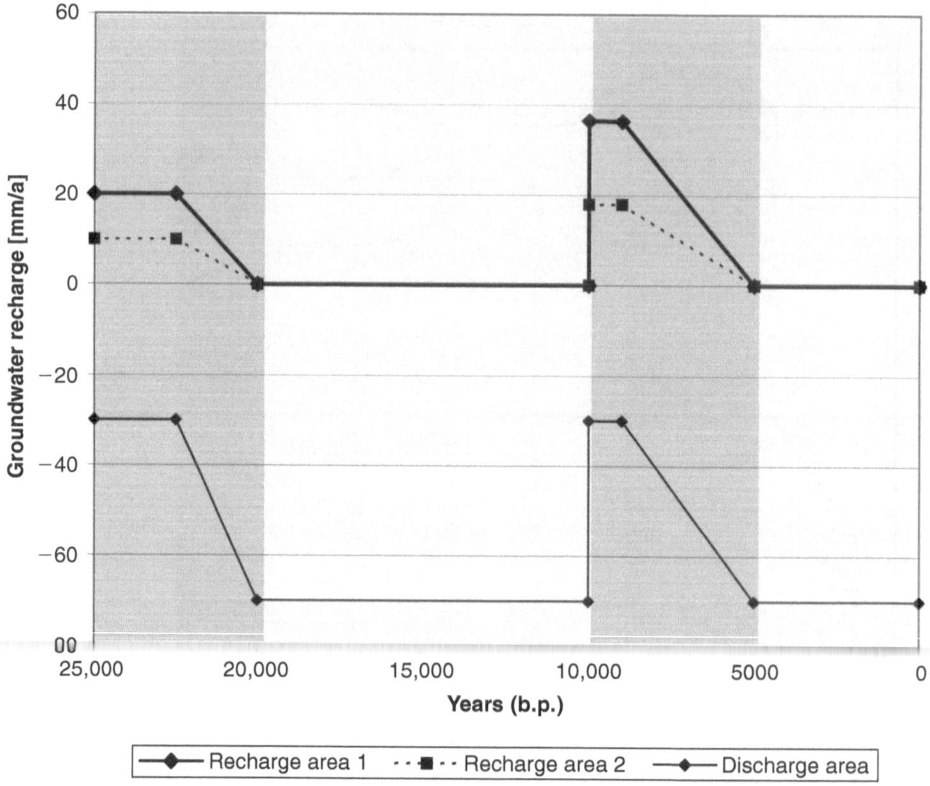

Figure 5. Time dependent groundwater recharge in the model area (see Figure 6 for the location of the different recharge areas).

depression areas and oasis varied from 70 mm/a during the dry periods and 35 mm/a in the wet periods (Figure 5).

Simulated groundwater levels are shown in Figure 7 for the finite element model. The pattern of the measured and the calculated groundwater isolines match in largest parts of the modeled area. To verify the model results in the past, the geological and geographical analysis of Kröpelin (1993) and Pachur (1999) were used. The model successfully simulated the behavior of known paleolakes during the period and in particular the behavior of the former Lake Ptolemy (Sudan) at 560 m above sea level. For more details see Gossel *et al.* (2004).

The modeled aquifer response to climate variations confirms that groundwater in the NAS most likely recharged during the more humid climatic periods by regional infiltration. The model indicates that the decline of groundwater levels started about 19,000 years ago, but was slowed down and reversed by regional infiltration during the last wet period (of approximately 10–5 ka BP). It took about 5500 years after the start of the second wet period for the water-levels to recover and the aquifer to become almost full, and water-levels have now been declining since then. Since the last wet period, natural discharge has not been balanced by recharge (Figure 8). Natural discharge does not depend directly on the climate

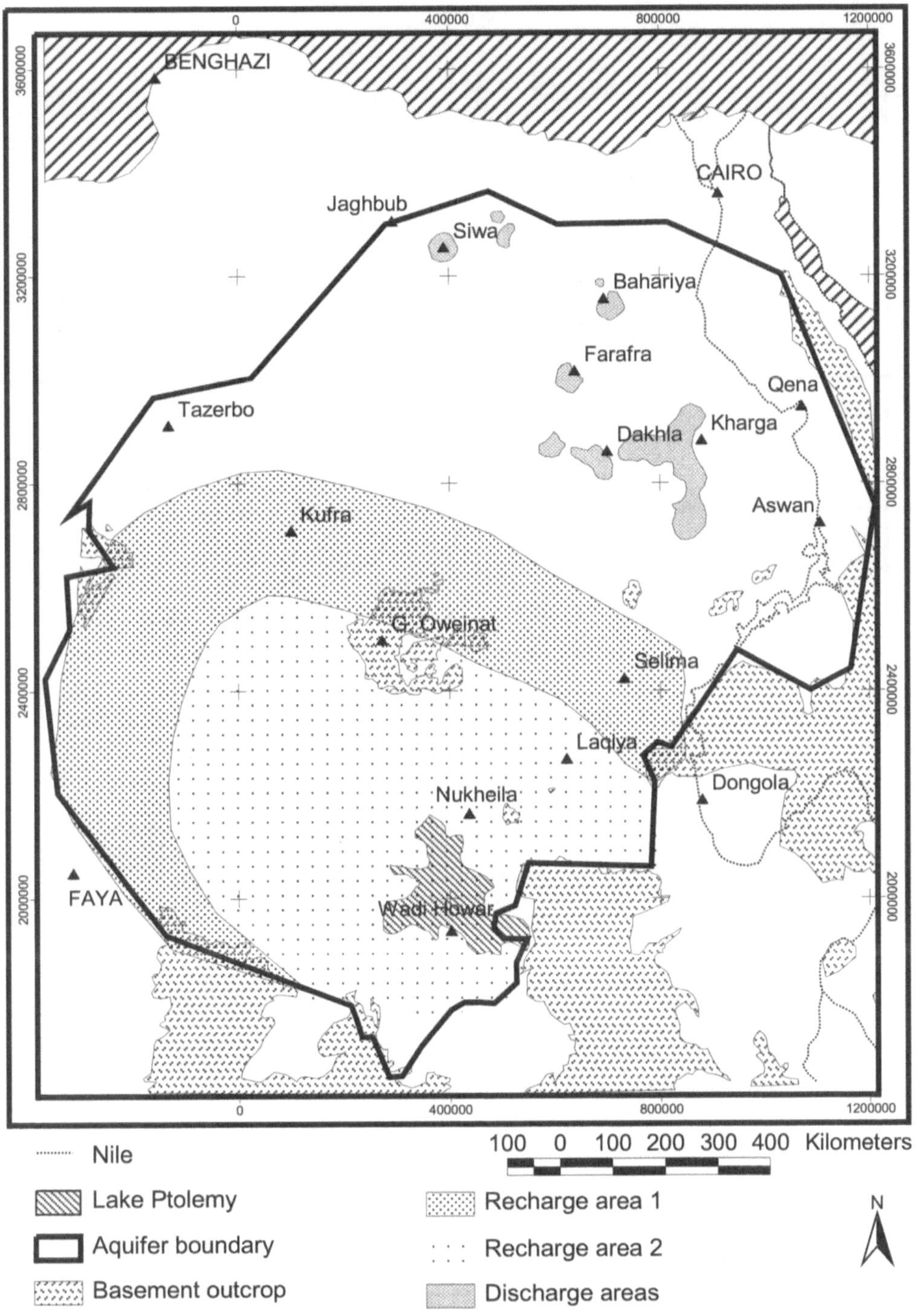

Figure 6. Recharge and discharge areas and the location of the former Lake Ptolemy (after Gossel *et al.*, 2004).

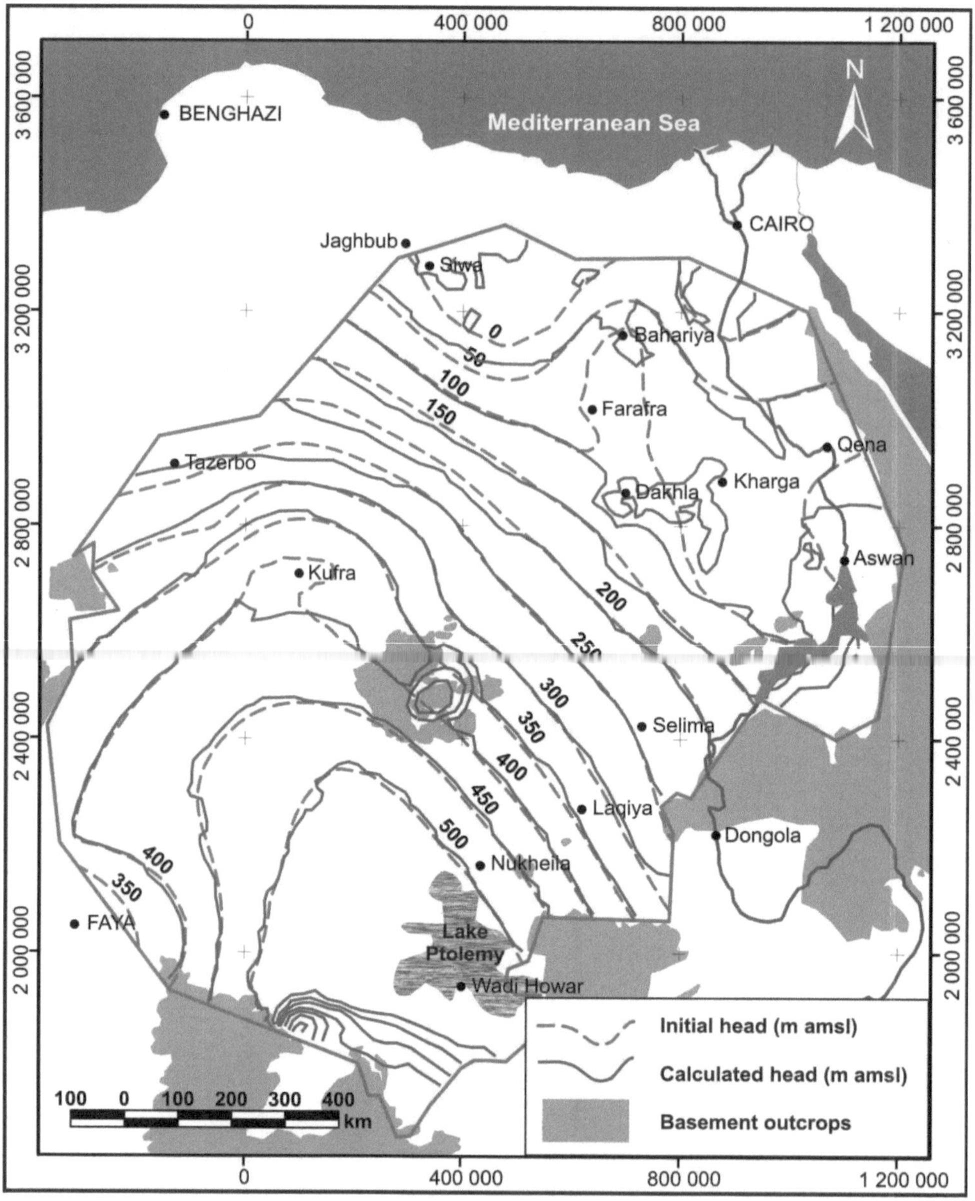

Figure 7. Simulated groundwater contours in 1960 after a simulation of climate change over the past 25,000 years. The groundwater contours of Ball's (1927) outside the Egyptian part of the aquifer were considered as the observed values for other areas (finite element model) (after Gossel *et al.*, 2004).

but rather the potential head distribution, therefore, discharge continues during the dry periods, meaning that the system is not in steady state.

Between 20–10 ka BP, infiltration stopped and the simulated groundwater contours indicate that groundwater level dropped, particularly in the higher areas. Within this time,

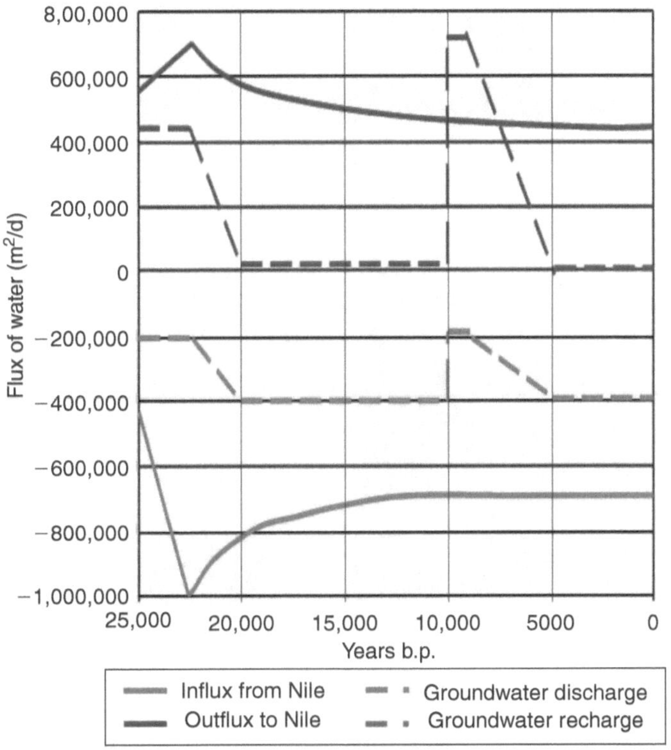

Figure 8. Groundwater balance for the Nubian Aquifer System in the last 25,000 years (finite element model). Note that groundwater discharge was imposed on the model to give evapotranspiration.

groundwater levels fell between 40 and 70 m in Gilf Kebir Plateau and about 20 to 30 m at the southern and western boundary of the model area. During the wet periods, however, the calculated depth to groundwater shows that in wide parts of the model area the aquifer comes to a nearly full condition. It is also interesting to note the time delays in the system. The highest groundwater levels are calculated about 50 to 100 years after the end of the wet periods in the southern parts of the model area. Groundwater levels have continued to decline since the end of the last wet period. Modeled depth to groundwater in 1960 is shown in Figure 10.

Even during wet periods a quasi steady state condition could have only existed for a short time for the whole aquifer area. Taking into account that the average distance between recharge areas in the south and discharge areas in the north is about 1000 km and the average groundwater velocity in sandstones is about 1 m/a, then the flow time between the recharge and development areas would be one million years. The ages of groundwater shown by (Thorweihe & Heinl, 1999) of about 20,000 to 40,000 years show that also in some northern parts recharge must have occurred in the wet periods. Older groundwater ages, found e.g. by Sturchio *et al.* (2004) can be found in deeper structures of the aquifers and in the northern parts of the NAS.

The main focus of the simulation was to decide whether or not flow in the Nubian Aquifer System was in a steady state prior to current abstraction. As shown above, the

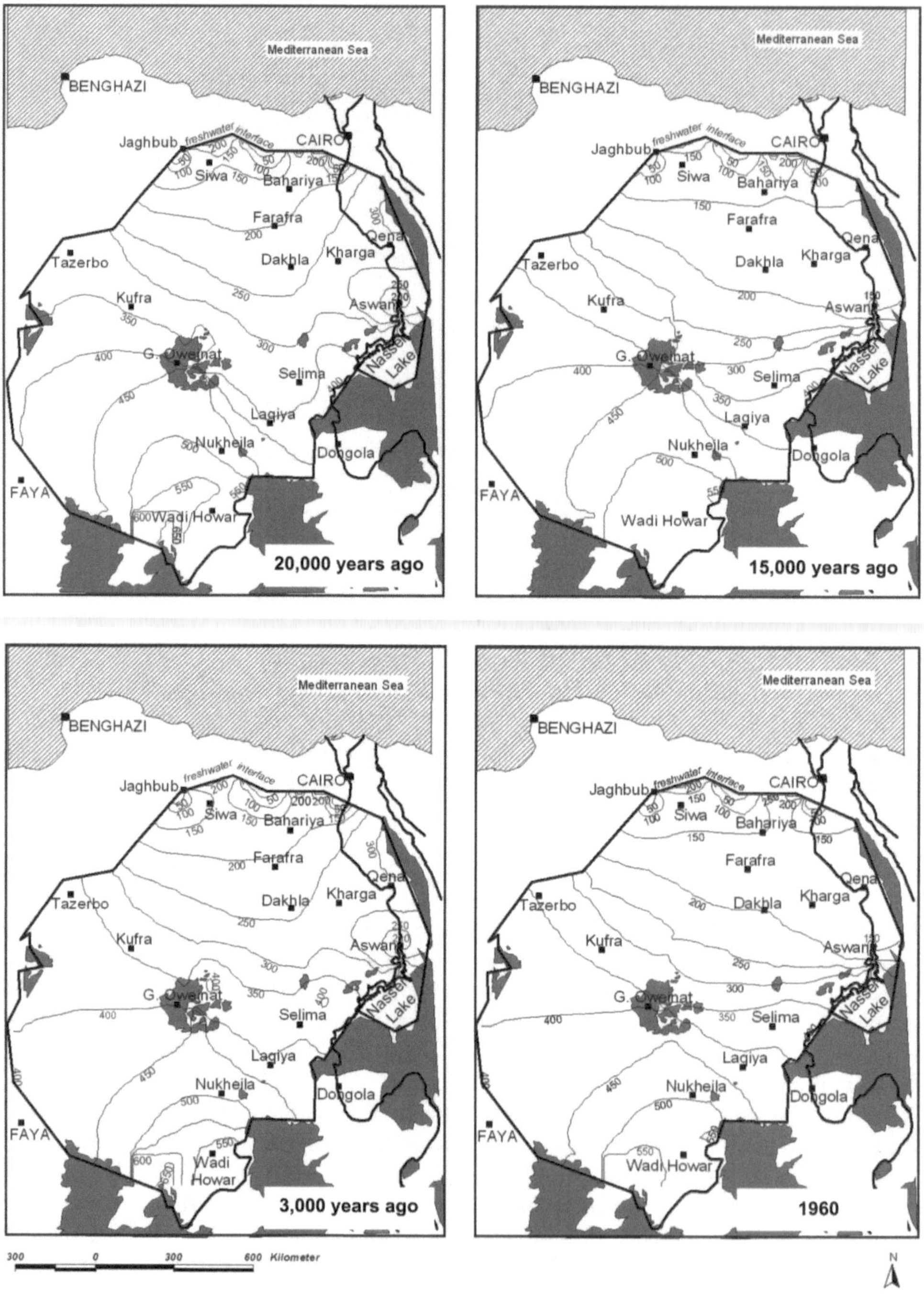

Figure 9. Simulated response of NAS to the climatic changes in the last 25,000 years.

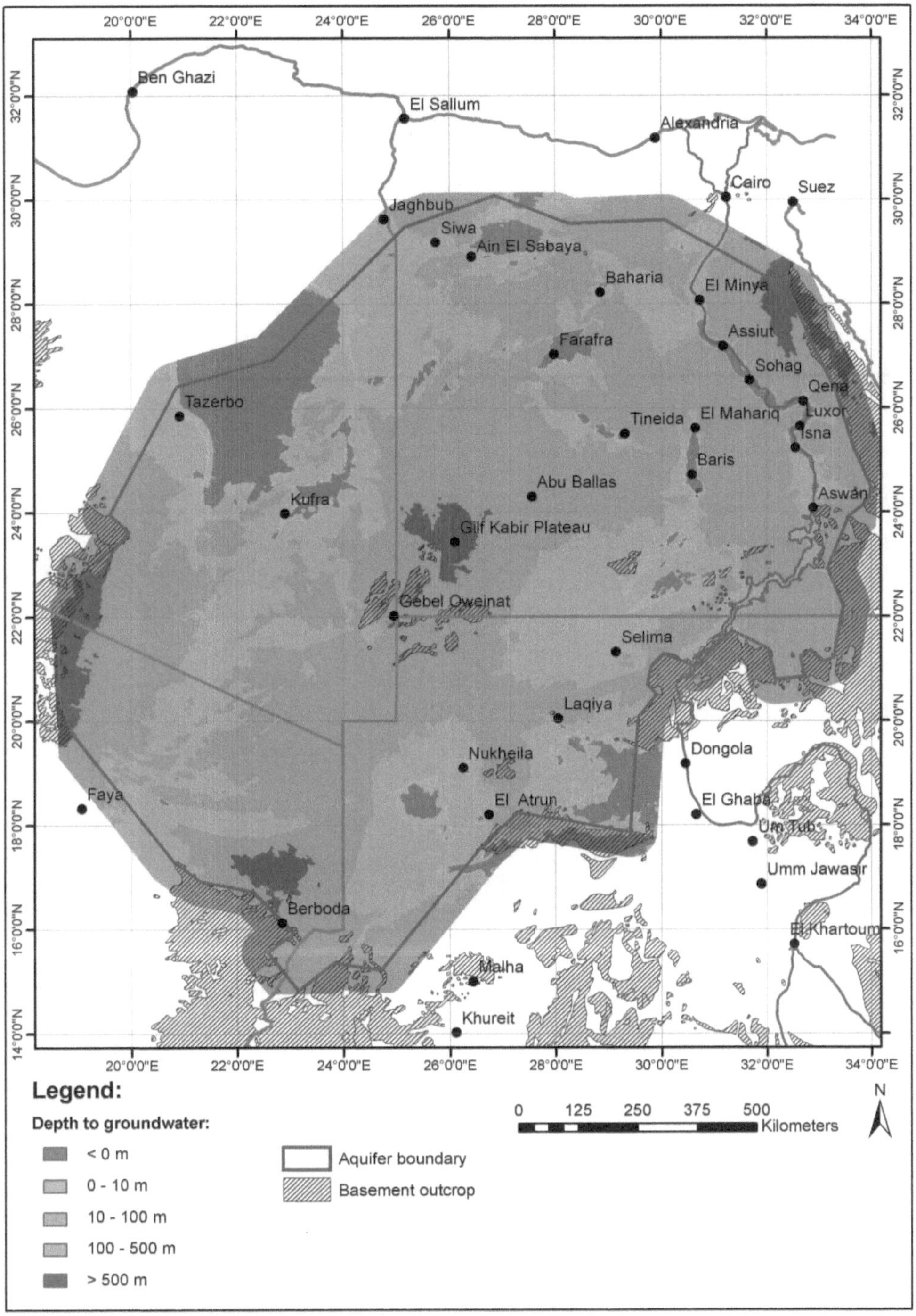

Figure 10. Depth to groundwater in 1960 (finite element model) (after Gossel *et al.*, 2004).

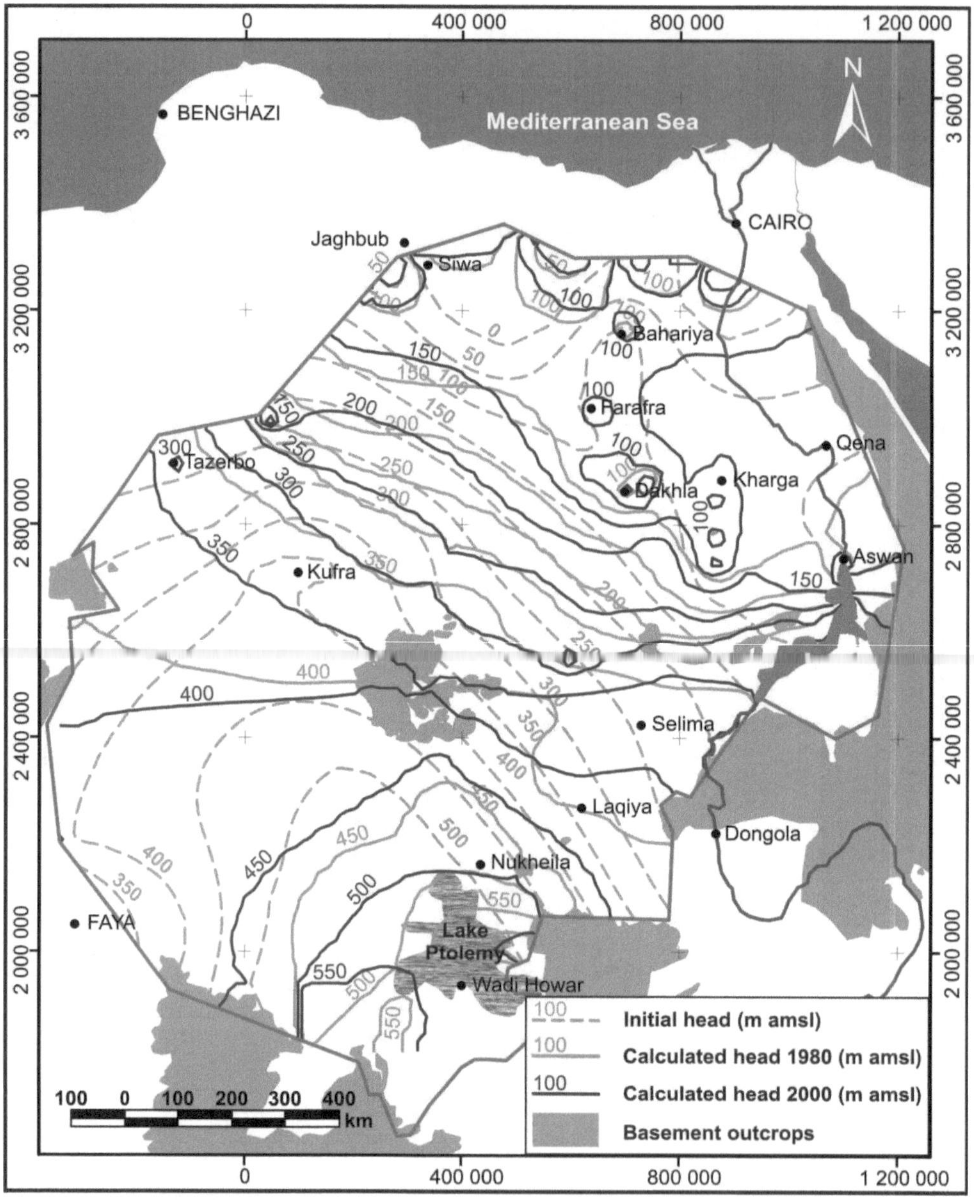

Figure 11. Simulated groundwater contours in 1980 imposing groundwater extraction in the period 1960–1980 after a very long simulation of climatic change. observed values are also shown between parentheses for the Egyptian Oasis. The groundwater contours of (Ball, 1927) outside the Egyptian part of the aquifer were considered as the observed values for other areas (finite difference model) (after Gossel *et al.*, 2004).

model indicates clearly that the water-levels were still declining since the end of the last wet period and had not reached the low levels of the time before 10,000 years BP. This indicates that the natural water-level decline is ongoing and could last for another few 1000 years.

Certain aspects of the research results reported in Pachur (1999) merit further investigation. The reported former lakes could be a result of higher runoff or elevated high groundwater levels. If they were formed as a result of a high groundwater level, the recharge period must have been a little bit earlier than the formation of lakes due to the time shift between groundwater recharge and rising groundwater levels in the concerned areas. The formation of lakes in consequence of a high runoff on the other hand leads to an increased infiltration in the gathering areas and thus to higher groundwater levels. Research into these aspects is ongoing.

4.2.3 *With groundwater abstraction*

The official records of groundwater abstraction for irrigation and domestic water supply were used to calculate the abstraction for each individual pumping centre in the NAS area. The abstraction of all boreholes in each development area were applied for each year in this period and considered as the abstraction rate of a single borehole. The modeling systems allow the location of well screens in more than one layer so that there was no need to classify boreholes as shallow and deep. Due to the intensive abstraction rates in the Dakhla, Kharga, and Kufra oases, wells in each of these oases were grouped into three pumping centers. The abstraction rates shown in Figure 9 are the formally reported rates for year 2000. However, the present abstraction rates may be higher at some places (e.g. Kufra oasis in Libya and Kharga and East Oweinat in Egypt). In a proposed abstraction scenario by CEDARE (2001), the present abstraction rate will probably be reduced in some areas.

Due to the close fit between the observed groundwater contours before the development time in 1960 and the computed groundwater contours at the end of the long simulation period (for 25,000 years before 1960), the computed groundwater contours were considered as the initial heads for the short simulation periods. Thus, to model the flow in the period 1960–1980, the model ran for a simulation period of 25,020 years involving the climatic changes in the first 25,000 years and groundwater abstraction in the last twenty years (1960–1980). In the Egyptian part of the aquifer (where groundwater development started in 1960), the simulated hydraulic heads in 1980 are in good agreement with the observed hydraulic heads for that year (Figure 9). In 1980, cones of depression started to develop in Kharga and Dakhla oases and become wider and deeper with time (Figure 9). Outside the Egyptian part of the aquifer, the simulated groundwater contours are also in a good agreement with the observed groundwater contours of the predevelopment time (Figure 9).

This simulation indicates that the natural groundwater level decline, ongoing since the end of the last wet period, is small compared to the effect of modern pumping (compare with Sefelnasr *et al.* (2006)).

5 PREDICTION RUNS

With the available evidence that the Kufra Basin is not separated from Dakhla Basin by uplift and even the small basins are not completely separated, the NAS has to be looked at as one aquifer. The Nubian Aquifer System in the development areas (Egyptian oases and Kufra oasis in Libya) is only a part of the whole NAS. Groundwater flow in the main aquifer layers is governed by conditions at the boundaries of the regional system. The developed and calibrated regional model for the whole NAS (Gossel *et al.*, 2004; Sefelnasr *et al.*, 2006) was used to model groundwater flow in the development area of the Egyptian oases

Table 1. The five possible future groundwater exploitation scenarios; abstraction in million cubic meters per year.

Area	Scenario 1	Scenario 2	Scenario 3	Scenario 4	Scenario 5
Bahariya	34	177	177	177	177
Farafra	283	471	471	471	471
Dakhla	439	625	625	625	625
Kharga	203	124	93	93	93
East Oweinat	164	300	600	900	1200
Kufra	200	300	400	400	500

and Kufra oasis in Libya by refining the grid in these areas. This modeling approach was used for the following reasons:

- it is easy and convenient to include all the local details of the development areas in the regional model by refining its grid in the development areas to the desired level (Leake & Claar, 1999);
- boundary conditions can only be well defined for these areas by using the connection to the large scale model;
- pumping from any of these development areas is affecting the whole system;
- records of hydraulic head observations needed for the model calibration procedure are only available for certain areas.

The grid cells in the regional model in these development areas were refined with a grid spacing of 5000 m in both directions. The regional model was used to calculate the fluxes across the boundary cells of the areas of the local scale model. These fluxes served as boundary conditions for the local scale model (areas with refined grid). This approach allowed precise analysis of pumping and the resulting drawdown in the development areas.

The simulation model was used to calculate five possible future exploitation/development plans (Table 1) for the NAS. The simulation was applied to the years 2000–2100. In the analysis of the simulation results, emphasis was given to the amount of decline in the hydraulic head, and consequently its effect on the depth to groundwater. The results of the modeling scenarios are summarized in Table 2. Some of the obtained results will be briefly discussed below whereas the detail is subject to ongoing research.

The simulation results of scenario 1 indicate that the major cones of depression are centered in the Kharga and Dakhla oases with maximum declines of 80, 50, and 20 m in the year 2100 for the Kharga, Dakhla, and East Oweinat areas, respectively.

In scenario 5, the consequences of expanding the groundwater abstraction rates in East Oweinat and Dakhla oasis to their planned rate (1200 and 625 Mm^3/a respectively) were investigated. The resulting declines in the water-levels (Table 2) indicate that the core of the cone of depression will cover the entire planned reclaimed area in East Oweinat in Egypt. The depth to groundwater in this area will be greater than 140 m. Also, a similar negative impact, but to a lesser extent, was observed for the area around the centre of Dakhla oasis by the year 2100. This indicates that an abstraction rate of 1200 Mm^3/year is not feasible for the East Oweinat area without excessive drawdown, and also it is safer if the rate in Dakhla oasis does not exceed 500 Mm^3/year. Therefore, Scenario 4 is the most

Table 2. The simulated depths to groundwater under the five possible future groundwater exploitation scenarios.

Area	Estimated economic pumping depth (m)	Scenario	Modelled depth to groundwater (m below ground surface)						
			1980	2000	2010	2020	2030	2060	2100
Bahariya	124	Scenario 1	−10*	−7	−6	−5.5	−5	0	3
		Scenario 2	−7	−5	−3	0	2	7	10
Farafra	80	Scenario 1	−60	−40	−35	−30	−25	−22	−18
		Scenario 2	−23	−15	−11	−6	0	1	3
Dakhla	63	Scenario 1	−25	−5	−2	2	5	18	37
		Scenario 2	−18	−1	12	25	39	45	57
Kharga	38	Scenario 1	12	20	24	28	40	50	53
		Scenario 2	−5	5	10	15	20	30	40
		Scenario 3	−7	4	8	13	17	26	36
East Oweinat	100	Scenario 1	−15	−4	0	7	15	20	23
		Scenario 2	16	20	22	24	25	30	34
		Scenario 3	26	30	32	35	37	45	58
		Scenario 4	28	35	39	44	53	64	95
		Scenario 5	12	40	55	70	80	95	145

*negative values indicate artesian flows.

efficient option for managing groundwater resource in the Nubian Sandstone Aquifer in Egypt.

6 CONCLUSIONS

The simulation results indicated that the total groundwater storage in the NAS is about 135,000 km^3, which is very close to the figure obtained by Heinl & Thorweihe (1993) by other means. The model also confirms the main recharge periods of the NAS as the wet periods at approximately 20–25 ka BP and 5–10 ka BP. Under present arid conditions, the sum of recharge from the River Nile and infiltration in the southern part is only 4% of the current discharge in the Libyan and Egyptian development areas.

A transient model simulation of the past 25,000 years shows results that correlate well with observed groundwater levels pre-development (Ball, 1927) and also a good correlation to geological research results for the lakes that existed about 6000 years BP.

To some extent, the groundwater contours from the transient model are similar to the simulated ones of a steady state condition obtained with infiltration on the highland at the southern edge and evaporation in the Egyptian oases. This similarity is caused by very slow groundwater fluctuations in the last 3000 years.

The simulation results of the regional model (without telescoping meshes) indicate that the water-levels were still declining in the aquifer in response to the end of the last wet period and had not reached the low levels of the time before 10,000 years BP. This indicates that the natural water-level decline was ongoing prior to development and could last for another few 1000 years.

The slope of the hydraulic gradient, which has the same direction as the gradient of precipitation from south to north, should not mislead some researchers to conclude that there is continuous recharge to the Egyptian part to account for a steady state condition before 1960.

After each transition to arid climate, natural discharge continues for the entire dry period with a slow decrease with time. After this arid depletion of the aquifer, the groundwater was replenished after the climatic transition to the subsequent wet period. During wet periods, a time span of only 3000 years and a few mm/a of recharge are sufficient to keep the groundwater level near the surface (filled up condition).

The simulation results of the regional model, with telescoping meshes to account for current abstraction, demonstrates that the groundwater reserves of the Nubian Aquifer System in Egypt and Libya are being mined. By expanding the presently established well fields to their full capacity by year 2020, the water levels will continuously decline.

To avoid groundwater depletion in the shallow aquifer and to ensure sustainable development of this precious natural resource in the Kharga oasis, the present abstraction rate in this oasis has to be limited to 93 Mm^3/year. The planned abstraction rate of approximately 500 and 1200 Mm^3/year in the Kufra oasis (Libya) and the East Oweinat area (Egypt), respectively, is not sustainable, and will have a negative impact on water-levels not only in these two areas but also in several areas within Dakhla oasis. However, rates of 300 and 900 Mm^3/year may be feasible in Kufra and East Oweinat areas respectively. The results of an ongoing detailed modeling study will provide a detailed picture about the sustainable extraction rates in these development areas.

The simulated potentiometric surface for the year 2100 for the most intensive development scenario in all development areas except Kharga oasis (scenario 5) did not drop below mean sea level in the northern part of the Nubian Aquifer System in Egypt. Therefore, the possibility of upconing of more saline water and invasion of seawater in the northern parts of the aquifer is low. However, further investigations to confirm or reject this possibility are needed.

REFERENCES

Ambroggi, R. P. 1966. *Water under Sahara.* Scientific American, **214**, 21–49.

Amer, A., Nour, S. & Meshriki, M. 1981. *A finite element model for the Nubian Aquifer System in Egypt.* In: Proceedings of the International Conference on Water Resources Management in Egypt, Cairo, 327–361.

Ball, J. 1927. *Problems of the Libyan Desert.* Geographical Journal, **70**, 21–38, 105–128, 209–224.

CEDARE 2001. *Nubian Sandstone Aquifer System Programme, Regional Maps.* CEDARE, Heliopolis, Cairo.

Churcher, C. S., Kleindienst, M. R. & Schwarcz, H. P. 1999. *Faunal remains from a middle Pleistocene lacustrine marl in Dakhleh Oasis, Egypt: Palaeoenvironmental reconstructions.* Palaeogeography, Palaeoclimatology, Palaeoecology, **154**, 301–312.

Dahab, K. A., Ebraheem, A. M. & El Sayed, E. 2003. *Hydrogeological and hydrogeochemical conditions of the Nubian Sandstone Aquifer in the area between Abu Simbel and Tushka depression, Western Desert, Egypt.* Neues Jahrbuch für Mineralogie, Geologie, und Paläontology, 228, 175–204.

Ebraheem, A. M., Riad, S., Wycisk, P. & Seif El Nasr, A. M. 2002. *Simulation of impact of present and future groundwater extraction from the non-replenished Nubian Sandstone Aquifer in SW Egypt.* Environmental Geology, **43**, 188–196.

Ebraheem, A. M, Riad, S., Wycisk, P. & Seif El Nasr, A. M. 2004. *A local-scale groundwater flow model for the management options in Dakhla Oasis, SW Egypt.* Hydrogeology Journal, **12**, 714–722.

Edmunds, W. M. & Wright, E. P. 1979. *Groundwater recharge and palaeoclimate in the Sirte and Kufra basins.* Journal of Hydrology, **40**, 215–241.

Edmunds, W. M. 1994. *Characterization of groundwaters in semi-arid and arid zones using minor elements.* In: Nash, H. and McCall, G. J. H. (eds.) Groundwater Quality, Chapman and Hall, London, 19–30.

ESRI 2005. ArcGIS 9. http://www.esri.com/software/arcgis/about/literature.html

Ezzat, M. A. 1974. *Groundwater Series in the Arab Republic of Egypt; Exploration of groundwater in El-Wadi El Gedid Project area.* Part 1 to IV, General Desert Development Authority, Ministry of Irrigation, Cairo.

Gossel, W., Ebraheem, A. M. & Wycisk, P. 2004. *A very large scale GIS-based groundwater flow model for the Nubian Sandstone Aquifer in Eastern Sahara (Egypt, northern Sudan, and Eastern Libya).* Hydrogeology Journal, **12**, 698–713.

Heinl, M. & Brinkmann, P. J. 1989. *A Ground Water Model for the Nubian Aquifer System.* Hydrological Science Journal, **34**, 425–447.

Heinl, M. & Thorweihe, U. 1993. *Groundwater Resources and Management in SW Egypt.* In: Meissner, B. & Wycisk, P. (eds.) Geopotential and Ecology, Catena Supplement, **26**, 99–121.

Hesse, K. H., Hissene, A., Kheir, O., Schnaecker, E., Schneider, M. & Thorweihe, U. 1987. *Hydrogeological investigations of the Nubian Aquifer System, Eastern Sahara.* Berliner Geowiss. Abh (A), **75**, 397–464.

Hellström, B. 1939. *The subterranean water in the Libyan Desert.* Geofisika Annaler, **21**.

Klitzsch, E., Harms, J. G., Lejal-Nicol, A. & List, F. K. 1979. *Major subdivision and depositional environments of the Nubian strata, Southwest Egypt.* Bulletin American Association of Petroleum Geologists, **63**, 967–974.

Klitzsch, E. & Lejal-Nicol, A. 1984. *Flora and fauna from strata in southern Egypt and northern Sudan.* Berliner Geowiss-Abh (A) **50**, 47–79.

Klitzsch, E. & Squyres, H. C. 1990. *Paleozoic and Mesozoic geological history of northeastern Africa based upon new interpretation of Nubian strata.* Bulletin American Association of Petroleum Geologists, **74**, 1203–1211.

Kröpelin, S. 1993. *Geomorphology, Landscape Evolution, and Palaeoclimates of southwest Egypt.* In: Meissner, B. & Wycisk, P. (eds.) Geoptential and Ecology. Catena Supplement **26**, 67–97.

Leake, S. A. & Claar, D. V. 1999. *Procedures and computer programs for telescopic mesh refinement using MODFLOW.* Open file report, USGS, **53**.

Meissner, B. & Wycisk, P. (eds.) 1993. *Geopotential and Ecology.* Catena Supplement, **26**.

NASA 2005. SRTM03 data. http://www2.jpl.nasa.gov/srtm.

Nour, S. 1996. *Groundwater potential for irrigation in the east Oweinat area, Western Desert, Egypt.* Environmental Geology, **27**, 143–154.

Pachur, H. J. 1999. *Paläo-Environment und Drainagesysteme der Ostsahara im Spätpleistozän und Holozän.* In: Klitzsch, E. & Thorweihe, U. (eds.) Nordost-Afrika: Strukturen und Ressourcen. Deutsche Forschungemeinschaft, WILEY-VCH, Weinheim, 366–445.

Sandford, K. S. 1935. *Sources of water in the northern-western Sudan.* Geographical Journal, **85**, 412–431.

Sefelnasr, A. M., Gossel, W. & Wycisk, P. 2006. *GIS-basierte Grundwasserströmungsmodellierung des Nubischen Aquifersystems, Westliche Wüste, Ägypten.* FH-DGG-Congress, Cottbus, Germany.

Sturchio, N. C., Du, X., Purtschert, R., Lehmann, B. E., Sultan, M., Patterson, L. J., Lu, Z. T., Müller, P., Bigler, T., Bailey, K., O'Connor, T. P., Young, L., Lorenzo, R., Becker, R., El Alfy, Z., El Kaliouby, B., Dawood, Y. & Abdallah, A. M. A. 2004. *One million year old groundwater in the Sahara revealed by krypton-81 and chlorine-36.* Geophysical Research Letters, **31** (5).

Thorweihe, U. 1986. *Isotopic identification and mass balance of the Nubian Aquifer System in Egypt.* In: Thorweihe, U. (ed.) Impact of climatic variations on East Saharian Groundwaters, Modeling of large scale flow regimes, Proceedings of "Workshop on hydrology", Berliner Geowiss. Abh. (A). **72**, 55–78.

Thorweihe, U. 1990. *Nubian Aquifer System.* In: Said R (ed.) The Geology of Egypt. 2nd Edition, Balkema, Rotterdam.

Thorweihe, U. & Heinl, M. 1999. *Grundwasserressourcen im Nubischen Aquifersystem.* In: Nordost-Afrika: Strukturen und Ressourcen. Deutsche Forschungemeinschaft. WILEY-VCH Verlag GmbH. D-69469 Weinheim, Germany.

Diersch, H.-J. 2003. Feflow reference manual. WASY GmbH, Berlin.

Waterloo Hydrogeologic Inc. 2002. Visual Modflow Pro User's Manual. Waterloo, Canada.

Wycisk, P. 1993. *Geology and Mineral Resources, Dakhla Basin, SW Egypt.* In: Meissner, B. & Wycisk, P. (eds.): Geopotential and Ecology. Catena Supplement, **26**, 67–97.

CHAPTER 5

Water resources management in the Lake Chad basin: Diagnosis and action plan

B. Ngounou Ngatcha
Department of Earth Sciences, Faculty of Science, University of Ngaoundéré, Ngaoundéré, Cameroon

J. Mudry
UMR Chrono-environnement, UFR Sciences & Techniques, University of Franche-Comté, Besancon, France

C. Leduc
IRD, UMR G-EAU, Université de Montpellier II, CC MSE, Montpellier, France

ABSTRACT: Scarcity of water resources constitutes a major problem for sustainable development in the Chad basin. By 1976, Lake Chad had lost nearly 90% of its water volume. Its surface area had shrunk from 25,000 km^2 to less than 3000 km^2, and its surface level had fallen by 4 m. Therefore, maintaining secure water supplies for drinking, industry and agriculture would be impossible without groundwater. To effect the sustainable management of the Lake Chad basin, the Lake Chad Basin Commission (LCBC) has been created. Discussions about the relationship between surface water and groundwater, sediment, water protection, social and economic benefits of water, and the use of water resources for irrigation have clearly outlined uncertainties in managing the system due to incomplete knowledge and lacking data. This study indicates the need for: (1) permanent monitoring research, which will provide the data essential to generate a quantitative perspective on the status of the water resources systems and to validate the conceptual understanding; (2) more international cooperation and coordination between the LCBC members within their jurisdiction; and (3) joint studies and research programmes in order to strengthen the capacity for strategic and integrated water resources management.

1 INTRODUCTION

The Lake Chad basin makes up an aquatic ecosystem with numerous important natural areas including wetlands or flood plains, lakes and swamps. It is of high environmental value, and of economic and social significance to the surrounding area: the Chad basin supports the drinking water supply, ecological, agriculture, industry, fishing, stock breeding, tourism, navigation, and the end disposal of waste waters. As a consequence of the diverse and competing uses for water, the Chad basin faces a challenge for integrated water resources management.

In the semi-arid area of the Chad basin, mean annual rainfall ranges between 100 and 800 mm. In this low rainfall region groundwater recharge is generally low and unpredictable. The high evaporation rate, about 2200 mm/a, accompanied by unreliable rainfall makes these areas prone to droughts, the frequency of which appears to range between 7 and 10 years. The severe droughts of the 1980s exposed major deficiencies in knowledge and data for the Chad basin. Information systems are disparate, poorly integrated, and not easily accessible for management purposes. Water resources are still managed in a fragmented way. Surface water and groundwater development and protection are generally considered separately and their interdependence is often not acknowledged. Moreover, land use practices are not coordinated with water resources management. Experience in other major river basins indicates that integrated management can increase benefits, reduce adverse impacts, and improve long-term resource development and management (Vadas, 1999).

From all indications, it is estimated that by 2025, the Chad basin will face severe pressure, either physical or economic. The population of 7 million in 1975 grew to 15 million in 2001 and is projected to be about 25 million by the year 2025. It is clear that stresses placed on the Chad basin environments and their water resources will be severe during the next few decades. The need to improve social conditions and reduce poverty for the inhabitants will lead to an increase in water supply and sanitation services. In addition, higher water demands are expected in the basin for economic activities, agricultural irrigation and fisheries. However, this water use will come at an environmental cost. Strategies for sustainable use need to take into account the characteristics of all components of the water cycle, and guarantee that full use is made of the scientific basis that can provide a much fuller understanding of groundwater.

The increasing recognition for the need for integrated water resource management (IWRM) in the area has led to a recent focus on: (1) the main factors which could influence water management in the Chad basin; and (2) what research is required during the next decade. The aim of this paper is to bring together the global experience in an accessible and helpful compendium of optimal approaches, to support the practical and effective development of IWRM in the Lake Chad basin.

The approach in formulating this work has been to gather all relevant hydrological data and information from previous studies and recent field investigation of water resource quality, irrigation, sediment, and groundwater recharge. These data provide a base for better understanding of the Lake Chad hydrological regime and groundwater framework. It is hoped that this paper will stimulate readers to re-think our strategy for long-term water resources management and to develop management strategies with special attention to the concepts of IWRM. In this context we take IWRM to mean focusing on the river basin (or sub-basin) scale and managing water allocations, water-use and water discharge on the basis of integrated and comprehensive plans developed by means of a broad-based participatory planning process (Kardoss, 1999).

2 OVERVIEW OF THE STUDY AREA

2.1 *Hydrological regime of the Lake Chad basin*

The Lake Chad basin (Figure 1) is a geographical area determined by the catchment area of both surface water and groundwater. It covers an area of about 2,335,000 km^2. The hydrological regime of the lake depends mainly upon the interaction between precipitation, surface

Figure 1. The Lake Chad catchment area and the Lake Chad Basin Commission area.

water and evaporation. For the moment, it is difficult to obtain an objective quantification of some parameters characterizing the drainage basins because of the lack of standard measurements and the high variability. Temperature and precipitation are characterized by high seasonal and daily variability. The basin is also characterized by rainfall deficits, southwards shift of the isohyets (Ngounou Ngatcha *et al.*, 2005), reduced runoff of the rivers discharging into Lake Chad and shrinkage of the Lake Chad from 25,000 km^2 in the sixties to less than 3000 km^2 at present (Olivry *et al.*, 1996). The uneven rainfall pattern naturally creates highly variable flows over time and space causing uncertainty in water availability to both domestic users and farmers alike. The high temperatures together with the low relative humidity and the active wind account together for the high rate of evaporation (2200 mm/a).

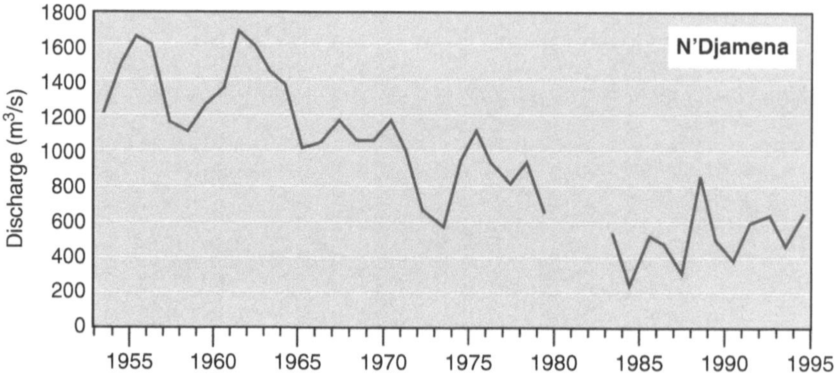

Figure 2. Annual mean discharge at N'Djamena from 1953 to 1994 (Olivry *et al.*, 1996).

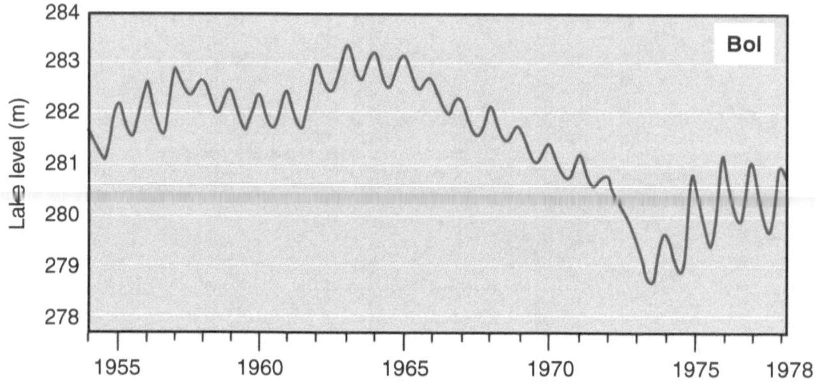

Figure 3. Monthly mean lake level at Bol, Chad from 1954 to 1978 (Olivry *et al.*, 1996).

The major tributaries of Lake Chad are Rivers Logone-Chari, Yedseram and Kamadougu Yobe. The Yedseram and Kamadougu Yobe are two Nigerian rivers essentially ephemeral and account for less than 10% of the total surface water input into the Lake Chad. The Rivers Logone-Chari is the main source of water (90%) into the lake. On the River Chari at a gauging station (N'Djamena) corresponding to a river basin of about 600 000 km^2, the average discharge over 59 years (1932–1992) is 33.5×10^9 m^3 (Olivry *et al.*, 1996). According to Figure 2, highest discharge is 54×10^9 m^3 (1955–1956) and least discharge is 6.7×10^9 m^3 (1984–1985). The elevation of the Lake Chad level ranges between 275 and 284 m above sea level and the water depth is between 2 and 6 m for a lake level of 282 m (Olivry *et al.*, 1996). Figure 3 shows the fluctuation of the mean monthly levels of the lake from 1954 to 1977. From long-term hydrological observations, the analysis of water balance (Olivry *et al.*, 1996) demonstrated that the volume of water stored in the lake decreased from 40–100 $\times 10^9$ m^3 (before 1964) to $7 - 45 \times 10^9$ (in 1990).

The declining water-levels are impacting on domestic economies such that the cumulative effects of all these different conditions lead the society of the Chad basin towards poverty. There has been a noticeable migration of the rural population to city areas, and consequently the economic losses according to IUCN (1996) and Akujieze *et al.* (2003) are likely to account for about 40% of the area's gross domestic product (GDP).

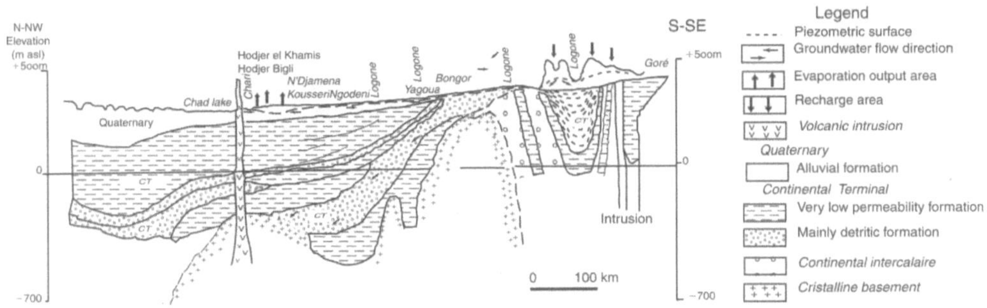

Figure 4. Schematic of the Chad lake-Gore aquifer systems (modified after Schneider, 1992).

The potential impacts of climate variability on water resources have long been recognised (Ngounou Ngatcha *et al.*, 2005; Favreau *et al.*, 2005). With the increased evidence of accelerated and permanent climate change there is a pressing need to increase field measurements of hydrological processes.

2.2 *Background of the shared aquifers systems*

Hydrogeological conditions and the aquifer system's characteristics have a direct bearing on groundwater management (Zektser & Everett, 2004). In the Chad basin, groundwater investigations mainly started at the local level within national boundaries and were rarely concerned with regional aquifer systems. The Chad basin is underlain, mainly, by Tertiary and Quaternary formations (Figure 4).

The Quaternary aquifer consists of sandy, deltaic and lacustrine deposits. It is a regional aquifer extending from Chad to eastern Niger, northwestern Nigeria, northern Cameroon and the Central African Republic. Groundwater in the unconfined alluvial aquifers is developed in a traditional manner by a large number of wells ranging between 5 and 50 m. Groundwater is classified as bicarbonate type with low mineralization (Eberschweiler, 1993). Sediment heterogeneity within the Quaternary aquifers strongly influences spatial variability of hydraulic conductivity and the paths and rates of groundwater flow. The productive levels in the aquifer consist of fine-grained sand and gravel. The thickness of the aquifer is variable, reaching 100 m around the Lake Chad.

In the Tertiary deposits, two regional units have been delineated: the Continental Terminal and the Pliocene aged deposits (Schneider, 1989). Since no evidence of a marine condition has been confirmed for the Pliocene deposits in the Chad basin, all could be termed Continental Terminal deposits, with the argument that after the Eocene, a continental environment was established in the Chad basin (Kilian, 1931; IGCP/UGIS/UNESCO, 1980; Ngounou Ngatcha, 1993; Ngounou Ngatcha *et al.*, 2006). Thus, the Continental Terminal (Ct) consists of continental sediments, ranging in age from Oligocene to Pliocene. Groundwater in the Ct aquifer is classified as bicarbonate type with high conductivity, TDS range between 75 and 700 mg/l, and high groundwater temperature (40–46°C). Groundwater from the Ct aquifer can pose a risk of salinization of irrigation fields (Oguntula, 2004).

The Chad basin suffers from serious chronic surface water shortage, and groundwater is increasingly being used to cover the demand. Since 1980, the advances in drilling well construction and pumping technologies – as well as increasing electrification in

Table 1. Annual groundwater use (million m^3) in the Lake Chad conventional basin in 1990 (after Eberschweiler, 1993)

	Chad	Cameroon	Nigeria	Niger	Total
Drinking water	48.7	32.2	87.9	2.8	171.6
Irrigation	25.5	1	6	15	47.5
Stock breeding	7.7	3	11	11.2	32.9
Total	81.9	36.2	104.9	29	252

rural areas (Cameroon and Nigeria) – mean that ever-increasing volumes of groundwater are being exploited without adequate planning. In 1990, groundwater consumption was estimated at 252 million m^3 with an additional loss from artesian flow of about 28 million m^3. The proportion of water used for drinking, irrigation and stock varies from country to country (Table 1). In the four countries, groundwater is the single most used natural resource. In spite of the widespread and increasing use of groundwater consumption during the last 15 years, the knowledge base concerning groundwater and its sustainable use is inadequate. Since groundwater flows very slowly, the consequences of over-exploitation may only become apparent after years or decades. Thus, future water strategies will have to include well planned monitoring of abstraction and quality. Water resources can be used sustainably only if their spatial extent and their variation through time are properly understood. There is still a critical need to improve the quality of hydrochemical and hydrodynamic data sets and organizing it on maps, in geographical information systems (GIS) and through mathematical models. However, models allow us to understand the data and analyse the effects of different management options.

Through the Lake Chad Basin Commission (LCBC), the study of regional aquifer systems received little attention. The LCBC collects climatological, hydrological and hydrogeological data from member states for purposes of carrying out regular assessment of the available resources, running the mathematical model and publishing the hydrological yearbook. Although more data from previous work are available, groundwater recharge assessment has not received enough attention in the scale of the whole basin.

Computer modelling techniques allow for a regional understanding of the groundwater system. The regional hydraulic continuity in the Chad basin aquifer system (Quaternary and Continental Terminal) permits the development of regional and sub-regional models which could be used to better understand and analyse the groundwater flow patterns, provide quantitative information on the regional assessment of the aquifer system, furnish boundary conditions for sub-regional and local simulation studies and contribute substantially to their socio-economic development.

At the regional level, a model called MARTHE has been developed (Eberschweiler, 1993). The discretization of the systems is relatively coarse (12.5 × 25 km, 25 × 25 km and 25 × 50 km). According to Ogontula (2004), a HYDRO-CHAD Model has been developed to investigate alternative water resources planning scenarios. This model is based on the St Venant equations. Field topographic data are being collected to improve the accuracy of the model which was initially calibrated with synthetic data on cross sections derived from stage and discharge measurements at hydrometric stations.

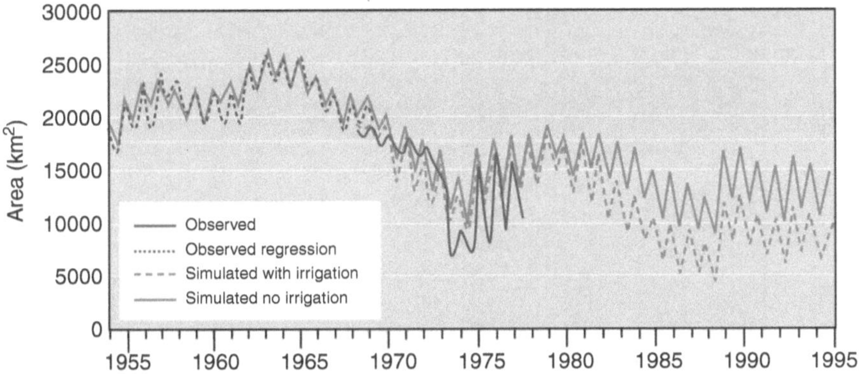

Figure 5. Monthly mean lake area from 1954 to 1994 and simulations of lake levels with and without irrigation. (Olivry *et al.*, 1996; Coe & Foley, 2001). Note that these simulations are contested (see text).

3 SOME KEY FACTORS OF WATER MANAGEMENT IN THE LAKE CHAD BASIN

Many activities influence hydrological regimes by their impact on soil moisture and subsurface water storage and movement. The abstraction and return of water for domestic, agricultural and industrial use, and the use of surface reservoirs also directly affects flow within rivers. Water quality changes also have a significant impact on water resource availability. Sedimentation in reservoirs due to erosion and subsequent transport of sediments is also a widely acknowledged water resources management problem. Changes in land use may lead to very important changes in the hydrological behaviour of a particular basin. The relative importance of all these attributes of a basin must be considered in any concept of regionally integrated water resources management.

3.1 *The hydrological effects of agricultural activities*

Throughout the world, agriculture is one of the most significant human activities affecting the hydrological cycle. Irrigation is reported to account for three-quarters of the world's water consumption. Hutchinson *et al.* (1992) envisage that future water demand in the Lake Chad basin is expected to increase, as the population becomes more dependent on irrigated agriculture. However, much confusion surrounds the impact of irrigation on lake levels.

With satellite altimetry data, Birkett (2000) has demonstrated that while irrigation losses are large compared to discharge, seasonal and annual variations in lake level are still predominantly determined by climate variability. However, Coe and Foley (2001) have estimated from an integrated biosphere model (IBIS) and a hydrological routing algorithm (HYDRA) that irrigation accounts for roughly 50% of the observed decrease in lake Chad area since the 1960s and 1970s (Figure 5).

Contrary to Coe and Foley (2001), agricultural development in the Logone-Chari plain did not depend on irrigation for the period 1954 to 1977. Areas under cultivation within the Logone-Chari plain changed little between 1970 and 1994, except at Maga (in Cameroon) where a dam went into operation in 1979 and 6000 ha of irrigated land were developed. Most irrigation potential in Nigeria is classified as small-scale or indigenous (Acharya, 2004). Farmers have neither the knowledge nor the means for correct agricultural practices and

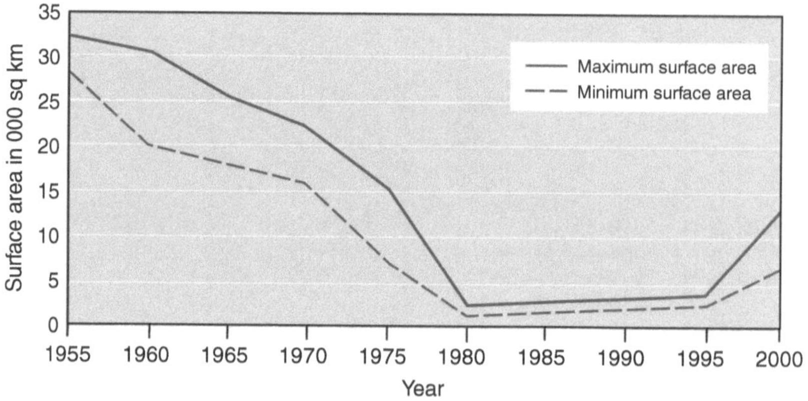

Figure 6. The Lake Chad surface water area during the past four decades.

soil conservation. In the early 1970s the Federal Government of Nigeria initiated a number of large-scale irrigation projects, including South Chad Irrigation Project (Acharya, 2004).

The environmental impacts due to the construction of the Maga dam were analysed within the IUCN Project (IUCN, 1996). Despite the fact that considerable amounts of Komadugu-Yobe water are used for irrigation, and apart from the significant impact of the Maga dam on the Cameroon flood plain itself, there is no evidence of high magnitude of irrigation influence on Lake Chad level as claimed by Coe and Foley (2001).

Although the irrigated area increased in the Logone-Chari and Komadugu flood plains since 1990, a pronounced increase in the lake level and area has taken place (Figure 6) with the trend towards increasing annual precipitation totals within the years 1990–1994. Previous investigations (Olivry *et al.*, 1996) have also indicated that the effect of irrigation is highly influenced by rainfall.

This review indicates that water use and demand for irrigation is poorly estimated and depends on the extent of the lake area; many irrigation schemes are presently too far from the available surface water. The impacts of climate variability, agriculture and land use have not yet been clearly and fully evaluated. The increasing use of groundwater for irrigating marginal farmland in the Chad basin is of particular concern. Studies on the groundwater regime in farmland are important for obtaining a better understanding of the irrigation influence on groundwater levels. In connection with irrigation, it could be concluded that irrigation may raise the groundwater table not only because of the infiltration increase over natural conditions, but also because of the higher moisture content produced in the topsoil, which reduces evaporation from the groundwater and contributes thus to raising the level thereof. A thorough and transparent assessment of the relative socio-economic value of groundwater in relation to surface water irrigation might contribute to mitigating or avoiding potential future conflicts (Garrido *et al.*, 2006).

3.2 *The effect of sediment deposits on water flow and lake storage*

The hydrological effects of sediment transport have been little studied in the Chad basin. Loss of Lake Chad storage due to sediment accumulation, if it were to occur quickly, would be truly catastrophic. The tendency of riverbeds to degrade causes changes in river regime. These changes sometimes bring difficulties in river management. Prolonged turbid water may influence the ecosystem of the river as well as river mouth area. The loss of reservoir

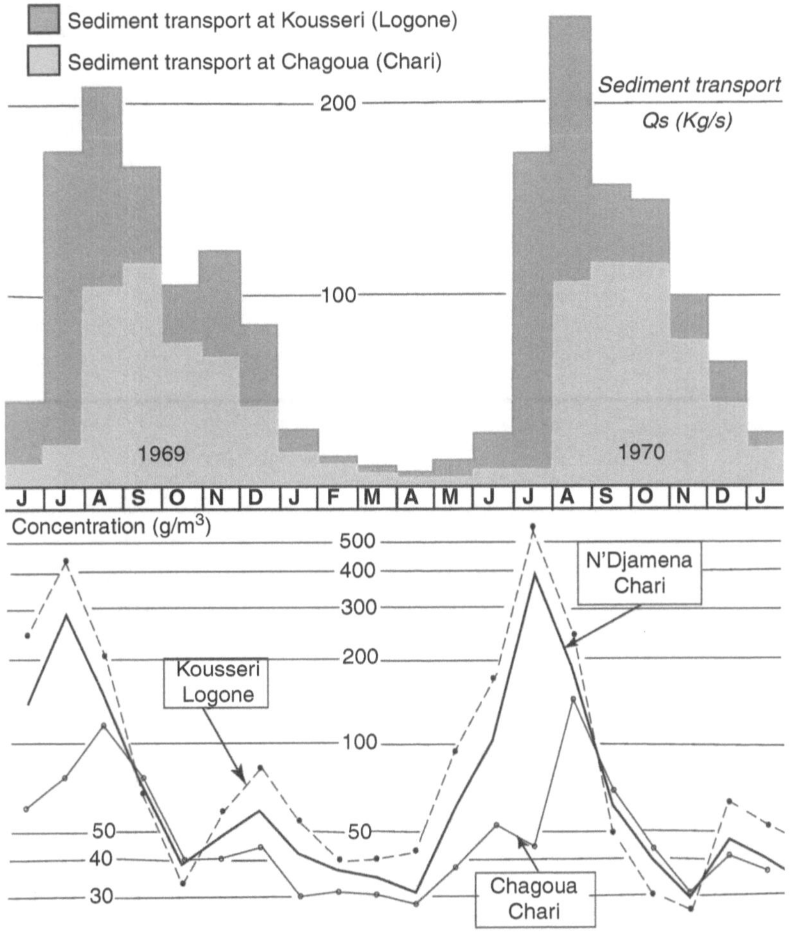

Figure 7. Monthly mean concentration and sediment transport of suspended material from Logone and Chari at Kousseri, Chagoua and N'Djamena (Modified from Olivry *et al.*, 1996).

storage could also reduce the effective life of the Lake Chad and diminish benefits for water supply, navigation, and recreation.

Only few data (Figure 7) are available which describe the quality of suspended material in the Chad basin (Olivry *et al.*, 1996). A short-term field observation along the Rivers Logone-Chari, and at the outlet of the River Chari, shows that there is considerable erosion of stream banks. Sediments consist of clay, sand and silt, but also organic materials. At N'Djamena, the Chari carries enormous quantities of clay and sand, the river has poor drainage resulting from the heavy sedimentation, and inundation occurs. The red colour of the river water demonstrates that clay is of great importance in the sediment transport.

The evaluation of sediment deposits is particularly important in the Chad basin as the rate of soil erosion and transportation is fairly high, mainly because of adverse hydrological factors, such as long periods of aridity but high runoff when it does rain. Sparse vegetative cover is also one of the main reasons for high quantities of suspended sediment.

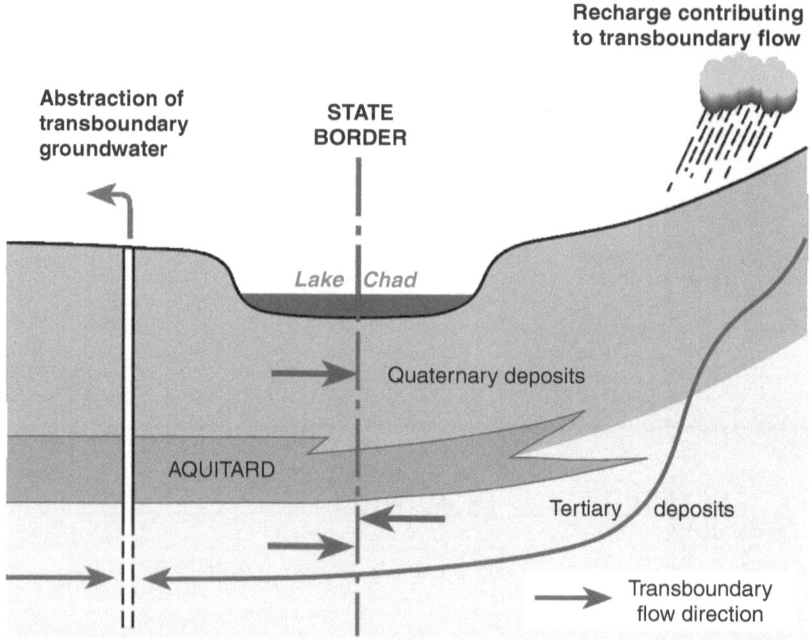

Figure 8. Schematic illustration of the transboundary flow in the Lake Chad basin (adapted from cross section on Figure 4).

An approximate evaluation of the amount of expected sediment deposit could supply designers with indications concerning the reservoir capacity. An adequate network of clay observation stations is essential to monitor the erosion. The method of combination of GPS (Global Positional System) and echo sounders to survey reservoir siltation seems to be an encouraging technique for field data collection.

3.3 *The dependence of groundwater upon surface water*

Groundwater is closely related to the surface water regimes of rivers and lake, and to atmospheric or climate processes. Most of the research established to help manage the Chad basin has focused on surface water (Olivry *et al.*, 1996) because of the fluctuations of Lake Chad. As scarcity becomes more pronounced and prolonged, it is apparent that proper management of surface water and groundwater cannot be accomplished separately. Because surface water and groundwater systems are interconnected, there is no doubt that any change in the surface water systems will affect the groundwater systems (Figure 8). Regional groundwater systems comprise an extensive set of aquifers and confining units that act on the regional scale as a single system. Groundwater does not stop at political borders. Pumping in one country can affect water in another. In such circumstances, groundwater management requires international cooperation and the existence of appropriate governmental and legal institutions.

The Lake Chad basin has ecosystems, landscape elements, or pre-existing water users that are dependent on current discharge or recharge patterns. Further development may require trading these dependencies in favour of new plans or policy. If dependencies are not well

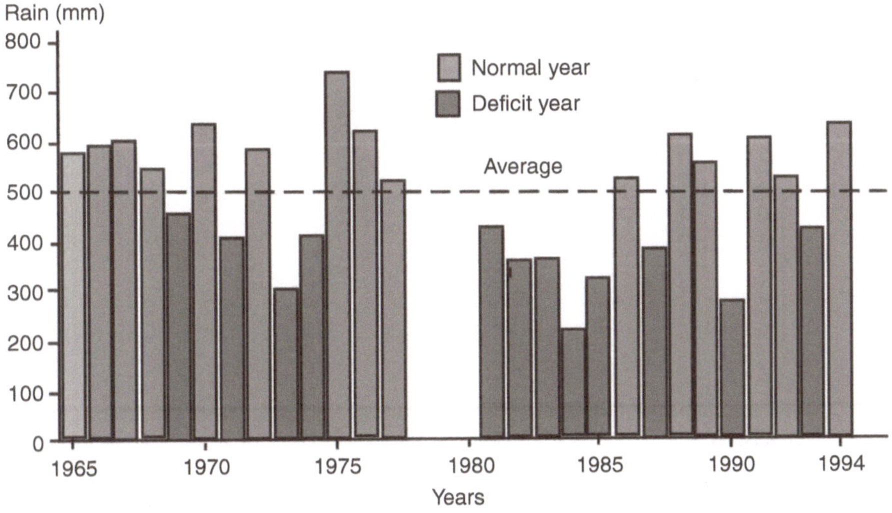

Figure 9. Annual rainfall variation in N'Djamena.

understood or considered, management changes may have major unanticipated impacts (Puri *et al.*, 2001).

Variation in precipitation (Figure 9) through time, and limited availability of groundwater present the greatest challenges to the management of water resources of the Lake Chad basin. Analysis of recharge conditions must relate not only to current climate, but also to previous climate conditions. Existing hydrogeological maps provide information, but water use, main recharge areas and general flow patterns are poorly known in the Chad basin. Isotopic investigations (Fontes *et al.*, 1970; Ketchemen, 1992; Ngounou Ngatcha, 1993; Njitchoua & Ngounou Ngatcha, 1997; Edmunds *et al.*, 1998; Leduc *et al.*, 1998; Goes, 1999; Djoret, 2000; Leduc *et al.*, 2000; Goni, 2002; Gaultier, 2004) have made an important contribution to develop understanding of the groundwater flow system and support better groundwater resources management.

In 2004, an examination of more than 20 wells in the N'Djamena region near the River Chari suggested that induced recharge from the River Chari may be an important source of groundwater recharge to this part of the Chad basin. Groundwater levels during the wet season are always higher than those during the dry season (Figure 10); but it is not yet clear how the losses of surface flow may be apportioned. Moreover, due to land use, the permeability and storage characteristics in the subsurface could be changed progressively and this may favour percolation of water from the flood plain into the aquifer. The variations of the groundwater level are naturally a measure of changes in groundwater storage but they can also be used as a measure of the soil moisture storage between the surface and the water table. This preliminary study illustrates the need for further and more detailed monitoring. The amount of water that enters and leaves the groundwater environment needs to be quantified in order to make a proper sustainable management program of groundwater resources.

Since the aquifer receives modern recharge, the global strategy consists of preserving natural outflows and of abstracting a volume equivalent to the average annual recharge. In any legal agreement to be drawn up for the equitable share of transboundary resources, the initial stage must be the correct identification of flow and movement of water, followed by its quantification. Identifying and mapping the boundaries and the various aquifers within

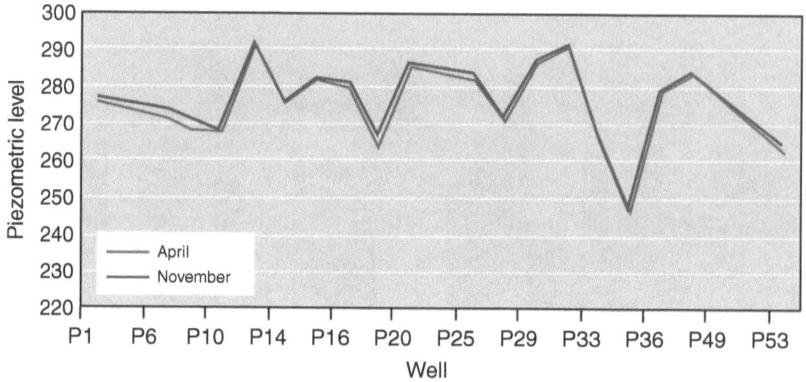

Figure 10. Piezometric level observed in some boreholes in N'Djamena in April (dry period) and November (wet period).

Table 2. Nitrate concentrations in groundwater in Cameroon and Chad

		Nitrate (NO_3 mg/l)		
Country	No of samples	Min	Max	Mean
Cameroon	285	1	300	14.7
Chad	31	8.8	167	35.3

a basin is a time consuming and expensive process, not to mention the time involved to develop complex hydrological and hydrogeological models to explain the dynamics of water resources inflows to and outflows from the basin (Schlager, 2006). When the recharge is assessed, then the management policy could be evaluated over a long period to consider the implications for future generations.

3.4 *Water resource protection*

One of the most important problems of natural resources management in the Lake Chad basin is the degradation of water resources caused by pollutants of agricultural, municipal and industrial origin. Nitrate concentrations of higher than 50 mg/l as NO_3 (Table 2) have been reported (Ngounou Ngatcha *et al.*, 2000; Djoret & Travi, 2001). The origin of these pollutants in groundwater remain largely uninvestigated. Furthermore, most surface and groundwater projects have no comprehensive environmental impact assessment (EIA) studies, and no reliable baseline studies or data to follow-up programmes for monitoring long-term environmental impacts on groundwater regimes and quality (Akujieze *et al.*, 2003). Water quality analyses are hardly done. The real danger in the Chad basin is not pollution but the ignorance of pollution problems.

Polluted water can transmit diseases and carry poisonous chemicals. Bako & Umaru (1998) have indicated threats to groundwater quality by chemical additives such as dextride and lingo-scale pollution from casing shoes at depths of 1200 m into the Lower aquifers of the Nigerian Chad basin as a result of petroleum exploration programmes. Other more common water related diseases are prevalent (Oguntola 2004): diarrhoea, cholera, typhoid fever, iodine deficiency and bilharzias.

Before appropriate protection measures can be designed and implemented, water resource pollution and its sources must be assessed, and the vulnerability of the environment to pollution evaluated. Monitoring the water vulnerability to contaminants will provide interesting views; in fact, if studies come to predict a significant increase in the human activities, then the consequences on water resource management will be high. A successful water resource protection program must include a combination of three basic alternatives: prevention (it is suggested that the states sharing the basin should enact a law to control the use of areas for agriculture, and establishing guidelines for delineation of protection zones around public groundwater supplies), remediation by natural attenuation, and remediation technology.

As the natural content of water varies considerably, we must define average conditions for natural and safe waters. Above a predefined threshold, water will be declared polluted. Any sustainable management approach will have to ensure that water can, by its quality, both satisfy the needs of human beings and maintain the natural functions of the ecosystem which shelters theme.

3.5 *Social and economic problems*

People in arid areas are uniquely vulnerable, not only to drought and other natural disasters, but also to economic and social changes. Achieving sustainable development has particularly significant implications for reducing poverty and hunger. In the Chad basin, natural and socio-economic conditions, culture and language often differ significantly between the different parts of the basin, and consequently upstream-downstream conflicts can occur easily.

Within the Lake Chad basin the population is approximately 15 million. The agriculture sector provides a subsistence livelihood (self-employment) for about 65–80% people in the rural areas. Problems related to economic development include poorly regulated exploitation of lands and natural resources for commercial purposes. The redistribution of population has led to increased urbanization throughout the basin. These populations have also created an increased demand for food which have been met by overfishing and by cultivating marginal lands. Superimposed on these factors are modifications of the natural hydrological regime of the Lake Chad basin. According to Acharya (2004), pastoralists use the wetlands seasonally, moving into the wetlands as the surrounding rangelands dry out. Grazing within the wetlands is crucial for the cattle and livestock owned by the nomadic populations and by some sedentary farmers. Conflicts over access to land and water are increasing between and among farmers and herdsmen since seasonal grazing grounds are now increasingly coming under dry-season groundwater-irrigation agriculture and there is increased jostling over land. Conflict resolution between users requires the gathering of a wide range of expertise and, of course, resources that are not available to national or regional institution (LCBC) responsible for these tasks.

The potential for ecotourism in the Chad basin has not been thoroughly studied. The free-tourism-area makes sense from an economic, political and environmental perspective and would serve as a significant model of cooperation and sustainable development within the Lake Chad basin.

In summary, cultural and ethical dimensions are likely to be as important as macro-economic dimensions in the evolution of approaches to address existing and emerging transboundary problems. Integrated assessment would allow the assessment of the impacts of future change on the water environment. In particular, the successful integration of socio-economic scenarios into the modelling of spatial land use could enable the indirect impacts, resulting from changing patterns of urbanization, flooding and cropping, to be assessed and quantified.

4 IMPROVEMENTS REQUIRED IN THE FUTURE

4.1 *The Lake Chad basin commission (LCBC): strategies and limits*

The countries sharing the Lake Chad basin took a strong integrated approach to its water resources management, particularly by creating in 1964 the Lake Chad Basin Commission (LCBC), with headquarters in N'Djamena (Chad). A series of principles, policies, general water laws, water administration regulations and basin oriented water laws have been made. They provide opportunities for information and risk sharing, facilitate pooling of technical and financial resources, thereby reducing the burden on individual countries. The LCBC also allows for joint mitigation and economic development programmes to maximize the benefits inherent in basin-wide management of water resources. The reforms undertaken so far in the LCBC offer some interesting insights and lessons. The LCBC water managers have also looked well into the future to secure new supplies, and although the different users and cities do compete for water supplies, they have been able to speak with one voice on interstate concerns over issues such as World Water Forum.

As discussed above, climate variability apparently reduced the water supply security for the population and economy in the Chad basin. In such circumstances, LCBC (with regards to other river basins in the world such as Dead Sea and the Aral Sea) has developed a project on the restoration of the Lake Chad level. The proposal for supplying the Lake Chad with additional water from Oubangui-Chari River (located in the Congo catchment, a humid area, 1350 km south of the lake) is still under investigation. Planning and project preparation, particularly the feasibility study, are vitally important for developing affordable financing plans and implementation plans. In order to achieve the above-mentioned objectives, a technical committee including Government, industry, agriculture, the public, and national and international scientific organisations will be welcome.

It is universally agreed that water resources must be managed at the basin or catchment level. The most serious constraints to LCBC performance in the water sector are: (1) the conventional basin boundaries (see Figure 1) do not conform to the principles of river catchment, geological unit and hydrogeological characteristics of the underlying rock formations; (2) the conventional basin lacks clear definition as a hydrogeological unit that could facilitate proficient groundwater management; and (3) the relatively small budget made available for water resources administration, infrastructure development and operations.

Given that hydrological network and data collection are prohibitively expensive in many circumstances, LCBC must therefore establish priorities and phase in their financial effort. The LCBC should: (1) do all in its power to reverse the trend of hydrological network decline and thus ensure that future generations will have the information they need to make wise decisions on Chad basin sustainable development, based on long time series of observational data; and (2) make water resources protection more visible through information dissemination and by introducing if possible, fees for water pollution or exploitation.

The main risk is that the legal mechanisms provided under the water law are not fully implemented. In fact, LCBC responsibility should be clearly defined within their jurisdiction. River basin managers must not only have the technical skill to manage a complex river system, but also the human interaction skill to facilitate resolution of conflict-ridden issues. Good skills in mediation and facilitation are critical to finding solutions to river basin problems.

4.2 *Need for a Lake Chad scientific research group*

Technology, knowledge transfer and research cooperation have not received sufficient attention at the regional scale. A scientific group should be recognised and integrated in river basin management. It will bring together, for the first time, researchers across the region (Cameroon, Central African Republic, Chad, Niger and Nigeria). Such an exchange of data and information is fundamental to regional co-operation and the mutual understanding and solution of regional problems.

The unrestricted exchange of data and knowledge is a prerequisite for efficient management and cooperation in international river basins. The data collected should be available and easily accessible internationally. The need for capacity building and innovative thinking is highly important in regional water sector development and management activities. A global multilateral interdisciplinary forum should be established to develop general principles and minimum standards for the sustainable management of international river basins.

4.3 *Need for data and information in support of water resource planning*

The first stage in the acquisition of hydrological knowledge in the region was the pioneer period of the 1950s, when many different kinds of water demand were emerging very fast but hydrological information was almost nonexistent (Olivry & Sircoulon, 1998). These networks consist of stations where variables such as precipitation, evaporation, soil moisture, water quality, groundwater level and river water-level and discharge were measured on a regular basis against agreed standards. Unfortunately, the severe droughts of the 1980s revealed major deficiencies in the hydrological services in the Chad basin: database and information systems are not efficient. They are disparate, poorly integrated, and not easily accessible for management purposes.

For all their economic, social and environmental benefits, groundwater basins are difficult to govern, partly because of poor information, and also because of poor visibility of the resource. According to Akujieze *et al.* (2003), the practice of hydrogeology in the region is influenced by the poor availability and status of hydrometric facilities such as hydrological, hydrometeorelogical and hydrogeological data acquisition instruments.

Due to the temporal and spatial variability of hydrological and climatic data, good sets of reliable data are essential for extending our understanding of hydrological processes, for the detection of significant trends in the relevant variables, for the verification of a wide number of models involving water behaviour or use, for the evaluation of the impacts of both variation and change in the hydrologic variables on ecosystems and socio-economic systems, and for the clear formulation of the alternative scenarios to be considered in key policy decisions (Rutashobya, 2003). Without these data, risks are substantially increased, projects can be underdesigned or overdesigned, costs can be inflated and substantial losses can occur, particularly in the event of a failure (Rodda, 2003). The available data should be compared with the requirements so that missing information can easily be identified.

4.4 *Key role of users, the public, planners and decision-makers*

Lessons from other international river basin organisations show that success can only be secured when interests of different stakeholders in the area are taken into account. The Lake Chad basin sustainable development can only be achieved when all multiple interests represented by farmers, herdsmen, fishermen and local communities are addressed. According to Mostert *et al.* (1999), public participation as a legal right is based on the notion that individuals and groups affected by decisions should have the opportunity to express their views and become involved in decision-making.

We should pay attention to the diversity of water uses, in order to project a more realistic image of the complex relationship between man and water within the basin. Collaborating with water users and providers is important in designing any management program, whether regulatory or not. Active user participation is the best way to solve conflicts in use; dialogue is the first step to wisdom. Public participation can be seen as a means of improving the quality and effectiveness of decision-making. Farmers need to be well informed on the benefits of adopting more-efficient irrigation methods. Because of the limited financial and administrative resources, governments should decentralise responsibilities, including financing, and involve users as much as possible.

5 CONCLUSIONS

Improvement of drinking water supply and sanitation is one of the priorities for the population of the Lake Chad basin. Groundwater is usually the only safe water source where surface water is limited. At present, there is very little international experience in the approaches needed for the integrated water resources management. The concept of IWRM poses a real challenge for its implementation in the Lake Chad basin. The water demand is rising as population, economic activities and agricultural irrigation grow. Available data on agriculture irrigation and sediment deposits are not adequate for extensive analysis. The role of water should be considered and managed in wider hydrological, ecological, economic trading and socio-political contexts. The Lake Chad conventional basin boundaries do not follow natural physical features and water resources can cross them. Institutional and political factors are the major obstacles to sustainable management of water resources, and in particular finding ways for the different countries to share information and experience. The primary challenge for LCBC lies in balancing formal regulations and delegated responsibilities at the local, national and international level. Given the current climate trends over the past 50 years and possible further reduction in runoff due to climate change, the future of the lake seems to be largely dependent on the availability of imported water from the River Oubangui (Congo basin). However, the behaviour of the groundwater systems, and how they interact with surface water, is poorly understood.

The above observations suggest that there are considerable gaps in data and knowledge about the system. The combination of lake, rivers and groundwater makes the Lake Chad basin an important, but complex, system to manage, particularly with climate variability and increased water demand in the region. Many strategies or actions are required with respect to the Lake Chad basin system to move from concept to reality with IWRM. But these strategies must be based on evidence, which requires systematic and prolonged data collection and rigorous analysis.

Here are some suggestions to improve integrated water resources management in the region and move towards a new water master plan for the Lake Chad basin:

- greater integration of the relevant information systems, e.g. hydrology, hydrogeology, water quality, land use, sediment transport;
- move away from the conventional basin and develop new approaches that will belong to a common geographical unit that does not recognise political boundaries;
- regional studies of hydrogeology, hydraulic properties, regional flow system and water quality that cross political boundaries;
- discussion and agreement on a common strategy for groundwater exploitation and equitable sharing across borders;
- protection of groundwater resources to safeguard long-term use and balance the demands of economic development with ecosystem conservation;
- charging an economic cost for groundwater exploitation, treatment and supply;
- re-introduce long-term hydrological observations (which have been discontinued in more ancient stations), and instigate new data collection on water use, irrigation and agriculture lands, water sediment deposits, industrial demands, urban development, recharge, hydraulic properties and groundwater/surface water interaction;
- land-use regulations to reduce the contamination of water resources, using the LCBC as a mechanism to help transboundary negotiations;
- public awareness raising programmes for water users, political decision makers and the general public;
- introduce measures to reduce land erosion, and consequent siltation in Lake Chad and the existing hydro-agricultural infrastructure;
- emphasising a participatory approach, involving users, planners and policy makers at all levels; women should be more involved in water management as important stakeholders.

Science and evidence should underpin IWRM in the Lake Chad basin. To achieve a common understanding there should be (1) joint studies and research programs for better understanding of the water resource mechanism, and (2) adequate research facilities (increase of public funding for research and innovation in the public interest) for data collection, analysis and interpretation.

ACKNOWLEDGEMENTS

Acknowledgement is due to the Ministère des Affaires Etrangères (MAE) for the facilities provided during the preparation of this work and to the Department of Geosciences, University of Franche-Comté, Besançon (France). Field work was funded by UNESCO Virtual laboratory Project.

REFERENCES

Acharya, G. 2004. *The role of economic analysis in groundwater management in semi-arid regions: the case of Nigeria*. Hydrogeology Journal, **12**, 33–39.

Akujieze, C. N., Coker, S. J. L. & Oteze, G. E. 2003. *Groundwater in Nigeria-a millennium experience distribution, practice, problems and solutions*. Hydrogeology Journal, **11**, 259–274.

Bako Adetole, B. A. & Umaru, A. F. 1998. *The Chad basin aquifers: new evidence from seismic reflection sections and wire line logs.* Water Resources Journal, Nigeria, **9**, 10–21.

Birkett, C. M. 2000. *Synergistic remote sensing of lake Chad: variability of basin inundation.* Remote Sens. Environ. **72**, 218–236.

Coe, T. M. & Foley, A. J. 2001. *Human and natural impacts on the water resources of the Lake Chad basin.* Journal of Geophysical Research, **106(D4)**, 3349–3356.

Coe, T. M. & Birkett, M. C. 2004. *Calculation of river discharge and prediction of lake height from satellite radar altimetry: Example for the lake Chad basin.* Water Resources Research, **40**, 1053–1065.

Djoret, D. 2000. *Etude de la recharge de la nappe du Chari Barguimi (Tchad) par les méthodes chimiques et isotopiques.* Thèse Doctorat, Université d'Avignon et des pays de Vaucluse.

Djoret, D. & Travi, Y. 2001. *Groundwater vulnerability and recharge or palaeorecharge in the south-eastern Chad basin, Chari Baguirmi aquifer.* In: Isotope techniques in water resource investigations in arid and semi-arid regions, TECDOC-1207, IAEA, Vienna, 33–40.

Eberschweiler, C. 1993. *Suivi et gestion des ressources en eaux souterraines dans le bassin du lac Tchad. Prémodélisation des systèmes aquifères, évaluation des ressources et simulations d'exploitation.* Rapport intermédiaire 2. BRGM/CBLT, R 35985.

Edmunds, W. M., Fellman, E., Goni, I., McNeill, G., & Harkness, D. D. 1998. *Groundwater, paleoclimate and paleorecharge in the SW Chad basin, Borno State, Nigeria.* In: Isotope techniques in the study of past and current environmental changes in the hydrosphere and the atmosphere. IAEA, Vienna, 693–707.

Favreau, G., Ardoin-Bardin, S., Goni, I., Boronita, A., Condom, T., Coudrain, A., Declaux, F., Dezetter, A., Gasse, F., Gaultier, G., Guero, A., Habou, L., Leblanc, M., Leduc, Ch., Lemoalle, J., Loubet, M., Ngounou Ngatcha, B., Niel, H., Razack, M., Seidel, J.L., Travi, Y., Vallet-Coulomb, Ch., Van-Exter, S. & Zaïri, R. 2005. *Impacts climatiques et anthropiques sur le fonctionnement hydrologique dans le basin du lac Tchad.* Actes premier colloque de restitution scientifique du Programme National coordonné ANR "Ecosphère continentale, risques environnementaux (ECCO), du 5 au 7 décembre 2005 au Centre International de Conférences de Météo France à Toulouse, France, 429–434.

Fontes, J-Ch., Gonfiantini, R. & Roche, M. A. 1970. *Deutérium et oxygène-18 dans les eaux du lac Tchad, Isotope Hydrology.* Proceedings of Symposium Vienna, 1970, 387pp.

Garrido, A., Martinez-Santos, P., & Llamas, R. M. 2006. *Groundwater irrigation and its implications for water policy in semiarid countries: the Spanish experience. Hydrogeology Journal*, **14**, 340–349.

Gaultier, G. 2004. *Recharge et paléorecharge d'une nappe libre en milieu sahelien (Niger Oriental: approches géochimique et hydrodynamique.* Thèse Doctorat. Université Paris Sud, Orsay, Paris.

Goes, B. J. M. 1999. *Estimate of shallow groundwater recharge in the Hadeija-Nguru Wetlands. Semi-arid northeastern Nigeria.* Hydrogeology Journal, **7**, 305–316.

Goni, I. B. 2002. *Realimentation des eaux souterraines dans le secteur Nigérian du bassin du lac Tchad: approche hydrogeochimique.* Thèse Doctorat, Université d'Avignon et des pays de Vaucluse, 123pp.

Hutchinson, C. F., Warshall, P., Arnould, E. J. & Kindler, J. 1992. *Development in Arid Lands: Lessons from Lake Chad*, Enviro, **34**, 16–43.

IGCP-IUGS-UNESCO, 1980. *Project no 127: Revision of the "Continental terminal" concept in Africa.* Report of the International Geological Correlation Programme (IGCP), **8**, 153–155.

IUCN, 1996. *Rehabilitation of the Waza-Logone floodplain, Republic of Cameroon.* Report IUCN and Ministry of the Environment and Forestry Yaoundé, Cameroon.

Kardoss, L. 1999. *Management of international river basins; the case of the Danube river.* In: Mostert, E. (ed.), River basin management, Proceeding of the International Workshop (The Hague, 27–29 October 1999). IHP-V, Technical Document in Hydrology, **31**, 81–95.

Ketchemen, B. 1992. *Etude hydrogéologique du Grand Yaéré (Extrême Nord du Cameroun). Synthèse hydrogéologique et étude de la recharge par les isotopes de l'environnement.* Thèse de Doctorat 3ème cycle, Université Cheikh Anta Diop Dakar, Sénégal.

Kilian, C. 1931. *Des principaux complexes continentaux du Sahara.* C. R. Somm. Soc. Géol. France, **9**, 109–111.

Leduc, C., Salifou, O. & Leblanc, M. 1998. *Evolution des ressources en eau dans le département de Diffa (bassin du lac Tchad, Sud-Est nigérien).* In: Servat, E., Hugues, D., Fritsch, J. M. & Hulme, M. (eds.), Water resources variability in Africa during the XXth century, IASH Publication, **252**, 281–288.

Leduc, C., Sabljak, S., Taupin, J. D., Marlin, C. & Favreau, G. 2000. *Estimation de la recharge de la nappe quaternaire dans le Nord-Ouest du bassin du lac Tchad (Niger oriental) à partir de mesures isotopiques.* C. R. Acad. Sci. Paris, **330**, 355–361.

Mostert, E., Beek, E. V., Bouman, N. W. M., Hey, E., Savenije, H. H. G. & Thissen, W. A. H. 1999. *River basin management and planning.* In: Mostert, E. (ed.), River basin management, Proceeding of the International Workshop (The Hague, 27–29 October 1999). IHP-V, Technical Document in Hydrology, **31**, 24–55.

Ngounou Ngatcha, B. 1993. *Hydrogéologie d'aquifères complexes en zone semi-aride. Les aquifères quaternaires du Grand Yaéré (Nord Cameroun).* Thèse de Doctorat, Université de Grenoble, France.

Ngounou Ngatcha, B., Njitchoua, R., Ekodeck, G. E., Naah, E., Mudry, J. & Sarrot-Reynauld, J. 2000. *Pollution par les nitrates des eaux souterraines de la partie septentrionale du Cameroun.* Actes Colloque ESRA'2000, Poitiers, 13 au 15 September 2000.

Ngounou Ngatcha, B., Mudry, J., Sigha Nkamdjou, L., Njitchoua, R. & Naah, E. 2005. *Climate variability and impacts on alluvial aquifer in a semiarid climate, the Logone-Chari plain (South of Lake Chad).* In: Regional impacts of climate change – Impact assessment and decision making. IAHS Publication, **295**, 94–100.

Ngounou Ngatcha, B., Mudry, J., Leduc, Ch. 2006. *A propos des aquifères du Continental terminal dans le bassin du lac Tchad.* 21st Colloquium on African Geology, Abstract book, Geoscience for Poverty Relief, Maputo, Mozambique, 3–5 juillet 2006.

Njitchoua, R. & Ngounou Ngatcha, B. 1997. *Hydrogeochemistry and environmental isotopic investigations of the North Diamaré plain, Extreme-North of Cameroon.* Journal of African Earth Sciences, **25**, 307–316.

Oguntola, J. A. 2004. *Management of transboundary aquifer systems in Africa. A case study of the Lake Chad Basin Commission.* In: Appelgren, B. O. (ed.), Managing shared aquifer resources in Africa, UNESCO, IHP-VI Series on Groundwater, **3**: 203–208.

Olivry, J-Cl., Chouret, A., Vuillaume, G., Lemoalle, J. & Bricquet, J. P. 1996. *Hydrologie du lac Tchad.* Monog. Hydrol. ORSTOM, Paris, France, **12**, 266pp.

Olivry, J-Cl. & Sircoulon, J. 1998. *Evolution des recherches hydrologiques en partenariat en Afrique sub-saharienne : l'exemple des pays francophones.* Rev Sci Eau Spécial, 61–75.

Puri, S., Appelgren, B., Arnold, G., Aureli, A., Burchi, S., Burke, J., Margat, J. & Pallas, P. 2001. *Internationally Shared (Transboundary) Aquifer Resources Management. Their significance and sustainable management.* IHP-VI, Series on Groundwater, **1**, 33–36.

Rodda, J. C. 2003. *Prioritising Hydrological Networks.* In: Hydrological networks for integrated and sustainable water resources management. IHP/OHP, 7–26.

Rutashobya, D. G. 2003. *Hydrological Networks in Africa.* In: Hydrological networks for integrated and sustainable water resources management. IHP/OHP, 51–65.

Schlager, E. 2006. *Challenges of governing groundwater in U.S. western states.* Hydrogeology Journal, **14**, 350–360.

Schneider, J. L. 1989. *Géologique et hydrogéologie de la République du Tchad.* Thèse de Doctorat d'Etat, Université d'Avignon, 818pp.

Vadas, R. G. 1999. *The Sao Francisco river basin.* In: Mostert, E. (ed.), River basin management, Proceeding of the International Workshop (The Hague, 27–29 October 1999). IHP-V, Technical Document in Hydrology, **31**, 97–114.

Zektser, I. S. & Everett, L. G. 2004. Groundwater resources of the world and their use. IHP-VI, Series on Groundwater, **6**, 346pp.

CHAPTER 6

The benefits of a scientific approach to sustainable development of groundwater in Sub-Saharan Africa

J.E. Cobbing
CSIR NRE Unit, Pretoria, South Africa

J. Davies
British Geological Survey, Wallingford, UK

ABSTRACT: With less than ten years to go before the deadline for the Millennium Development Goals (MDGs) falls due, there is an increasing urgency behind the supply of safe drinking water and sanitation facilities to African countries. Although groundwater will form a substantial part of the water used in water supply schemes, particularly in rural areas, the resource is poorly understood in many parts of the continent. Careful and appropriate data collection during project implementation, together with data interpretation and knowledge dissemination can prevent past mistakes being repeated, and reduce the ultimate cost of water supply schemes both from a human and a financial point of view. Hydrogeologists are familiar with this argument, but are not always consulted when water supply schemes are planned. As funding agencies prepare to increase water supply and sanitation implementation in Sub-Saharan Africa, it is vital that a scientific approach to groundwater development is more widely adopted, and incorporated at the planning stage of new projects.

1 INTRODUCTION

Improved water supply and sanitation services are recognised as essential to addressing poverty and underdevelopment in Africa. As with other topics in the field of international development, there is no lasting consensus on the best way to proceed in installing or supporting the necessary infrastructure, managing and maintaining it, and replacing it when no longer functional. The debate has moved from models advocating centrally funded, top-down implementation run by national governments, to the more recent emphasis on community funded and managed systems in which non-governmental organisations (NGOs) and the private sector play an important role. Today, partnerships between the government and the private sector ('public-private partnerships'), or between government agencies, NGOs and communities, are frequently discussed. One of the unchanging realities of water-related infrastructure, however, is that the available water resource must be understood and managed prudently if water supply and sanitation schemes are to be sustainable (Robins *et al.*, 2006).

2 GROUNDWATER AND RURAL WATER SUPPLY

2.1 *The new impetus for water supply in Africa*

On the 8th of September 2000, at the largest gathering of heads of state in history, the United Nations General Assembly adopted General Assembly Resolution 55/2, or the 'Millennium Declaration', in which member countries endorsed a series of values and principles designed to advance global development and reduce poverty. Eight Millennium Development Goals (MDGs) were adopted, each with specific targets (Table 1). The Millennium Development Goals have since become a focus for international development activities, with the targets providing a way to measure progress since adoption of the resolution.

The supply of safe drinking water, and the related provision of adequate sanitation, is critical to poverty alleviation and economic growth in the world's poorest countries, and Target 10 of Goal 7 commits the signatories to halving, by 2015, the proportion of people worldwide without access to safe drinking water compared to 1990 levels. At the World Summit on Sustainable Development in Johannesburg in August 2002, the target of halving the proportion of people worldwide without access to sanitation by 2015 (compared with 1990 levels) was added to existing agreements. Whilst the provision of sanitation and clean water is a target in itself, it is integral to the other goals. For example, people (usually women and girls) are relieved of the burden of collecting water from distant sources, and can devote more time to economic activities (Goal 1) or education (Goal 2). Lifting this burden contributes to the empowerment of women (Goal 3). Sanitation and clean water are vital in reducing infant mortality and disease (Goals 4 and 6), and improving maternal health (Goal 5) and resistance to diseases such as HIV/AIDS, and these outcomes all promote economic development (Goal 8). Bettering water and sanitation services is one of the cheapest ways of improving people's health (World Bank, 2005). In development terminology, water supply and sanitation is truly a 'cross-cutting' issue.

2.2 *Providing water: the task for the next ten years*

World progress towards meeting the MDGs, five years after they were adopted, is variable. Whilst some developing regions are on track to reach the safe water supply and sanitation targets, Africa south of the Sahara is lagging behind (World Bank, 2005), and if the current slow rates of improvement in Sub-Saharan Africa continue, the region will not even come

Table 1. Summary of the Millennium Development Goals (after UNDP 2003).

Goal	Summary
Goal One	Eradicate extreme poverty and hunger.
Goal Two	Achieve universal primary education.
Goal Three	Promote gender equality and empower women.
Goal Four	Reduce child mortality.
Goal Five	Improve maternal health.
Goal Six	Combat HIV/AIDS, malaria and other diseases.
Goal Seven	Ensure environmental sustainability.
Goal Eight	Develop a global partnership for development.

close to meeting the water supply and sanitation targets. Other indicators are equally dire; for example, rates of child malnutrition in Sub-Saharan Africa are rising (World Bank, 2005). The scale of the task facing Africa is daunting. Over the next ten years it is estimated that between 150 and 300 million people will need to gain access to a water supply, and over 200 million to sanitation. If the MDG targets are to be met, a massive improvement in water supply and sanitation coverage is called for.

Development bodies, national governments and funding institutions are well aware of the improvement that is needed in water supply and sanitation coverage in Sub-Saharan Africa, and recent initiatives promise to give a boost to the task of providing clean water and sanitation to all Africans. For example, in April 2005 The African Development Bank (ADB) announced a 'Rural Water Supply and Sanitation Initiative (RWSSI)', in which it pledged to provide an extra half a billion dollars (US) per year for the next 3 years, a sum that almost doubles current funding (Boucher, 2005). The donor community was asked to match the ADB commitment, and indications are that donors are supportive. African countries have been encouraged to scale up their rural water supply and sanitation operations accordingly. Both the Commission for Africa Report (2005) and the EU Water Initiative emphasise the importance of water supply and sanitation in addressing poverty in Africa. Worldwide, there is currently a momentum by donor countries towards increasing aid budgets, in some cases towards 0.7% of GDP, and this has led to substantial increases in aid in recent years. For example, the annual UK official development assistance, which amounted to more than GBP 4.1 billion in 2004/2005 is set to reach about GBP 6.5 billion by 2006/2007 (House of Commons, 2004). Although much of this money will not be allocated to Africa, some of these increases can be expected to translate to better funding for water-related development initiatives in Sub-Saharan Africa.

2.3 *The role of groundwater*

It is now broadly accepted that if the MDG targets are to be met, groundwater will have a central role to play (see for instance Pietersen, 2005 or MacDonald, *et al.*, 2005a). Groundwater is often the most appropriate water source for rural water supply in Africa because it is generally found close to where it is needed, the natural quality is usually good, and it is resistant to even prolonged droughts. In contrast, surface water must be piped from dams or rivers, requires relatively expensive treatment, and can be vulnerable to even short periods of dry weather. In Sub-Saharan Africa, more than 80% of people without access to a safe water supply live in rural areas (WHO/UNICEF, 2007). The funds and expertise to supply all deprived areas of the subcontinent with viable surface water schemes will not be available in the foreseeable future, and this means, quite simply, that if the MDG targets are to be met a large increase in groundwater development will be essential.

There is a risk inherent in proceeding with large increases in groundwater use without an adequate understanding of the state of the resource (Robins *et al.*, 2003). Questions that need to be answered include:

- How much groundwater is available?
- What is the groundwater quality, and how might this change with time?
- How will abstractions affect the environment?
- Will abstractions be sustainable?
- What is the best way to protect the groundwater from contamination?

Historically, poor data holdings on water resources (both surface water and groundwater) in Africa have compounded the difficulties in developing further resources, and this has been recognised as an important regional issue by the World Water Council and others (World Water Forum, 2000).

Current aid policies increasingly favour the direct transfer of money to African governments ('budget support'), and funding for rural water supply and sanitation schemes is often dependent on the recipient government having an acceptable water policy in its Poverty Reduction Strategy Paper (PRSP). Thus, an incentive exists for governments to include water supply strategies in PRSPs and obtain funding although the capacity to implement and monitor the schemes may still be lacking. Whilst the benefits of this type of direct funding by donors are many, there is a possible risk of groundwater projects being implemented without sufficient regard for and a scientific understanding of the groundwater resource. Treatment of water and sanitation in PRSPs is generally neither comprehensive nor consistent (SDTF, 2003; ODI, 2004).

Groundwater is often called a 'hidden resource' because it cannot be seen in the same way as water in a river, lake or reservoir. The volumes of groundwater are large, however – it is estimated that there is about one hundred times more fresh groundwater on earth than all the fresh water in rivers and lakes (Shiklomanov, 1998). The management of groundwater has frequently been overlooked in Africa for a number of reasons.

The lack of capacity for the centralised or organised collection, storing and dissemination of groundwater data in many African countries. Economic 'Structural Adjustment' policies applied in many African countries since the early 1980s led to a decline in public sector funding and consequent shrinking of public sector ability. The centralised coordination of groundwater development was eroded, and not always replaced adequately by private sector alternatives. Viable institutions underpin national policy and legal frameworks necessary for efficient service expansion (Carter, *et al.*, 1993).

The lack of skilled staff. In many countries, expatriate hydrogeologists working in the post-independence years were not always succeeded by skilled local replacements; it can also be difficult to retain skilled personnel working on rural water supply issues. HIV/AIDS has worsened staff shortages in recent years (Ashton & Ramasar, 2002).

The fact that there are relatively few high yielding regional aquifers that can be monitored or tested in a conventional way. Much of Africa is underlain by low-yielding basement type aquifers, which cannot be assessed in the same way as regional aquifers (Clark, 1985).

The small quantities abstracted for community supply are often believed to be inherently sustainable (both from a quality and a quantity point of view). This is in part due to a perception which arose in the 1980s 'water decade' that groundwater suitable for small water supplies was ubiquitous across most of Africa, and that the challenges were primarily technical and logistical such as mobilisation, drilling, and operation and maintenance. This misconception persists in some quarters today, even where higher yielding groundwater sources are considered.

Essentially, there is a danger that groundwater development for basic rural needs will be seen only in terms of the engineering problems: e.g. drilling, pump installation, maintenance and the like, together with the software aspects of community participation, management and ownership. Rural water supply will be a matter of 'tanks and taps', rather than the prudent use of a sometimes-complex resource. This approach neglects understanding the resource base on which water supply depends, and thus minimises the importance of data collection during installation as well as some form of monitoring after the scheme is in place. In comparison to the importance given to monitoring and regulating surface water resources, and the expense of surface water treatment, it is surprising that groundwater is often assumed to need little or no assessment or monitoring.

Finally, without accurate data, it is difficult to assess the success of groundwater supply schemes beyond simple 'wet or dry' criteria. This approach ignores the nature of the resource (geology, recharge, etc.), so that when comparing the relative success of water schemes we may be comparing apples with pears. One scheme might be in a good alluvial environment, where success rates of more than 98% and very good water quality might be expected, whilst another scheme might be in a hydrogeologically difficult area where success rates of 50 or 60% and water of adequate quality would be considered exceptional. This leads to the danger of implementing agencies steering clear of difficult areas for fear of affecting their statistics, and concentrating on areas where they know wet boreholes are easy to site. This has happened on the Afram Plains in Ghana (see later), and conversations with hydrogeological technicians in Nigeria in 2005 confirm that this is also happening there, to the obvious detriment of those unfortunate communities who happen to live in areas underlain by (for example) soft mudstones (MacDonald *et al.*, 2005b).

Apart from the huge increase in groundwater development that is necessary to meet the MDGs, additional demands will be made on groundwater in many areas in the coming years. For example, many rural people in South Africa consider a water supply from a handpump an adequate facility only, and aspire to a water supply from a tap, preferably inside their household. In addition, many regard on-site or 'dry' sanitation as backward, and desire a water-borne waste removal system (Harvey & Reed, 2004). Both piped water supplies and waterborne sewage systems require water resources larger than required by a handpump system. As economic growth takes place, it is inevitable that greater demands will be placed on groundwater resources.

Today, some 41% of arable land in southeast Asia is irrigated, compared with less than 4% in Africa (NEPAD, 2003). The 'green revolution' in south Asia was partly brought about by a large increase in the area of irrigated land. Niaz (1985) estimated that the World Bank had loaned about 1.4 billion US dollars up until 1983 for the purposes of groundwater development, much of it for agriculture, and that this assistance was provided mainly to Asia. Some experts see a substantial increase in irrigation as a fundamental requirement for Africa to become sufficient in food, for rural growth, and for rural livelihoods to be sustained (NEPAD, 2003; Commission for Africa, 2005). However, whilst much of the irrigated land area in Asia is underlain by very productive regional aquifers (for example the alluvial aquifer systems of the Gangetic plain), much of Africa's irrigable land is underlain by lower-yielding basement aquifers. Boreholes in the alluvial aquifers of Asia frequently yield 40 or 50 l/s, but successful boreholes in many African environments yield less than 10%, or more often 1% of that (MacDonald *et al.*, 2005). Groundwater resources in large areas of Africa will need to be carefully assessed and managed if even a modest increase in irrigation by groundwater is to be sustainable. High yielding aquifers do exist in places in Sub-Saharan Africa (for example the Witwatersrand Dolomites of South Africa or the coastal alluvial aquifers of Mozambique), but it is likely that innovative solutions will be required, and new approaches tried in pilot projects, to exploit groundwater effectively for irrigation purposes in many countries.

3 CASE STUDIES

Two case studies have been selected illustrating the value gained by adopting a scientific approach to finding, developing and managing groundwater. Both examples are drawn from Africa, but the principles could equally be applied to any developing region in which groundwater is important.

3.1 *Afram Plains, Ghana*

The Afram Plains area is located in the Eastern Region of Ghana (Figure 1), in the Volta River basin, and receives variable rainfall during a six-month wet season. Although the average annual rainfall is high (around 1200 mm/a), during the dry season water is scarce – surface water sources such as ponds that form along the courses of the ephemeral drainage system soon dry up, and springs are rare. Reliance on unprotected pools and dugouts for water supply in the dry season results in diarrhoeal disease, and much time and effort in water collection. The area is also one of the few remaining in Africa where guinea worm infections in humans are still found. The Afram Plains is underlain by rocks of the Voltaian Sedimentary Basin, deposited unconformably upon older Precambrian rocks (Shackleton, 1976; Black & Liegeois 1993). These rocks have minimal primary permeability and porosity, and groundwater storage and movement is consequently via fractures, bedding planes and other discontinuities. Groundwater exploration is regarded as particularly difficult due to the low primary permeabilities and the variable geology, and serious problems are experienced by water development agencies active in the area.

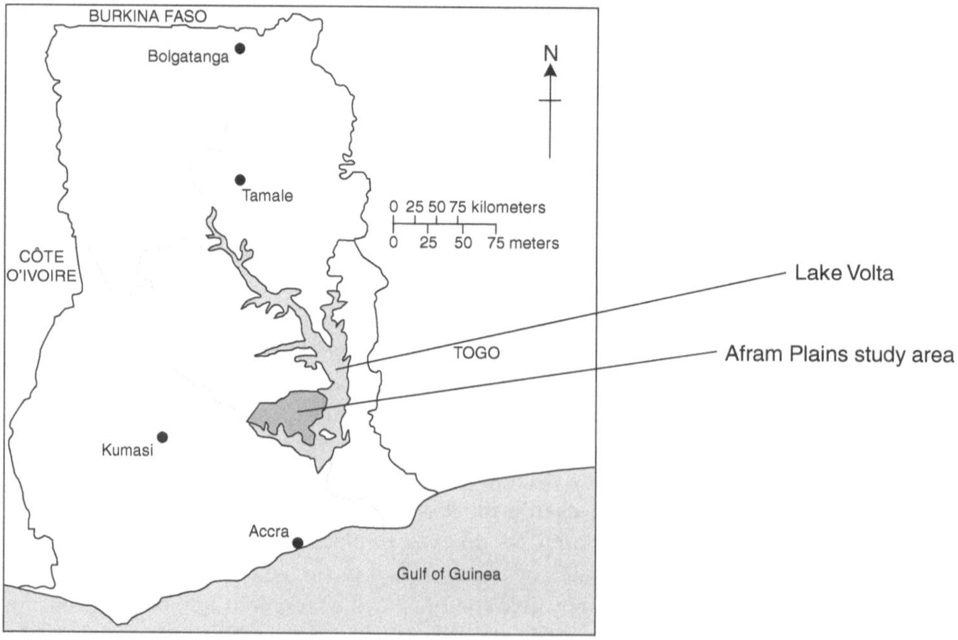

Figure 1. Map of Ghana showing the Afram Plains study area.

Development of the area's groundwater resources began in 1963 when the Geological Survey of Ghana and the Volta River Authority (VRA) drilled several boreholes to serve populations displaced by the rising waters of Lake Volta following the completion of the Akosombo Dam on the Volta River. Since that time, the population has grown considerably, and census data show a 250% increase in the farming population between 1970 and 1984, attracted by fertile soils and improving infrastructure. There are now more than 140 villages on the Afram Plains.

Table 2 shows the substantial numbers of dry boreholes drilled in this area. In addition, some boreholes which were initially classed as successful have failed after two to three years of use. This is highly significant, as communities come to rely on their groundwater sources, and populations attracted to the area by the groundwater supply have no effective alternative water source if boreholes fail. Generally, boreholes are classed as successful if, in the opinion of the driller or site supervisor, they will support a handpump.

A British Department of International Development (DfID) funded study by the British Geological Survey working together with WaterAid, the Danish aid agency DANIDA, and the Afram Plains Development (Davies & Cobbing, 2002; Cobbing & Davies, 2004) collected the following data during a programme of drilling in the area: EM-34 surface geophysics was carried out at the sites of new boreholes; borehole chip-samples were logged and photographed; drilling penetration rates, water strikes, blow yields, final water-levels and total borehole depths were recorded; borehole construction details were recorded; basic water quality measurements were made on site, and water samples taken for geochemical analysis; and boreholes were geophysically logged (conductivity, temperature, natural gamma, resistivity, and point resistance).

Based on these and other data, guidelines for groundwater exploration for the Afram Plains were subsequently compiled, and a basic groundwater exploration map was constructed to assist future groundwater development (see Figure 2 and Table 3). Initial use of these resources has been encouraging – for example, the five deep exploration boreholes sited towards the end of the project were successful, and one of the outcomes of the work has been a realisation that deeper (> 100 m) boreholes may well have yields > 2 l/s, possibly because they intercept fractures which are recharged by the lake. It is very likely that the careful use of this information will increase borehole success rate in the area. Prior to this project, the drilling success rate had not improved significantly over more than 30 years of groundwater exploration efforts. This is partly because data from each drilling project were

Table 2. Summary of available information on borehole drilling success on the Afram Plains, 1963 to 2001 (after Cobbing & Davies (2004)).

Organisation	Period	Total boreholes	Successful boreholes	Unsuccessful boreholes
Volta River Authority	1963–1965	10	6	4
Catholic Church group	1984	47	28	19
World Vision International	1990–1995	152	92	60
WaterAid/Afram Plains Dev. Org.	1996–2001	101	67	34
DANIDA exploration boreholes	2001	5	5	0
Total		315	198 (63%)	117 (37%)

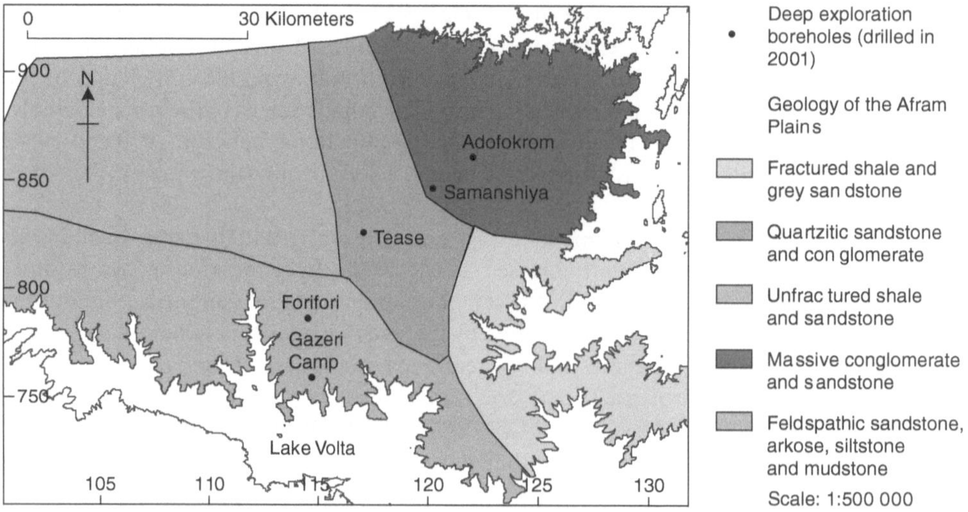

Figure 2. Map of the Afram Plains showing simplified hydrogeological units (compiled using a variety of sources reported in Cobbing & Davies (2004)).

either not collected and interpreted, or were not readily available to future workers. In effect, lessons were not learned from one project to the next, resulting in mistakes being repeated.

A typical example of the difficulties of sharing groundwater data was observed in Ghana. The drilling contractor employed a geophysical team to site the village water supply boreholes that were drilled on the Afram Plains in 2001. The team used conductivity (EM-34) and resistivity methods, and also took topography and vegetation into account. At least two sites were identified at each village in the event of the first site being dry, and this produced a lot of data. Large amounts of similar data are held by the geophysicist, gained from a variety of other projects in West Africa. It was suggested that these data could be used by other projects in the Afram Plains and in similar areas, and could also form the basis of a valuable paper or report since it reflected considerable local knowledge and experience gained over many years. The geophysicist responded by stressing that the data had value not only to him but also to others working in the groundwater field. Since he was only paid for the time that he actually spent working, making his data more accessible could lead to a reduction in contracts and therefore income for him. He gave the example of a previous contract that required him to locate three potential sites for each of a series of boreholes. Many of the sites remained "unused". Some years later his data were used by another geophysical contractor to site more boreholes in the area for little extra cost or effort.

In other words the geophysicist perceived that making his data widely available helped his competition and thus harmed his interests. There was also a lack of access to equipment, such as GPS receivers, copiers and laptop computers, to record and compile field data. Data often remained in a rough form as they were recorded in the field (and as a single copy), since analysis was often performed in the field and the results not formally recorded with an outside audience or agency in mind. A great amount of potentially valuable data have been lost in this way. It is clear that incentives may be needed in many parts of Africa if private contractors are to make 'their' data accessible, or alternatively the handover of data needs to be specified in contracts.

Table 3. Summary of notes on groundwater potential accompanying the map.

Geological area (see Figure 2)	Summary of groundwater potential	Notes on groundwater occurrence
Massive conglomerate and sandstone	Moderate	Weathered conglomerate gravel often visible at surface. Good recharge, best borehole sites located in valleys. Boreholes should be drilled to below present day lake level for best yields, since flow may be induced from the lake along fracture zones. Problems with pollution in villages, otherwise groundwater quality good. Success rate ~66%; 38% ≥30 l/min.
Fractured shale and grey sandstone	Low to very low	Low altitude lakeside areas. Little information has been recorded in this area; it is possible that few boreholes have been attempted. Groundwater quality thought to be poor to saline. Rainwater harvesting may be necessary in some villages.
Quartzitic sandstone and conglomerate	Moderate	Quartzitic sands often visible at surface. Moderate recharge, best sites located in valleys. Boreholes should be drilled to below present day lake level. May be able to induce flow from the lake along fracture zones. Problems with pollution in villages, otherwise groundwater quality good. Success rate ~67%; 40% ≥30 l/min.
Feldspathic sandstone, arkose, siltstone and mudstone	Low to moderate	Weathered purple brown sandstone platform surface beneath thin ferricrete. Very poor recharge potential due to re-cemented layer down to ~60 m. Deep holes may intercept weathered zones. Direct recharge from lake along fractures unlikely. Fractures poorly defined. Groundwater quality good. Success rate ~66%; 39% ≥30 l/min, however boreholes known to fail after two or three years of use. May need to consider artificial recharge of boreholes.
Unfractured shale and sandstone	Low	Poor to moderate recharge to tight formation except where conglomeratic bands are present. Boreholes should be drilled to below present day lake level. Groundwater quality may be poor to saline. Success rate ~50%; t. 14% ≥30 l/min.

The Afram Plains is perhaps an extreme case, in that the hydrogeology is particularly difficult (i.e. successful boreholes are difficult to site, water quality is sometimes poor, and some boreholes fail after a few years due to inadequate recharge), there is an acute lack of skilled staff and other resources in this remote area, and institutions in Ghana responsible for the collection and assessment of drilling data lack funding and capacity.

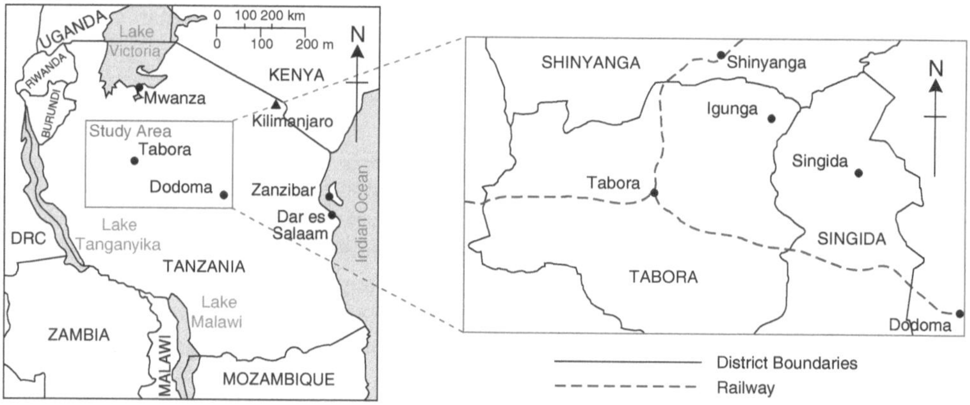

Figure 3. Map of Tanzania showing the general study area.

3.2 *North central Tanzania*

The Tabora, Singida and Manyara regions of Tanzania (Figure 3) are amongst the most deprived in Sub-Saharan Africa in terms of water supply and sanitation coverage, and substantial efforts have been made over the past few years to improve the water supply situation. The occurrence of groundwater within the Tabora, Singida and Arusha (including Manyara) regions is described in Regional Water Master Development Plans produced during the 1980s and 1990s (Arusha Regional Water Master Plan, 2000). Since the compilation of the Regional Water Development Plans, rural groundwater development has proceeded primarily as a partnership between the Ministry of Water, its former nominally privatised component parts (such as the Drilling and Dam Construction Agency (DDCA) and international NGOs such as WaterAid and World Vision. The UK has provided scientific advice over a number of years on the sustainable development of the groundwater resources of the area in cooperation with these partners (e.g. Smedley *et al.*, 2002; Davies & Ó Dochartaigh 2002; Davies, 2005).

The area is underlain by granitic and metasedimentary basement rocks of the Archean Granite Shield. Sandy soils occur with ferricretes on hillsides and heavy "Mbuga" clays (black vertisol soils) in valleys. Cenozoic and later lacustrine and alluvial deposits are found in some areas (Davies, 2005). Recognised technical problems in these areas include:

1. water quality problems, including the widespread occurrence of high fluoride concentrations in groundwater;
2. difficulties with the location of wells and boreholes capable of yielding sustainable quantities of water sufficient to meet community water supply needs.

These problems are exacerbated by problems common to many developing areas, including a lack of trained personnel, a lack of good quality groundwater data, and limited resources with which to carry out groundwater development work. An increasing population and better water infrastructure is placing growing demands on already limited groundwater resources, which in some cases are unable to deliver water of the desirable quantity and quality. However, a scientific approach to the difficulties of developing groundwater resources has helped to identify the following important points, which have proved valuable in on-going projects.

- The development of databases by using GPS equipment to locate boreholes and other water sources (including the locations of dry boreholes) on topographic and geological maps has proved to be extremely useful in planning and costing future projects, and arriving at a realistic assessment of the water resources of an area.
- Experienced ex-government hydrogeologists are employed by WaterAid to supervise and acquire geological and hydrogeological data during borehole drilling and testing, data which have proved to be of considerable benefit in planning future projects, and in the monitoring of existing boreholes.
- Data on the groundwater resources, gathered both during implementation and in follow-up monitoring phases, have proved that in some areas the water supply needs of rural communities cannot be met by groundwater alone, and that other solutions (such as conjunctive use of groundwater and rainwater harvesting) are needed. The data which underpin the groundwater development makes this less of a 'guessing game'.
- In some areas, hand-dug-wells are feasible. These avoid high drilling costs whilst at the same time allowing a much greater degree of community participation. However, in other areas traditional methods of accessing groundwater are not adequate. The choice of technology is frequently determined by hydrogeological conditions (e.g. knowledge of water table fluctuations or quality changes with depth) and scientific advice allows planners to make a much more realistic assessment of what is possible.

Data collected during recent BGS/WaterAid project work (Davies & Ó Dochartaigh, 2002; Smedley, *et al.*, 2000) and the collation of available existing data, provide a framework for the development of groundwater resources in the region (Davies, 2005). The hydrogeological characteristics of the various low permeability rock types that underlie the Tabora Region are complex. Groundwater potential depends on many factors, including geology, structure (particularly fracture patterns), geomorphology, and past climates. The different geological units have different hydrogeological characteristics, but all are low yielding. Groundwater occurs in zones of weathering and in discrete fracture zones within bedrock. Shallower aquifer units, often on hillslopes, contain young, recently recharged water, which flows rapidly downslope to discharge in valleys. This rapid movement of water can also lead to rapid transport of contaminants in shallow zones, which is seen in the high levels of nitrate in some of the shallow wells tested. Older, more mineralised water is often present in fracture zones. In Nyanzian rocks, water bearing fracture zones are often buried too deep beneath Mesozoic and Quaternary sediments to be determined using geophysical survey methods. Water from most of the boreholes and hand-dug-wells tested had high levels of iron, aluminium, fluoride and/or barium, all of which associated with health problems.

Data from this project and other work have been interpreted to provide, amongst other things, a summary diagram of the hydrogeological potential of the Nzega and Tabora areas, together with a table describing the groundwater characteristics and potential of each unit (see Figure 4 and Table 4). These resources were designed to be laminated as a single A4 sheet for use in the field by technicians and others.

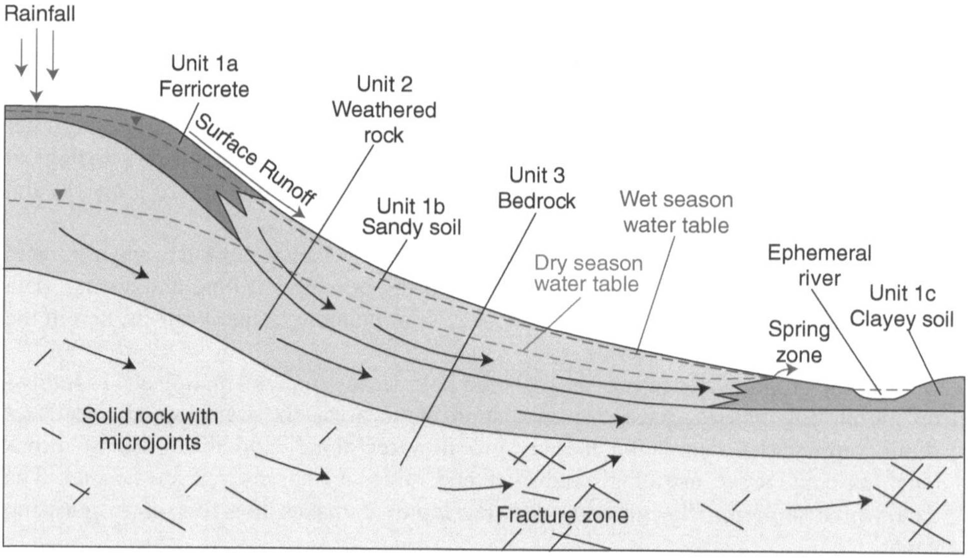

Figure 4. Simplified diagram of groundwater occurrence in the Nzega and Tabora areas, Tanzania (Davies, 2005).

In addition to the single-sheet summary provided above, a list of key issues for the development of groundwater in north-central Tanzania was compiled:

The Importance of Accurate Site Recording and Location: Both wet and dry boreholes are a valuable source of data. In the past, water supply development programmes have failed to locate boreholes and other survey sites accurately, or have relied on sometimes-ambiguous place names. Inexpensive GPS systems now provide a simple means of accurately locating boreholes, villages, rivers, roads and other data points.

Use of Geophysical Surveys for Borehole Siting: Electrical resistivity surveys are used to determine the apparent thickness of the weathered zone. Borehole sites are then located where this zone appears to be thickest. Geophysical surveys need to be undertaken in conjunction with the interpretation of aerial photography (used to locate linear target structures such as fault zones) with sites located on a 1:50 000 topographic map using a GPS if the optimum amount of information is to be obtained from the correlation of drilling data and geophysical survey results. EM34 equipment can be used for the rapid determination of lineament locations. Ideally target lineations should be located using conductivity (EM-34) traversing; and electrical resistivity used to investigate depths of weathering on the fault zone.

Data Gathered During Borehole Drilling: At little added expense, much useful geological and hydrogeological data can be gathered during the drilling of a borehole, including detailed geological descriptions, penetration rates, and flow rates. These data can be used to make a more objective assessment of the borehole potential, so that low yielding boreholes can be abandoned, and higher yielding boreholes identified and perhaps targeted for further development. Photographs have been used to record rock colour changes with depth that can be correlated with patterns of weathering, fracturing and water struck zones.

Test Pumping: A bail test (a modification of a slug test) demonstrated during the field study in Tabora provided field personnel with a rapid procedure to assess the yield potential

Table 4. Summary of the groundwater potential of the units identified in Fig. 3 for the Nzega and Tabora areas, Tanzania (after Davies, 2005).

Hydrogeological unit	Groundwater potential	Groundwater Quality	Notes on groundwater occurrence
Unit 1a: Upper slope soils – Nodular, hard, red-brown ferricrete with white clay at base. 2–10 m thick.	Moderate	Low pH (4.5–5.5); SEC <200 μS/cm; may be high iron. Risk of nitrate and e-coli pollution.	Ferricrete seen at surface and exposed in wells. Moderate-high storage, short groundwater residence time: high yield in wet season when ferricrete is saturated and first weeks/months of dry season; low/no yield available for remainder of dry season. Wells 10–15 m deep.
Unit 1b: Lower slope soils – Light brown sands with interbedded clay and clay at base. 2–10 m thick.	Moderate	Neutral pH and low SEC. May be clay in suspension and/or high fluoride.	Sands seen at surface and in wells, investigate with hand auger. Unlined wells prone to collapse. Saturated sands have moderate storage and residence times: moderate yield; low yield at end of dry season; low/no yield after long droughts. Spring zones at junction with valley bottom clays. Wells 10–15 m deep or spring boxes.
Unit 1c: Valley bottom soils dark grey, cracking mbuga clays with thin sands and calcrete nodules; gravels at base. 2–10 m thick.	Low	May be brackish.	Gravel seen in wells, or investigate with hand auger. Gravels beneath surface clays have low storage, very slow groundwater throughflow: low/no yield. Wells 10–15 m deep.

(*Continued*)

Table 4. (Continued)

Hydrogeological unit	Groundwater potential	Groundwater Quality	Notes on groundwater occurrence
Unit 2 (Nzega): Weathered Nyanzian meta-sediments and ash. Silvery-grey, red-brown or brown; moderately hard to very soft. 2–30 m thick.	Low to moderate	TDS and SEC increase with depth	Upper weathered zone and weathered fracture zones. Low groundwater storage, very long residence times: low yields available year-round. Wells 10–15 m deep, boreholes <60 m deep.
Unit 3 (Nzega): Unweathered hard grey, red or brown Nyanzian metasediments, some hard black ash, fractures near top, quartz or calcite veins, >50 m thick.	Low to high	TDS and SEC decrease with depth	Fractures near top have low-moderate groundwater storage and short residence times: high yields for short periods following borehole construction. Main rock body has low storage & very long residence times in microjoints: very low or no yield. Boreholes <60 m deep.
Unit 2 (Tabora): Weathered brown, grey and red-brown Dodoman igneous and metamorphic rocks, medium to coarse grained, hard to very soft. 2–20 m thick.	Low to moderate	TDS and SEC increase with depth, may contain white suspended clays and/or high iron.	Upper weathered fracture zones, especially pegmatites. Low groundwater storage, very long residence times: low yields available year-round. Wells 10–15 m deep, boreholes <60 m deep.
Unit 3 (Tabora): Unweathered hard white, black, pink, dark green or black Dodoman igneous and metamorphic rocks, fine to coarse grained, fractures near top >50 m thick.	Low to high	TDS and SEC decrease with depth	Fractures near top have low-moderate groundwater storage and short residence times: high yields for short periods following borehole construction. Main rock body has low storage and very long residence times in microjoints: low or no yield. Boreholes <60 m.

of a borehole, without undertaking a standard pumping test (see MacDonald *et al.*, 2005a). Simple procedures were provided to help interpret these tests. However, in some fractured aquifers interpretation can be more difficult, especially if the tests are carried out over short time periods (e.g. less than 5 hours). Fractured aquifer systems can initially give high yields, but if pumping continues they can be dewatered and may suddenly fail. If this behaviour is common in a project area, a longer-term pumping test should be carried out to allow a more accurate assessment of borehole potential.

Hydrochemical Sampling: Hydrochemical sampling and analysis is used to establish inorganic groundwater quality, and provide information about recharge and contamination. The routine measurement of borehole water specific electrical conductance (SEC) by field staff provides indications of changes in water quality. The results obtained can be used to define areas or depths of different water quality that can provide information on aquifer recharge, dewatering or anthropogenic contamination, especially within village environments.

Borehole Monitoring: Some boreholes in the Tabora Region have experienced declining yields and some have failed, after periods of abstraction lasting from some months to a several years. The simple monitoring of borehole yields, which can be undertaken informally by the borehole users, would provide a warning of this problem. Informal monitoring can also be used to identify construction problems such as pump failures.

Hydrogeological Database: The construction of a hydrogeological database, accurately geo-referenced, forms an outcome of data collection activities. This will inform future workers, and help to improve drilling success rates. It will also serve to inform the expectations of development workers and communities. A good database can also be the basis for a useful groundwater potential map showing average yields for the different aquifer types, likely depths to groundwater, modes of groundwater occurrence, and water quality information.

4 CONCLUSIONS

None of the general points made above are new. For example, Grey *et al.* (1985) demonstrated the importance of hydrogeological expertise in rural water supply work in Malawi more than twenty years ago. These authors showed that costs were lower when a hydrogeologist was involved, especially when he or she is present at the project planning stage. In particular, the hydrogeologist is able to advise on appropriate drilling equipment, borehole siting methods and borehole design. More recently, Robins *et al.* (2003) have shown the importance of data collection and interpretation to the success of groundwater development projects. Donors, implementers, national governments and others now broadly agree on the need for a more scientific approach to development work, which includes the exploitation of groundwater resources.

The challenge is to make this a reality, since groundwater development in many parts of Africa continues today with very little hydrogeological input. The case studies discussed above show the value of scientific data collection and interpretation, which should ideally be explicitly specified in contract documents. Common standards for the collection, storage and sharing of groundwater data need to be agreed on. African groundwater institutions (universities, geological surveys, research organisations, and others) are beginning to cooperate more closely in developing better ways to develop and manage groundwater resources, particularly trans-boundary resources (note for instance the recent International

Workshop on Groundwater Protection in Africa, hosted by the University of the Western Cape in Capetown, South Africa, in November 2005). It is necessary to imbed the kind of cooperative, scientific approach demonstrated at this workshop into the discourse of water development in Africa more generally. Most groundwater scientists are now aware of the critical importance of the social, political and institutional context in which water development work is carried out, but it is essential that development managers and fund holders incorporate a scientific assessment of the groundwater resource into water supply strategies.

REFERENCES

Arusha Regional Water Master Plan 2000. Part III of the Plan: Methods, Data and Analysis: Volume 14: Groundwater Resources (Hydrogeology), Volume 15 Water Quality and Volume 16: Water Engineering, Final Report, December 2000.

Ashton, P. & Ramasar, V. 2002. *Water and HIV/AIDS: Some strategic considerations in Southern Africa*. In: Turton, A. R. & Henwood, R. (eds.) Hydropolitics in the Developing World: A Southern African Perspective. University of Pretoria, Pretoria.

Black, R. & Liegeois, J. P. 1993. *Cratons, mobile belts, alkaline rocks and continental lithospheric mantle: the Pan-African testimony*. Journal of the Geological Society, **150**, 89–98.

Boucher, C. 2005. *The Rural Water Supply and Sanitation Initiative*. (Chanel Boucher, Vice President, Policy, Planning and Research.) African Development Bank Group, New York, United Nations.

Carter, R. C., Tyrrel, S. F & Howsam, P. 1993. *Lessons learned from the UN Water Decade*. Journal of the Institution of Water and Environmental Management, **7**, 646–650.

CIA, 2006. *CIA World Fact Book*. World Wide Web Address http://www.cia.gov/cia/publications/factbook/index.html (accessed April 2006).

Clark, L. 1985. *Groundwater abstraction from Basement Complex areas of Africa*. Quarterly Journal of Engineering Geology, **8**, 25–34.

Cobbing, J. E. & Davies, J. 2004. *Understanding problems of low recharge and low yield in boreholes: an example from Ghana*. In: Proceedings of the International Conference on Water Resources of Arid and Semi-arid Regions of Africa (WRASRA), August 3–6 2004, Gaborone, Botswana.

Commission for Africa 2005. *Report of the Commission for Africa: March 2005*. World Wide Web Address: http://www.commissionforafrica.org/english/report/introduction.htm (accessed April 2006).

Davies, J. & Cobbing, J. E. 2002. *An assessment of the hydrogeology of the Afram Plains, Eastern Region, Ghana*. British Geological Survey Technical Report, **CR/02/137N**.

Davies, J. & Ó Dochartaigh, B. É. 2002. *Low Permeability Rocks in Sub-Saharan Africa. Groundwater development in the Tabora Region, Tanzania*. British Geological Survey Commissioned Report, **CR/02/191N**.

Davies, J. 2005. *Scoping Study to Assess Hydrogeological Support to WaterAid Tanzania*. British Geological Survey Technical Report **CR/05/174C**.

Grey, D. R. C, Chilton, P. J., Smith-Carrington, A. K. & Wright, E. P. 1985. The expanding role of the hydrogeologist in the provision of village water supplies: an African perspective. Quarterly Journal of Engineering Geology, **18**, 13–24.

Harvey, P. & Reed, B. 2004. *Rural Water Supply in Africa. Building Blocks for Handpump Sustainability*. WEDC, Loughborough University, UK.

House of Commons Science and Technology Committee 2004. *HC 133-1 The Use of Science in UK International Development Policy*. Thirteenth Report of Session 2003-04. Volume 1. The Stationery Office, London.

MacDonald, A. M., Davies. J, Calow R. C. & Chilton, P. J. 2005a. *Developing Groundwater: a guide for rural water supply*. ITDG Publishing, Rugby, UK, Warwickshire, 384 pp.

MacDonald. A. M., Kemp, S. J. & Davies, J. 2005b. *Transmissivity variations in mudstones*. Ground Water, **43**, 259–269.

New Partnership for Africa's Development (NEPAD) 2003. Comprehensive Africa Agriculture Development Program (CAADP). NEPAD, Midrand, South Africa.

Niaz, S. M. 1985. *International funding of groundwater development schemes*. Quarterly Journal of Engineering Geology, **18**, 3–12.

Overseas Development Institute (ODI) 2004. *From Plan to Action: Water Supply and Sanitation for the poor in Africa*. ODI Briefing Paper July 2004, Overseas Development Institute, London.

Pietersen, K. 2005. *Groundwater Crucial to Rural Development*. The Water Wheel March/April 2005. Water Research Commission, Pretoria.

Robins, N. S., Davies, J., Farr, J. L. & Calow, R. C. 2006. *The changing role of hydrogeology in semi-arid southern and eastern Africa*. Hydrogeology Journal, **14**, 1483–1492.

Robins, N. S., Davies, J., Hankin, P. & Sauer, D. 2003. *Groundwater and data – an African experience*. Waterlines, **21**, 19–21.

SDTF3 2003. *Sustainable Development Task Force. Overview of DfID Activities in the Water Sector SDTF3-1003-1*. World Wide Web Address: http://www.sustainable-development.gov.uk/government/task-forces/word/10water.doc, (accessed April 2006).

Shackleton, R. M. 1976. Pan-African structures. Philosophical Transactions of the Royal Society, London, **280**, 491–497.

Shiklomanov, I. A. 1998. *World Water Resources – A New Appraisal and Assessment for the 21st Century*. UNESCO, Paris.

Smedley, P. L., Nkotagu, H., Pelig-Ba, K., MacDonald, A. M., Tyler-Whittle, R., Whitehead, E. J. & Kinniburgh, D. G. 2002. *Fluoride in groundwater from high-fluoride areas of Ghana and Tanzania*. British Geological Survey Commissioned Report, **CR/02/316**.

UNDP 2003. *Human Development Report 2003*. United Nations Development Program (UNDP), New York.

WHO/UNICEF 2007. *JMP Report 2006*. World Health Organization, UNICEF, Geneva.

World Bank 2005. *Global Monitoring Report 2005. Millennium Development Goals: From Consensus to Momentum*. The World Bank, Washington.

World Water Forum 2000. *The Africa Water Vision for 2025: Equitable and Sustainable Use of Water for Socioeconomic Development*. World Water Forum, The Hague, Netherlands.

CHAPTER 7

Sustainable groundwater development in Nigeria

A. Onugba & O.O. Yaya
Department of Hydrogeology, National Water Resources Institute, Kaduna, Nigeria

ABSTRACT: Developing groundwater is generally an excellent option for sustainable water supplies in Nigeria, despite some challenges. However, to achieve a sustainable supply, planning is required which needs hydrological and hydrogeological data as well information on water demand and general socioeconomic conditions. Data requirements include: the quantity of water required per year, intended rate of abstraction, the use of water and the amount of money available; detailed geology of area – type, extent, structure and variability of rocks, aquifer properties; rainfall & surface water sources; the cost of drilling, pumps and other materials; and potential incomes from water sales. Development of groundwater resources involves a sequential process with three phases: exploration, evaluation and exploitation. Often, groundwater supply projects concentrate on exploitation to the neglect of the evaluation phase. Groundwater supply projects can fail if this sequential development process is not followed. Adequate data cannot be available for meaningful planning if no proper evaluation is carried out before exploitation. Equally, to ensure sustainability there is need for periodic re-evaluation of demand, performance and changes in hydrological and hydrogeological conditions. The paper examines groundwater resources development and water supply programmes in Nigeria with emphasis on technical and social constraints to achieving sustainable groundwater development.

1 INTRODUCTION

Approximately 60 million Nigerians live in rural communities of less than 5000 and half of these in even smaller communities of less than 1500. The main economic activities in the rural areas are agriculture and livestock rearing with about two-thirds of the population engaged in subsistence farming. Nigeria as a whole is well endowed with water resources. The country is well-drained with a reasonably close network of rivers and streams (Figure 1). Some of these rivers, particularly the smaller ones, are seasonal, particularly in the northern part of the country where the rainy season is only three or four months in duration. The average rainfall ranges from <500 mm/a in the north to <3000 mm/a in the south. Despite the abundant water resources in the country, the current water-supply coverage in rural areas is 40–50% for which groundwater is the most widely used option.

In Nigeria groundwater provides the only realistic water supply option for meeting dispersed rural demand. Alternative water sources can be unreliable and expensive to

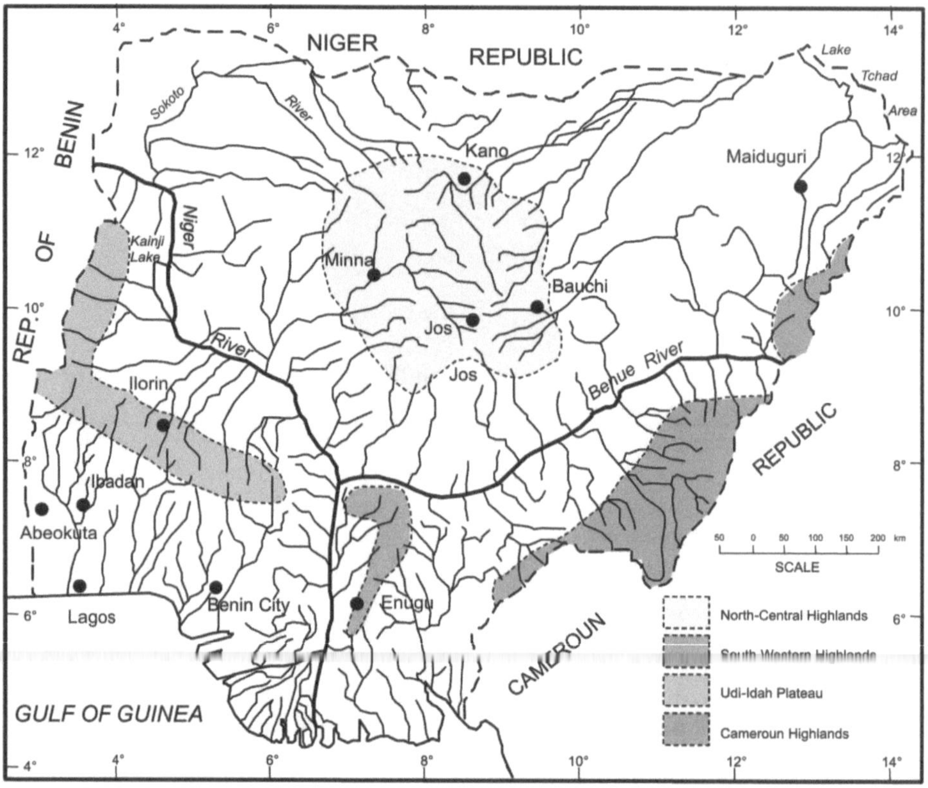

Figure 1. Map of Nigeria showing the locations discussed in this paper.

develop: surface water is prone to contamination and often seasonal; rainwater harvesting can be expensive and requires good rainfall throughout the year. Groundwater, however, can be found in most environments with the appropriate expertise. Wells give reasonably good yields in most areas although saline intrusion is a problem in the coastal areas. Artesian water occurs in the Chad Basin and parts of Sokoto basin as well as in a few locations of Paleocene-Eocene sedimentary rocks of the southwestern areas. Groundwater generally requires no prior treatment since it is naturally protected from contamination; it does not vary significantly seasonally and is often drought resistant. Also it lends itself to the principles of community management – it can be found close to the point of demand and be developed incrementally (MacDonald *et al.*, 2005).

Some of the benefits of groundwater development are linked to the inherent characteristics of groundwater resources: most aquifers provide large water storage space and help stabilize water supply during peak of dry season and droughts; the sluggish flow of groundwater through small voids helps in purifying water, necessitating lower or no treatment costs prior to its use as drinking water; the general availability of groundwater makes it a resource easy to access; and in areas of extensive aquifers, groundwater development can increase recharge and also decrease flood intensity.

This paper examines groundwater development water supply programmes in Nigeria with emphasis on technical and social constraints to achievement of sustainability for water supply in the country.

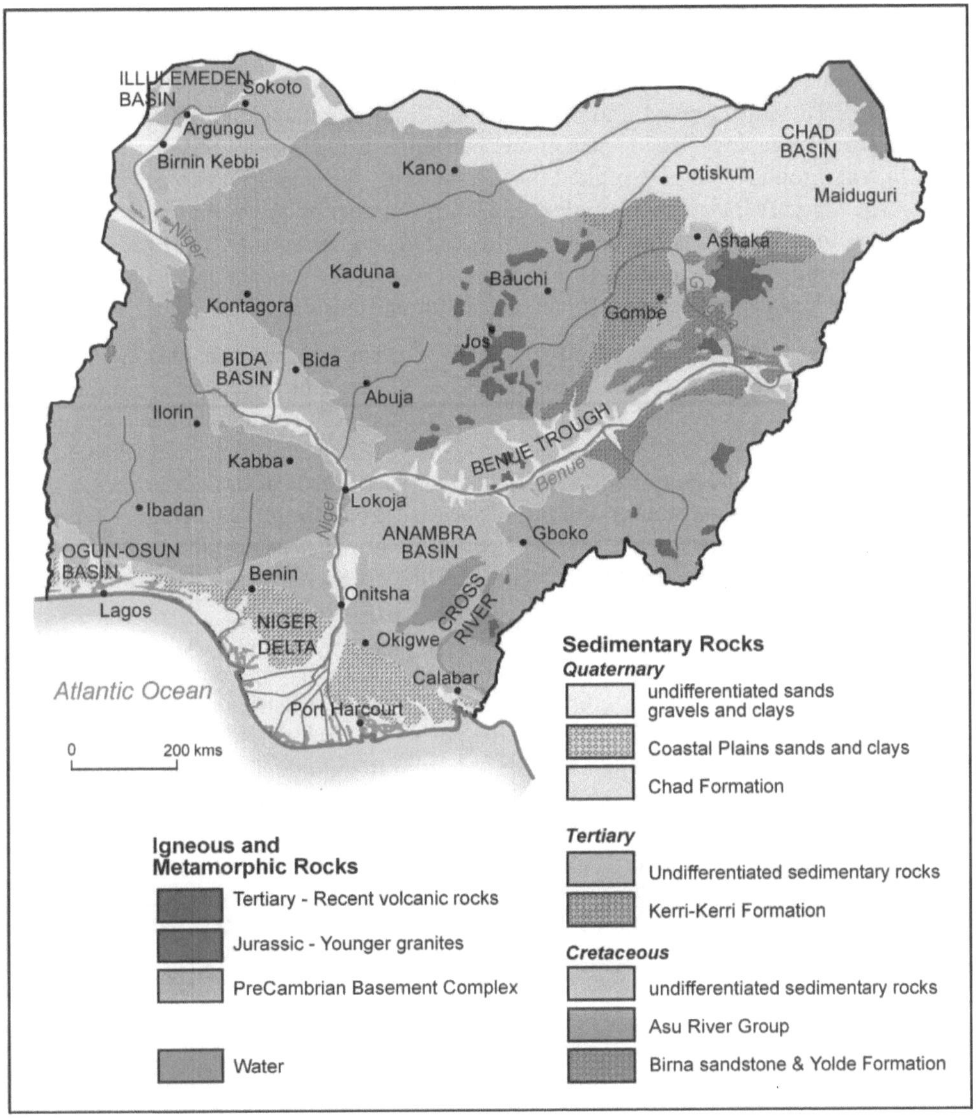

Figure 2. A generalised geological map of Nigeria showing the location of the main basins (see Table 1). Diagram adapted from MacDonald *et al.* (2005).

2 GEOLOGY AND HYDROGEOLOGY

The hydrogeology of Nigeria is discussed in detail in Adelana *et al.* (2008); a summary is given here. There are three main aquifer types in Nigeria: crystalline rocks, consolidated sedimentary rocks and unconsolidated sedimentary deposits (Figure 2). In the crystalline rock areas, folds, faults, joints and shear zones are common but localized. The weathered mantle renders the normally impermeable crystalline rocks suitable for ingress and storage of water (Ayoade, 1975; Wright & Burgess, 1992). In prospecting for groundwater in these areas, one needs therefore to determine the lateral and vertical limits of the faults, fractures, joints and zones and determine the extent and thickness of the weathered mantle.

Thus in Nigeria, all the major aquifers have already been located and are being used to some extent. They are divided into the following eight regions as hydrogeological areas according to their geology, basin and aquifer occurrence and nature as presented in Table 1.

In Nigeria, groundwater is exploited through springs, wells and boreholes. Springs are common in Nigeria's Basement Complex area. Large number of dug-wells in the rural areas supply water to communities and/or household located far from springs; there are numerous boreholes equipped with handpumps. The number of mechanized boreholes is increasing, as demand for town supplies, industry and agriculture grows.

(i) Drinking water. The Federal Government of Nigeria has constructed dug-wells and boreholes through the 'Small Town and Rural Water Supply Project', and the State Water Agencies. Individual households drill mechanized private boreholes depending on affordability.
(ii) Industrial use includes boreholes for industrial water needs and water selling bodies. Water marketing organisations bottle and package groundwater resources for commercial purposes. This wealth and job creation avenue provides clean and safe water to the people who can pay moderate prices, e.g. Spring Waters of Nigeria, Yankari, Ragolis etc. who tap spring water.
(iii) Agricultural supplies are sourced from irrigation farming in the Fadamas with subsidized support by the World Bank Assisted-Agricultural Development Projects (ADPs), River Basin Development Authorities (RBDAs) and Ministries of Agriculture and Rural Development (MARDs) for private individual farmers and communities in the arid and semi-arid areas of the north extending from latitude 8°N northward.
(iv) In rural areas, dug-wells (lined/unlined) are often shallow, unprotected and seasonal water sources mainly in shallow aquifers (crystalline rocks, alluvium). Spring development is now encouraged as an alternative safe water supply sources. Spring are natural sources but are being developed for the protection from contaminants and environmental pollution sources.

The handpump option for the rural areas has emerged as one of the most cost effective abstraction means of sustainable safe water supply. It is estimated that over 90% of safe water sources in rural Nigeria are handpump operated sources. Conventional water supply schemes (water treatment plants, reticulated borehole schemes), have proved not to be sustainable for the rural Nigeria over the years due to sophisticated technology associated with their operation and maintenance. Other abstraction mechanisms include motorized and solar powered schemes for semi-urban and urban centres.

3 IMPACTS OF GROUNDWATER DEVELOPMENT (PHYSICAL AND SOCIAL)

Most frequently, at least in the short and medium term, the impact of groundwater use is positive and include such benefits as improved health, reduced time to collect water, food security, job creation, livelihood diversification and general economic and social improvement. In the long term depending on the factors as mentioned above, the impact

Table 1. Major hydrogeological regions and aquifer occurrence in Nigeria.

Region	Area (km^2)	Aquifers	Bore depths	Yields (l/s)	Sources of data
Sokoto Basin Area (sedimentary) Illulemeden trans boundary aquifer system	63, 700	Multilayered, unconfined, confined and artesian conditions exist, pressure as high as 25 m but declining		Yield highly variable	Egboka, 1983 Oteze, 1976 Du Preez & Barber, 1965 Anderson *et al.*, 1973
		Gwandu	250		
		Sokoto group	360	6.3–8.8 l/s artesian	
		Rima group	140–210	0.3–12 l/s	
		Gundumi	300	3–8 l/s	
Chad Basin Aquifers	120, 400	Water table, unified and artesian but declining pressure head.			
		Upper Aquifer	0–180	2.5–30 l/s	Du Preez & Barber, 1965 Miller *et al.*, 1968
		Middle Aquifer	150–580	5–12 l/s (artesian flow 0.35–3.4 l/s)	
		Lower Aquifer	>500	2–25 l/s	
Bida Basin	38, 300	Bida sandstone		0.8–2.5 l/s	–
		Nupe sandstone		0.8–36.7 l/s	
Benue Basin (Benue Trough)	116, 300	Bima, Yolde	50–250	1–8 l/s	
		Gombe		1.9–5 l/s	
		Makurdi Sandstone		0.1–6.8 l/s	
		Lower Benue Ajali formation			
Anambra Basin		Ajali Formation	90–370	1–13 l/s	Offodile, 2002
		Nsukka			Egboka, 1983
Imo – Cross River Basins		Low permeability shale groups	20–50 m	<1 l/s	MacDonald *et al.*, 2005
Ogun – Osun		Abeokuta formation	77	10 l/s	Idowu *et al.*, 1999
Basement complex areas	>510, 000	Weathered overburden	20–80	<2.0 l/s	–
		Fractured/Fissured Basement			

might be negative, such as lowering the water-table, deterioration of water quality, saline intrusion in coastal areas.

In cases of fossil or compacting aquifers (noted from isotope aided studies in the Chad Basin and Iullemeden Aquifer systems) where recharge is either unavailable or unable to refill drained pore spaces, depletion constitutes groundwater mining. Hence the need for resource protection.

Furthermore, in renewable aquifers, depletion is indicated by persistent yield declines. The scope of the problem of decreasing yield or groundwater depletion has not been documented in Nigeria in recent time, however there are reports that the pressure head in artesian boreholes has declined in the Iullemeden aquifer in the Sokoto/Rima basin in northwest Nigeria. Also in the Maiduguri area in northeast Nigeria, the groundwater reserves in the upper and middle zones of the multilayered Chad aquifer system are being depleted. Rural dwellers are migrating to urban and semi-urban centres creating additional pressure on water supply systems in the area.

In some cases, removing the most easily recoverable fresh groundwater leaves residents with inferior water quality. This is due in part to induced leakage from the land surface, confining layers or adjacent aquifers that contain saline or contaminated water. As depletion continues, impact can worsen, making it imperative to carry out an objective analysis of the problem and its possible solutions.

Groundwater development is crucial for livelihoods and food security of millions of Nigerians. The impact of intensive groundwater use may be either positive or negative, depending on the factors, such as the nature of the aquifer, pressure on the aquifer, recharge rates, types of use, climate and so on. The key challenge then becomes to manage risk in such a way to minimize chances of long-term negative impacts without seriously damaging short and medium term benefit flow. Groundwater irrigation has also ensured food security and helped alleviate poverty. With increase in areas under groundwater irrigation, the contribution of groundwater to total agricultural output and rural wealth has been on the increase. There is no firm and accurate estimate of the total volume of groundwater used for irrigation in Nigeria. This is a challenge in groundwater development, as we have no knowledge of how much groundwater is used. We need the information for planning and development as in Nigeria, we have no evidence of over-exploitation due to lack of study.

4 GROUNDWATER DEVELOPMENT IN NIGERIA (PROCESS AND HISTORY)

The basic steps in groundwater development in Nigeria involve the search for productive aquifers, their exploration, the construction and the testing of water wells and boreholes, water quality and pollution considerations, groundwater management and modelling. It is viewed as a sequential process with three major phases (Freeze & Cherry, 1979).

- exploration stage in which surface and subsurface geological and geophysical techniques are brought to bear on the search for suitable aquifers;
- evaluation stage that encompasses the measurement of hydrogeological parameters, the design and analysis of wells/boreholes and the calculation of aquifer yields; and
- exploitation and/or management stage in which there is consideration of optimal development strategies and an assessment of the interaction between groundwater exploitation and the regional hydrologic system.

During the pre-independence period, The Nigerian Geological Survey, at its inception in 1917, had as one of its main objectives, to search for groundwater in the semi-arid areas of the former northern Nigeria (Offodile, 2002). These activities culminated in the commencement, in 1928, of a systematic investigation of towns and villages for digging of hand-dug-wells. The concrete lined well sinking technique designed and developed by the Geological Survey is still in use all over the country. Later in 1938, a water drilling section of the Survey was set up. By 1947 the engineering responsibility of the water supply section was transferred to the Public Works Department, the forerunner of today's Ministry of Works. The Geologists of the Geological Surveys collected data on groundwater occurrence, distribution, quality and quantity. During those days, the Geological Survey provided the only indigenous Consultancy Services available on the specific problems of groundwater supply.

There were numerous published and unpublished reports of the Geological Survey on the subject notable among were: Du Perez & Barber (1965) 'The Distribution and Chemical Quality of groundwater in Northern Nigeria'; Miller *et al.* (1968) 'Groundwater hydrology of the Chad Basin in Bornu and Dikwa Emirates'; Anderson *et al.* (1973) 'Aquifers in Sokoto Basin, North-western Nigeria'; Offodile (1975) 'Ground Water level fluctuation in the East Chad Basin of Nigeria'; Oteze (1975) 'The Hydrogeology of the North-Western Nigeria Basin'; Hazel (1960) 'The Hydrogeology of Eastern Nigeria'; and Carter (1963) 'Ground Water Resources of Western Nigeria'. For more information see Hazell (2004).

More recently, with the inauguration of the Nigerian Association of Hydrogeologists in 1986 and yearly publication of the Water Resources Journal from 1988, valuable reports on research carried out by members of the Association on groundwater resources of the country are published periodically.

The joint programme of the Nigerian Geological Survey and the United States Geological Survey marked a striking development in the approach to groundwater studies in Nigeria in 1963. The programme was designed to explore the artesian basins of the Sokoto and Chad Basins. The successful execution of this investigation, underscored the need for systematic studies of the Nigerian hydrogeological basins. A third programme for the Anambra basin was unfortunately interrupted by the Nigerian civil war.

The Government of Nigeria with the assistance of External Support Agencies (ESAs) has undertaken several massive water supply development projects through the following agencies:

- National Borehole Programme (1981–1986) of President Shagari's Civil Administration;
- Directorate of Food, Roads and Rural Infrastructure (DFRRI) Rural Water and Sanitation Programme (1986–1992) of General Babangida's Military Government;
- Petroleum Trust Fund Rural (PTF) Rural Water Supply and Sanitation Programme (1996–1999) of General Abacha's Military Government;
- Improved Access to Water Supply and Sanitation Programme (2000–2001) of President Obasanjo's Civilian Administration;
- National Rural Water Supply and Sanitation Programme (2001–2010) of the present administration.

These are mainly borehole projects with installed handpumps. These programmes depended on the Government in power. Hence there has not been any continuity as each is separate from the other. Programmes supported by External Support Agencies include:

- UNICEF Assisted State Water and Sanitation Projects (1981–2010) on-going;
- UNDP's RUSAFIYA (An acronym in local language) Project (1988–1993);
- World Bank Assisted-Agricultural Development Projects (1983–1992);
- European Union (EU) Water & Sanitation Programme (2002–2009);
- Department for International Development's (DFID) Water and Sanitation Pilot Project (2002–2008);
- WaterAid's Rural Water Supply and Sanitation Programme (1996–2010);
- Japanese International Cooperation Agency's (JICA) Rural Water Supply Project (1992–1994);
- Development of local manufacture of handpumps (1988–2010).

5 ISSUES RESPONSIBLE FOR GROUNDWATER PROJECT FAILURES

Project reviews and stakeholders consultation has revealed low level of sustainability of rural water supplies as boreholes and handpumps fail (Hodges, 2001; Jawara, *et al.*, 2001; Habila 2002). The key causes of low level of sustainability include inappropriate policy or legislation; insufficient institutional support; unsustainable financing mechanisms; ineffective management systems; and lack of technical support. The studies indicate that the problem will only be solved by adopting a holistic approach to planning and implementation rather than focusing on one issue (such as community management of spare parts supply) in isolation. In Nigeria, workshops have been held to explore the reasons behind the apparent failure to consistently and efficiently deliver water supply and sanitation services to its citizens. The important issues identified include: the policy and systems; coordination, planning and management; funding; community empowerment and management; public-private partnership; and technical issues. These are summarized as follows (Fellows *et al.*, 2003):

Policy and systems issues: (1) no clear long-term sector programme at all levels; (2) no specific agency with role of sanitation; (3) no clear monitoring and database system/inadequate and unreliable data; (4) involvement of the FMWR in direct execution of water supply projects; (5) politicization of implementation process;

Co-ordination, planning and management issues: (1) no co-ordination between service providers resulting in duplication of efforts and resources; (2) poor collaboration between Local, States and Federal Governments; (3) poor planning; (4) poor policy implementation by government; (5) inadequate training of water delivery operators; (6) poor maintenance culture.

Funding Issues: (1) low investment by government in WSS; (2) inadequate funding of maintenance of WSS scheme and equipment.

Community empowerment and management issues: (1) lack of sense of ownership by communities; (2) inadequate participation of women in rural water supply.

Public-private partnership issues: (1) inadequate private sector participation, particularly professionals in WSS (2) incapable contractors involved in the execution of water supply projects as corruption contributes to high costs of WSS projects.

Technical issues: (1) inappropriate technology to find and develop sustainable supplies; (2) poor yields or poor quality water; (3) erratic power supply and high diesel cost associated with operations/high costs in production of water.

6 MAIN CHALLENGES FOR SUSTAINABLE GROUNDWATER DEVELOPMENT IN NIGERIA

The oil-rich countries of the arid and semi-arid areas of North Africa and the Persian Gulf have learnt from experience that water is as important a resource as oil. Efficiently sustainable groundwater development should be taken as seriously in Nigeria as we take the establishment of Nigerian National Petroleum Corporation (NNPC), Power Holding Company of Nigeria (PHCN) and Federal Road Management Agency (FERMA). These issues and challenges require institutional reforms; the objective of which should be the achievement of efficiency and effectiveness in service delivery management through decentralization of management functions and provision of mechanisms for enhancing co-ordination, partnership and accountability.

Global concerns with the increased use of groundwater in agriculture and the associated problems of sustainability and resource depletion has necessitated renewed attention towards issues of groundwater governance (Konikow & Kendy, 2005). In doing so, direct regulation through licensing and registration of boreholes and aggressive pricing policies is advocated.

Small land holdings and overdependence on groundwater. In the urban centres – Abuja, Kano, Kaduna, Lagos, Enugu etc., high population density, small landholdings and overwhelming dependence on groundwater is an issue of concern for groundwater development in Nigeria. A landowner can extract as much water as he desires without any kind of restriction, except that the investment in drilling and pumping equipment is expensive and not all can afford it.

The institution of water market (*water sellers*) popularly known as pure water in sachets and bottled water from springs and boreholes is a big market in Nigeria. Water market has been very crucial in alleviating poverty, generating employment and improving welfare service. However, in regions where the groundwater resource base is scarce, water market could hasten resource depletion.

Politcalisation of groundwater development. This constitutes real constraints and it affects both contract award and distribution of borehole projects. As a result, professionals are not utilized and projects not professionally executed and this often leads to rampant project failure and waste of funds and resources.

7 RECOMMENDATIONS

It would appear that in most Nigerian drilling programmes, groundwater management is still not seriously considered. Hundreds of boreholes were, and are still being drilled with little or no hydrogeological control. Use of professionals to undertake systematic studies and prepare management guidelines for the entire country is hereby advocated. Revisit Water Resources Decree no. 101 with a view to amending, and enforce as appropriate. In its present form it fails to address, emphasize and lay down criteria for the development of groundwater resources in Nigeria. The following are recommendations for progressive and sustainable development of water resources in Nigeria:

(i) Revisit Local Manufacturing of development Facilities. In early 1960s – 1980s the Naira was strong and importing was cheap. It is cheaper today to patronize local goods and create employment.

(ii) Set up a coordinating body, to be known as Groundwater Development Commission, and should be an autonomous body, saddled with the responsibility of groundwater development , similar in nature to the bodies governing the petroleum industry, power and solid minerals sector.

(iii) Review and conduct evaluation or assessment/appraisal of the RBDA for improved performance and efficiency in service delivery.

(iv) Improvement in capacity building programme (in knowledge, skills and attitude) for groundwater development as groundwater has become the major source of water supply schemes for both rural and semi-urban communities in Nigeria. Presently, government (Federal, States, and LGAs) and ESAs (UN Agencies and JICA) retain the substantial number of drilling crews and drilling machines in Nigeria. Equally they comprise by far the majority of implementing forces for groundwater resources development in Nigeria.

(v) Organize stakeholders meeting / workshop periodically for preparation of practical agenda for judicious development and management of groundwater resources of Nigeria.

8 CONCLUSIONS

Groundwater development is crucial for livelihoods and food security of millions of Nigerians. In the short and medium term, impact of groundwater development is positive and include such benefits as increased productivity, food security, job creation, livelihood diversification, and general economic and social improvement. In the long term, the impact might be negative, such as uncontrolled depletion, which might result in deterioration of quality, land subsidence, increased cost of abstraction and total drying of water points. The key challenge is to manage risk in such a way as to minimize changes of long term negative impacts without seriously damaging short and medium-term benefits flow. Nigeria has abundant groundwater resources but no adequate reliable data on quantity and quality of reserve to work with. There is serious effort on all sides to improve water supplies to meet the Millenium Development Goals. The sector needs reforms and groundwater development should be programmed to outlast political cycles while participation in its implementation by all stakeholders must be encouraged and coordinated.

Groundwater is a resource and should be valued appropriately even though abstraction is carried out as social service.

ACKNOWLEDGEMENT

The authors are grateful to the anonymous reviewers and to Dr. Petr Vrbka and Dr Alan MacDonald who helped with the figures.

REFERENCES

Adelana, S. M. A., Olasehinde, P. I., Bale, R. B., Vrbka, P., Edet, A. E. & Goni, I.B. 2008. *An overview of the geology and hydrogeology of Nigeria*. In: Adelana, S. M. A. & MacDonald, A. M. Applied Groundwater Studies in Africa. IAH Selected Papers on Hydrogeology, Volume 13, CRCPress/Balkema, Leiden, The Netherlands.

Anderson, H. R. & Ogilbee, W. 1973. *Aquifers in Sokoto Basin N.W. Nigeria with a description of general hydrogeology of the region.* Geological Survey Water Supply Paper, **1757.**

Ayoade, J. D. 1975. *Water Resources and their Development in Nigeria.* Hydrological Sciences Bulletin, **20,** 581–591.

Carter, J. D., Barber, W. & Tait, E. A. 1963. *The geology of parts of Adamawa, Bauchi and Bornu Provinces in north-eastern Nigeria.* Bulletin of the Geological Survey Nigeria, **30**.

Du Perez, J. W. & Barber, W. 1965. *The Distribution and Chemical Quality of Groundwater in Northern Nigeria.* Nigerian Geological Survey Bulletin, **36.**

Egboka, B. C. E., 1983. *Analysis of the Water Resources of Nsukka area and environs, Anambra State Nigeria.* Nigerian Journal of Mining and Geology, **20**, 1–16.

Fellows, W., Habila, O. N., Kida, H. M., Metibaiye, J., Mbonu, M. C. & Duret, M. 2003. *Reforming the Nigerian Water and Sanitation Sector.* In: Proc. of the 29th WEDC International Conference, Abuja, Nigeria. WEDC, Loughborough, UK.

Freeze, R. A. & Cherry, J. A. 1979. *Groundwater.* Prentice-Hall, Englewood Cliffs. New Jersey, USA.

Habila, O. N. 2002. *Rural Water and Sanitation Development in Nigeria.* In: proceedings of the 28th WEDC International Conference, Kolkata, India. WEDC, Loughborough, UK.

Hazell, J. R. T. 1960. *Groundwater in the Eastern Region of Nigeria.* Geological Survey of Nigeria Report, **5198**.

Hazell, J. R. T. 2004. *British hydrogeologists in West Africa – an historical evaluation of their role and contribution.* Geological Society, London, Special Publications, **225**, 229–237.

Hodges, A. (ed.) 2001. *Children's and Women's rights in Nigeria: A wake-up call.* Published by National Planning Commission and UNICEF Nigeria.

Idowu, O. A., Ajayi, O. & Martins, O. 1999. *Occurrence of groundwater in parts of Dahomey Basin, SW Nigeria.* Nigerian Journal of Mining and Geology, **35,** 229–236.

Jawara, D., Colin, J., Hellandendu, J., Okuofu, C., Parry-Jones, S., Yakubu, M. & Wedgwood, A. 2001. FGN/UNICEF Water and Environmental Sanitation Programme, Nigeria – Evaluation Report. WELL, DFID, London, UK.

Konikow, L. F. & Kendy, E. 2005. *Groundwater Depletion: a global problem.* Hydrogeology Journal, **13**, 317–320.

MacDonald, A. M., Cobbing, J. & Davies J. 2005. Developing groundwater for rural water supply in Nigeria. British Geological Survey Commissioned Report **CR/05/219N**.

MacDonald, A. M., Davies, J, Calow, R. C. & Chilton, P. J. 2005. *Developing Groundwater: a guide for rural water supply.* ITDG Publishing, Rugby, UK.

MacDonald, A.M., Kemp, S.J. & Davies J. 2005. *Transmissivity variations in mudstones.* Ground Water, **43**, 259–269.

Miller, R. E., Johnston, R. H., Olowu, I. & Uzoma, J. 1968. *Groundwater Hydrology of the Chad Basin in Bornu and Dikwa Emirates.* Nigerian Geological Survey Water Supply Paper, **1757**.

Offodile, M. E. 1975. *Groundwater Level Fluctuation in East Chad Basin of Nigeria.* Nigerian Journal of Mining and Geology, **1**.

Offodile, M. E. 2002 *Groundwater Study and Development in Nigeria.* Mecon Geology and Engineering Service, Jos, Nigeria.

Oteze, G. E. 1976. *The Hydrogeology of North Western Nigerian Basin.* In: C. A. Kogbe (ed.) Geology of Nigeria. Elizabeth Pub. Co, Ibadan, Nigeria .

Wright, E. P. & Burgess, W. G. (eds.) 1992. *The hydrogeology of crystalline basement aquifers in Africa.* Geological Society, London, Special Publications, **66**.

CHAPTER 8

A remedial approach for soil and groundwater impacted by crude oil in the Niger delta

G.A. Bolaji
Department of Civil Engineering, University of Agriculture, Abeokuta, Nigeria

ABSTRACT: Oil spillage is a common problem in oil producing communities in Africa, and has been a particular problem in the Niger delta, since oil production began. The effectiveness of bioremediation and phytoremediation was investigated at an oil spill site close to Eriemu in Nigeria. The impacted area was divided into three blocks, A and B represent heavily and moderately polluted areas respectively and block C an unimpacted block. Much of the contamination was within the top 3 m of the soil. Excavation was carried out in block A and B to 2 m and the excavated materials from each depth interval were laid on PVC sheets as biopiles. Seeded microbes (from block C) were applied according to nutrient formulation needed to degrade the level of TPH in each biopile. TPH reduced from 20,000 ppm to close to 0 ppm within 8 weeks. The effect of solar energy, though not investigated, but could not be ruled out as contributory to TPH reduction under this circumstance. Specially trained water hyacinth was used in aqua cells to treat pumped polluted groundwater at the site before re-injection into the ground. It was concluded that bioremediation and phytoremediation are alternative cheap methods of ameliorating soil and groundwater polluted by crude oil spillage.

1 INTRODUCTION

Oil spillage is a common phenomenon in the oil producing communities in Africa, especially in the Niger delta area of Nigeria. The environmental consequences of this problem have raised much concern. When an oil spillage accident occurs, several thousands of hectares of land are usually affected before any action is taken. The ecology of the area affected suffers as forest wildlife, and water resources are impacted. In most cases, the oil spilled also runs off into water bodies of the Niger delta. In addition to environmental losses, there are economic impacts: farmland and plantations (rice, rubber and oil palm) are lost, and fishing deteriorates. Spilled crude oil also percolates into the ground to pollute groundwater, where large quantities are retained to cause lasting damage (Brown & Donnelly, 1983).

Crude oil is composed of different grades of complex hydrocarbons and trace metals. Prior to the advent of the oil industry, enzymes had evolved in parallel with natural occurring organic compounds (Brown & Donnelly, 1983). Consequently, micro-organisms were capable of degrading almost all organic compounds in the quantities that they were found on the earth's surface. However, the oil industry has introduced many structurally novel compounds into the environment over a relatively short time span which has not allowed the organisms sufficient time to develop appropriate mechanisms to rapidly degrade these compounds (Alexander, 1981; McEdowney *et al.*, 1993). As a result, many new compounds are resistant to microbial breakdown in the environment, thereby allowing pollutants to exist in natural ecosystems (such as watercourses) through which they enter the food chain and ultimately threatening the health of other organisms.

Concern about environmental pollution caused by oil industry has led to significant changes in government legislation. In Nigeria, the Federal Ministry of Environment and Federal Ministry of Petroleum Resources through their implementation agency, Federal Environmental Protection Agency (FEPA) and Department of Petroleum Resources (DPR) have set up integrated guidelines to provide control measures and have resulted in strict control of oil spillage and remediation programmes. The petroleum companies operating in the Niger delta area are made to comply with stipulated guidelines through monitoring programmes. The operations of environmental monitoring covered land, swamp (fresh and brackish water) and shallow offshore facilities including flow stations, compressor stations, gas plants, sewage treatment plants, jetties, loading terminals and power plants. As a consequence of the regulation, oil companies (through consultants) are monitoring the effect of all forms of discharges they release into the environment during their operations and taking immediate action when significant fluxes are observed.

Bioremediation involves the use of micro-organisms and to remove oil based pollutants from the environment. The by-products of effective biodegradation of complex hydrocarbons, such as water and carbon dioxide are non-toxic and can be accommodated without harm to the living organisms and the environment. It is a cheap alternative to most other methods (Loehr *et al.*, 1992). Biodegradation is the most significant natural attenuation process for petroleum hydrocarbons and is also being used and studied for chlorinated solvents such as trichloroethylene (TCE) and perchloroethylene (PCE). These solvents, used for degreasing and dry-cleaning, are commonly found as groundwater contaminants in urban areas (Bower & Zehnder, 1993).

Phytoremediation is a new technology that involves the use of aquatic plants for treating contaminated groundwater and soils. The uptake and detoxification of contamination by plant species has been used successfully in the treatment of sewage. Certain plants (such as water hyacinth) are known as hyper-accumulators and have been used in the bioremediation of polluted water. Particularly isolated species have been genetically engineered to increase capacity to biodegrade various forms of pollutants found in water. They have the capacity to absorb dissolved toxic substances from water and deposit them around roots cell vacuoles. The success of bioremediation is controlled by the availability of adequate amount of nutrients (nitrogen, phosphorus and oxygen) which stimulate the quick growth and multiplication of microbes and hyper-accumulator plants such as water hyacinth (Bull, 1992; Kovalick, 1999). The major objective of this paper is to demonstrate the potentials of bioremediation and phytoremediation in to clean up oil spill sites in the Niger Delta, Nigeria.

2 MATERIALS AND METHODS

2.1 *Study area*

The study site is located along Eriemu to UQSS (400 mm diameter pipe) trunk line, about 5 km from UQSS station. It is located approximately on latitude $5^\circ\ 31''$N and longitude

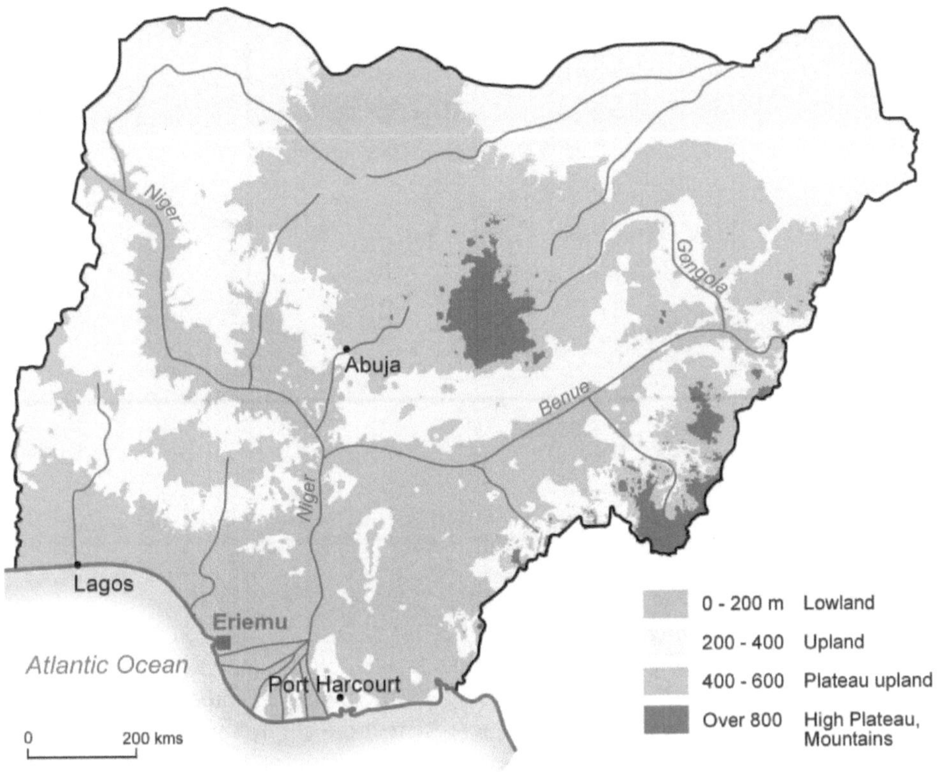

Figure 1. Map of Nigeria showing the sample site.

5° 45′E (see Figure 1) about 3.5 kilometres from Igwemaro community. The climate of Eriemu is hot and humid. The annual rainfall ranges between 2500 mm and 3500 mm and has a bimodal distribution with peaks occurring in the months of July and September. The wet season covers a period of ten months (February to November), and the highest daily (mean) temperature is recorded during the short dry season. The vegetation around Eriemu is a secondary forest dominated by shrub, hedges and grasses. Some primary forests are maintained as forest reserves and rubber plantations with a few oil palm trees dotting the terrain. Another form of vegetation present is the fresh water swamp forest. The geological formation of the area is generally classified as part of Warri deltaic plain, which comprises unconsolidated sands and gravels.

The oil spillage occurred as a result of underground leakage on a 400 mm diameter pipe delivery line from Eriemu flow station to UQSS. An estimated volume of 350 barrels (70,000 litres) was recovered during initial cleanup. The area covered by the spill was approximately 1050 m^2 and the soil profile was saturated with crude oil up to average depth of 4 m. The seasonal rise and fall in shallow water table ranges between <1 m during the wet season falling to 4–12 m during the dry season.

2.2 *Soil sampling*

The site was cordoned off with wooden stakes painted red and white and red/white stripped polythene tape was tied to the stakes to mark the polluted area. A health and safety plan

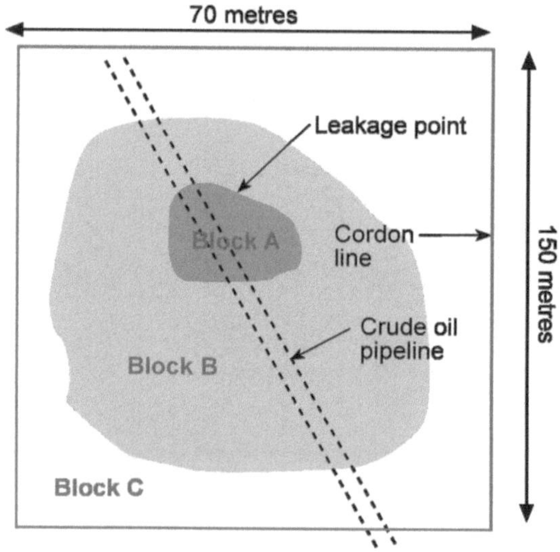

Figure 2. Site of the crude oil spill, and the location of the sample blocks.

was included in the sampling plan (Boulding, 1995; NIOSH/OSHA/USGC/EPA, 1985). The cordoned area was divided into three blocks labelled A, B, and C (Figure 2). Block A represented a heavily polluted area (50 m by 15 m); block B represented moderately polluted area (50 m by 15 m); and block C represented the control area (50 m by 15 m). A graduated stainless steel auger was used to collect soil sample at depths 0, 30, 60, 90 and 120 cm on each block. The bulk samples were homogenized, kept in glass bottles, labelled and transported to the laboratory under stabilized conditions (preserved and refrigerated) within 48 hours. The parameters determined in the laboratory were the total petroleum hydrocarbon (TPH) and the heavy metals present in the sample, according to recommended procedures (API, 1996).

The results of analyses for TPH content in each sample was used as a guide for dividing the soil profiles into horizons. This was also used as a guide for excavating from each horizon and the application of nutrients. The contaminated soil in blocks A and B, were excavated according to the horizon and laid out on PVC sheets in heaps labelled as biopiles BP1A, BP2A, BP3A, BP4A and; BP1B, BP2B, BP3B, BP4B for the depth intervals 0–50 cm, 50–100 cm, 100–150 cm and 150–200 cm in each block respectively. Samples were obtained from each biopile and sent to the laboratory for analysis.

2.3 *Identification and isolation of indigenous oil degrading bacteria*

Soil samples were collected from 0.5 m depth from the control area (Block C) for microbial examination and identification of heterotrophic/hydrocarbon degrading bacteria. Approximately 80 g of soil sample was homogenized and put inside each sterile glass container, corked, stored in ice packed coolers and transported to the microbiology laboratory within 24 hours. Soil samples were logged; mixed thoroughly using a sterile spatula and 1 g of soil from each sample was put inside an autoclaved and sterile McArthney bottle. A saline solution (0.85% NaCl) of 900 ml was added to the soil in the McArtney bottle for total heterotrophic count. Another 1 g of soil from the sample stock was put inside a separate autoclaved McArthney bottle and 900 ml of mineral salt solution was added for crude oil degrading bacteria count. The contents in each McArthney bottle was thoroughly agitated for 30 minutes and allowed to stand for 2 minutes before transferring 100 ml into separate fresh McArthney bottles, making a ten-fold serial dilution which were then plated based on the condition required for each group of micro-organisms (Mara, 1974). About 1 ml of each dilution was introduced into cooled sterile Petri dish containing 15 ml of nutrient medium cooled to room temperature.

2.4 *Laboratory multiplication of oil degrading bacteria*

Confirmed indigenous oil degrading bacteria were multiplied in nutrient/yeast extract broth medium in the laboratory. Measured quantities (5 ml each) of bacterial suspension in normal saline solution were added to 2 litres of broth in techord tubes and incubated at ambient room temperature. Total colony forming unit (CFU) in each tube was determined at 48 hours interval until the microbial population reached 10^8–10^{10} CFU. On attainment of the desired population, the content in each tube was aseptically transferred into the bioreactor containing half strength nutrient/yeast extract (Chio & Burkhead, 1982; Gonzalez-Martinez & Dugue–Luciato, 1992). The bioreactor was continuously aerated using oxychrome. At this stage, they were ready to be used for inoculation.

2.5 *Training of halophytes*

The modified principle of formulated hydrocarbon degraders was considered in the training of young water hyacinth plants. Freshwater hyacinth plants were made to grow in a 20% toxicant concentration of hydrocarbon contaminated water. Lead powder was added to the water after 3 weeks. Observation after another 4 weeks of exposure showed that most of the plants died and few survivors were left. The survivors had brown patches, fragile petioles and reduced leaf size. These survivors were transferred into 40% toxicant concentration hydrocarbon then 60% and 80% successively. Further training of the plants were carried out by adding formulated hydrocarbon degraders to the water polluted at 70% and 100% toxicant concentration of hydrocarbon and observed for another 4 weeks. The trained water hyacinth were assumed to have been prepared to withstand highly toxic conditions and developed capacity for accelerated uptake of hydrocarbon from contaminated medium with the seeded microbes.

2.6 *Treatment of biopiles*

Each contaminated soil block was excavated to a depth of 2 m and heaped to form biopiles on thick polythene sheet laid on the ground. As the biopiles were made ready for treatment by bioremediation, samples were taken from each biopile and analyzed to determine the amount of nutrient present. Nutrient deficiency was used to calculate the quantity of nitrogen (N), phosphorus (P) and potassium mineral fertilizer that should be applied to each biopile (Lynch & Genes, 1989). The quantity of lime ($Ca(OH)_2$) to be added to each biopile was determined through the analysis for the presence of magnesium that may have been seriously depleted as a result of the presence of petroleum hydrocarbon in the impacted area. Each biopile was bio-stimulated by addiction of mineral fertilizer (NPK) at defined proportions. Polluted soil was mixed with virgin topsoil (obtained from nearby burrow pits) at ratio 1:2 using mechanical concrete mixer at 250 rounds per minute and discharged into a wheelbarrow from where it was spread thinly (5–10 cm) on the biopiles. A CP15 knap sack sprayer was used to apply the "seeded" bacteria into the bio-stimulated biopiles. Adequate aeration of the biopiles was achieved through regular turning and winnowing with rake after which they were covered with PVC sheets. Every week, the conditioned soil in the biopile were sampled, seeded with microbes and then aerated. The samples were tested on gas chromatography to determine the rate of degradation of the hydrocarbons and the quantity of hydrocarbon that was yet to be degraded. The population of oil degrading microbes was also determined so as to guide the quantity of microbes to be added.

2.7 *Treatment of impacted groundwater*

Shallow boreholes were drilled around the blocks and water samples were collected from the boreholes using a Teflon water sampler. Rapid analysis for total hydrocarbon (TPH) was carried out on site using TPH kits and duplicate samples were sent to the laboratory for duplicate to allow the results to be compared. Fourteen piezometers were installed in the aquifer at depths varying from 3 m to 11 m, well distributed around the three block areas.

Contaminated groundwater was pumped out into aqua cell tanks of 15,000 litres each, arranged in parallel and connected in series with operating valves (Figure 3). Groundwater

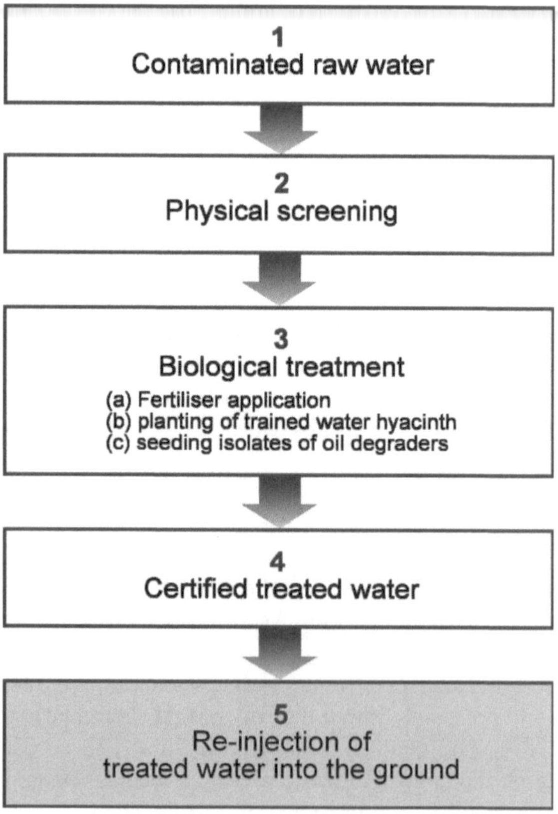

Figure 3. Flow diagram of phytoremediation in aqua cells.

was first aerated using an oxychrome air compressor; crude oil that rose to the water surface was continuously skimmed off. Cultured indigenous strains of heterotrophic bacteria and nutrients were introduced into groundwater through excavated test pits, the trained water hyacinth plants were used as bioreactors to treat the contaminated groundwater. Treated water for re-injection into the ground was taken from the last tank.

Water samples were taken from inlets to aqua cell 1 and labelled against the depth and block area of extraction. The water samples were collected inside clean containers, labelled, preserved and transported to the laboratory under refrigeration. Both water and soil samples were analyzed to determine the total petroleum hydrocarbon (TPH) and some other physiochemical parameters.

3 RESULTS AND DISCUSSION

The concentration of total petroleum hydrocarbon in the soil profile from core samples in Block A is shown in Figure 4. The TPH concentration reduces with depth and this may be attributed to the high cation exchange capacity and nature of the topsoil layer with high affinity for retention of crude oil in soil colloids.

The pH and total petroleum hydrocarbon in the soil biopiles (from Blocks A and B) were monitored weekly for two months. The TPH concentrations in the biopiles reduced markedly to less than 100 ppm after eight weeks (see Figure 5). The rates of TPH depletion after reconditioning the soils were mainly due to the addition of optimum soil nutrients and high loads of heterotrophic hydrocarbon degrading bacteria and the attainment of higher pH. However, the total percentage depletion may not solely be due to microbes, as the low molecular weight fractions of some hydrocarbons are readily volatized by the solar energy. In a period of six weeks soil pH gradually increased (Figure 6). The pH of the re-conditioned soil of the biopiles averaged 6.0 (5.9–6.3), which is close to the optimum range (6–8) for normal bacteria physiological activity.

The physical screening and aeration of pumped groundwater into the aqua cell containing trained water hyacinth proved effective in the treatment of groundwater. The seeded microbes break down the complex hydrocarbons into simple organic molecules in their metabolic activity to energy and biomass while the water hyacinth also absorb, metabolize and bioaccumulate heavy metals present the water. The end products of these reactions are non-toxic to the environment and there is an excellent symbiotic association between

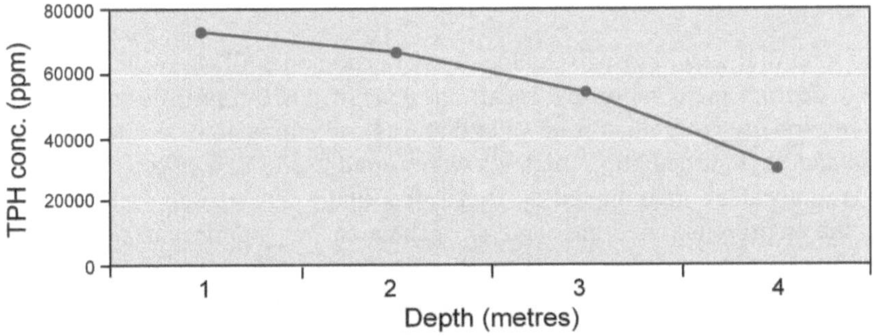

Figure 4. Average TPH concentration versus depth for block A.

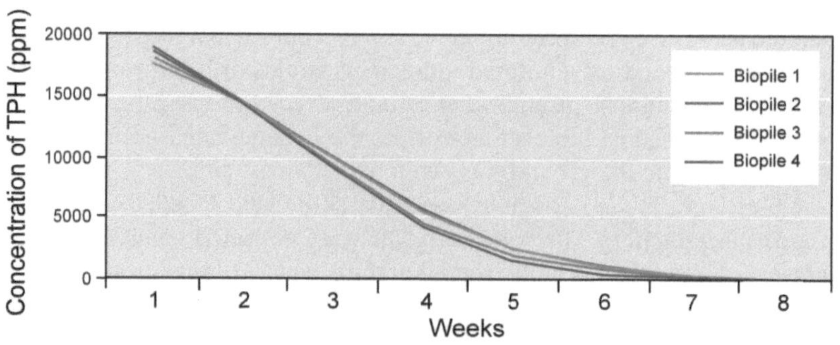

Figure 5. TPH concentration in the biopiles formed from Block A and B.

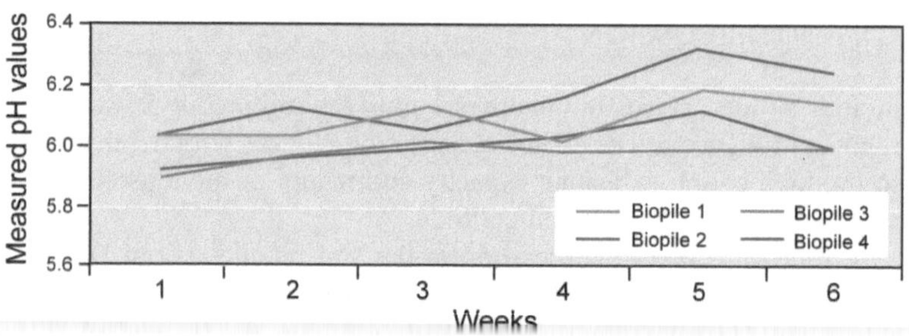

Figure 6. Adjusted pH of conditioned soil in the biopiles from Blocks A and B.

the microbes and the water hyacinth. The end product (i.e. carbon dioxide) in the aerobic catabolic activity on the complex hydrocarbons by the microbes is used up by the water hyacinth in anabolic activity to produce oxygen that is used up by the microbes as terminal electron acceptor. The water hyacinth also provides nutrients to microbes by the decaying spongy roots.

The quality of contaminated water continued to improve as it moves from one aqua cell to the other until about 90–95% reduction is achieved at the final aqua cell tank. Re-injection was only from the base of the final aqua cell tank leaving traces of crude oil film on the water surface.

4 CONCLUSIONS

Bacteria found in the soil close to the location of crude oil spill site were isolated, seeded and used to degrade the complex hydrocarbons deposited in the soil. Water hyacinth plants were trained and used to treat groundwater that was contaminated by crude oil. Both bioremediation and phytoremediation methods were found useful and effective in remediating soil and groundwater contaminated by crude oil spillage. The exercise left no undesirable effect on the environment and the ecology of the area was significantly improved within 3 months.

The risk of contamination is too great to rely on natural attenuation alone, especially in densely populated urban areas (Nyer & Gearhart, 1997). An additional useful measure to increase effectiveness would to increase air circulation through air sparging or the injection of air into biopiles or directly into aquifers (Rabideau & Blayden 1998). The bacteria being used in the remediation process will then have enough oxygen at its disposal to transform the contaminant and or volatilize organic contaminant for removal or venting into the atmosphere. This will enable faster and economical cleanup.

ACKNOWLEDGEMENTS

The author wish is acknowledge the efforts of Miss Esosa I. Oyawale (my former student) and the Global Environmental Consultant that collaborated with me for the research.

REFERENCES

Alexander, M. 1981. *Biodegradation of chemicals of environmental concern.* Science **211**, 132–138.

API 1996. *Manual on disposal of refinery waters.* American Petroleum Institute, New York, USA.

Boulding, J. R. 1995. *Practical handbook of soil, vadose zone and groundwater contamination: assessment, prevention and remediation.* CRC Press, Florida, USA.

Bouwer, E. J. & Zehnder, A. J. B. 1993. *Bioremediation of organic compounds-putting microbial metabolism to work.* Trends in Biotechnology, **11**, 360–367.

Brown, K. W. & Donnelly, K. C. 1983. *Influence of soil environment on biodegradation of refinery and petroleum sludge.* Environmental Pollution (Series B) **6**, 119–132.

Bull, A. T. 1992. *Degradation of hazardous wastes.* In: Bradshaw, A. D., Southwood, R. & Wemer, F. (eds.). The treatment and handling of wastes. The Royal Society, Chapman and Hall, London, UK.

Chio, E. & Burkhead, C. E. 1982. *The hydrodynamic evaluation of fixed media biological process.* Kings Island Pub, Ohio, USA.

Gonzalez-Martinez, S. & Dugue–Luciato, J. 1992. *Aerobic submerged biofilm reactors for wastewater treatment.* Water Research, **26**, 825–833.

Kovalick, W. 1999. *The Perspectives for Clean up.* Paper presented at EPA Innovative Clean-up Approaches Conference, Bloomingdale. Illinois, USA.

Loehr, R. C., Martini Jr, J. H. & Neuhauser, E. F. 1992. *Land treatment of an aged oily sludge-organic loss and change in characteristics.* Water Research, **26**, 805–815.

Lynch, J. & Genes, B. R. 1989. *Land treatment of hydrocarbon contaminated soils.* In: Kostecki, P. T. & Calabrese, E. J. (eds.). Petroleum contaminated soils, (Vol. 1). Lewis Publishers, Chelsea, USA.

Mara, D. D. 1974. *Bacteriology for Sanitary Engineers.* Churchill Livingstone, Edinburgh, UK.

McEldowney, S., Hardman, D. & Walte, S. 1993. *Pollution, Ecology and Biotreatment.* John Wiley and Sons, New York, USA.

NIOSH/OSHA/USCG/EPA 1985. *Occupational safety and health guidance manual for hazardous waste site activities.* DHHS(NIOSH) Publication No 85-115, US Government Printing Office, Washinton DC.

Nyer, E. & Gearhart, M. J. 1997. *Treatment Technology: Plumes don't move.* Groundwater Monitoring and Remediation, **17**, 52–55.

Rabidaeu, A. J. & Blayden, J. M. 1998. *Analytical Model for Contaminant Mass Removal by Air Sparging.* Groundwater Monitoring and Remediation, **18**, 120–130.

Groundwater and Rural Water Supply

CHAPTER 9

African hydrogeology and rural water supply

A.M. MacDonald
British Geological Survey, Edinburgh, UK

J. Davies & R.C. Calow
British Geological Survey, MacLean Building, Wallingford, Oxfordshire, UK

ABSTRACT: The widespread development of groundwater is the only affordable and sustainable way of improving access to clean water and meeting the Millennium Development Goals for water supply by 2015. Current approaches to rural water supply, in particular demand driven approaches and decentralisation of service delivery have many benefits to the overall efficacy and sustainability of water supplies, however, problems arise when projects do not take into consideration the nature of the groundwater resources. Different approaches and technologies are required depending on the hydrogeological environment. Sub-Saharan Africa (SSA) can be divided into four hydrogeological provinces: (1) The crystalline basement occupies 40% of the land area of SSA and supports 235 million rural inhabitants. (2) Volcanic rocks occupy 6% of the land area of SSA, and sustain a rural population of 45 million, many of whom live in the drought stricken areas of the Horn of Africa; (3) Consolidated sedimentary rocks occupy 32% of the land area of SSA and sustain a rural population of 110 million: (4) Unconsolidated sediments occupy 22% of the land area of SSA and sustain a rural population of 60 million. Hydrogeological expertise can have significant benefit to rural water supplies by increasing capacity throughout projects by effectively transferring knowledge; by providing authoritative benchmarking, by focused research and by providing accessible advice to planners and policy makers.

1 INTRODUCTION

At least 44% of the population in Sub-Saharan Africa (some 320 million people) do not have access to clean reliable water supplies (JMP, 2004). The majority of those without access (approx 85%) live in rural areas where the consequent poverty and ill health disproportionately affect women and children (DFID, 2001; JMP, 2004). In response, the international community has set the Millennium Development Goals (MDGs) which commit the UN membership to reduce by half the proportion of people who are unable to reach, or afford, safe drinking water by the year 2015 (United Nations, 2000). Poverty reduction and sustainable development are now given highest priority.

In this context, the need for sustainable development and management of groundwater cannot be overstated. Across large swathes of Africa, South America and Asia, groundwater provides the only realistic water supply option for meeting dispersed rural demand – alternative water resources can be unreliable and expensive to develop (Foster *et al.*, 2000; MacDonald *et al.*, 2005a).

Yet many projects spend large amounts of money installing water sources without trying to understand the groundwater resources on which these sources depend. As a result, many supplies are unsuccessful or perform poorly (Robins *et al.*, 2006). Successfully developing groundwater resources sustainably and cost-effectively on the scale required to help achieve the Millennium Development Goals is not trivial. The challenge is more than just providing extra drilling rigs to the worst-affected countries: technology, software and hardware must all be appropriate to the nature of the groundwater resources in the project area.

In this paper we discuss the groundwater resources in Sub-Saharan Africa (SSA) in the context of rural water supply: finding and developing groundwater resources to sustain community hand pumps. An extended reference list is given to help follow up technical aspects of the hydrogeology of Africa, which are not discussed in detail here.

2 GROUNDWATER RESOURCES IN SUB-SAHARAN AFRICA

The availability of groundwater resources in Sub-Saharan Africa depends critically on the geology, the history of weathering faulting, and the recharge to groundwater.

Figure 1 shows the average annual rainfall across Sub-Saharan Africa based on data from New & Hulme (1997). This clearly demonstrates the arid areas, where groundwater recharge is limited and erratic. However, there is no simple direct relationship between average annual rainfall and recharge, and significant recharge (10–50 mm) can occur where annual rainfall is less than 500 mm (Edmunds & Gaye, 1994; Butterworth, 1999; Edmunds *et al.*, 2002). Climate change will significantly alter patterns of rainfall and recharge across Africa. Climate models predict that the number of drought episodes in Africa will increase, particularly in Sahel areas, and the number of people affected by sever drought will grow (Hulme *et al.*, 2001; Magrath & Simms, 2007). Rural water supply, however, does not require large quantities of recharge, and a simple mass balance indicates that recharge of 10 mm per annum would support community boreholes (5 m^3/d or 0.17 l/s) with hand pumps at a spacing of 500 m across Africa.

A simplified hydrogeological map is shown in Figure 1 based on a synthesis of studies (Foster, 1984; Guiraud, 1988; UNTCD, 1988; UNTCD, 1989; MacDonald & Davies, 2000) and using the 1:5,000,000 scale geological map of Africa as a base (UNESCO, 1991; Persits *et al.*, 1997). The classifications reflect the different manner in which groundwater occurs, constrained by the geological information available at this scale throughout SSA. The four different environments are: Precambrian "basement" rocks; volcanic rocks; unconsolidated sediments; and consolidated sedimentary rocks. Basement rocks form the largest hydrogeological environment, occupying 40% of the 23.6 million square kilometres and volcanic rocks are the smallest hydrogeological environment with only 6% of the land area (see Table 1). This basic division forms the basis for the rest of this paper.

The potential of each hydrogeological environment to contribute to rural water supply is best indicated by the rural population living in each one. As discussed above, the rural communities are most dependent on local resources for water supply, since transportation is often prohibitively expensive and difficult to manage. Using spatial data from ESRI (ESRI, 1996) and statistics from the World Bank and the WHO/UNICEF Joint Monitoring

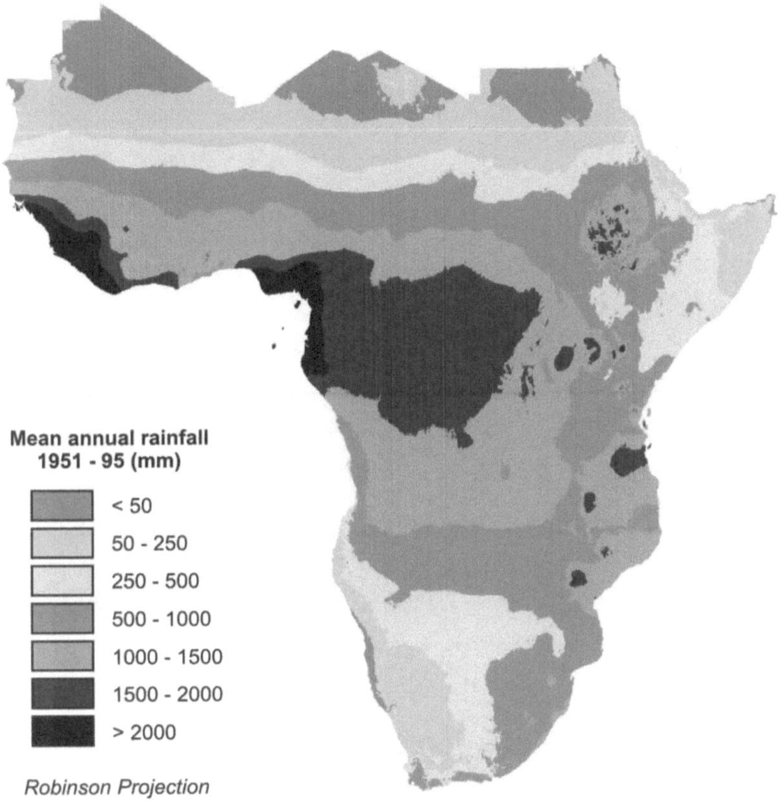

Figure 1. Average annual rainfall for Sub-Saharan Africa for the period 1951–1995. Produced from data from New & Hulme (1997).

Table 1. The land area of each hydrogeological environment shown in Figure 2, and and estimate of the rural population living on each (ESRI, 1996; World Bank, 2000; JMP, 2004).

Hydrogeological Environment	Land area (%)	Rural population (millions)
Basement rocks	40	235
Consolidated sedimentary rocks	32	120
Volcanic rocks	6	45
Unconsolidated sediments	22	70
Total	**100%**	**470**

Programme (JMP, 2004), an approximation was made of the distribution of rural population throughout Sub-Saharan Africa. The results are shown in Table 1. Basement rocks support the largest population (235 million) and volcanic rocks the least (45 million). However, despite, the relatively low number of people living in hydrogeological volcanic areas, they are particularly important since they are home to some of the poorest and most drought prone people in Africa.

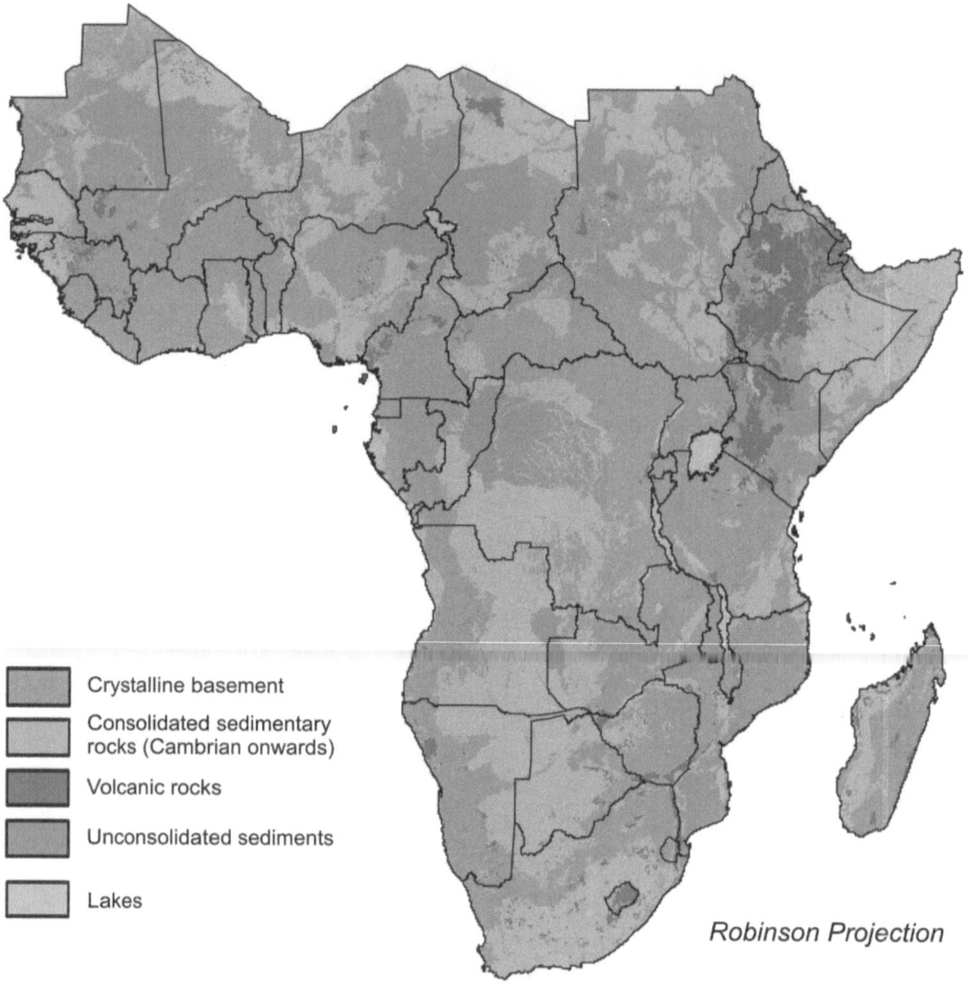

Figure 2. The hydrogeological environments of Sub-Saharan Africa (from MacDonald & Davies, 2000; MacDonald *et al.*, 2005a).

3 HYDROGEOLOGICAL ENVIRONMENTS

In this section each of the hydrogeological environments are described with references to key publications on each. Table 2 provides a summary of the groundwater potential for each environment.

Before discussing each hydrogeological environment it is important to define what constitutes an aquifer in the context of rural water supply. If we assume that a borehole is to supply a minimum of 5 m^3/d to be successful, and that recharge is not a constraint (see above), then the minimum aquifer properties can be estimated that would give a successful source. Modelling indicates that transmissivity is the key limiting factor and that generally transmissivity >1 m^2/d will give a successful borehole (MacDonald *et al.*, 2008).

Table 2. A summary of the groundwater potential of the African hydrogeological environments (modified from MacDonald *et al.*, 2005a).

	Hydrogeological Sub-Environment	GW potential & average yields	Groundwater Targets
Crystalline basement rocks	Highly weathered and/or fractured basement	Moderate 0.1–1 l/s	Fractures at the base of the deep weathered zone. Sub-vertical fracture zones.
	Poorly weathered or sparsely fractured basement	Low 0.1 l/s–1/s	Widely spaced fractures and localised pockets of deep weathering.
Consolidated sedimentary rocks	Sandstone	Moderate – High 1–20 l/s	Coarse porous or fractured sandstone.
	Mudstone and shale	Low 0–0.5 l/s	Hard fractured mudstones Igneous intrusions or thin limestone/sandstone layers.
	Limestones	Moderate – High 1–100 l/s	Fractures and solution enhanced fractures (dry valleys)
	Recent Coastal and Calcareous Island formations	High 10–100 l/s	Proximity of saline water limits depth of boreholes or galleries. High permeability results in water table being only slightly above sea level
Unconsolidated sediments	Major alluvial and coastal basins	High 1–40 l/s	Sand and gravel layers
	Small dispersed deposits, such as river valley alluvium and coastal dunes deposits.	Moderate 1–20 l/s	Thicker, well-sorted sandy/ gravel deposits. Coastal aquifers need to be managed to control saline intrusion.
	Loess	Low – Moderate 0.1–1 l/s	Areas where the loess is thick and saturated, or drains down to a more permeable receiving bed
	Valley deposits in mountain areas	Moderate – High 1–10 l/s	Stable areas of sand and gravel; river-reworked volcanic rocks; blocky lava flows
Volcanic Rocks	Extensive volcanic terrains	Low – High Lavas 0.1 – 100 l/s Ashes and pyroclastic rocks 0.5–5 l/s	Generally little porosity or permeability within the lava flows, but the edges and flow tops/bottom can be rubbly and fractured; flow tubes can also be fractured. Ashes are generally poorly permeable but have high storage and can drain water into underlying layers.

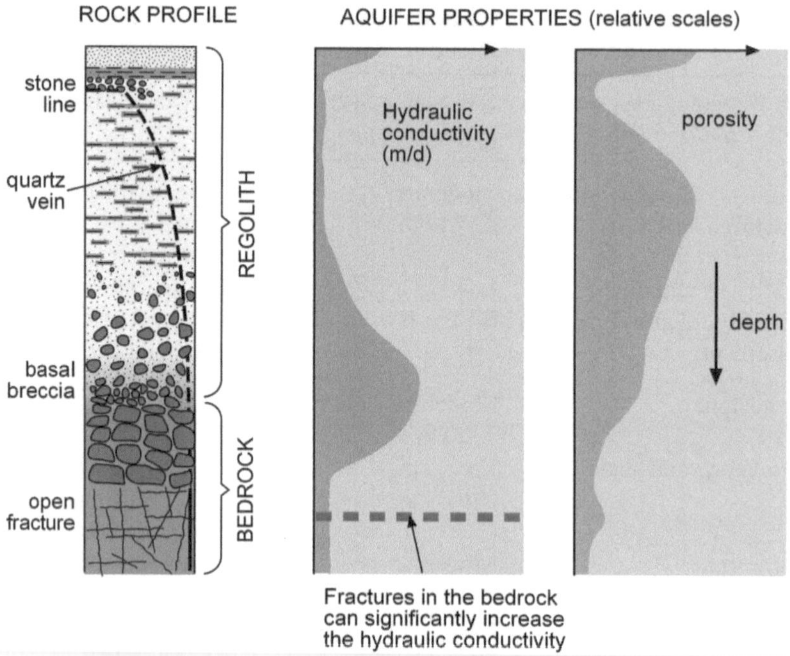

Figure 3. Schematic diagram of the variation of permeability and porosity with depth in the crystalline basement (adapted from Acworth, 1987; Chilton & Foster, 1995).

3.1 *Precambrian basement*

Precambrian basement rocks occupy 40% of the land area of SSA and support approximately 235 million rural inhabitants. They comprise crystalline igneous and metamorphic rocks over 550 million years old (Key, 1992). Unweathered and non-fractured basement rocks contain negligible quantities of groundwater. Significant aquifers however, develop within the weathered overburden and fractured bedrock.

The geology of Precambrian Basement areas is complex, reflecting the long history that the environment has been subjected to. Although Precambrian basement terrains largely comprise crystalline igneous and metamorphic rocks, there are also areas of metamorphosed consolidated sediments comprising sandstones, conglomerates, shales and mudstones in west. Two notable examples are the Voltaian Sediments of Ghana and the Transvaal, Waterberg and Ventersdorp Groups in southern Africa.

Five factors contribute to the weathering of basement rocks (Jones, 1985; Acworth, 1987):

- tension and stress fractures;
- geomorphology of the terrain; e.g. weathering along fracture controlled valleys, formation of inselbergs;
- temperature and occurrence of groundwater controlling the depth and nature of weathering;
- mineral content of the basement rock;
- the palaeo-climates experienced by the near surface deposits.

The resulting weathered zone can vary in thickness from just a few metres in arid areas to over 90 m in the humid tropics. Historical erosion surfaces may also be important in preserving ancient weathered surfaces. Figure 3 summarise the permeability and porosity profiles for the weathered zone. Porosity generally decreases with depth; permeability however, has a more complicated relationship, depending on the extent of fracturing and the clay content (Wright & Burgess, 1992; Chilton & Foster, 1995). In the soil zone, permeability is usually high, but groundwater does not exist throughout the year and dries out soon after the rains end. Beneath the soil zone, the rock is often highly weathered and clay rich, therefore permeability is low. Towards the base of the weathered zone, near the fresh rock interface, the proportion of clay significantly reduces. This horizon, which consists of fractured rock, is often permeable, allowing water to move freely. Wells or boreholes that penetrate this horizon can usually provide sufficient water to sustain a handpump.

Deeper fractures within the basement rocks are also an important source of groundwater, particularly where the weathered zone is thin or absent. These deep fractures are tectonically controlled and can sometimes provide supplies of 1–5 l/s. The groundwater resources within the regolith and deeper fracture zones depend on the thickness of the water-bearing zone and the relative depth of the water table. In general terms, the deeper the weathering, the more sustainable the groundwater. However, due to the complex interactions of the various factors affecting weathering (an in particular the presence of clay in the weathered zone), water-bearing horizons may not be present at all at some locations.

Various techniques have been developed to locate favourable sites for the exploitation of groundwater resources within basement rocks. These include remote sensing (Lillesand & Kiefer, 1994) geophysical methods (Beeson & Jones, 1988; McNeill, 1991; Carruthers & Smith, 1992) and geomorphological studies (Taylor & Howard, 2000). Geophysical surveys using combined resistivity and ground conductivity (EM) surveys have often been found useful in siting wells and boreholes (see Table 3). These can often be successfully interpreted by using simple guidelines, (MacDonald *et al.*, 2005a). Although groundwater is generally abstracted through boreholes and wells, more sophisticated systems, such as collector wells have also been used with success, (Ball & Herbert, 1992; Lovell, 2000).

3.2 *Volcanic rocks*

Volcanic rocks occupy only 6% of the land area of SSA and are found in east and southern Africa where they can form important aquifer systems. In total, about 45 million people are dependent on volcanic rocks for rural groundwater supplies, and they underlie much of the poorest and drought stricken areas of SSA. The groundwater potential of volcanic rocks varies considerably, reflecting the complexity of the geology. There have been few systematic studies of the hydrogeology of volcanic rocks in Africa, although good site studies are given by Aberra (1990), Vernier (1993), Demlie *et al.* (2007) Kebede *et al.* (2007). Volcanic rocks are important aquifers in India and have been extensively studied there (e.g. Kulkarni *et al.*, 2000).

The volcanic rocks in SSA were formed during three phases of activity during Cenozoic times, associated with the opening of the East African rift valley and an earlier Late Karoo (Jurassic) phase. These events gave rise to a thick complex sequence of lava flows, sheet basalts and pyroclastic rocks such as agglomerate and ash. Thick basalt lava flows are often interbedded with ash layers and palaeosoils. The potential for groundwater depends largely on the presence of fractures. The top and bottom of lava flows, particularly where associated with palaeosoils, are often highly fractured and weathered; towards the middle of the lava flows, the basalt tends to be more competent and less fractured. Figure 4 shows aspects of groundwater flow in highland volcanic areas in Ethiopia. In southern Africa, large volumes of flood basalts erupted from centres in present day Lesotho, SE South Africa,

Table 3. A summary of common geophysical investigations used in rural water supply projects (see Milsom (2002) for more details).

Geophysical technique	What it measures	Output	Usual maximum depth of penetration	Comments
Resistivity	Apparent electrical resistivity of ground	1-D vertical geoelectric section; more complex equipment gives 2-D or 3-D geoelectric sections	100 m	Can locate changes in the thickness and nature of the weathered zone and differences in geology. Also useful for identifying thickness of sand or gravel within superficial deposits. Often used to calibrate FEM surveys (see below). Slow survey method and requires careful interpretation.
Frequency domain Electro-Magnetic methods (FEM)	Apparent terrain electrical conductivity (calculated from the ratio of secondary to primary electromagnetic fields)	Single traverse lines or 2D contoured surfaces of bulk ground conductivity	50 m	Quick and easy method for determining changes in thickness of weathered zones or alluvium. Interpretation is non-unique and requires careful geological control. Can also be used in basement rocks to help identify fracture zones.
Transient Electro-Magnetic methods (TEM)	Apparent electrical resistance of ground (calculated from the transient decay of induced secondary electromagnetic fields)	Output generally interpreted to give 1D resistivity sounding	150 m	Better at locating targets through conductive overburden than FEM, also better depth of penetration. Expensive and can be difficult to operate.

Very Low Frequency (VLF)	Secondary magnetic fields induced in the ground by military communications transmitters.	Single traverse lines, or 2D contoured surfaces.	40 m	Can locate vertical fracture zones and dykes within basement rocks or major aquifers
Ground penetrating radar (GPR)	Reflections from boundaries between bodies of different dielectric constant	2D section showing time for EM waves to reach reflectors	10 m	Accurate method for determining thickness of sand and gravel. The technique will not penetrate clay, however, and has a depth of penetration of about 10 m in saturated sand or gravel.
Seismic refraction	P-wave velocity through the ground	2-D vertical section of P-wave velocity	30 m	Can locate fracture zones in basement rock and also thickness of drift deposits. Not particularly suited to measuring variations in composition of drift. Fairly slow and difficult to interpret.
Magnetic	Intensity (and sometimes direction) of earth's magnetic field	Variations in the earth's magnetic field either along a traverse or on a contoured grid	100 m	Can locate magnetic bodies such as dykes or sills. Susceptible to noise from any metallic objects or power cables.

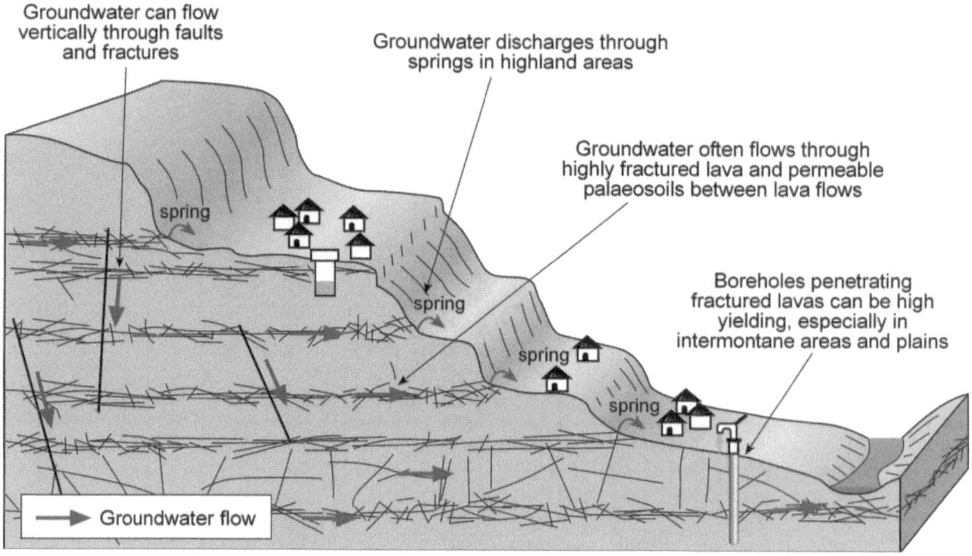

Figure 4. Groundwater occurrence in volcanic rocks.

Mozambique, western Zimbabwe and north eastern Botswana. These thick extrusive sequences of basalts and ash deposits were associated with the intrusion of extensive dolerite ring and dyke swarm complexes within the present day regions of south-eastern South Africa, southern Zimbabwe and north eastern Botswana (TAMS, 1996; Woodford & Chevallier, 2002).

The most important factors for the development of aquifers within volcanic rocks are given below (Kehinde & Loenhert, 1989; Vernier, 1993):

- Thick palaeosoils or loose pyroclastic material between lava flows are often highly permeable;
- joints and fractures due to the rapid cooling of the tops of lava flows provide important flow pathways;
- contact between lava flows and sedimentary rocks or earlier volcanic material such as domes etc. are often highly fractured and contain much groundwater;
- gas bubbles within lava flows, and porosity within ashes and agglomerates can provide significant groundwater storage.

Fractured lava flows can have very high permeability, but yields exhibit large variations with average values from boreholes about 2 l/s (UNTCD, 1989), which is more than adequate for rural domestic water supplies. Boreholes are generally more suitable than hand dug wells, since the fracture zones with significant groundwater are often deep. However, in Kenya, where the volcanic rocks form vast tablelands, the groundwater can be shallow, and sometimes exploited by dug wells. Dug wells can also be used in mountainous areas, where aquifers are small and water-levels sometimes shallow. Springs are common in volcanic rocks, particularly in highland areas. The interconnected fractures and cavities found in the lava flows provide rapid discrete flow paths for groundwater, which often discharge as springs at impermeable boundaries. Springs, particularly at higher altitudes can be more susceptible to drought failure than boreholes (Calow *et al.*, 2002).

The quality of groundwater can be a problem in volcanic rocks. Fluoride concentrations are sometimes elevated and concentrations in excess of 1.5 mg/l can lead to health problems such as dental or skeletal fluorosis. High fluoride groundwaters are common in the rift valley regions of Kenya and Tanzania (e.g. Ashley & Burley, 1995; Reimann, 2003).

Geophysical techniques have sometimes been used in volcanic terrain to site boreholes, but few general guidelines have been developed. Remote sensing techniques may be valuable for detecting different geological units and identifying fracture zones but have not been widely applied. Boundaries between volcanic rocks and sedimentary rocks could be easily identified with magnetic methods. Resistivity methods have been used to locate vertical and horizontal fracture zones in East Africa (Drury *et al.*, 2001). However locating deep horizontal fracture zones (such as the boundary between lava flows) can be difficult using geophysics, and boreholes may have to be drilled relying solely on experience from previous drilling in the area.

3.3 *Consolidated sedimentary rocks*

Consolidated sedimentary rocks occupy 32% of the land area of SSA (Figure 2). Approximately 135 million people live in rural areas underlain by these rocks. Sedimentary basins can store considerable volumes of groundwater, but in arid regions, much of the groundwater can be non-renewable, having been recharged when the area received considerably more rainfall. Also, sedimentary rocks are highly variable and can comprise low permeability mudstone and shale as well as more permeable sandstones and limestones Examples of large sedimentary basins in Sub-Saharan Africa (Figure 5) are the Karoo (tillite, mudstone, sandstone and conglomerate), and Kalahari Basin sediments (mudstone and sandstone with uncompressed cover sediments of poorly consolidated muds, silts and sands with associated consolidated evaporate sediments such as calcrete and silcrete) of Southern and Central Africa (Truswell, 1970), sediments within the Somali basin of East Africa and the Benue Trough of West Africa (Selley, 1997).

For the purposes of creating the simplified map shown in Figure 2, sedimentary rocks deposited before Quaternary times are assumed to be mainly consolidated. Figure 6 illustrates how groundwater occurs in consolidated sedimentary rocks.

Consolidated sandstone and limestone contain significant groundwater. Shallow limestone aquifers are often vulnerable to saline intrusion and pollution (e.g. the limestone aquifers along the East African coast). Carefully constructed deep boreholes into thick sandstone aquifers can provide high yields (e.g. Middle and Upper Karoo sandstones of South Africa, Botswana and western Zimbabwe (Interconsult, 1985; Woodford & Chevallier, 2002; Cheney *et al.*, 2006). Yields are highest where the sandstones are weakly cemented or fractured. This makes the aquifers highly suited to large-scale development for reticulated urban supply, industrial uses and agricultural irrigation. However, rural water supply generally relies on shallow boreholes or wells close to communities. Only rocks immediately surrounding the community and to a depth of less than 100 m are usually considered. In large sedimentary basins, where modern recharge is limited, water-levels can be deeper than 50 m, and therefore difficult to abstract using a handpump. More sophisticated approaches may be required for rural water supply in such areas requiring deeper boreholes, header tanks and distribution systems.

Although mudstone and siltstone are poor aquifers, groundwater can often be found in these environments with careful exploration. Studies in Nigeria showed that where

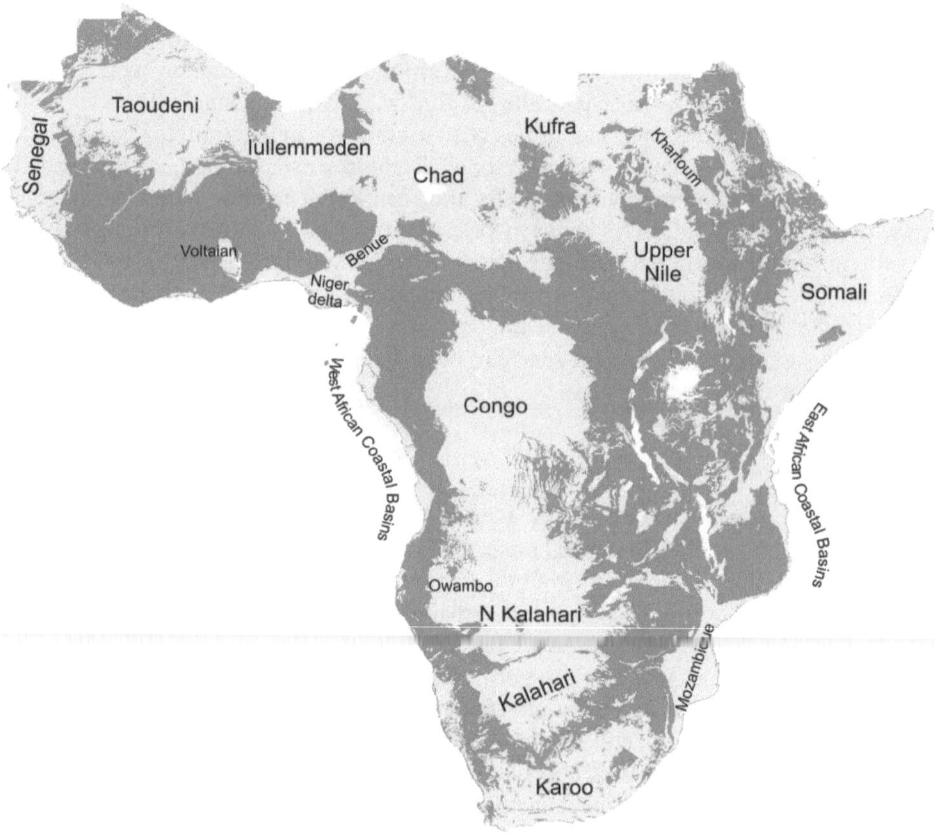

Figure 5. The location of large sedimentary basins in Sub-Saharan Africa.

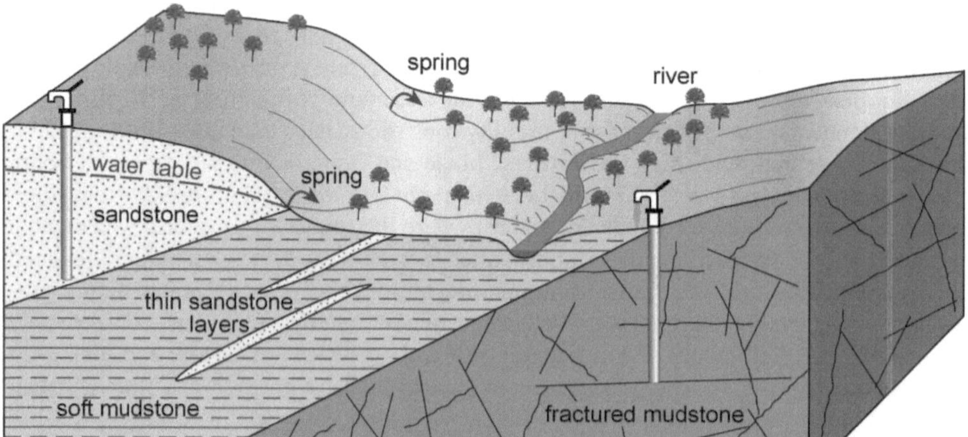

Figure 6. A schematic diagram illustrating how groundwater for rural water supply can exist within sedimentary rocks.

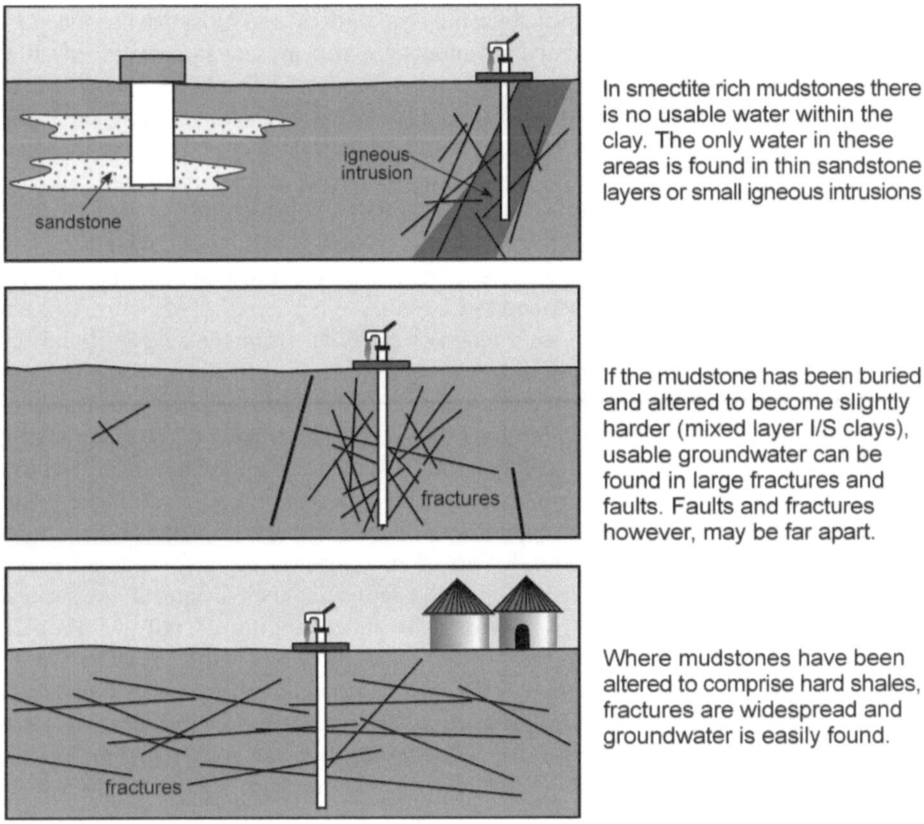

Figure 7. Groundwater occurrence in mudstone areas (after MacDonald *et al.*, 2005b).

mudstone is soft and dominated by smectite, negligible groundwater exists; in slightly metamorphosed mudstone, where the rocks have been altered to become harder, fractures can remain open and usable groundwater can be found (MacDonald *et al.*, 2005b). Similar problems have been encountered in the fine grained aeolian and fluvially deposited Karoo age sediments (as in south Africa and Lesotho (Sami, 1996)). It is estimated that 65% of all sediments are mudstone (Aplin *et al.*, 1999); therefore, up to 75 million people may live directly on these mudstone areas. Figure 7 illustrate how groundwater exists in mudstone areas.

Geophysical techniques can be used to identify good aquifers. Sandstone can easily be distinguished from mudstone using ground conductivity or resistivity methods (e.g. Interconsult, 1985; Bromley *et al.*, 1996;. Similarly, harder mudstones can also be distinguished from soft mudstone (MacDonald *et al.*, 2001). In areas where large sandstone or limestone aquifers are present, little or no detailed siting is required for rural domestic supply; boreholes can be drilled anywhere. Occasionally, if the aquifers and groundwater levels are shallow, dug wells can be constructed.

3.4 *Unconsolidated sediments*

Unconsolidated sediments form some of the most productive aquifers in Africa. They cover approximately 22% of the land surface of SSA (Figure 2). However, this is probably an underestimate of their true importance since only the thickest and most extensive deposits are shown on the map. Unconsolidated sediments are also present in many river valleys throughout Africa. Examples of extensive deposits of unconsolidated sediments are found in Chad, Congo and Mozambique (the amalgamated deltas of the Save, Zambezi and the Limpopo) and in the coastal areas of Nigeria (Niger Delta), Ghana, Somalia, Namibia, Madagascar and Kenya. There is no clear dividing line between unconsolidated sediments and consolidated sedimentary rocks, as the time taken for consolidation can vary. However, for most purposes it can be assumed that sediments deposited in the past few million years (during Quaternary and late Neogene times) will remain unconsolidated. Unconsolidated Sedimentary Aquifers are often described as UNSAs.

Unconsolidated sediments comprise a range of material, from coarse gravel and sand to silt and clay. They are deposited in different environments such as rivers and deltas by various combinations of physical processes. Large unconsolidated sedimentary basins can store large amounts of groundwater. Guiraud (1988) describes several of the major UNSAs in Africa. As with consolidated sedimentary rocks, where the basins are now in arid regions, the water they contain may not be currently renewable. The size and physical characteristics of the aquifer depend on how the sediment was deposited. Sand and gravel beds can be continuous over hundreds of kilometres, but are often multi-layered, with sands and gravels interbedded with silts and clays. Depending on the depositional environment, the structure of the aquifers can be highly complex, with sediments changing laterally within a few metres (see Figure 8).

Small UNSAs are found throughout SSA. On basement, volcanic and consolidated sedimentary rocks, UNSAs can be found in valleys- deposited by present day rivers. Here, groundwater is close to the surface, so pumping lifts are small; also the proximity to the rivers offers a reliable source of recharge. In southern Africa, sand-rivers are important sources of water for domestic and stock watering use. Research into the occurrence of groundwater in sand rivers has been undertaken in Botswana (e.g. Wikner, 1980; Herbert *et al.*, 1997; Davies *et al.*, 1998), Namibia (Jacobson, *et al.*, 1995), South Africa (Clanahan & Jonch, 2005) and Zimbabwe (Owen, 1989). These rivers rarely contain surface water, but the thick sediment within the river channel can contain significant groundwater. In northern Nigeria, shallow floodplains known as fadamas, are important sources of groundwater (Carter & Alkali, 1996). These floodplains may be several kilometres wide and can contain 10 m of sands and gravels. They rely on annual flooding for recharge.

Where the structure of UNSAs is complex, geophysical techniques can be used to distinguish sand and gravel from clay. Ground penetrating radar, shallow conductivity and resistivity surveys are all routinely used in groundwater exploration in UNSAs. Ekstrom *et al.* (1996) describe the application of resistivity to find groundwater in river alluvium in SW Zimbabwe; Davies *et al.* (1998) used shallow seismic refraction to investigate sand rivers in NE Botswana and MacDonald *et al.* (2000) describe the use of ground conductivity and ground penetrating radar for locating groundwater in alluvium and blown sands. Remote sensing techniques such as satellite imagery and aerial photography can also be used to provide information on the distribution of sedimentary systems.

UNSAs are easy to dig and drill, so exploration is rapid and inexpensive. Where groundwater is shallow, simple hand drilling is often effective. Where boreholes have to be deeper, drilling can be more problematic. Groundwater quality problems can occur in UNSAs due to natural geochemistry and contamination. Problems can arise where groundwater is developed from such sediments with little regard to the water chemistry. High arsenic

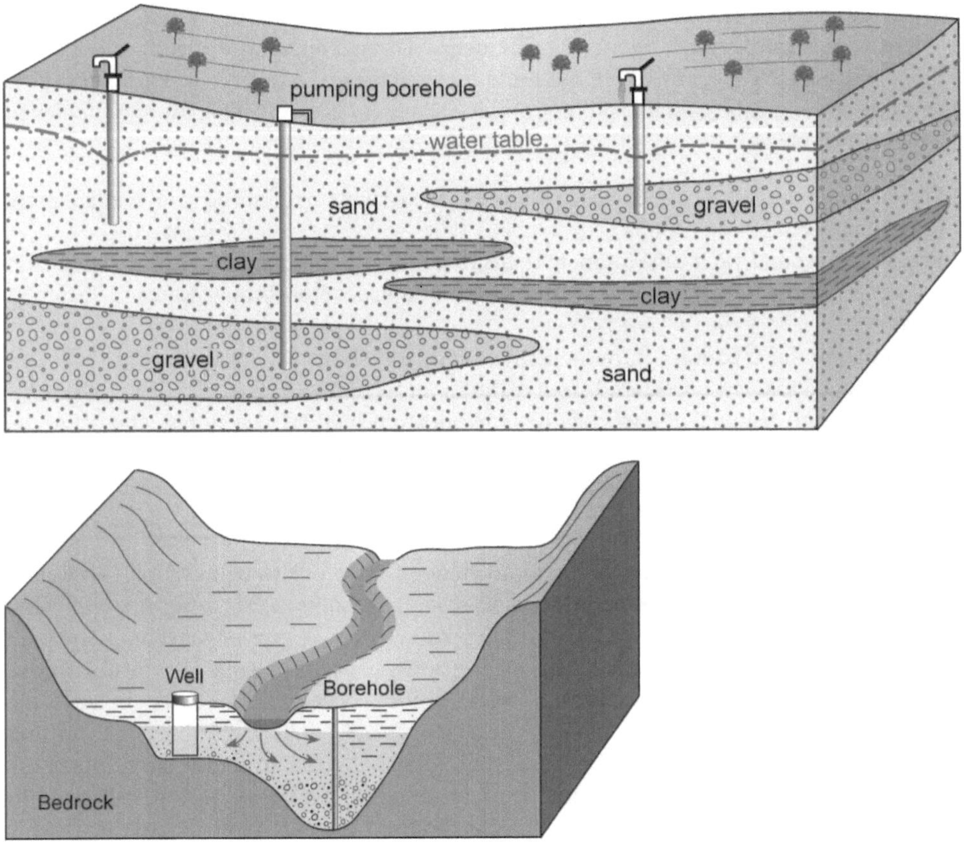

Figure 8. Groundwater occurrence in large unconsolidated sedimentary basins (top) and small unconsolidated valley deposits (bottom).

concentrations in groundwater within Bangladesh and India were undetected until the local population developed symptoms of arsenic poisoning (Kinniburgh *et al.*, 2003).

4 DISCUSSION: USING APPROPRIATE GROUNDWATER KNOWLEDGE

The premise of this paper is that knowledge of groundwater resources and hydrogeological expertise is fundamental to successful and sustainable rural water supplies. The technical capacity required to develop groundwater resources differs with the hydrogeology: in some environments little expertise is required, while in others considerable research and money is required to develop groundwater. In this discussion we highlight four areas where hydrogeological expertise can be focussed to maximise the benefit to rural water supplies. (1) existing knowledge transfer; (2) benchmarking; (3) researching difficult areas; and (4) providing accessible advice to planners and policy makers.

4.1 *Knowledge Transfer*

Implementers of rural water supply programmes (those tasked with siting water points, designing boreholes/wells, assessing yield, quality and sustainability) form a varied group with a wide range of skills and disciplines. They may be water engineers, geologists, geophysicists, hydrogeologists or general technicians. Trained and experienced groundwater specialists are rare. Therefore, there is a pressing need for groundwater development skills to be made more widely available within rural water supply projects. This could be achieved in a number of ways:

1. Developing manuals designed specifically for the issues surrounding rural water supply (e.g. MacDonald *et al.*, 2005a).
2. Producing groundwater development maps that indicate the availability of groundwater resources, and the techniques required to find and develop groundwater in different areas.
3. Running regular incountry training courses and workshops, where the useful skills can be developed and lessons shared from different projects.
4. Creating local, regional and worldwide networks to provide a pyramid of support to the project staff undertaking the work.
5. Investing research into developing new simple but effective techniques for use by project staff on rural water supply.

To be able to transfer knowledge, that knowledge first needs to be available. However, decentralisation and the promotion of demand-responsive approaches to service provision have had significant implications for building knowledge of groundwater in Africa. In particular, local institutions – including local government and NGOs. While this move has many benefits and promises greater sustainability, decentralisation has been to the detriment of national databases, national knowledge and control over borehole drilling and construction standards. As a consequence, knowledge of groundwater resources is not growing or even being maintained in much of SSA. This new reality will need to be embraced and new methods and possibly institutions developed to ensure that data is captured from ongoing projects and transformed to information which can be assimulated as knowledge by those who need it.

4.2 *Benchmarking*

As the provision of rural water supply becomes increasingly decentralised, budget holders (who are often based in district or local government) have little knowledge about the complexities of hydrogeology and groundwater investigations. This makes it difficult for them to judge whether a drilling success rate is justified due to difficult terrain, or whether a project is over specified in easy terrain. Knowledge of local hydrogeology and the business of groundwater development is required to be able to make informed decisions about designing and managing projects.

Simple cost-benefit analysis can help if data are available on drilling costs and success rates 'with' and 'without' different levels of investigation. As noted in Farr *et al.* (1982) the use of a particular search technique is only justified if it increases the chances of subsequent boreholes being successful, such that the overall saving in drilling costs (through drilling fewer unsuccessful boreholes) is greater than the cost of the search. In some environments, where groundwater is readily available, expensive methods may not be justified. In other environments, however, seemingly expensive methods or studies may be entirely justified by long term savings in drilling costs.

Therefore, it is fundamental to the success of decentralised programmes that the local governments have sufficient hydrogeological expertise. Hydrogeologists can be of value in a number of ways:

1. Facilitating the acquisition of the knowledge and skills required by the decentralised bodies to manage contractors. This may be through the various methods described in the section on "knowledge transfer" above.
2. Providing authoritative guidelines against which proposals, projects and contractors can be assessed. This "benchmarking" role is increasing in importance and visibility within the development community.

4.3 *Research in complex areas*

Areas where sustainable groundwater sources are hard to find often have the greatest problems with health and poverty. In these areas, women have to walk further to find water and waterborne diseases such as guinea worm are more common. Helping to solve water problems in these difficult areas may have greater impact on reducing poverty in Sub-Saharan Africa than drilling many more boreholes in areas where it is relatively easy to find water.

By effectively disseminating techniques to project staff in areas where it is relatively easy to find groundwater, hydrogeological expertise and research budgets can focus on more difficult areas where groundwater occurrence is not well understood and rural water supplies rarely effective. Some issues that demand more research are:

- the age, recharge and sustainability of groundwater supplies in basement areas, particularly during drought; this will become increasingly important with a changing climate;
- the existence of groundwater in areas where groundwater is difficult to find (e.g. poorly weathered crystalline basement and mudstone areas), and developing techniques that can be used by project staff in these areas to find groundwater and develop rural water supplies;
- identify and understanding the constraints on rural water supply caused by natural groundwater contaminants, such as fluoride and arsenic;
- matching more closely technologies for accessing groundwater (wells, boreholes, springs, collector wells) with the hydrogeological environment and socio-economic conditions to maximise yield and sustainability, and minimise costs;
- examine the risks to rural water supply caused by the increase in onsite sanitation.

4.4 *Informing policy*

Many of the current pillars of rural water supply policy stem from a change in thinking about the value of water and a recognition that centralised approaches to service delivery are unsustainable. A key objective of policies is the provision of potable water on a continuous basis: security of supply across seasons and between wet and dry years is essential if health and poverty alleviation benefits are to be met and sustained. Central pillars of modern policy include: treating water as a social and economic good; using demand responsive approaches which allows consumers to guide investment decisions; moving from community participation to community management; embedding rural water supply in larger initiatives which include sanitation and hygiene promotion; decentralizing service delivery; and recognising the broader livelihood benefits of rural water supply rather than concentrating only on public health.

These approaches have many benefits for improving rural water supply in Africa, however, they rely on informed decisions being made on technology choices and approaches. Unfortunately, this is rarely true. As discussed above, capacity rarely exists within projects to advise communities on the most appropriate approach to use in their particular community. Therefore, hydrogeologists have much to offer in influence how policies are translated into workable approaches on the ground. Only with determined interdisciplinary approaches will interventions be effective and sustainable.

5 CONCLUSIONS

Groundwater is central to helping Sub-Saharan Africa meet the Millennium Development Goals for water supply. Rocks with poor aquifer properties ($T \sim 1\ m^2/d$ and $S \sim 0.001$) will generally still support a village borehole with a handpump. The current approaches to rural water supply, in particular demand driven approaches, community participation and poverty focus have many benefits to the overall efficacy and sustainability of water supplies. However the underlying presumption that groundwater is ubiquitous, or can easily be found at each site is dangerous and my lead to many failures.

Crystalline basement occupies 40% of the land area of SSA; 235 million people live in rural areas underlain by crystalline basement rocks. Volcanic rocks occupy 6% of the land area of SSA, and sustain a rural population of 45 million, many of whom live in the drought stricken areas of the Horn of Africa. Consolidated sedimentary rocks occupy 32% of the land area of SSA and sustain a rural population of 110 million. Unconsolidated sediments occupy 22% of the land area of SSA and sustain a rural population of 60 million. They are probably more important than these statistics suggest since they are present in most river valleys throughout Africa.

Hydrogeologists have a key role in helping to meet the Millennium Development Goals, but must learn to work within the existing policy framework which gives social and economic factors a higher priority than groundwater resources. The four main areas of work for hydrogeologists are: 1) communicating techniques and knowledge to those responsible for siting and developing groundwater supplies; 2) benchmarking the expected expertise and quality for different hydrogeological environments; 3) researching complex areas, where little is known about groundwater resources; and 4) providing accessible and appropriate advice to policy makers. The hydrogeological community can help to reduce poverty in Sub-Saharan Africa. The most pressing task is to communicate what we know to the people who need to know it.

ACKNOWLEDGEMENTS

This paper is published with the permission of the Executive Director of the BGS (NERC). Much of the work on which the paper is based was funded by the UK Department of International Development (DFID). The views expressed, however, are not necessarily those of DFID.

REFERENCES

Aberra, T. 1990. *The hydrogeology and water resources of the Ansokia highland springs, Ethiopia.* Memoires of the 22nd Congress of IAH, Vol XXII, Lausanne.

Acworth, R. I. 1987. *The development of crystalline basement aquifers in a tropical Environment.* Quarterly Journal of Engineering Geology, **20**, 265–272.

Aplin, A. C., Fleet, A. J. & MacQuaker, J. H. S. 1999. *Muds and mudstones: physical and fluid flow properties.* In: Aplin, A. C., Fleet, A. J. & MacQuaker, J. H. S. (eds.) Muds and Mudstones: Physical and Fluid Flow Properties. Geological Society London Special Publications, **158**, 1–8.

Ashley, R. P. & Burley, M. J. 1995. *Controls on the occurrence of fluoride in groundwater in the Rift valley of Ethiopia.* In: Nash, H. & McCall, G. J. H. (eds.) Groundwater Quality. Chapman & Hall, London.

Ball, D. F. & Herbert, R. 1992. *The use and performance of collector wells within the regolith aquifer of Sri Lanka.* Ground Water, **30**, 683–689.

Beeson, S. & Jones, C. R. C. 1988. *The combined EMT/VES geophysical method for siting boreholes.* Ground Water, **26**, 54–63.

Bromley, J., Mannstrom, B., Nisca, D. & Jamtlid, A. 1994. *Airborne geophysics: Application to a ground-water study in Botswana.* Ground Water, **32**, 79–90.

Butterworth, J. A., Macdonald, D. M. J., Bromley, J., Simmonds, L. P., Lovell, C. J. & Mugabe, F. 1999. *Hydrological processes and water resources management in a dryland environment III: groundwater recharge and recession in a shallow weathered aquifer.* Hydrology and Earth System Sciences, **3**, 345–352.

Calow, R. C, MacDonald, A. M, Nicol, A. Robins, N. S. & Kebede, S. 2002. *The struggle for water: drought, water security and rural livelihoods.* British Geological Survey Commissioned Report **CR/02/226N**.

Carruthers, R. M. & Smith, I. F. 1992. *The use of ground electrical methods for siting water supply boreholes in shallow crystalline basement terrains.* In: Wright, E. P. & Burgess, W. G. (eds.) The Hydrogeology of Crystalline Basement Aquifers in Africa. Geological Society London Special Publications, **66**, 203–220.

Carter, R. C. & Alkali, A. G. 1996. *Shallow groundwater in the northeast arid zone of Nigeria.* Quarterly Journal of Engineering Geology, **29**, 341–356.

Cheney, C. S., Rutter, H. K., Farr, J. & Phofuetsile, P. 2006. *Hydrogeological potential of the deep Ecca aquifer of the Kalahari, southwestern Botswana.* Quarterly Journal of Engineering Geology and Hydrogeology, **39**, 303–312.

Chilton, P. J. & Foster, S. S. D. 1995. *Hydrogeological characterisation and water-supply potential of basement aquifers in tropical Africa.* Hydrogeology Journal, **3**, 36–49.

Clanahan, M. J. & Jonck, J. L. 2004. A critical evaluation of sand abstraction systems in Southern Africa. Water Research Commission of South Africa Report **829/3/05**, Pretoria, RSA.

Demlie, M., Wohnlich, S., Wisotzky. F. & Gizaw, B. 2007. *Groundwater recharge, flow and hydrogeochemical evolution in a complex volcanic aquifer system, central Ethiopia.* Hydrogeology Journal, **15**, 1169–1181.

Davies, J., Rastall, P. & Herbert, R. 1998. *Final report on the application of collector well systems to sand rivers pilot project.* British Geological Survey Technical Report **WD/98/2C.**

DFID 2001. *Addressing the water crisis: healthier and more productive lives for poor people.* Department for International Development, London.

Drury, S. A, Peart, R. J. & Andrews Deller, M. E. 2001. *Hydrogeological potential of major fractures in Eritrea.* Journal of African Earth Sciences, **32**, 163–177.

Edmunds, W. M. & Gaye, C. B. 1994. *Estimating the spatial variability of groundwater recharge in the Sahel using chloride.* Journal of Hydrology, **156**, 47–59.

Edmunds, W. M., Fellman, E. Goni, I. & Prudhomme C. 2002. *Spatial and temporal distribution of groundwater recharge in northern Nigeria.* Hydrogeology Journal, **10**, 205–215.

Ekstrom, K., Prenning, C. & Dladla, Z. 1996. *Geophysical Investigation of Alluvial Aquifers in Zimbabwe.* MSc Thesis. Department of Geotechnology, Institute of Technology, Lund University, Sweden.

ESRI 1996. *ArcAtlas: Our Earth.* Environmental Systems Research Institute, USA.

Farr, J. L., Spray, P. R. & Foster, S. S. D. 1982. *Groundwater supply exploration in semi-arid regions for livestock extension – a technical and economic appraisal.* Water Supply and Management, **6**, 343–353.

Foster, S. S. D. 1984. *African groundwater development – the challenges for hydrogeological science.* Challenges in African Hydrology and Water Resources (Proceedings of the Harare symposium, July 1994), IAHS Publication **144**.

Foster, S. S. D., Chilton, P. J., Moench, M., Cardy, F. & Schiffler, M. 2000. *Groundwater in rural development.* World Bank Technical Paper **463**, The World Bank, Washington DC.

Guiraud, R. 1988. *L'hydrogeologie de l'Afrique.* Journal of African Earth Sciences, **7**, 519–543.

Herbert, R., Barker, J. A., Davies, J. & Katai, O. T. 1997. *Exploiting ground water from sand rivers in Botswana using collector wells.* In: Fei Jin, Krothe, NC (eds) Proceedings of the 30th International Geological Congress China, **22**, Hydrogeology, 235–257.

Hulme, M., Doherty, R., Ngara, T., New, M. & Lister D. 2001. *African climate change: 1900–2100.* Climate Research, **17**, 145–168.

Interconsult 1985. National *Master Plan for Rural Water Supply and Sanitation. Volume 2/2 Hydrogeology.* Ministry of Energy and Water Resources and Development, Republic of Zimbabwe.

Jacobson, P. J., Jacobson, K. M. & Seely, M. K. 1995. *Ephemeral rivers and their catchments: Sustaining people and development in Western Namibia.* Desert Research Foundation of Namibia, Windhoek, 160pp.

JMP 2004. Global water supply and sanitation 2004 report. Joint monitoring programme WHO/UNICEF, World Health Organisation, Geneva.

Jones, M. J. 1985. *The Weathered Zone Aquifers of the Basement Complex Areas of Africa.* Quarterly Journal of Engineering Geology, **18**, 35–46.

Kebede, S., Travi, Y., Asrat, A., Alemayehu, T. & Tessema, Z. 2007. *Groundwater origin and flow along selected transects in Ethiopian rift volcanic aquifers.* Hydrogeology Journal, DOI: 10.1007/s10040-007-0210-0.

Kehinde, M. O. & Loehnert, E. P. 1989. *Review of African groundwater resources.* Journal of African Earth Sciences, **9**, 179–185.

Key, R. M. 1992. *An introduction to the crystalline basement of Africa.* In: Wright E. P. & Burgess, W. G. (eds.) The Hydrogeology of Crystalline Basement Aquifers in Africa. Geological Society London Special Publications, **66**, 29–58.

Kinniburgh, D. G. , Smedley, P. L., Davies, J., Milne, C. J. , Gaus, I., Trafford, J. M., Burden, S., Huq, S. M. I., Ahmad, N. & Ahmed, K. M. 2003. *The scale and causes of the groundwater arsenic problem in Bangladesh.* In: Welch, A. H. & Stollenwerk, K. G. (eds.), Arsenic in Groundwater, Kluwer Academic Publishers, pp 211–257.

Kulkarni, H., Deolankar, S. B., Lalwani, A., Josep, B., Pawar, S. 2000. *Hydrogeological framework of the Deccan basalt groundwater systems, west-central India.* Hydrogeology Journal, **8**, 368–378.

Lillesand, T. M. & Kiefer, R. W. 1994. Remote Sensing and Image Interpretation (3rd Edition). John Wiley & Sons, New York.

Lovell, C. 2000. *Productive water points in dryland areas – guidelines on integrated planning for rural water supply.* ITDG Publishing, Rugby, UK.

MacDonald, A. M, Ball, D. F. & McCann, D. M. 2000. *Groundwater exploration in rural Scotland using geophysical techniques.* In: Robins, N. S. & Misstear, B. D. R. (eds.) Groundwater in the Celtic Regions: studies in hard rock and Quaternary Hydrogeology. Geological Society London Special Publications, **182**, 205–217.

MacDonald, A. M. & Davies, J. 2000. *A brief review of groundwater for rural water supply in Sub-Saharan Africa.* British Geological Survey Technical Report **WC/00/33**.

MacDonald, A. M., Davies, J. & Peart, R. J. 2001. *Geophysical methods for locating groundwater in low permeability sedimentary rocks: examples from southeast Nigeria.* Journal of African Earth Sciences, **32**, 115–131.

MacDonald, A., Davies, J., Calow, R. & Chilton, J. 2005. *Developing groundwater: a guide for rural water supply.* ITDG Publishing, Rugby, UK, 358 pp.

MacDonald A. M., Barker, J. A. & Davies, J. 2008. *The bailer test: a simple effective pumping test for assessing borehole success.* Hydrogeology Journal, DOI10.1007/s10040-008-0286-1.

MacDonald A. M., Kemp, S. J. & Davies, J. 2005b. *Transmissivity variations in the mudstones.* Ground Water, **43**, 259–269.

Macdonald, D. M. J., Thompson, D. M. & Herbert, R. 1995. *Sustainability of yield from wells and boreholes in crystalline basement aquifers.* British Geological Survey Technical Report **WC/95/50.**

Simms, A., Magrath, J. & Reid, H. 2004. *Africa: up in Smoke?* Oxfam, UK.

McNeill, J. D. 1991. *Advances in electromagnetic methods for groundwater studies.* Geoexploration, **27**, 65–80.

Milsom, J. 2002. *Field geophysics, 3rd Edition.* John Wiley and Sons Ltd, London.

New, M. & Hulme, M. 1997. *A monthly rainfall dataset for Africa for 1951 to 1995.* IPCC. University of east Anglia.

Owen, R. J. S. 1989. T*he use of shallow alluvial aquifers for small scale irrigation; with reference to Zimbabwe.* Final report of ODA Project R4239. University of Zimbabwe and Southampton University, UK.

Persits, F., Ahlbrandt, T., Tuttle, M. Charpientier, R., Brownfield, M. & Takahashi, K. 1997. *Maps showing geology, oil and gas fields and geological provinces of Africa.* USGS Open-file report **97-470A**.

Reimann, C., Bjorvatn, K., Frengstad, B., Melaku, Z., Tekle-Haimanot, R., & Siewers, U. 2003. *Drinking water quality in the Ethiopian section of the East African Rift Valley I - data and health aspects.* Science of the Total Environment, **311**, 65–80.

Robins, N. S., Davies, J., Farr, J. L. & Calow, R. C. 2006. *The changing role of hydrogeology in semi-arid southern and eastern Africa.* Hydrogeology Journal, **14**, 1483–1492.

Sami, K., 1996. *Evaluation of the variations in borehole yield from a fractured Karoo aquifer, south Africa.* Ground Water, **34**, 114–120.

Selley, R. C. 1997. *African Basins.* Sedimentary Basins of the World 3 (Series editor: K. J. Hsu) Elsevier, Amsterdam.

TAMS 1996. Water *Resources Management: Policy and Strategies, Final Repor*t. Department of Water Affairs, Ministry of Natural Resources, Lesotho. (2.4.24), Section 1, Part 4 Groundwater Resources.

Taylor, R. G., Howard, K. 2000. *A tectono-geomorphic model of the hydrogeology of deeply weathered crystalline rock: Evidence from Uganda.* Hydrogeology Journal, **8**, 279–294.

Truswell, J. F. 1970. *An introduction to the historical geology of South Africa.* Purnell, Cape Town, RSA.

UNESCO 1991. *Africa Geological Map Scale (1:5,000,000).* 6 Sheets. UNESCO, Paris.

United Nations 2000. United Nations Millennium Declaration. United Nations General Assembly, A/RES/55/2. United Nations. New York.

UNTCD 1988. *Groundwater in North and West Africa.* Natural Resources/Water Series, **18**, United Nations.

UNTCD 1989. *Groundwater in Eastern, Central and Southern Africa.* Natural Resources/Water Series, **19**, United Nations.

Vernier, A. 1993. *Aspects of Ethiopian Hydrogeology.* In: Geology and mineral resources of Somalia and surrounding regions, Ist Agron. Oltremare, Firenze, Relaz e Monogr. **113**, 687–698.

Wikner, T. 1980. *Sand rivers of Botswana.* Results from phase 1 of the Sand Rivers Project sponsored by the Swedish International Development Authority (SIDA).

Woodford, A. C. & Chevallier, L. 2002. *Regional characterisation and mapping of Karoo fractured aquifer systems – an integrated approach using geographical information system and digital image.* Water Research Commission report **653/1/02**, 192 pp.

World Bank, 2000. *African development Indicators 2000.* World Bank, Washington DC.

Wright, E. P. 1992. *The hydrogeology of crystalline basement aquifers in Africa.* In: Wright E. P. & Burgess, W. G. (eds.) The hydrogeology of crystalline basement aquifers in Africa. Geological Society London Special Publications, **66**, 1–27.

CHAPTER 10

Groundwater as a vital resource for rural development: An example from Ghana

P. Gyau-Boakye, K. Kankam-Yeboah, P.K. Darko & S. Dapaah-Siakwan
Water Research Institute, Accra, Ghana

A.A. Duah
Groundwater Group, Department of Earth Sciences, University of the Western Cape, Bellville, Republic of South Africa

ABSTRACT: Rural communities in Ghana are defined as those with less than 5000 inhabitants. Rural people make up 56.6% of Ghana's total population of 18.9 million. Most of these rural communities have traditionally relied on a wide variety of water sources for their water supply: dug-wells, ponds, dug-outs, dam impoundments, streams, rivers and rainwater harvesting. Many of these sources, particularly those based on surface water resources, are contaminated and a major contributor to the disease and poverty endemic in many rural communities. Presently only about 52% of the rural inhabitants have access to improved safe water supplies – which are mainly derived from groundwater sources. In Ghana, developing groundwater is seen as the most economic way of providing safe water supplies to the dispersed rural settlements. To improve the standard of living and boost economic activities in the rural areas, the government has drawn up a policy of supplying most of the rural communities with potable water. This is in line with the United Nations Millennium Development Goal of reducing by half the population of the world without access to potable water by the year 2015. The government of Ghana has set itself a more ambitions task of covering about 85% of the rural communities with potable water by 2015.

1 INTRODUCTION

The population of Ghana increased from 12.3 million in 1984 to 18.9 million in 2000. The rural population, i.e. people living in communities with less than 5000 inhabitants also rose from 8.4 million to 10.7 million within the same period. Based on the year 2000 population census about 56.6% of the population of Ghana lives in rural communities. Most rural settlements have traditionally relied on raw waters from such sources as streams, rivers, lakes, ponds, dug-outs and impoundment reservoirs. Some of these surface water sources are heavily polluted, resulting in water-borne and water-related diseases such as diarrhoea, guinea worm, bilharzias, etc.

Following the 'Water Resources Sector Studies', commissioned by the Government of Ghana in 1969–70 (Nathan Consortium for Sector Studies, 1970), it became the official policy that for the supply of potable water, communities below 500 inhabitants are to be helped to construct hand-dug-wells. Supplies to communities with 500–2000 inhabitants are to be assisted by means of hand-dug-wells or boreholes fitted with handpumps, whereas communities of population between 2000 and 5000 are to be supplied with reticulated systems based mostly on groundwater sources. Thus, the policy relied heavily on developing groundwater resources. Only where groundwater was not readily available, were rainwater harvesting, spring sources and simple techniques for obtaining surface water from dam-impoundments to be tried.

Certain features make groundwater attractive as a source of potable water supply in Ghana (Quist *et al.*, 1988). Firstly, there are aquifers in several areas of the country that can frequently be tapped at shallow depths close to the water demand centres in response to the dispersed nature of rural settlements. Some of these aquifers (e.g. the Birimian System) have been moderately assessed and their characteristics are fairly well known. Secondly, water stored in aquifers is for most part protected naturally from evaporation and pollution, and well yields are in many cases adequate, offering water supply security in regions prone to protracted droughts as is common in the northern parts of Ghana. Thirdly, with adequate aquifer protection, groundwater has excellent microbiological and chemical quality, and requires minimal or no treatment at all. Lastly, the capital cost of groundwater development as opposed to the conventional treatment of surface waters is relatively modest and the resource lends itself to flexible development capable of being phased in with rising demand.

In the late 1980s, the Government developed a policy (Ghana Vision 2020, 1998), which aimed at supplying all the rural communities with potable water mainly from groundwater sources by the year 2020. Also, at the United Nations Millennium Summit held between 6th and 8th September 2000 at the United Nations Headquarters in New York, 189 Heads of State adopted the Millennium Development Goals (MDGs), which set clear, numerical, time-bound targets for making real progress by 2015 in tackling the most pressing issues facing developing countries. One of the MGDs is to cut by half the proportion of people without sustainable access to safe drinking water and sanitation by 2015. In line with this goal, the Strategic Investment Programme (SIP) of the Ghana Poverty Reduction Strategy (II) is aiming at an ambitious target of achieving 85% coverage in both water supply and sanitation by the year 2015. In Ghana, most of the population targeted by these goals are in the rural communities.

According to the Community Water and Sanitation Agency (CWSA), the national coverage for potable water supply in both rural communities and small towns was 51.7% at the end of 2004 (CWSA, 2004). At the end 2004, 13,196 boreholes and 1344 hand-dug-wells had been constructed and were providing sources of potable water to rural communities and small towns. The available information also indicated that the national coverage for potable water supply to the rural communities increased from 41% in 1984 to 52% in 1998 and declined to 44% in 2000 and increased slightly to 46.3% at the end of 2003 and 51.7% at the end of 2004 (CWSA, 2004). The decline in the national coverage figures for potable water supply to the rural communities from 1998 to 2000 was due to the reclassification of some small towns (some with populations of 15,000) as rural. The low rural potable water coverage of 52% helps to explain the endemic diseases and inherent poverty in rural areas.

Since the new rural water supply schemes are to be based mainly on groundwater resources, it is necessary to have a proper assessment of the country's groundwater potential to ensure efficient utilization of the groundwater resources and the management of aquifers, hand-dug-wells and boreholes.

2 BACKGROUND TO GHANA

Ghana lies on the west coast of Africa with a total land area of 238,000 km^2 between latitude 04°30'N and 11°25'N and longitude 01°E and 03°30'W. The physiographic features are made up of the coastal plains in the south, a forest dissected plateau in the southwest, savannah plains in the north, the Voltaian basin covering large portions of the northeast to the central parts and the Togo-Buem mountain ranges stretching from the coast to the north-east.

The climate of Ghana is influenced by three air masses namely, the South-West Monsoon, the North-East Trade Wind (Tropical Continental Air Mass) and the Equatorial Eastern. The warm but moist South-West Monsoon which originate from the Atlantic Ocean and the warm, dry and dusty Tropical Continental Air Mass (Harmattan) from the Saharan Desert approach the tropics from opposite sides of the equator and flow towards each other into a low pressure belt known as the Inter Tropical Convergence Zone (ITCZ) (Ojo, 1977). The slow and irregular north-south oscillations of the ITCZ gives rise to the regime of wet and dry seasons. The wet season in the southern sections of Ghana is characterized by two main rainfall regimes, i.e., double maxima whilst the northern sections experience single rainfall regime in a year. The extreme southwestern portion of Ghana is the wettest part of the country with more than 2000 mm/a (Figure 1). Rainfall, which mainly recharges the aquifers, generally decreases towards the north and southeastern sections of the country. The driest part of the country is found in the southeast coastland plains where the mean annual rainfall is approximately 800 mm.

Mean monthly temperature over the country never falls below about 25°C and open water (pan) evaporation is generally high and ranges from about 1200 mm per year in the southwest to more than 2600 mm in the north. Relative humidity is high along the coast (>95% during the night and early morning), but lower in the north (<30%) where the area comes under the influence of the dry Harmattan.

There are three river basin systems in Ghana: the Southwestern Rivers System, which is comprised of Pra, Ankobra, Tano, and Bia rivers; the Coastal Rivers System, which includes the Kakum, Ochi-Nakwa, Ochi-Amisa, Ayensu, Densu, and Tordzie rivers and; the Volta Rivers System, made up of the Black and White Volta, and the Oti rivers (Figure 2).

The Volta Rivers System covers about 70% of the country and contributes about 64.7% of the actual runoff from Ghana. The Southwestern Rivers System covers about 22% of Ghana and contributes 29.2% of the actual runoff, while the Coastal Rivers System covers about 8% of the country and contributes 6.1% of the actual runoff. The total annual mean runoff for the country is 54.4 billion (10^9) m^3 of which the runoff originating in Ghana alone is 39.4 billion m^{3}representing 68.6% of the total annual runoff. The rest of 15.0 billion m^3 originate from outside Ghana's territory (WARM, 1998).

3 GEOLOGY AND HYDROGEOLOGY

3.1 *Hydrogeological provinces – an overview*

Two major hydrogeological provinces exist in Ghana (Figure 3). These are: (1) the Basement Complex composed of Precambrian crystalline igneous and metamorphic rocks, and (2) Palaeozoic sedimentary formations. Minor provinces consist of the Cenozoic, Mesozoic

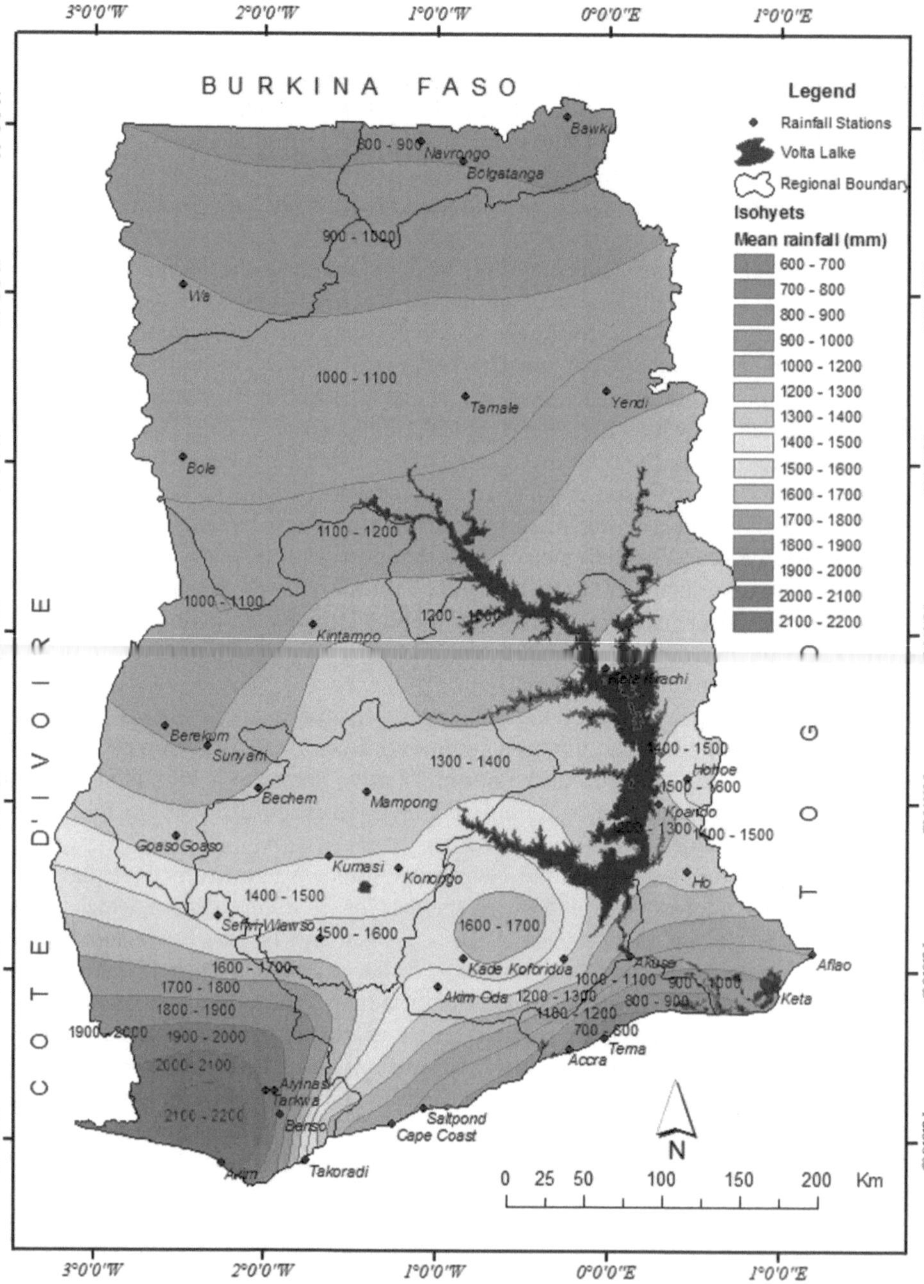

Figure 1. Mean annual rainfall in Ghana.

and Palaeozoic sedimentary strata along narrow belts on the coast; and Quaternary alluvium along major stream courses.

The Basement Complex underlies about 54% of the country and is further divided into sub-provinces on the basis of geologic and groundwater conditions (Gill, 1969). Generally,

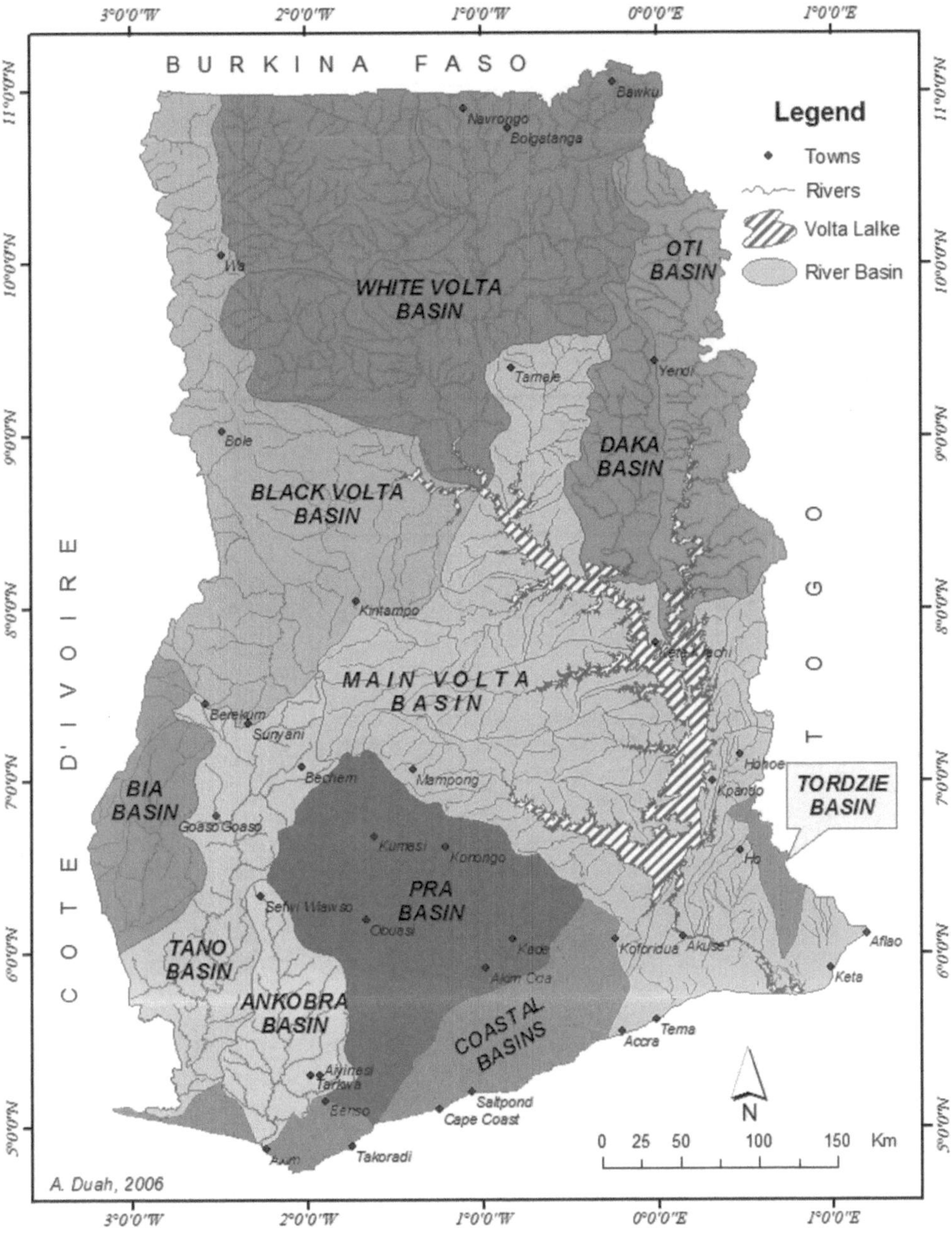

Figure 2. The main river basins in Ghana.

these sub-provinces include metamorphosed and folded rocks of the Birimian System, Dahomeyan System, Tarkwaian System, Togo Series and Buem Formation (Figure 4) which comprise mainly gneiss, phyllite, schist, migmatite, granite-gneiss and quartzite. Large masses of granite have intruded the Birimian rocks.

The Palaeozoic sedimentary formations, locally referred to as the Voltaian Formation, underlie about 45% of the country and consist mainly of sandstone, shale, arkose,

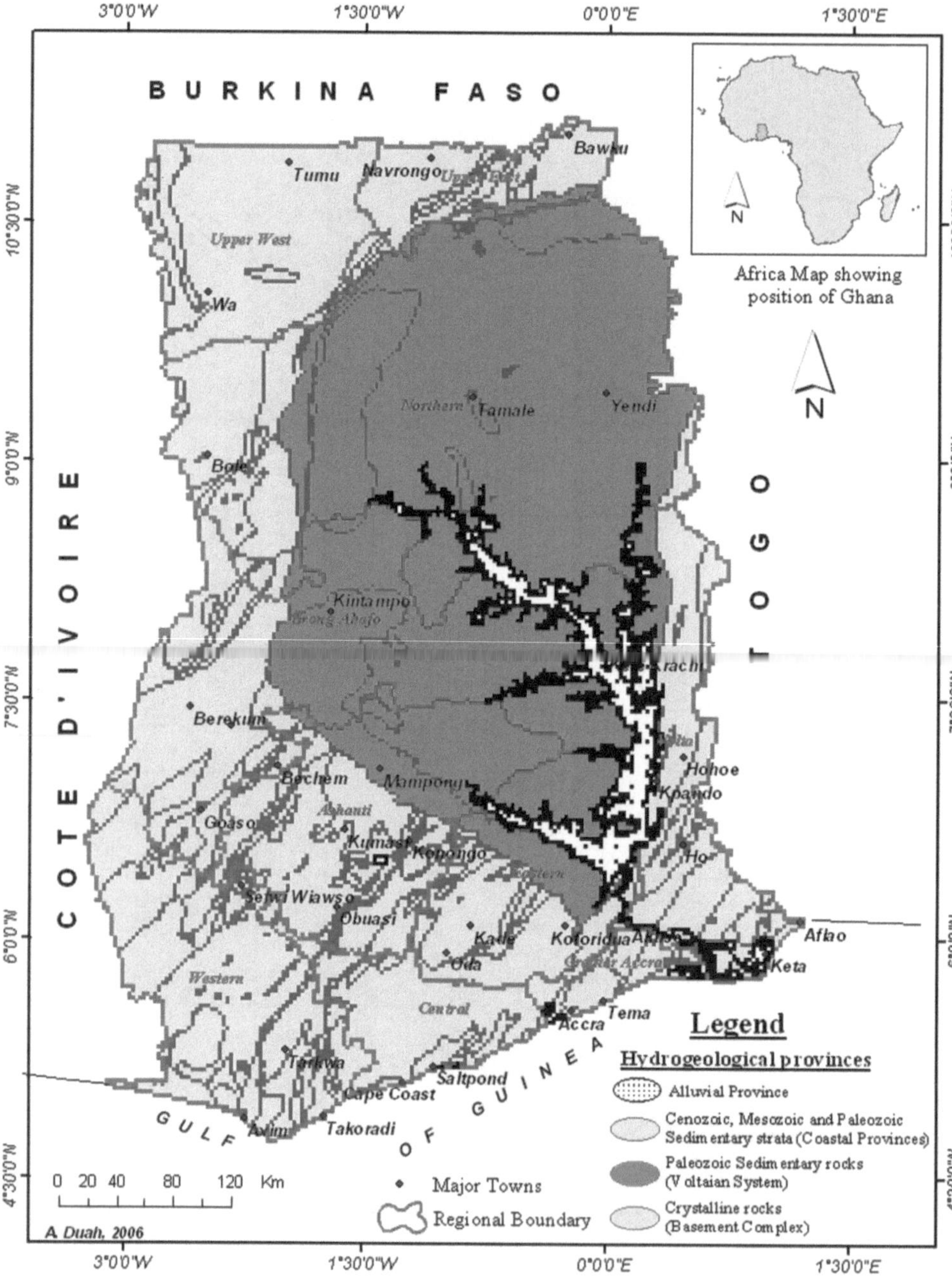

Figure 3. Hydrogeological provinces and river systems.

mudstone, sandy and pebbly beds and limestone. The Voltaian Formation is further subdivided on the basis of lithology and field relationships into the following sub-provinces (Junner & Hirst, 1946; Soviet Geological Survey Team, 1964–1966): (1) Upper Voltaian (massive sandstone and thin-bedded sandstone); (2) Middle Voltaian

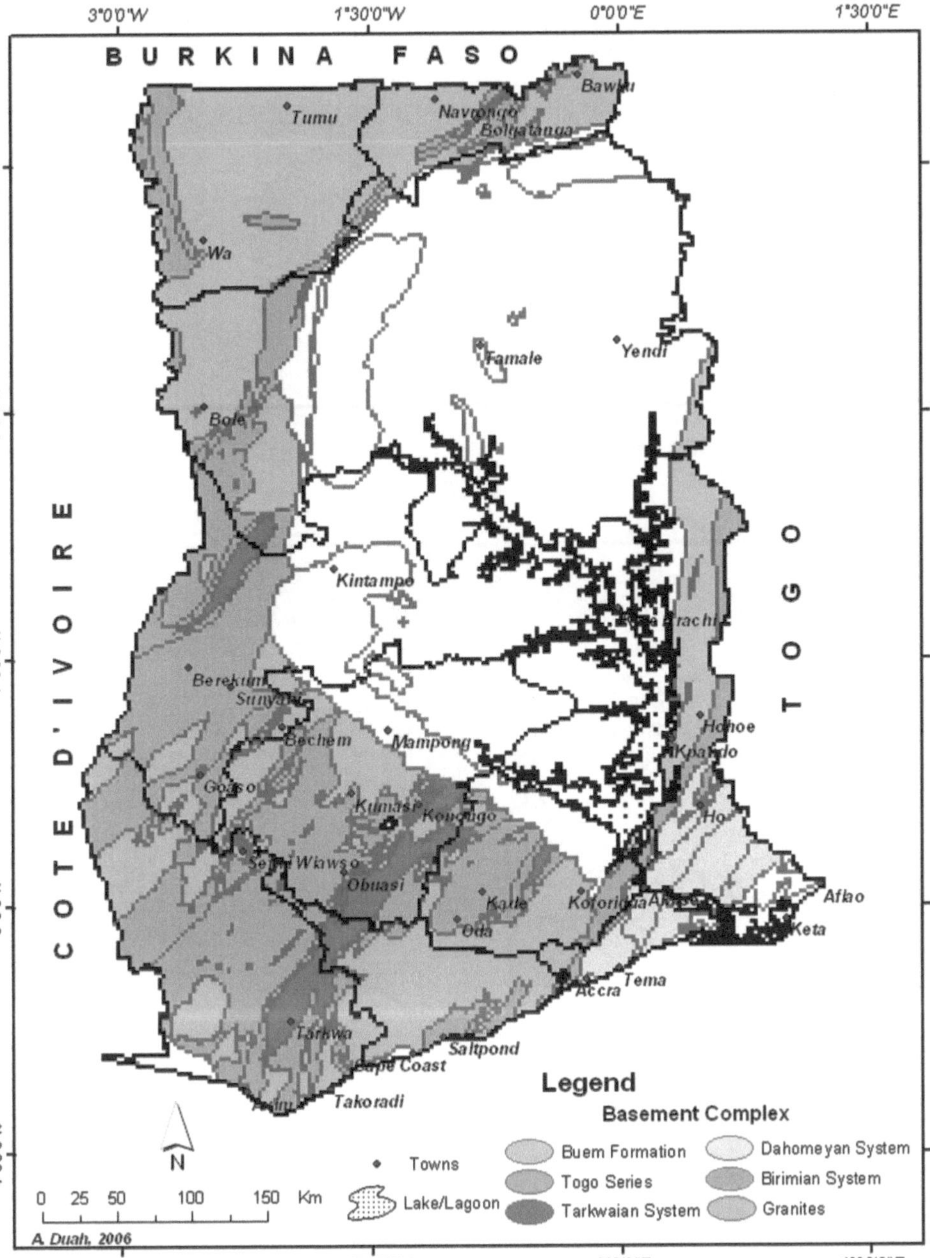

Figure 4. Hydrogeological sub-provinces of the Basement Complex (modified from Geological Survey of Ghana 1969).

(Obosum and Oti Beds); and (3) Lower Voltaian. Their distribution is shown in Figure 5.

The remaining 1% of the rock formations are made up of two coastal provinces (the Coastal Block-Fault Province and the Coastal-Plain Province) and the Alluvial Province.

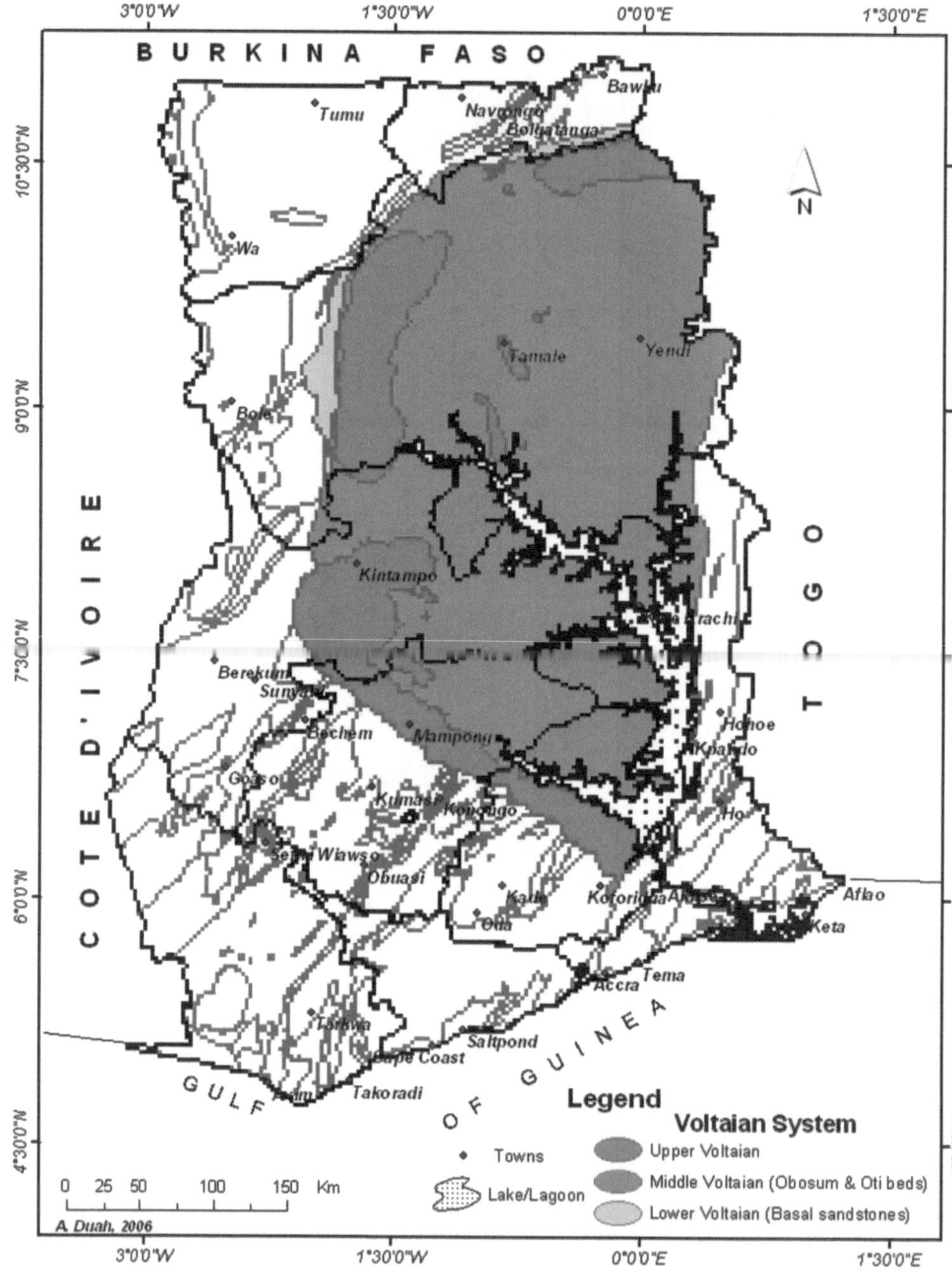

Figure 5. Hydrogeological sub-provinces of the Voltaian System.

The Coastal Block-Fault Province consists of a narrow discontinuous belt of Devonian and Jurassic sedimentary rocks that have been broken into numerous fault blocks and are transected by minor intrusions (Kesse, 1985). The Coastal-Plain hydrogeological Province is underlain by semi-consolidated to unconsolidated sediments ranging from Cretaceous to

Holocene in age in southeastern Ghana and in a relatively small isolated area in the extreme southwestern part of the country. The Alluvial hydrogeological Province includes narrow bands of alluvium of Quaternary age, occurring principally adjacent to the Volta River and its major tributaries and in the Volta delta.

3.2 *Precambrian crystalline igneous and metamorphic rocks (basement complex)*

3.2.1 *Upper and lower Birimian system and associated granite*

This region, which extends from the north through the mid-west to the south-western parts of the country, is made up of rocks of the Birimian System and associated intrusions which are of Paleoproterozoic era (1800–2100 Ma) (Figure 4). The Birimian System consists of a great thickness of isoclinally folded, metamorphosed sediments intercalated with metamorphosed tuff and lava. The tuff and lava are predominant in the upper part of the System, whereas the sediments are predominant in the lower part. These dominantly argillaceous sediments were metamorphosed to schist, slate and phyllite. The entire sequence is intruded by batholithic masses of granite and gneiss.

The granite and gneiss associated with the Birimian rocks are of considerable importance in the water economy of Ghana because they underlie extensive and usually well-populated areas. The rocks are not inherently permeable, but permeability and porosity have developed as a result of fracturing and weathering. In some areas, weathered granite and gneiss form fractured rock aquifers. Major fault zones also are favourable locations for groundwater storage. Depending on the degree of local weathering, the granite and gneiss can form low-lying areas or massive, poorly jointed inselbergs that rise above the surrounding lowlands. In the northwestern sections, around the Wa District (Figure 4), the zone of weathering can be greater than 130 m thick, but in the eastern districts, the zone of weathering is much thinner.

In the Wa Granite area, the proportion of successful boreholes yielding >0.12 l/s is about 85%. The yields obtained from these boreholes average 1.5 l/s but some have been recorded greater than 6 l/s.

Boreholes tapping the granites in the Winneba area of the southwestern section of the hydrogeological sub-province have lower yields. The success rate is about 68% and the average yield of successful boreholes is around 0.11 l/s.

In the mid-sections around Kumasi and surrounding areas, the boreholes have higher yields than in other areas; the average is 2.6 l/s. The rainfall around Kumasi and surrounding areas is greater than around the Wa and Winneba areas; hence the zone of weathering is thicker and well yields are greater. Most of the boreholes in the Birimian System are fitted with handpumps and have an average depth of about 35 m. In the granite, where it is more difficult to find water, boreholes are drilled to an average depth of 60 m.

The phyllite, schist, slate, greywacke, tuff, and lava of the Birimian System are generally strongly foliated and fractured. Where these rocks crop out or are near the surface, considerable water may percolate through them. Boreholes tapping the Upper and Lower Birimian rocks have an average yield of about 3.5 l/s. In the mid-western part of the Brong-Ahafo Region (Figure 3), boreholes in the Birimian rocks have similar yields (mean 3.2 l/s). The success rate for boreholes tapping the Upper Birimian rocks in the Western Region is about 75%. There are few boreholes in the Upper Birimian rocks; the greatest number of high-yielding boreholes is in the Enchi and Bogoso areas (see Figure 3 for locations).

Around Axim, near the coast in the Western Region, the boreholes have much lower yields (WRRI, 1996a).

3.2.2 *Dahomeyan system*

Rocks of the Dahomeyan System of Neoproterozoic era (550 Ma) underlie the Accra Plains and southern parts of the Eastern and Volta Regions and extend from Ho to Accra (Figure 3). The rocks consist mainly of crystalline gneiss and migmatite, with subordinate quartz schist, biotite schist and sedimentary rocks. The gneiss is generally massive with few fractures. The two main varieties are silicic and mafic gneisses, which weather, respectively, to slightly permeable clayey sand and nearly impermeable calcareous clay. The generally impervious nature of the weathered zone and massive crystalline structure of the rocks limit the yields that can be obtained from hand-dug-wells or boreholes.

An analysis of about 200 boreholes drilled in the Accra plains and elsewhere in this hydrogeological province indicates that the success rate for developing wells is about 36%, based on the use of geophysical surveys as an aid to site selection. Borehole yields average 0.3–0.8 l/s. With improved site-selection methods, including the detailed study of aerial photographs and the use of more sophisticated geophysical surveys, borehole yields could probably be increased to as much as 3 l/s.

3.2.3 *Tarkwaian system, Togo series and Buem formation*

Although different in age, these three rock formations are similar in lithology. The Tarkwaian rocks of Paleoproterozoic era (1650–1850 Ma) comprise slightly metamorphosed, shallow-water, sedimentary strata, chiefly sandstone, quartzite, shale and conglomerate, resting unconformably on and derived from rocks of the Birimian System. The rocks are intruded by thick laccoliths or dykes and sills of epidiorite and, like the Birimian rocks, are folded along axes that trend northeast. The rocks of the Tarkwaian System are not as extensive as the Birimian rocks. The largest area lies in a band (Figure 4) that extends from Konongo to Tarkwa (Figure 3).

The Togo Series of Neoproterozoic era (550 Ma) consists of metamorphosed arenaceous and argillaceous sedimentary strata. The rock types include indurated sandstone, quartzite, quartz schist, shale, phyllite, sericite schist and some limestone. These rocks are highly folded and form the chain of hills known as the Akwapim-Togo Ranges that extend northeast from the coast near Accra to the Togo border (Figure 4). The quartzites and related rocks commonly form hills, and the shale and phyllite occur in intervening valleys.

The Buem Formation also of Neoproterozoic era (550 Ma) consists of a thick sequence of shale, sandstone and volcanic rocks with subordinate limestone, tillite, grit and conglomerate. The rocks underlie an elongated area of very considerable size on the western side of the Akwapim-Togo Ranges, including the areas around Kpandu, Jasikan and Hohoe and extending north-east to the Togo frontier (Figure 4). The sandstone overlies the basal beds of shale and the conglomerate and the tillite overlie the sandstone. Rocks of volcanic origin form the upper part of the Buem Formation and include lava, tuff and agglomerate interbedded with shale, limestone and sandstone.

The three rock groups, the Tarkwaian System, Togo Series and Buem Formation have negligible primary porosity or permeability but contain openings along joints, bedding, and cleavage planes. Where these openings are extensive, good supplies of groundwater can be developed from boreholes. Springs frequently occur along the flanks of hills where quartzites are in contact with argillaceous rocks of the valleys, such as in the Akwapim-Togo

Ranges. Weathering of quartzites yields an unconsolidated alluvium of sand and quartzite fragments that are a source of good supplies of groundwater from shallow wells. Generally, these Neoproterozoic rocks have a relatively good potential for groundwater development, the most favourable areas being the valleys where rocks are highly fractured.

In the Volta Region (Figure 3), the success rate for obtaining water from boreholes in the Togo Series and Buem Formation is about 88%. The average yield from these boreholes is about 2.5 l/s although can be greater than 6 l/s. The higher yielding boreholes in this area probably tap large fractured systems or fault zones. Boreholes are drilled to an average depth of 65 m in sandstone.

3.3 *Palaeozoic sedimentary formations (Voltaian system)*

This hydrogeological province is underlain by rocks of probable Cambrian to Silurian age. The province occupies approximately 45% of the country and extends northeastward beyond the borders of Ghana almost to the Niger River. The rocks underlie the central and eastern parts of the Northern Region, the central and eastern parts of the Brong-Ahafo Region, the northeastern parts of the Ashanti and Eastern Regions and northern part of the Volta Region. (Figure 3). The Voltaian System consists of Lower, Middle, and Upper sub-provinces (Figure 5).

3.3.1 *Lower Voltaian*

The Lower Voltaian sub-province occupies a narrow band in the western and northern parts of the area (Figure 5). This sub-province is underlain by the Basal sandstone, consisting mainly of quartz-sandstone and pebbly grits, and grits with ripple marks and galls. Even though weathering of the Basal sandstones produces sandy superficial deposits and is probably well jointed in many places, it has virtually not been explored for groundwater because the area is relatively sparsely inhabited.

3.3.2 *Middle Voltaian*

In the Middle Voltaian sub-province, the rocks, which occur in a large sedimentary basin and include the Obosum and Oti Beds, form the most extensive sedimentary sequence in Ghana. The sub-province consists of interbedded mudstone, sandstone, arkose, conglomerate and some sandstone. The rocks are mainly flat lying or gently dipping. The rocks are generally well consolidated and are not inherently permeable except for a few places, such as the long belt between Kete Krachi and Sang (Figure 5), where the strata may be permeable locally. Shale crops out in the central part of the sub-province. Where sandstone crops out and the relief is low, the shale lies at a shallow depth and is generally capped by a few metres of laterite.

In the wet season, large areas are covered by shallow ephemeral lakes or ponds that disappear during the dry season. The lack of springs on permanent tributary streams indicates the absence of shallow groundwater. The success rate for obtaining water from boreholes in the Northern Region is about 56% and the average borehole yield is 1 l/s. Saline water is fairly extensive in the northern part of the basin. Salt beds are known to crop out in the Tamale and Daboya areas. The boreholes in the Voltaian Sandstones of the Kete Krachi area have much higher yields than those in the north. The average yield obtained in this area is 2 l/s. Also, some of the shallow boreholes in this area have artesian flows. The hydrogeology of the Voltaian Rocks in the Afram Plains in the south of the basin is described in Cobbing & Davies (2008).

3.3.3 *Upper Voltaian*

The Upper Voltaian sub-province lies to the south, west and the north of the Middle Voltaian sub-province (Figure 5). The sandstones in the Upper Voltaian sub-province, particularly those to the west and south of the Middle Voltaian sub-province, store considerable quantities of groundwater, which discharges in springs along joints and the bedding planes at many localities. These springs maintain many permanent streams that rise in the sandstone hills. Along the escarpment between Wenchi and Anyaboni (Figure 5), the average yield of boreholes is about 2 l/s.

3.4 *Cenozoic, Mesozoic and Palaeozoic sedimentary strata (Coastal provinces)*

3.4.1 *Coastal block-fault province*

The Coastal Block-Fault hydrogeological province is underlain by rocks of the Accraian and Sekondian Formations of Devonian age and the Amisian Formation of probable Jurassic age. The rocks have been subjected to post-depositional igneous activity and major block faulting. The Devonian rocks, which underlie Accra, Takoradi and Sekondi, crop out along the coast between Sekondi and Cape Coast. The rocks at Accra include sandstone, grit, and shale, whereas the Sekondian Formation near Sekondi and Takoradi consists mainly of sandstone and shale with conglomerate, pebble beds, grit and mudstone. The rocks of both formations unconformably overlie a complex of granite, gneiss and schist of Precambrian age. The Amisian Formation, which crops out near the mouth of the Amisa River, is composed of poorly sorted, semi-consolidated sedimentary rocks, largely pebbly and bouldery shale and sandstone deposited in a freshwater environment.

Boreholes tapping the Accraian, Sekondian, and Amisian Formations yield on the average about 1 l/s.

3.4.2 *Coastal plain province*

In southwestern Ghana, from near Esiama to the Ivory Coast frontier (also known locally as the Tano Basin; Figure 3), Cretaceous to Lower Tertiary sedimentary rocks of the Coastal Plain extend inland 8–24 km. The sedimentary sequence includes a thick section of alternating sand and clay with occasional thin beds of gravel and fossiliferous limestone. The limestones are known to have an oil and gas potential. Seepages of oil and gas have been reported at several places along the coast near Bonyeri, Techinta, Tobo and Nauli (Figure 3). Groundwater supplies from boreholes in the Cretaceous to Lower Tertiary sediments between Esiama and Half-Assini are obtained largely from the upper 100 m of the section with an average yield of 3.5 l/s.

Southeastern Ghana, in the vicinity of Keta and other districts, is also underlain by Cretaceous to Lower Tertiary consolidated and semi-consolidated marine sedimentary strata. In this area, locally known as the Keta Basin, these strata are covered by younger continental deposits. Two limestone horizons have been traced in the subsurface of the Keta Basin. The upper limestone is equivalent to limestone penetrated by boreholes further inland at Anyako (Figure 3), northwest of Keta. In boreholes at Anloga, Anyanui, and Ada, however, this limestone was apparently not penetrated, even at depths of 600 m. These lateral discontinuities have been considered by Akpati (1975).

The lower limestone aquifer in the Keta area probably represents a single hydrogeological unit that is recharged from intake areas at higher altitudes farther inland. Two recharge areas have been identified: one is centred at Avenopedu and Agbodrafo on the west, and the other at Dzodze and Ehi on the east (Figure 3). These recharge areas are separated by an area of negative head from Wuti to Afife, where flowing wells are common, as typified by the Anyako Well (Akiti, 1977). A third possible recharge area is at the Mono River in Togo, which flows from north to south along the limestone outcrop (Akiti, 1977) and at least 112 km northeast of the Keta area.

Records of boreholes in the limestone aquifer in the Keta area indicate that the average yield of boreholes is 3.5 l/s. Boreholes tapping limestone aquifers along the coast from Aflao to Keta and around Anyako yield 6 l/s on average, and range from 1–15 l/s.

3.5 *Alluvial province*

Surface deposits of Quaternary age are generally not extensive in Ghana. Locally, however, relatively thick deposits of permeable water-bearing alluvium are present in valleys of the larger streams, such as the Volta River and its tributaries (Figure 3). Extensive but relatively thin alluvial deposits also occur on the delta of the Volta River in southeastern Ghana. Although the alluvial deposits have not yet been developed to any significant extent for water supply, they have considerable potential. In areas where the deposits are permeable, relatively thick, and located adjacent to perennial streams, they offer desirable locations for shallow wells, particularly for irrigation purposes.

4 SUMMARY OF BOREHOLE YIELDS

Analyses of borehole yields have been made for the various geological formations in the country (WRRI, 1996a) using recorded borehole yields at the time of drilling. The least-explored geological unit is the Voltaian System, due to low population density in the area where it occurs. Based on that data, the Water Resources Research Institute (WRRI) of the Council for Scientific and Industrial Research (CSIR) prepared a borehole-yield map of Ghana (Figure 6) (WRRI, 1996b). This map indicates the borehole yields to be expected in any particular area and Table 1 gives a summary of data on borehole yields for the various hydrogeological units. The estimates of borehole yields (Table 1) sometimes differ from those of the map (Figure 6) for the following reasons: (1) the data sets are not based on the same number of boreholes, and because the degree of weathering and fracturing differs within each geological formation, the choice of boreholes could make a difference in yield analysis; and (2) the boreholes are not evenly and uniformly distributed in the country, resulting in the need to interpolate and extrapolate some of the data. The map and table are based on estimated yields from time of drilling and may not represent the true long-term sustainable yield of the boreholes.

In 1998 the Ministry of Works and Housing who are also in charge of water resources, conducted a study on the number of boreholes drilled in the country and found out that about 11,500 boreholes had been drilled nationwide. The yields, static water-levels and other vital information from these boreholes were well documented. As shown in Table 1 the yields of successful boreholes in the Basement Complex range from 0.1 to 8 l/s with an average of 0.8 to 3.5 l/s for the various sub-provinces. Borehole yields in the Voltaian System range

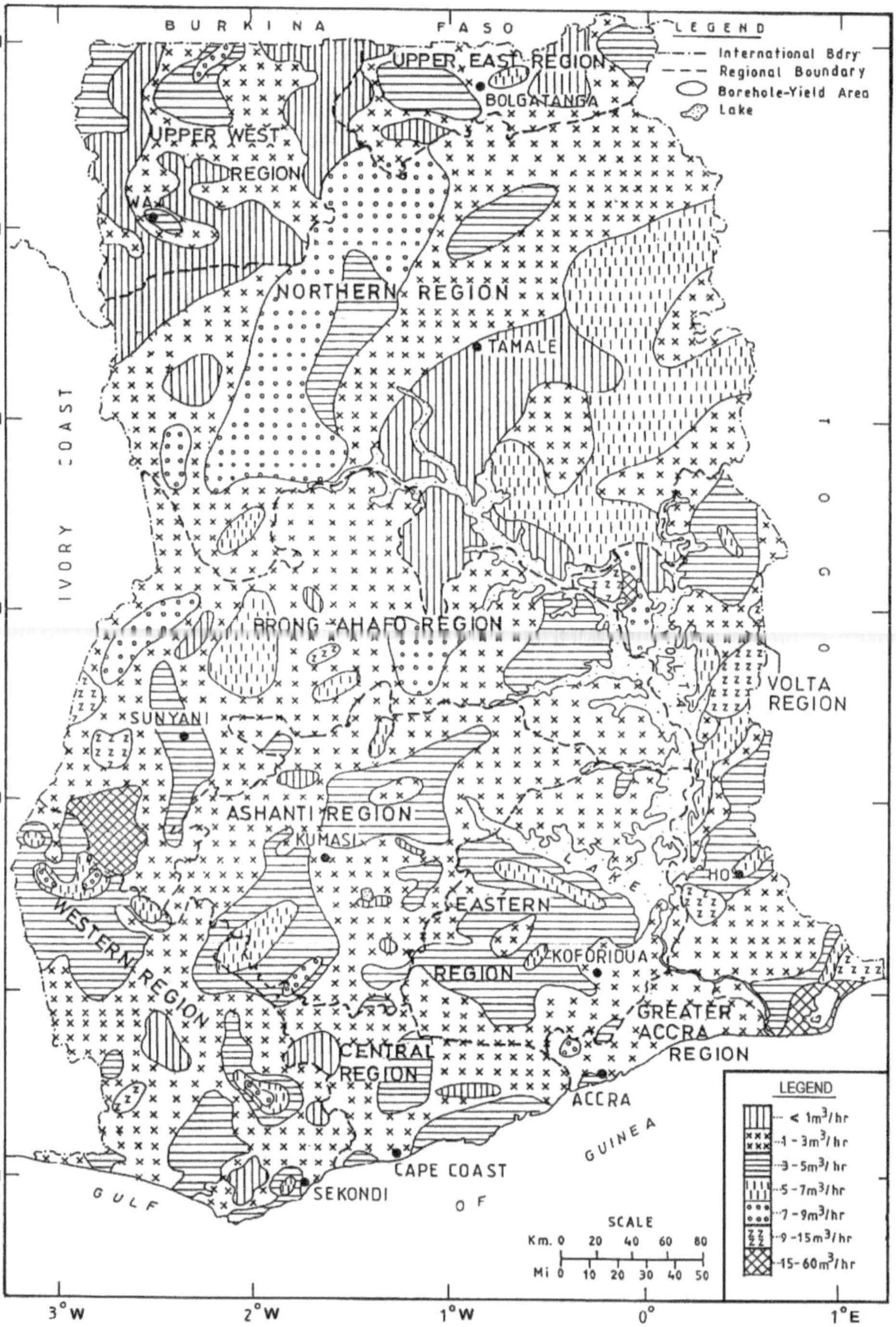

Figure 6. Distribution of borehole yields (WRRI, 1996b).

Table 1. Summary of borehole yields of hydrogeological provinces and sub-provinces.

Hydrogeological province and sub-province	Borehole completion success rate (%)	Yield of successful boreholes (l/s)	
		Range	Mean
Basement Complex			
Lower Birimian System	75	0.1–8	3.5
Upper Birimian System	77	0.1–7	2.1
Dahomeyan System	36	0.3–0.8	0.8
Tarkwaian System	83	0.3–6	2.4
Togo Series	88	0.2–7	2.6
Buem Formation	88	0.2–7	2.6
Voltaian System			
Lower Voltaian	55	0.3–2.5	2.4
Middle Voltaian	56	0.1–2.5	1.7
Upper Voltaian	56	0.3–2.5	2.4
Cenozoic, Mesozoic, and Palaeozoic			
Coastal Block-Fault Province	36	0.3–1.4	1.1
Coastal-Plain Province	78	1.3–15	4.3
Alluvial Province	67	0.3–4	3.3

between 0.1 and 2.4 l/s, with an average of 1.7–2.4 l/s for the various sub-provinces. In the Coastal Provinces, the recorded yield of successful boreholes vary from 0.1 to 15 l/s, with an average of 1.1–4.3 l/s for the sub-provinces. There are over 15,000 boreholes drilled in the country currently.

There are few reliable pumping tests in Ghana to estimate aquifer properties. The following is a summary of the available data. The Voltaian sedimentary have mean transmissivity 12 m^2/d (range 0.3–270 m^2/d) (Darko, 2005). Granites have mean transmissivity 6.6 m^2/d with a (range of 0.3–110 m^2/d). The Birimian and Tarkwaian rocks are characterized by mean transmissivity of 7.4 m^2/d (range $<$0.1–2119 m^2/d) and 8 m^2/d (range 0.9–43 m^2/d), respectively. The Dahomeyan rocks have mean transmissivity of 4.5 m^2/d with a range of 0.3–42 m^2/d.

Because most of the boreholes in Ghana are fitted with handpumps and are meant to supply rural communities, a successful borehole yield (airlift) is considered to be at least 13 l/m (0.22 l/s) or more. This minimum yield per borehole is designed to meet the demand of rural communities with populations of 200–2000 in accordance with the Government policy of providing about 25 l/person/day. In a few localities where sufficiently high yields are difficult to obtain close to rural settlements, borehole yields as low as 0.1 l/s are developed for rural communities to replace the polluted traditional sources such as streams, rivers, lakes, ponds, dug-outs and impoundment reservoirs. In some low yielding boreholes, the technology of hydro-fracturing is sometimes applied, where identified fractures in these boreholes are subjected to pressurized air and water to help open up the fractures to improve the yield. This technology has, however yielded mixed results and further work needs to be done to ascertain its viability or otherwise.

5 GROUNDWATER EXPLORATION AND EXPLOITATION

Groundwater in Ghana is abstracted for supply through either hand-dug-wells with or without handpumps, boreholes with hand or electric pumps depending on the yield, or protected springs. The available statistics (CWSA, 2004) showed that by the end 2004, a total of 13,196 boreholes and 1344 hand-dug-wells had been constructed and were providing sources of potable water to 51.7% of the rural population in Ghana. The Government's rural water supply policy of reliance on groundwater, is due to feasibility and economic value. Studies in Ghana have indicated that the cost of potable water supply based on surface water sources was about twice that of groundwater based systems for rural communities (Bannerman, 1975). Similar studies undertaken in the Imo state of Nigeria arrived at the same conclusion (Uma & Egboka, 1988). This does not mean all rural communities need to be supplied with groundwater. Some of the rural communities within the urban environs may benefit from extension of urban water supply systems, while due to geological limitations others may have to resort to conjunctive use of both surface and groundwater resources.

In many parts of the country, particularly where the nature of the rocks make it difficult to locate aquifers, there are problems in selecting suitable sites for drilling. The main methods used for groundwater investigation, often used collectively, in Ghana include:

- review of archival reports;
- interpretation of topographic maps;
- interpretation of geological and structural maps;
- survey of existing boreholes and other water sources, including terrain evaluation; and
- discussions with residents of communities.

In addition, certain ecological and physical indicators are helpful in locating aquifers. Some of these indicators include:

- the presence of baobab and other large trees in the north of the country;
- alignment of big trees or clusters of trees in the forest areas may indicate either a fracture or localized aquifer;
- ant hills in the Accra plains may be taken as probable areas of shallow groundwater, which can be developed for hand-dug-wells; and
- valleys and low-lying areas.

Geophysical investigations are used in conjunction with the above factors to locate fractures within the sub-surface. The common geophysical methods used in Ghana are electrical resistivity survey, electromagnetic survey and vertical electrical sounding.

Because groundwater prospecting, until the late 1980s, had been done without the benefit of detailed studies of aerial photographs and the use of geophysical surveys, development of groundwater was restricted to the areas where it was most likely to achieve success, such as the weathered parts of the Birimian Systems and limestone aquifers of the southeastern part of the country. Consultants and groundwater developers avoided the areas where it was more difficult to develop adequate water supplies such as the rocks of the Dahomeyan and the Voltaian Systems. These formations are among the least explored in the country probably because of the low population densities where they occur and their hydrogeological characteristics. As has been pointed out earlier, in many parts of the country, particularly where the lithological nature of the formations makes it difficult to locate aquifers, there is a problem in selecting suitable sites for drilling. These constraints together with lack of information on aquifer geometry and recharge characteristics of aquifers are manifested in the low yields and poor success of some boreholes.

Part of this problem is being overcome by the increased use of detailed aerial photographs and geophysical surveys to interpret hydrogeological conditions and a more scientific approach to borehole siting (see Cobbing & Davies, this volume). These studies improve the chances of identifying fractures and weathered zones in the impermeable rocks, thereby significantly helping to improve the likelihood of identifying aquifers in the more massive and less permeable formations.

Various additional problems are associated with siting of boreholes in Ghana. Many times, potentially suitable sites are ignored because of difficulty in accessibility of heavy–duty drilling rigs. Some potentially favorable sites are also rejected because of their proximity to rubbish dumps, cemeteries and pit latrines. Many potentially favourable sites, particularly in formations of the Voltaian and the Dahomeyan Systems, are located in swampy areas, and intermittently dry valleys where, during the rainy season, polluted surface water could enter the boreholes.

6 MANAGEMENT AND UTILIZATION OF GROUNDWATER

6.1 *The development of the current status*

The Act 310 (of 1965) established the Ghana Water and Sewerage Corporation (GWSC), which later after 1998 by Act 461 as amended under statutory corporation Legislative Instrument (L.I.) 1648 (1998) was converted to a Limited Liability Company known as the Ghana Water Company Limited (GWCL). Act 310 empowered the GWSC to be responsible for the provision, distribution, conservation and management of water supply and to set the criteria for all water supply development and installation, and to coordinate all activities related to the water supply industry. Within the GWSC was established in 1994 a section known as Community Water and Sanitation Division (CWSD), which was later in 1998 (Act 564, 1998) transformed into an autonomous Community Water and Sanitation Agency (CWSA), to be responsible for rural water supply and sanitation. This mandate of CWSA has been expanded to include small towns, some of which have populations of up to and sometimes more than 15,000.

A new policy of CWSA requires that the supply of water to the rural communities should be demand driven and community managed. Hence the communities are required to make a contribution (currently 5%) of the capital cost of providing the facility to inculcate in the communities the idea of ownership of the project. This policy has recently been criticised due to the endemic poverty in the rural areas, which makes it almost impossible for the rural communities to raise their contribution, and accordingly the policy is currently under review.

Until 1994, the GWSC did not have a specialized division that dealt with groundwater resources, hence some individuals, agencies and Non Governmental Organisations (NGOs) planned, implemented and operated their own schemes which did not, at all times, meet the criteria set by GWSC. Whilst some of these schemes have sometimes disregarded priorities in the implementation of National Water Supply Programme, others have supplemented Government efforts in areas of great need. Presently, there is better co-ordination between the NGOs and the other interested parties and the CWSA. In the past, there was no proper 'handing over' of new water supplies to the rural communities as the locals were not trained to take over the running of the systems. To address such a situation, a community participation programme which involves the beneficiaries in the planning, implementation, operation and maintenance of water supply schemes, has been put in place.

6.2 *Groundwater utilization in Ghana*

Groundwater in Ghana is used primarily for domestic purposes, i.e. about 95% of all groundwater use. All major groundwater development projects that have contributed to more than 20 or more boreholes have been done to produce water for domestic consumption. The few boreholes that have been constructed for the purposes of irrigation water or cottage industries and other agricultural uses such as animal rearing were done by individuals or companies that could afford only a few boreholes or some number of hand-dug-wells. Many hand-dug-wells may be used for irrigation purposes but these are only for subsistence farming or minor commercial vegetable farming. These dug wells often do not have motorized pumps fitted on them, water is generally lifted by bucket and rope systems. Most large scale irrigated agriculture is done using surface water. In recent times (last decade) however, a number of boreholes are being drilled for large scale commercial bottled water industry in southern Ghana. In a recent survey of groundwater use in the Densu basin in southern Ghana, it was found out that all the major commercial bottled water industries in Ghana use groundwater sources for the water production (Darko *et al.*, 2003). These industries constitute about 85% of groundwater use in the Densu basin.

6.3 *Decline in groundwater levels*

Some decline in groundwater levels have been observed in certain parts of the country particularly in the semi-arid northern parts of the country and some areas in the south, i.e. some wells in the Ho area in the Volta Region. The excessive reliance on groundwater has led to the gradual depletion of the groundwater resources. Studies conducted on the Wa (Upper West Region) and Bawku well fields in the Upper East Region indicated that in some cases, water-levels have dropped by 5 m (Quist *et al.*, 1988). This may be due to poor spacing of the wells which manifest themselves in low yields, and non-sustainable exploitation of aquifers due to lack of information on recharge mechanisms. To mitigate these effects, refinement of well field design criteria, detailed recharge studies and new recommendations on pumping to address the overdraft, are required.

6.4 *Rehabilitation of boreholes*

Currently more than 40% of boreholes in the country are between 15 and 35 years old, and have hardly been maintained or rehabilitated, and in some cases leading to disuse. The Canadian International Development Agency (CIDA) in the late 1970s and early 1980s undertook rehabilitation of some boreholes in the Upper Regions. Currently, some rehabilitation works are being undertaking on boreholes in certain District capitals under the Small Towns Water Supply Projects.

6.5 *Identification and development of springs*

With the exception of the Volta Region where almost all the springs have been identified and some are being developed, the national development of springs as sources of rural water supply is still rudimentary (WRRI, 1992). There is a great need to identify and document the locations of springs in the country in order to promote their large-scale use. This is because springs are relatively cheap to develop and the technology required can easily be adapted by local water supply engineers.

6.6 *Pollution of hand-dug-wells and boreholes*

The water in many hand-dug-wells tends to be turbid and polluted as a result of high levels of nitrate content (30–60 mg/l) and abundant coliform (WARM, 1998). This could be avoided to some extent through improved construction and adequate protection of wells from surface runoff and animal droppings by providing, for example, covers for the wells. Boreholes and hand-dug-wells are normally recommended to be sited upstream and away from pit or KVIP (Kumasi Ventilated Improved Pit) latrines, cemeteries and refuse dumps to avoid pollution, and also from mining activities to avoid contamination from heavy metals.

6.7 *Quality of groundwater*

In general, the quality of groundwater in Ghana is considered good for multi-purpose use (including for drinking), except in a few cases where there are low pH (3.5–6.0) waters in the forest zones of southern Ghana. In addition, high levels of iron, manganese and fluoride occur in certain localities as well as high mineralization with total dissolved solids (TDS) in the range 1458–2000 mg/L in some coastal aquifers particularly in the Accra plains. High fluoride levels in the range 1.5–5.0 mg/l are found in some boreholes located in the granitic formations of the Upper East (e.g. Bongo and Bolgatanga Districts) and Upper West Regions.

The most prominent water quality problem with groundwater supply in Ghana is excessive iron concentrations. About 30% of all boreholes in Ghana have high iron content. High iron concentration in the range 1–64 mg/l has been observed in boreholes in all geological formations. Iron originates partly from the attack of low pH waters on corrosive pump parts and partly from the aquifers. There is therefore the need to incorporate some simple iron removal plants through which iron contaminated water from the outlets of handpumps can flow to render the water acceptable for domestic use. The Water Research Institute (WRI) has developed and tested successfully prototypes of such iron removal plants (Amuzu, 1987). Unfortunately, due to inadequate funding to transfer this technology, very few rural communities are benefiting from the iron removal technology.

6.8 *Recharge studies*

Recharge studies in Ghana have been carried out in isolated cases, usually directed at few river basins. Direct groundwater recharge in the various aquifers is generally low (Atobra, 1983; Frempong & Kortatsi, 1995; Kankam-Yeboah, *et al.*, 2005) ranging from about 8% to about 20% of mean annual rainfall over different formations.

In the absence of detailed recharge studies, groundwater resources have to be efficiently managed to minimize the risk and vulnerability of this resource in certain geological formations. For example, in the Voltaian System and parts of the Basement Complex, like the Dahomeyan where the groundwater resources are considered vulnerable and borehole yields are insufficient to meet the water needs of some rural communities, it is necessary to resort to conjunctive use of both groundwater and surface waters.

7 TOWARDS POVERTY REDUCTION

The Ghana Poverty Reduction Strategy (GPRS 2002–2004) is the foundation on which Ghana's development is based. The GPRS outlines a number of measures to be introduced with the objective of achieving an accelerated reduction in the overall poverty. The GPRS is expected to achieve its objectives through a series of interventions including prudent fiscal policies, private sector-led growth, sound and sustainable management of the environment, promotion of agriculture and agro-based industrial expansion, export promotion, increased investment in social services, and accelerated decentralization.

In the restructuring of the water sector, CWSA has been made a facilitator to ensure the acquisition of drinking water and sanitation facilities, through community participation, ownership, management and cost recovery of operation and maintenance. The GPRS had set targets for access to potable water and sanitation (latrines) in the rural areas by 2004 at 54% and 25% respectively. Also one of the objectives of the Millennium Development Goals (MDGs) is to reduce by half the population of the world without access to potable water and sanitation by the year 2015. In line with this objective of the MGD, the Government of Ghana has set itself a more ambitious target of providing 85% of the rural population with potable water supply mainly from groundwater sources by 2015.

Groundwater development is therefore being pursued vigorously in the rural areas of Ghana to replace the polluted traditional surface water sources in order to reduce incidence of water-borne and water-related diseases common in rural areas. The decline in the incidence of water-borne and water-related diseases in rural areas is expected to reflect in lower patronage of health institutions and lower spending on drugs. The savings made through lower spending on drugs and hospital attendance can thus be channeled to other productive ventures. After all, it is often said that 'water is life' and safe water guarantees good health, which is an important condition for increased production.

The construction of boreholes and protected hand-dug-wells near the rural settlements also allows women enough time to undertake economic activities instead of the usual long treks and time consuming search for water. Children, who usually fetch water together with their mothers are also spared the long treks for water and can thus attend school regularly and punctually.

The provision of potable water in rural areas among others will encourage cottage industries and hence provide a catalyst for the reduction of rural poverty and enhancement of quality of life through employment generation. In addition the utilization of hand-dug-well and borehole water assures all year round crop production, increased yield and productivity in agriculture particularly in the rural areas.

8 CONCLUSION

The groundwater potential of the various hydrogeological provinces in Ghana and the groundwater quality has been assessed. This is necessary since groundwater has been found to be both feasible and the most economic source of water supply for the socio-economic development of rural communities. Groundwater investigations and exploration in Ghana have been limited mostly to areas of the more concentrated human settlements. It has been observed that the various geological formations have varying degrees of groundwater potential (between 36 and 88% of borehole success rate in the Dahomeyan System and the Togo/Buem Series respectively). Groundwater quality has been observed (generally TDS <1000 mg/l) to be generally suitable for potable use, provided the boreholes and hand-dug-wells are properly protected from sources of pollution and contamination.

However, this overview clearly shows that there are many hurdles in the way of the desired process. There is an incomplete study of the Nation's hydrogeological settings; inadequate protection of many hand-dug-wells; lack of detailed studies on the aquifer conceptualisation and characterisation as well as aquifer hydrogeochemistry will need to be timely addressed.

More than half of Ghana's population lives in the rural areas where only about 52% are served with potable water mainly from groundwater resources. Most of the rural population that is not provided with potable water relies on polluted surface water sources. These are usually sources of water borne and water related diseases, e.g. guinea worm and bilharzias, which are common in the rural areas. The government has a policy of providing 85% of the rural population with potable water supply mainly from groundwater sources by 2015. Groundwater supplies have been found to be very feasible and the most economic source of potable water for the rising rural population. The government's policy is aimed at eradicating water borne and water related diseases in the rural areas in order to raise a healthy and more productive people. The provision of groundwater supplies for agriculture and cottage industries will raise the rural standard of living and reduce poverty.

Where the groundwater potential is found to be low and the yields of boreholes or hand-dug-wells are not sufficient to meet the water supply needs of the rural communities, a conjunctive use of both ground and surface water resources is recommended. To this end, it is further recommended that proper conservation techniques and efficient management of the country's groundwater and surface water resources have to be put in place.

REFERENCES

Act 564 1998. *Community Water and Sanitation Agency Act.* The 564 Act of the Parliament of Ghana Accra, Ghana.

Akiti, T. T. 1977. *Groundwater from the Keta-Anlo artesian basin.* Proceedings of the 10th Biennial Ghana Science Association Conference, Cape Coast, Ghana.

Akpati, B. N. 1975. *Geological structure and evolution of Keta basin.* Ghana Geological Survey Report **75/3**, Accra, Ghana.

Amuzu, A. T. 1987. *Prototype iron removal system suitable for small-scale groundwater supplies.* Technical Report, WRRI (CSIR), Accra, Ghana.

Atobra, K. 1983. *Groundwater Flow in the Crystalline Rocks of the Accra Plains, West Africa.* PhD, Princeton University, USA.

Bannerman, R. R. 1975. *The role of groundwater in rural water supplies in Ghana.* Hydrological Sciences Bulletin, **20**, 191–201.

Cobbing, J. E. & Davies, J. 2008. *The benefits of a scientific approach to sustainable development of groundwater in Sub-Saharan Africa.* In: Adelana, S. M. A. & MacDonald, A. M. Applied Groundwater Studies in Africa. IAH Selected Papers on Hydrogeology, Volume 13, CRCPress/Balkema, Leiden, The Netherlands.

CWSA 2004. *Strategic Investment Plan 2000–2015.* Community Water and Sanitation Agency, Ministry of Works and Housing, Accra, Ghana.

Darko, P. K. 2005. *Prevailing transmissivity in hard rocks of Ghana.* Journal of the Ghana Science Association, **4**, (2), 99–107.

Darko, P. K., Duah, A. A. & Dapaah-Siakwan, S. 2003. *Groundwater Assessment: An Element of Integrated Water Resources Management. The Case of Densu River Basin.* Technical Report for the Water Resources Commission, Accra, Ghana.

Frempong, D. G. & Kortatsi, B. K. 1995. *Groundwater Modelling in Ghana.* Presented at Geological Society Africa, International Conference, Oct. 9–13, 1995, Nairobi, Kenya.

Geological Survey of Ghana 1969. *Geological map of Ghana.* Survey of Ghana, Accra, Ghana.

Ghana Vision 2020 1998. *The First Medium Term Development Plan (1997–2000).* National Development Planning Commission, Accra, Ghana.

Gill, H. E. 1969. *A groundwater reconnaissance of the Republic of Ghana, with a description of geohydrologic provinces.* USGS Water-Supply Paper **1757-K**.

Junner, N. R. & Hirst, T. 1946. *The geology and hydrogeology of the Voltaian basin.* Gold Coast Geological Survey Memoir **8**, Gold Coast Geological Survey, Accra, Ghana.

Kankam-Yeboah, K., Darko, P. K. & Nishigaki, M. 2005. *Sustainable groundwater exploitation under natural conditions in southwest Ghana.* J Fac Environ Sc Tech, **10**, 83–88.

Kesse, G. O. 1985. *The mineral and rock resources of Ghana.* AA Balkema, Rotterdam, The Netherlands.

Legislative Instrument 1998. Statutory Corporations (Conversion to Companies) (Schedule Amendment) Instrument 1998, Date of Gazette notification: 10th November 1998, Entry into force: 16th December, 1998, Accra, Ghana.

Nathan Consortium for Sector Studies 1970. *Water Sector Studies Reports-Ghana.* Report prepared for the Ministry of Finance and Economic Planning. Nathan Consortium/Ministry of Finance & Economic Planning, Accra, Ghana.

Ojo, O. 1977. *The Climates of West Africa.* Heinemann, UK.

Quist, L.G., Bannerman, R.R. & Owusu, S. 1988. *Groundwater in rural water supply in Ghana.* In: Ground Water in Rural Water Supply, Report of the West African Sub-Regional Workshop held in Accra, Ghana, 20–24 October 1986, UNESCO Technical Documents in Hydrology, Paris. 101–126.

Soviet Geological Survey Team 1964–1966. Unpublished reports, Geological Survey Department, Accra, Ghana.

Uma, K. O. & Egboka, B. C. E. 1988. *Problems of rural water supplies in a Developing Economy: Case studies of Anambra and Imo States of Nigeria.* Water International, **13**, 33–45.

WARM 1998. *Ghana's water resources, management challenges and opportunities.* Prepared for the Ministry of Works and Housing, Accra, Ghana.

WRRI 1992. *Preliminary report on the inventory and assessment of potential for hand-dug-wells in the Volta Region.* WRRI publication, Accra, Ghana.

WRRI 1996a. *Groundwater resources assessment of Ghana.* WRRI publication, Accra, Ghana

WRRI 1996b. *Borehole yield map of Ghana*, WRRI publication, Accra, Ghana.

CHAPTER 11

An overview of the geology and hydrogeology of Nigeria

S.M.A. Adelana, P.I. Olasehinde & R.B. Bale
Geology Department, University of Ilorin, Ilorin, Nigeria

P. Vrbka
Dieburger Str. 108, Groß-Zimmern, Germany (formerly Geology Institute, Technical, University, Darmstadt, Germany)

A.E. Edet
Geology Department, University of Calabar, Calabar, Cross-River State, Nigeria

I.B. Goni
Geology Department, University of Maiduguri, Maiduguri, Borno State, Nigeria

ABSTRACT: There are considerable groundwater and surface water resources available in Nigeria; however, much less information is available on groundwater resources than the surface water. This paper reviews the groundwater resources of Nigeria, in line with recent research across the country. The occurrence of groundwater in Nigeria is dependant primarily on the geology, and groundwater regions can be classified accordingly. The extent in any given groundwater region is defined by the limits of a group of rock types with similar hydrogeological conditions. The primary aim is to produce a synoptic overview of the hydrogeological character in each groundwater basin to guide an overall management and protection of groundwater resources in Nigeria. Groundwater management and aquifer protection must be taken seriously in Nigeria to ensure the long-term sustainable use of groundwater in future generations.

1 INTRODUCTION

Nigeria is faced with increasing demands for water resources due to high population growth rate and growing prosperity. The importance of groundwater in meeting a substantial percentage of this water need, and in the overall development of Nigeria's economy cannot be over-emphasized. Hence, a country-wide inventory of groundwater resources is desirable. In order to set the context for sustainable groundwater use and management options across the country there is need to develop better predictive models, which are dependent on a thorough understanding of geological and hydrogeological settings. With the objective to present a clear picture of the hydrogeological conditions of the various groundwater regions, this paper attempts to integrate results of local and scattered studies into a comprehensive, regional description. It begins with the review of the hydrogeology of northwestern,

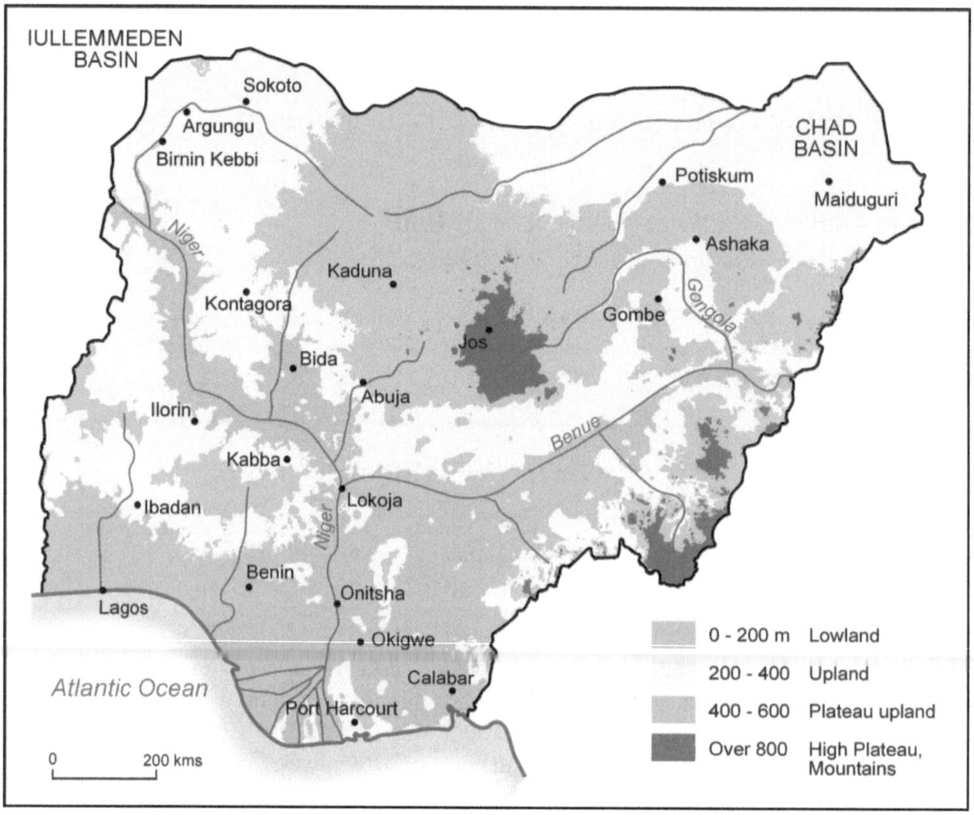

Figure 1. Location and base map morphology.

northeastern, southeastern, southwestern and central Nigeria based on recent information available and concludes with defining strategies and setting priorities to achieve sustainable management as well as a protected aquifer system in Nigeria.

2 LOCATION AND CLIMATE

Nigeria lies south of the Sahara within West Africa, with the Atlantic Ocean bordering the southern coastal region (Figure 1). The entire country lies on longitudes 2° 50′ to 14° 20′ E and from latitudes 4° 10′ to 13° 48′ N, occupying a land area of 923,768 km^2. Covering an average distance of 1120 km from south to north, Nigeria displays physiographic regions of varying characters in relief, nature and spatial distribution. Nigeria is the most populous country in Africa with a population of over 140 million. Just over half of the population (52%) is rural (WHO/UNICEF, 2007). Annual demographic growth rate currently exceeds 3%.

The climate in Nigeria is semi-arid in the north, and humid in the south. Due to its location, Nigeria has a tropical climate characterized by the hot and wet conditions linked with the movement of the Inter-Tropical Convergence Zone (ITCZ) north and south of the equator. The country experiences consistently high temperatures throughout the year. There are, however, wide diurnal ranges in temperature particularly in the very hot months. The mean monthly temperatures during the day sometimes exceed 36°C while monthly average temperatures at night fall below 22°C. Since temperature varies only slightly, rainfall distribution, over space and time, becomes the single most important factor in differentiating the seasons and climatic regions. Except for the coastal zone, where it rains all year round, rainfall is seasonal with distinct wet and dry seasons.

The mean annual rainfall along the coast in the southeast is 4800 mm while it is less than 500 mm in the northeast (although there are considerable spatial and annual variations to the long-term mean (Goni, 2002; Adelana *et al.*, 2003a).

Surface water flowing from Nigeria to the sea is estimated at 263 km^3/a (FAO, 2005). There are four main river systems: the Niger-Benue, the Chad basin area, the southwestern littoral and the southeastern littoral, with transborder flows from Cameroon. Nigeria receives annually an estimated 30 km^3 from River Niger (although in the 1980s the average flow of River Niger at the border was about 18 km^3), and 29 km^3 from Cameroon in the Benue and tributaries. Groundwater recharge is estimated at 87 km^3/a, of which a great part (about 80 km^3) discharges as base flow to major rivers. Total active surface water reservoir capacity in 1993 was 30.3 km^3 and a total of 60 large dams and 100 small dams have been built so far. Water allocation from reservoirs is shared as follows: 10.9 km^3 (36%) for irrigation, 0.8 km^3 (3%) for water supply, and 18.6 km^3 (61%) for hydropower (FAO, 2005). Agricultural, domestic and industrial water withdrawal was estimated at 3.6 km^3 in 1987.

3 GEOLOGICAL SETTING

The geology of Nigeria, as detailed in Kogbe (1989), is made up of three main rock groups: mainly Precambrian basement crystalline metamorphic-igneous-volcanic rocks; Mesozoic to Tertiary sediments, granites and volcanics; and Quaternary alluvial deposits (Figure 2).

3.1 *Precambrian basement complex rocks*

Precambrian Basement Complex rocks underlie three areas of Nigeria: North-central area including the Jos Plateau; South-west area adjacent to Benin; and south-east area adjacent to Cameroon.

The rocks of the North-central area are composed of gneisses, migmatites, granites, schists, phyllites and quartzites. The narrow, tightly folded north-south trending schist belts of northwestern Nigeria include igneous rocks, pelitic schists, phyllites and banded ironstones. The migmatite-gneiss complex of amphibolites, diorites, gabbros, marbles and pegmatites form a transition zone between the schist belt of NW Nigeria and the granites of the Jos Plateau to the east. There, extensive Precambrian age Older Granites crop out extensively. These have been intruded by Jurassic age Younger Granites that are characteristic ring complex structures.

The Precambrian Basement rocks of southwestern Nigeria, as found in the Dahomeyan (Benin) Basin, consist of migmatites, banded gneisses and granite gneisses, with low grade metasedimentary and metavolcanic schists, intruded by Pan-African age granites and charnockites (Oyawoye, 1972). The migmatites and gneissic metasediments are often intruded by pegmatite veins and dykes (Oluyide *et al.*, 1998). Older granites, granodiorites and syenites, with dolerite dykes, also form part of the Precambrian basement of

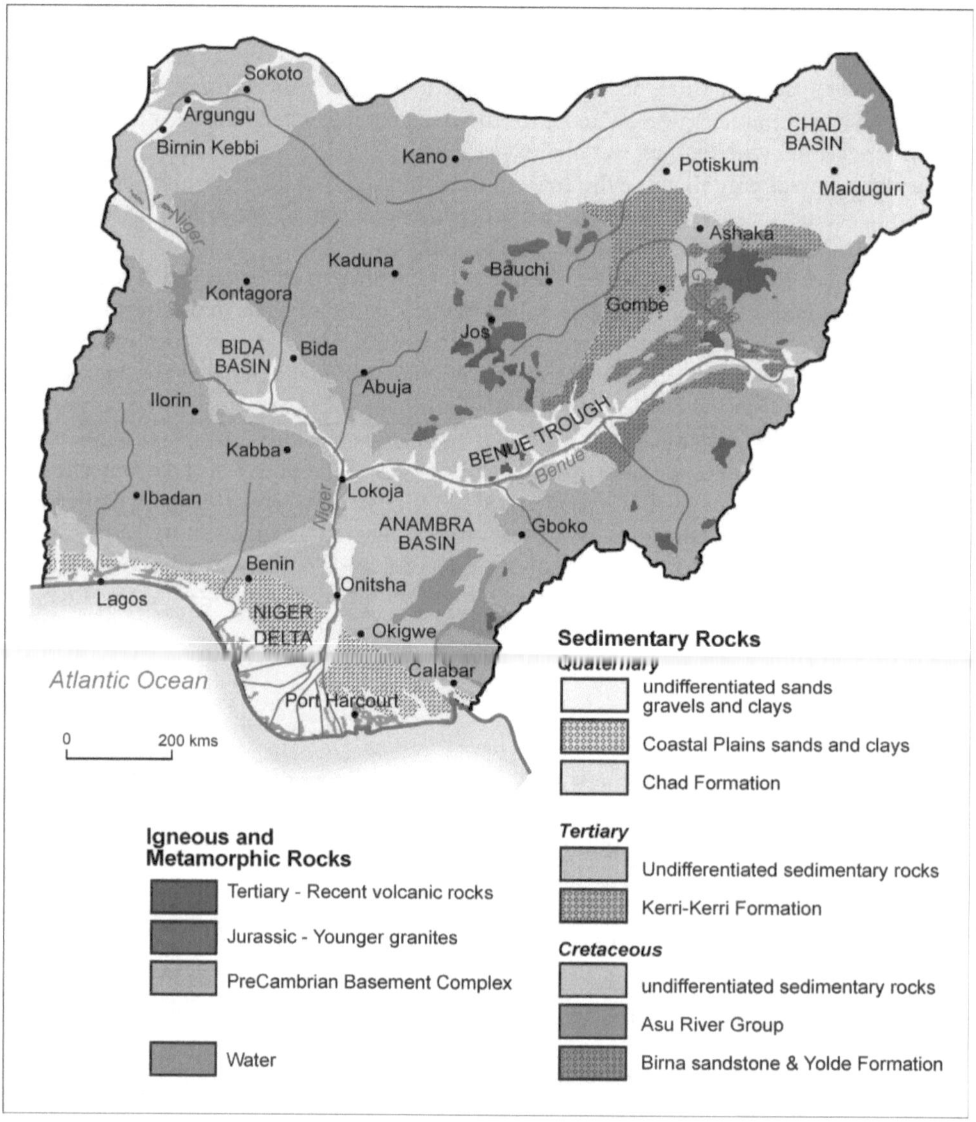

Figure 2. Generalized geological map of Nigeria (from MacDonald *et al.*, 2005).

SW Nigeria. The Precambrian Basement rocks of south-eastern Nigeria occur in three blocks along the border with Cameroon (Figure 2). The crystalline basement rocks include biotite-hornblende gneiss, kyanite gneiss, migmatite gneiss and granites and are well fractured (Ekwueme, 1987).

3.2 *Mesozoic to tertiary rocks*

Tertiary age olivine basalts, trachytes, rhyolites and tuffs , overlying or interbedded with coarse grained alluvial sediments, occur in the Jos Plateau and adjacent plateau areas. The surfaces of these volcanic lava flows have been weathered to form succession of laterite

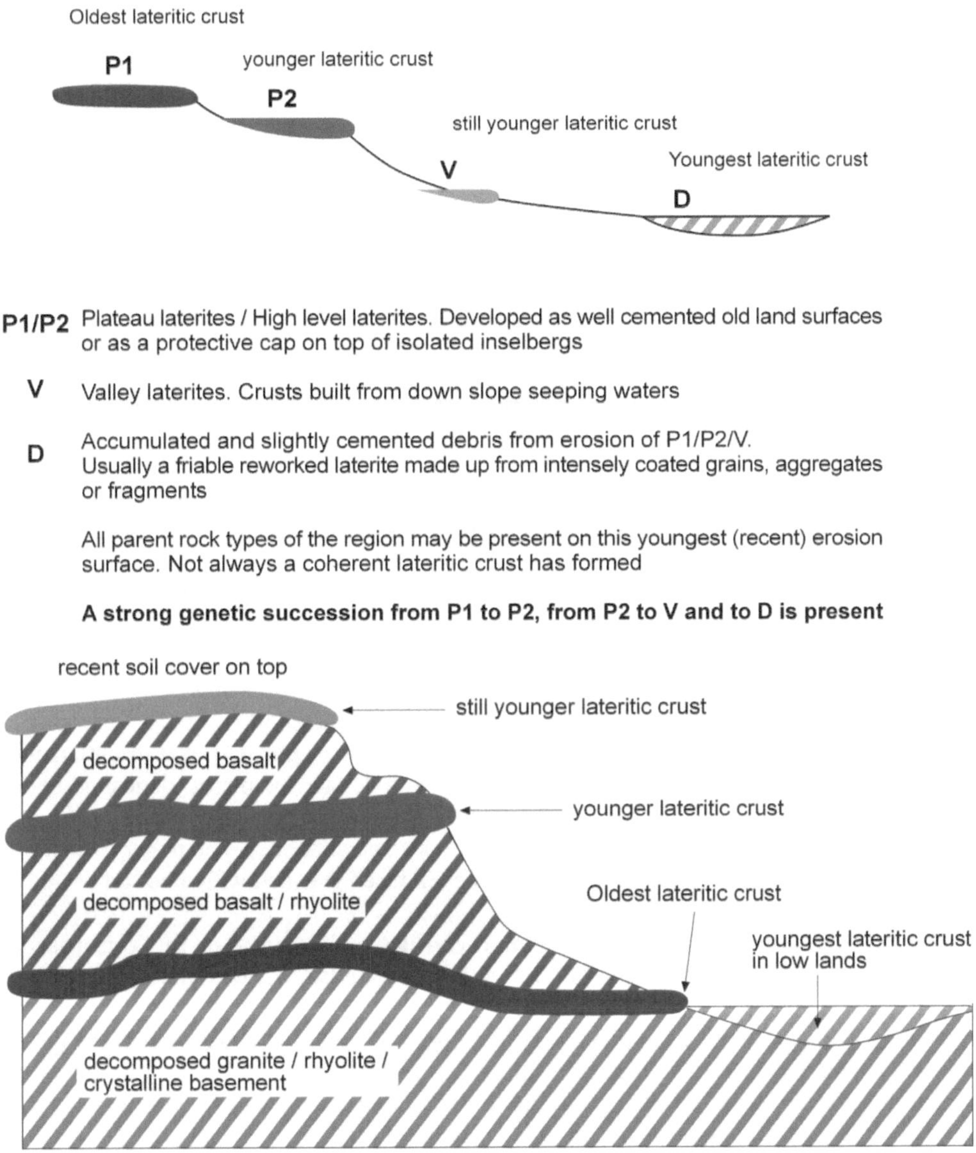

Figure 3. Genetic succession and lateritization in the Jos Plateau (after Farnbauer & Tietz, 2000).

palaeosols on the Jos Plateau related to the uplift of the plateau (Figure 3). (Farnbauer & Tietz, 2000).

The Mesozoic and Tertiary strata, of the Sokoto part of the Iullemmeden Basin in NW Nigeria, comprise interbedded sandstones, clays, and limestones that dip to the north-west. These formations are capped by laterite. The sedimentary sequence includes the late Jurassic to early Cretaceous Illo and Gundumi Formations, the Maastrichtian Rima Group, the late Paleocene Sokoto Group and the Eocene-Miocene Gwandu Formation (Figure 4). These were deposited during a series of overlapping marine transgressions. Over 1250 m

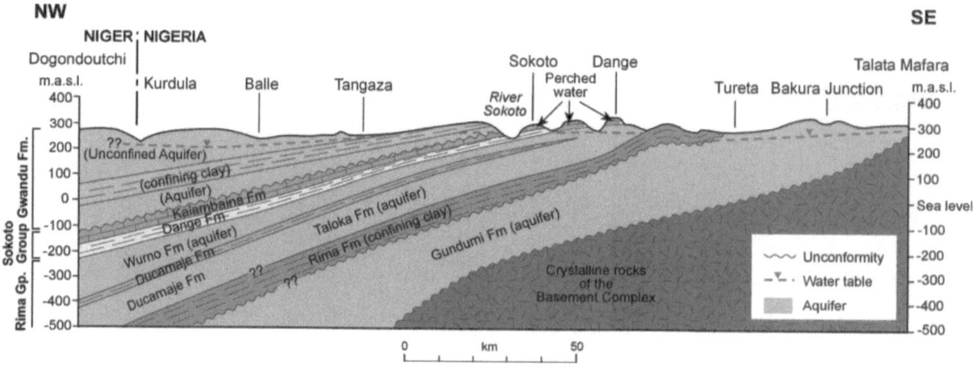

Figure 4. Hydrogeological cross-section through Sokoto Basin, North-West Nigeria (modified from Anderson & Ogilbee, 1973).

of sediments occur in the down-warped Sokoto Basin, unconformably overlying Precambrian Basement rocks. Quaternary age alluvial deposits occur along the course of the River Sokoto.

In the Chad Basin of NE Nigeria, Cretaceous sediments include the Albian-Cenomanian Bima Sandstone Formation continental poorly sorted and thickly bedded feldspathic sandstones and conglomerates to fluviatile and deltaic sediments. The early Turonian age Gongila Formation, up to 500 m thick, includes marine limestones, sandstones and shales. The Senonian-Maastrichtian age Fika shale Formation, 100 to 500 m thick, consists of gypsiferous shales and limestone of marine and continental origin. The Maastrichtian age Gombe Sandstone Formation consists of estuarine and deltaic sediments, deposited upon marine shales with sandstone/shale intercalations. The lower deposits of siltstone, mudstone and ironstone are overlain by well-bedded sandstones and siltstones. The upper formation contains coals and cross-bedded sandstones. The Cretaceous ended with a period of uplift and erosion.

The Palaeocene age Kerri-Kerri Formation consists of lacustrine or fluvio-lacustrine loosely cemented cross-bedded coarse- to fine-grained sandstones, with locally occurring claystones, siltstones, ironstones, lignites and conglomerates. The Kerri-Kerri Formation rests unconformably on the Gombe Sandstone Formation and thickens towards the basin center where it is overlain by Chad Formation (Figure 5).

The Pleistocene age Chad Formation (up to 840 m thick) consists of poorly sorted fine to coarse-grained sand, with sandy clay, clay and diatomite.

The southern Nigeria sedimentary basin includes the Lagos-Osse and the Niger Delta Basins that are separated by the Benin Hinge Line (the Okitipupa ridge). These basins have a common basal formation of marine Albian age, arkosic, gravelly, poorly-sorted, cross-bedded sandstones and sandy limestones.

The Lagos-Osse Basin, the eastern sector of the Benin Basin, underlies the western low-lying coastal zone, where rock exposures are poor due to thick soil cover. The Tertiary geology of the area comprises the basal Araromi/Ewekoro Formation, consisting of shelly and sandy black shale with thin sandstones and limestones that is overlain by the Palaeocene age Imo Shale Formation (Okosun, 1998). These are overlain by the Oshosun Formation of shales, clays and sandstones. The succeeding Miocene age Benin Formation consists of up to 200 m of sands with shales, clays and lignite. The near surface Quaternary geology includes recent littoral sandy alluvium and lagoon/coastal plain sands (Jones & Hockey, 1964; Longe *et al.*, 1987).

In the Niger Delta Basin, Quaternary age sediments underlying the Delta Plain consist of coarse to medium grained unconsolidated sands and gravels with thin peats, silts, clays, and shales, forming units of old deltas. The underlying Miocene age Benin Formation is composed of gravels and sands with shales and clays. This multi-aquifer system formation crops out to the northeast of the coastal belt.

The Cretaceous sediments of the down faulted and failed rift that is the Benue Trough occur in a series of sedimentary basins that extend north east of the confluence of the Niger and Benue Rivers, bounded by the Basement Complex strata to the north and south of the Benue River (Figure 2) (Reyment, 1965). The Lower Benue Basin consists of shales, silts and silty shales with subordinate sandstones and limestones intruded by dolerite dykes. The Upper Benue Basin consists of a thick succession of continental sandstones overlain by marine and estuarine deposits (Carter *et al.*, 1963). The basal formation is the Bima Sandstone.

The Bida Basin runs north west from the confluence of Niger and Benue rivers from the Anambra Basin in the southeast and to the northwest towards the Sokoto Basin (Figure 2). The basin contains Cretaceous age mainly continental sandstones, siltstones, claystones and conglomerates. The Middle Niger Basin at the confluence of the Niger and Benue rivers, contains 500 to 1000 m of increasingly marine sediments (Ladipo, 1988). The main unit, the Lokoja Formation, consists of alluvial to deltaic coarse to medium-grained cross-bedded to massive sandstones with subordinate siltstones, kaolinitic claystones and shales indicate.

Quaternary to Recent age alluvial deposits occur along the main river valleys. These deposits range from thin discontinuous sands to thick alluvial deposits up to 15 km wide and 15 to 30 m thick along the Niger and Benue rivers. The alluvial deposits include gravel, coarse and fine sand, silt and clay. Thin deposits of unconsolidated and mixed sands and gravels occur along the courses of ephemeral fadamas in northern Nigeria.

4 GROUNDWATER AREAS AND REGIONAL HYDROGEOLOGY

4.1 *Summary of the groundwater areas of Nigeria*

The geological framework described in the previous section has, to a large extent, defined the distribution of groundwater in Nigeria. There is a large groundwater potential in Nigeria, which by far exceeds the surface water resources. Details of the various hydrogeological basins are presented in Offodile (2002). Little is known about aquifer boundaries in some regions, so the aquifers are simply classified by geographical location. Moreover, the poor quality of drilling records prevents some detailed study in certain aquifers. Table 1 shows borehole information based on available data from UNICEF (Nigeria) on wells drilled within the Middle Belt and Northern Nigeria. Note that failed boreholes often go unreported, therefore, the true success rate is likely to be much lower.

The distribution and occurrence of groundwater in Nigeria described below iis based on the various geological units identified above, on which basis the groundwater areas have been classified. The extent in any given area is defined by the limits of a group of rock types, geological formations or group of formation with similar hydrogeological conditions.

Table 1. Summary of borehole information (Data source: UNICEF, unpublished).

Period	No. of borehole drilled	No. of completed boreholes	Success rate (%)	Borehole Depth (m)		Static Water Level (m)		Yield (m^3/h)	
				Mean	Range	Mean	Range	Mean	Range
1989–2002	16,827	13,903	93	38.6	2–240	7.8	0.1–120	54.6	1–3333

4.2 *Regional hydrogeology: northwestern Nigeria*

In northwestern Nigeria, i.e. Sokoto Basin, up to 2000 m of clastic sequences rest upon the basement. Sequences of semi-consolidated gravels, sands, clay, some limestone and ironstone are found. The analysis of pumping test data carried out in the shallow aquifer indicates transmissivity in the range of 200–5000 m^2/d and storage coefficients of 10^{-2} to 10^{-5} indicating semi-unconfined to confined conditions (Bassey *et al.*, 1999). The yield of boreholes up to 20 m depth is generally >2 l/s. The fluctuation of the water table in fadama (low lying) areas is about 2–3 m throughout the year. The water table is lowest in June and highest in September at the end of the wet season.

Precipitation in the area occurs within 3–5 months giving short-lived, but strong, runoff. Stream discharge has been measured on daily basis at several gauging stations on the Sokoto-Rima Drainage Basin, and values range from 9.64×10^8 m^3/a to 7.13×10^9 m^3/a (Anderson & Ogilbee, 1973). The recharge characteristics of the Rima Group Aquifer are described in Oteze (1989a).

Artesian aquifers occur at depth in the Gundumi Formation, the Rima Group, and the Gwandu Formation. A perched groundwater body occurs locally at shallow depth in the limestone of the Kalambaina Formation (Anderson & Ogilbee 1973; Oteze 1989b). This aquifer sustains numerous dug wells, springs and ponds.

Five major aquifers in the Sokoto Basin are distinguished based on the geography and geological setting (Figure 4).

1. The Gundumi Formation (Lower Cretaceous) includes river and lacustrine deposits, which contain comparatively coarser materials than any of the younger overlying formations of the Sokoto Basin.
2. The Illo Group (Cretaceous), similar in lithology to the Gundumi Formation, includes non-marine cross-bedded pebbly sand and clay that underlie an area of about 6400 km^2 in the southwestern part of the Sokoto Basin.
3. The Rima Group (Upper Cretaceous) consists of a marine transgressive series of fine-grained sand and friable sandstone, mudstone, and some marly limestone and shale. North of the River Sokoto, the group is divided into three formations, the Taloka at the base, the Dukamaje in the middle, and the Wurno Formation at the top (Jones, 1948; Oteze, 1989a, b).
4. The Sokoto Group (Paleocene) consists of a lower unit (the Dange Formation, which is basically a marine clay shale), an upper unit of light-grey and white-clayey limestone with modular crystalline limestone, known as the Kalambaina Formation.
5. The Gwandu Formation (Eocene) crops out over 13,600 km^2 in the western third of the Sokoto Basin with sediments made up of interbedded semi-consolidated sand and

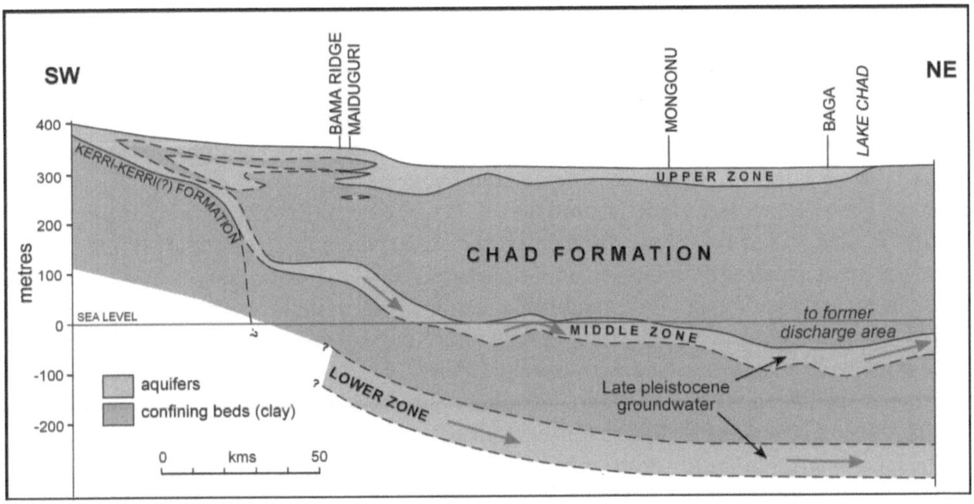

Figure 5. Geological cross section showing the aquifers of the Chad Formation.

clay. The Gwandu Formation unconformably overlies the Kalambaina Formation in the northern and central parts of the basin. These aquifers remain the sole source of drinking water in most parts of the basin, as rainfall is sparse. Generally, of all the water-bearing formations in the basin, the Eocene-Miocene Gwandu Formation is the most prolific aquifer with an annual recharge exceeding 6.6×10^7 m^3 and estimated groundwater storage of about 8.17×10^{12} m^3 (Oteze, 1989b).

Although it has not been possible to determine the recharge area accurately, about 20 km^2 of the Kalambaina Formation is exposed to recharge around Sokoto town. Annual recharge of the Wurno Aquifer from precipitation is estimated at 4.26×10^6 m^3 while total storage in this aquifer is 2.25×10^{10} m^3. Whereas, the total annual recharge into the Taloka aquifer is estimated at 7.3×10^7 m^3 with groundwater storage of 2.55×10^{11} m^3 (Oteze, 1989a).

Also, from hydrological data, it has been estimated that the Rima River system between Sabon-Birni and Sokoto town loses an average of 4×10^8 m^3/a. A part of this water may be evaporated while the remaining infiltrates into the alluvium, and in turn recharges the aquifers. Recent quantitative estimation of groundwater recharge using chloride method indicates mean recharge is in the order of 19.6 mm/a based on an annual rainfall average of 670 mm from 1916–1993 in the Sokoto area, although rates can be highly variable in space and time. Results further show recharge around the Wurno and Goronyo areas as representing $<1\%$ of annual rainfall while areas outside this region (Argungu/Gwandu areas) 3.2% of annual rainfall recharges the groundwater. This sharp variation was attributed to local conditions of climate and lithology (Adelana *et al.*, 2006).

4.3 *Regional hydrogeology: northeastern Nigeria*

In the Plio-Pleistocene Chad Formation, groundwater occurs under water table condition, in perched condition, in semi-confined and confined conditions. Three well-defined arenaceous horizons within the argillaceous Chad Formation constitute the aquifers, and were named by Barber & Jones (1960) as the Upper, Middle and Lower Aquifers (Figure 5).

The Upper Aquifer in most of the study area is within the superficial deposits, and extends across the entire outcrop of the Chad Formation. It is composed of alluvium and aeolian sands and gravel deposited during recent times.

However, around the type locality (Maiduguri) the Upper Aquifer includes not only a surface zone of recent sands with an unconfined water table, but deeper layers of sands of the Chad Formation complexly intercalated between clays, and are partially confined by these clays. Beacon Services Limited (1979) working in Maiduguri called the Upper Aquifer a system, because of the three definable units, which they termed A, B and C unit: (A) is the unconfined aquifer largely within the superficial deposits; (B) is artesian to leaky artesian; and (C) is artesian with negligible leakage. These units seem to be restricted to Maiduguri and environs, which is an ancient shoreline of Mega-Chad Lake – deposits of these units have been associated with beach or deltas. The units exhibit extremely variable thickness. Argillaceous deposits (100–150 m) confine the Middle and Lower Zone Aquifers (Figure 5), and boreholes drilled to these zones give rise to artesian wells.

The Kerri-Kerri Formation crops out towards the western part of the northeastern Nigeria. The arenaceous beds in this formation are coarse grained and highly permeable, with little or no water at the top. This poor water occurrence is as a result of the highly permeable nature of the sands, which makes it possible for the infiltrating water to run through the aquifer very easily and rapidly to considerable depths, except where clays, shale and silts are encountered (Offodile, 1972). This explains the variable water-level in wells tapping this formation. It is also possible that infiltrating water moves laterally to recharge the Lower Zone Aquifer of the Chad Formation. This is a possibility in view of the fact that the Kerri-Kerri Formation gently dips towards the northeast, and dovetails with the formation being a lateral equivalent of the Lower Zone of the Chad Formation (Oteze & Fayose, 1988). Fairly good yields have been obtained from deep boreholes tapping this formation (Offodile, 2002).

4.4 *Regional hydrogeology: the Middle Niger Basin and central Nigeria*

The area of the Middle Niger Basin extends northwestwards along the River Niger from the confluence with the River Benue in Lokoja. Crystalline basement rocks adjoin the area on all sides except to the southeast where it has a common boundary with the Cretaceous sediments of the Lower Benue Basin (Du Preez & Barber, 1965). The Lower Niger Basin area is drained by minor rivers flowing SW-NE, and N-S from the Kabba and Jos Highlands, in the south and the north respectively. The Patti Formation has a maximum thickness of about 100 m, and is composed of fine- to medium-sized sandstones, some claystones and carbonaceous siltstones. Oolithic inclusions of ironstones were observed towards the top. The samples have distinctively lower hydraulic conductivities than those of the deeper Lokoja Formation; hence not a primary target for groundwater exploration.

Additionally, intercalated clay and shale layers may reduce the productivity of wells, if vertical recharge plays a major role. The sandstones of the Lokoja Formation rest directly on the granitic to quartzitic basement rocks. The whole sequence consists of pebbly, clayey grits and sandstones with a minimum thickness of 250 m (Figure 6). Due to their higher hydraulic conductivity values, groundwater exploration tend to focus on the Lokoja Formation. Mands (1992) reported on three pumping tests situated within the sedimentary series of the Bida Basin, which were performed at Shashi Dama, Gaba and Etsugaie (near Bida). Transmissivity values are in the range of 5.5–29.3 m^2/d. The resulting hydraulic conductivities were reported as 6×10^{-2} to 3×10^{-5} m/s (0.5–2.6 m/d), which are similar to values obtained by Vrbka *et al.* (1999). The work of Kehinde (1990) indicates lower transmissivity for sediments of the Bida Basin (mainly <1–$50\,m^2/d$) presented in Vrbka *et al.* (1999). This discrepancy may partly be due to the fact that the latter examined mainly sandy sediments.

Furthermore, pumping wells usually are not fully penetrating and the screens may have been positioned in zones with relatively low hydraulic conductivities. Thus, transmissivity values from pumping tests are based on relatively short screen sections of some meters, whereas we calculated the bulk transmissivity for a sediment thickness of 500 m. Based on the mean value for screen lengths of 3–10 m, a transmissivity in the range of 10–33 m^2/d would result (Vrbka *et al.*, 1999), to some extent confirming the findings of Kehinde (1990) and Mands (1992). Based on the thickness data of Ojo & Ajakaiye (1989) of 500 m for the Lokoja-Abaji section and a void space of 12–18%, the entire groundwater resource (sedimentary cover about $60\,km \times 80\,km = 4{,}800\,km^2$) is estimated to be in the range of 290–430 km^3.

The areas underlain by Basement Complex rocks in west-central Nigeria include part of Niger, Kwara, Kaduna, Kogi, Benue and Plateau States. This is composed mainly of granitic and migmatitic gneisses and granites. There are also extensive areas of schists, phyllites occurring in Kaduna and Kwara states. Both the Older and Younger Granites of these areas are poor aquifers. However, over most of these areas occur thin, discontinuous mantle of weathered rocks or joint and fracture systems in the unweathered basement provide secondary reservoir. The decomposed mantle is sometimes too thin to harbour large quantities of water and is usually too clayey to be highly permeable. The crystalline basement of Nigeria generally represents the deeper, fractured aquifer and is noted as a poor source of ground water. This is partly overlain by a shallow, porous aquifer within the lateritic soil cover as described by Annor & Olasehinde (1996). Streams and rivers rise quickly during and after precipitation events, thus indicating a rather low storage capacity of the soil and underlying rocks. From a borehole situated on the southern outskirts of Ilorin on a major fracture in migmatitic granite, Offodile (1992) reports a yield of about 7.2 m^3/h (2 l/s). Yields between 1–2 l/s are considered good, and yields are more commonly below 1 l/s.

4.5 *Regional hydrogeology: southwestern Nigeria*

Groundwater occurrence in much of southwestern Nigeria is essentially semi-confined to unconfined. Due to the crystalline nature of the rock types in this region, the porosity and permeability necessary for the occurrence of large groundwater resources is lacking. However, appreciable porosity and permeability may have developed through fracturing and weathering processes. In the southern end of Kwara State (southeastern of Ilorin) the aquifer has a fairly heterogeneous geological structure, being covered by approximately 3–7 m of unsaturated soil composed of a humus layer and a sandy loam overlying a lateritic and clay layer (Adelana & Olasehinde, 2004). The highly permeable weathered/fractured zone varies locally and is sometimes composed of quartzo-feldspathic and amphibolitic materials.

Further information from the available borehole lithological logs in the region revealed that weathering is fairly deep and that the rocks have been jointed and fractured severely occurring at between 30–68 m below the surface (Adelana & Olasehinde, 2004). These joints and fractures have, to a large extent, controlled the flow direction of most rivers in this region. The weathered zone immediately overlying this fresh basement rocks are made

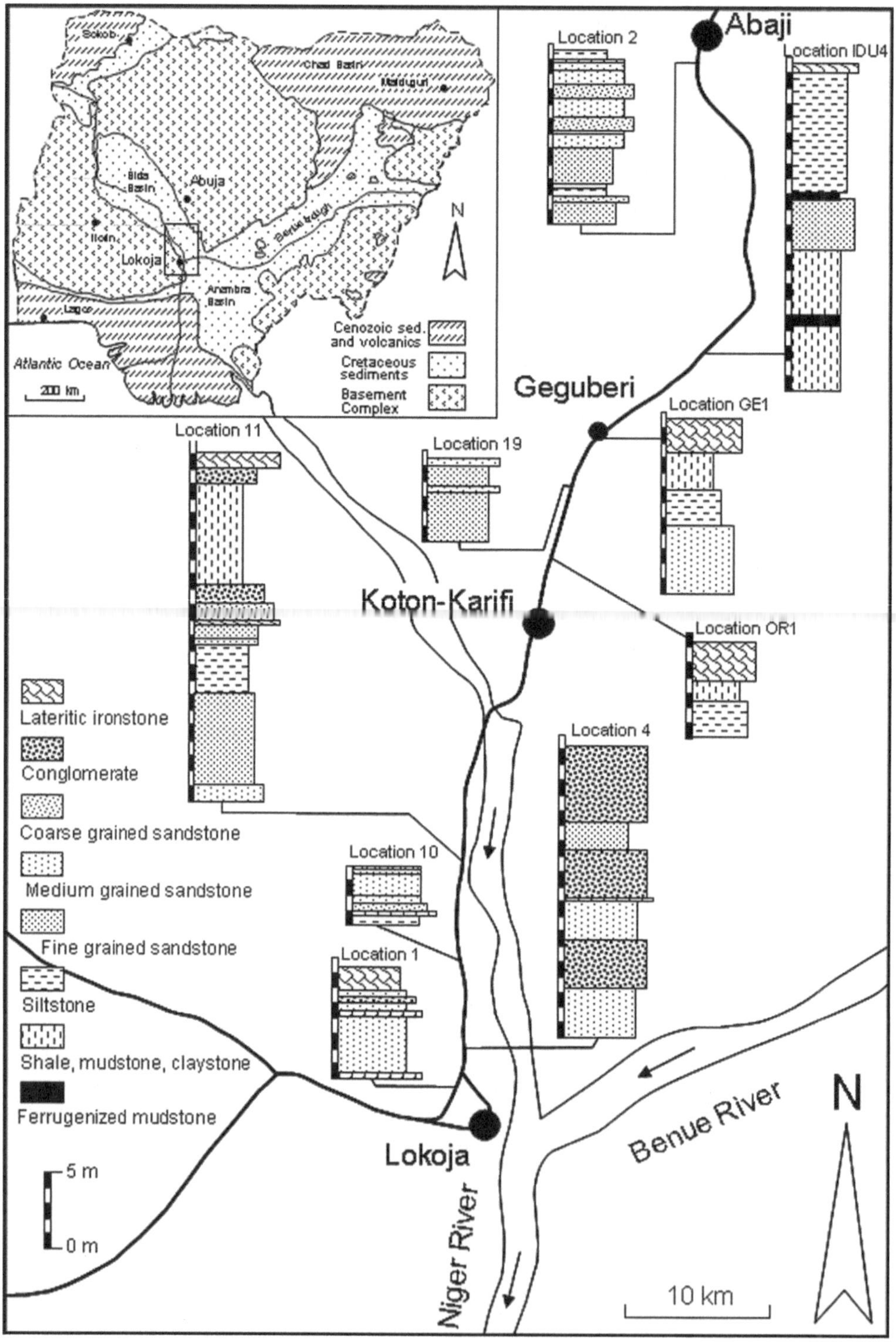

Figure 6. Hydrostratigraphic section through the Middle Niger Basin (after Vrbka *et al.*, 1999).

of similar lithology but slightly brittle as a result of intense chemical weathering of the primary rock-forming minerals, such as feldspars and ferromagnesian, into secondary clay and iron oxides (i.e. laterite).

Water resources inventory of southwestern Nigeria shows that water supplies are mostly from surface sources, such as dams and weirs in streams and rivers. Borehole and dug-wells, tapping groundwater, are used to supplement the short supply from surface water. Available data from UNICEF-water assisted projects have shown that boreholes in southwestern Nigeria are designed to tap water from the weathered zone aquifer or the jointed/fractured basement rock aquifer. This is in accordance with the earlier proposals of Oteze (1981) that groundwater occurs in localized and disconnected aquifers in the crystalline Basement Complex of Nigeria.

The boreholes are either single-screened or multiple-screened, and sometimes open wells are drilled through fractured basement rocks yielding substantial amount of water for private stakeholders. Depth to water level rarely exceeds 24 m, while the mean dry season water level in this region is generally between 10 and 15 m. Most aquifers in this region occur within 40 m from the surface under unconfined conditions, and very few wells tap water below 60 m. An average yield of 0.4 l/s and borehole depth of 40–80 m is estimated for the crystalline basement rocks in Nigeria (Oteze, 1983). Records from available boreholes (in the southern end of Kwara State extending to Osun State) show the borehole depth between 25–68 m while overburden thickness is varying between 3–24 m. In the area around Ibadan (Oyo State) overburden thickness and borehole depth are within the same range. Eniola *et al.* (2006) found that the thickness of the overburden aquifer in the rural areas of Ibadan related to tectonic fractures rather than weathering, influenced the long-term yield of the wells but not on the short-term yield. Hence yields of boreholes are generally low, <2.5 l/s (Adelana & Olasehinde 2004). Yields between 1–2 l/s in the Nigerian basement rocks are considered good enough for the installation of motorized submersible pumps, while values of borehole yield <0.5 l/s are still adequate for handpumps (Offodile, 2002). The recharge into the weathered aquifer is predominantly through infiltration of rainwater (Adelana *et al.*, 2005; Adelana *et al.*, 2006), and prolonged yields from motorized pumps may not be sustainable.

Major aquifers in the coastal zone of western Nigeria occur in sands and overburden/superficial deposits while shales and clays form the impermeable horizons (Longe *et al.*, 1987). The aquifers have variable thickness with first and third horizons attaining thickness of about 200 m and 250 m respectively at Lekki peninsula while the second horizon is approximately 100 m thick (Figure 7). Calculations of groundwater stored in the first aquifer horizon are at a mean of $2.87 \times 10^3\ m^3$ (Asiwaju-Bello & Oladeji 2001). The water table is generally close to the surface, ranging 0.4–21 m below ground level, with a relatively annual fluctuation of less than 5 m. The major aquifer is within coastal plain sand, sometimes underlain by impermeable horizons of shale and lenses of clay. Several high-yielding boreholes provide >30% of the water supply in Lagos and its environs.

4.6 *Regional hydrogeology: The Niger Delta*

The most important aquifers in the Niger Delta are the Deltaic and Benin Formations. Most of the boreholes in the northern parts of the Niger Delta tap unconfined aquifers. In most of these boreholes the geological sequence consists of continuous sandy formations from

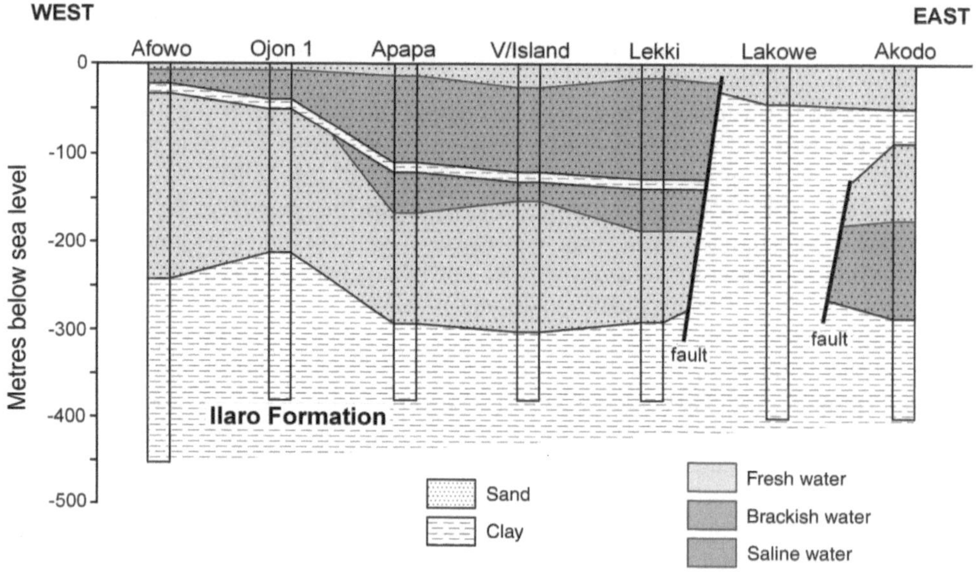

Figure 7. Schematic hydrogeological cross-section along the coastal area of Lagos State, south-western Nigeria.

top to the bottom. However, some aquifers occur under confined conditions resulting in artesian flows. The marked distinction in this area is discussed below.

4.6.1 *Unconfined aquifers*

(1) Deltaic Formation: The water-table in the Niger Delta area is very close to the ground surface, ranging from 0 to 9 m below ground level. The aquifers in this area obtain steady recharge through direct precipitation and major rivers. Rainfall in the Delta is heavy, varying from about 2400 mm a year inland to 4800 mm near the coast. Some proportion of the rainfall is lost by runoff and evapotranspiration. NEDECO (1961) recorded 720–960 mm/a of evapotranspiration along the coast, which is mainly thick mangrove swamps, to 1200 mm/a in the northern parts of the delta which is less thickly vegetated. Hence, very limited water-table fluctuation is expected in these areas where there is heavy rainfall nearly all the year round. According to Offodile (1992), the Deltaic Plains have specific capacities in the range 160–320 m^3/d/m.

(2) Benin Formation: The sediments of the Benin Formation are more permeable than those of the Deltaic Plains. The depth to water table ranges between 3 and 15 m below ground surface. A few values for seasonal fluctuations obtained from the area, indicate seasonal differences between 2.1 and 3.6 m. The Benin Formation, is sandy and highly permeable, with specific capacity 150 and 1400 m^3/d/m (Offodile 1992).

4.6.2 *Confined aquifers*

Confined aquifers occur within both the Deltaic Formation and Benin Formation. These formations are characterized by moderately high yielding artesian flows. In some areas the aquifers are confined by a shale or clay bed up to 36 m thick. The total depth of the aquifers below this shale bed is not yet determined, however, borehole data indicate a depth

Table 2. Lithostratigraphic and hydrostratigraphic correlation between Niger Delta, Calabar Flank, Abakaliki Trough and Anambra Basin in southern Nigeria.

AGE/BASIN	NIGER DELTA	CALABAR FLANK	ABAKALIKI TROUGH		ANAMBRA BASIN	Hydrostratigraphic Units	Hydrogeological groups
PLIOCENE	Benin Formation				Benin Formation	Benin Formation aquifer	Upper
MIOCENE OLIGOCENE EOCENE	Agbada Shale Formation				Ogwashi-Asaba Formation Bende-Ameki Formation	Ogwashi-Asaba aquitard Bende-Ameki Aquifer	
PALEOCENE	Akata Shale				Imo Shale Group	Imo Shale aquitard	
MAASTRICHTIAN	Nkporo Shale	Nkporo Shale			Nsukka Formation Ajali Sandstone Mamu Formation	Nsukka aquitard Ajali Sandstone aquifer Mamu aquiclude	Middle
CAMPANIAN					Enugu Shale Nkporo Shale	Enugu shales aquitard Nkporo shales aquitard	
SANTONIAN							
CONIACIAN	Nkalagu Formation	New Netim Marl	Nkalagu Formation		Agwu Shale	Agwu aquitard	Lower
TURONIAN		Ekenkpon Shale Formation	Eze Aku Group	Agu-Ojo Sandstones Nara Shale	Eze Aku Group	Eze Aku aquitard	
CENOMANIAN			Asu River Group	Ezillo Ibir/Aglla Sandstone	Asu River Group	Asu River aquitard	
ALBIAN		Mfamosing Limestone Formation		Ngbo Ekebeligwe			
APTIAN ?		Awi Formation					
OLDER	Basement Complex					Basement aquifer	

[a] Modified from Murat (1972), Ojoh (1990), Petters (1982), Ekwueme *et al.*, (1995)

of approximately 100 m. Hydrogeological information indicate a hydrological connection between the confined aquifers along the coastline and the unconfined aquifers of the Benin Formation to the north, inland. The aquifers increase in thickness towards the mainland, while the confining clays thin out. The specific capacity for this formation varies from 90–320 m^3/d/m. In the area underlain by the Benin Formation, the confined aquifers occur in the southeastern part of the Niger Delta. The aquifer was confined by several shale and clay beds. The confined aquifers consist mainly of very coarse to medium-grained sands. The specific capacity for this formation varies between 140 and 180 m^3/d/m, and artesian flows have been measured up to 6 l/s.

4.7 *Regional hydrogeology: southeastern Nigeria*

Generally the area falls into three broad hydrogeological groups; the first occurs within the area underlain by the predominantly shaly formations. These are the Asu River Group, Mfamosing Limestone Formation, Ekenkpon Shale, New Ntetim Marl, the Eze-Aku Shale, the Awgu Shale, Nkporo Shale, and Enugu Shale (Lower Hydrogeological Group) while the second group develops within the Mamu, Ajali and Nsukka Formations (Middle Hydrogeological Group) which contain prominent sandy horizons, and the Imo Shale, Bendi-Ameki, Ogwashi-Asaba and the Benin Formations constitutes the Upper Hydrogeological Group (see Table 2).

(1) Lower Hydrogeological Group

The major feature of the first hydrogeological group is the occurrence of a thin shallow but extensive unconfined aquifer. The aquifer is formed by the top weathered horizon within the fractured shales and sandy horizons. The aquifer is extensively exploited by hand-dug wells. The depth to the water-table below the surface varies from place to place but is generally less than 20 m.

Borehole yields can be highly variable, and in many locations it can be difficult to find sustainable water supplies. Groundwater is found in open fractures at shallow depth, between 10 and 40 m. Within the shale units, the degree of metamorphism of the mudstone has been found to be directly related to transmissivity: mudstones that are unaltered (such as the Awgu Shale) have low transmissivity, which mudstones that have been subject to burial and low grade metamorphism (such as the Asu River Group) have much higher transmissivity and a greater chance of finding successful boreholes (MacDonald *et al.*, 2001; MacDonald *et al.*, 2005) The saturated thickness is generally less than 50 m. Yields of successful boreholes are generally less than 0.3 l/s.

Another major hindrance to groundwater development is the occurrence of saline groundwater. The occurrence of saline water in these areas appears to be controlled by the major tectonic event in the area. It has been shown (Uma, 1998) that the saline waters originate from artesian and subartesian groundwater upfluxing through near vertical fractures to the surface. The problem is further compounded by the complex tectonic history (complex folding/faulting) which makes it difficult to reconstruct the hydrostratigraphy of the area. Other poor quality water can be found at shallow depths due to the presence of gypsum ($CaSO_4$)

(2) Middle Hydrogeological Group

The major feature of the Middle Hydrogeological Group is the occurrence of a deep and thick unconfined aquifer. The aquifer is found mostly within the outcrop area of the Ajali Sandstone and has been encountered in much of the area. The total depths of the boreholes range from 70–240 m with an average of 140 m. The yield of the boreholes vary from 3 to >30 l/s and specific capacities are in the range 200–500 m^3/d. Available drilling information indicates an average saturated thickness of 78 m.

Despite the regional extent for the unconfined aquifer, confined situations also occur in some case with pressures sometimes above the ground surface. Pumping test data gave transmissivity values ranging from 36–62 m^2/d and storativity in the order of 2×10^{-2}. The potential of the aquifer decreases towards the west where it is confined beneath the shales of the Nsukka Formation. The decreasing potential is probably related to the decreasing thickness of the aquifer in this region (Uma & Onuoha, 1991).

A striking feature within the second hydrogeological group is the occurrence of perched aquifer systems. The perched aquifers develop mostly in areas where the Nsukka Formation occurs as outliers on the Ajali Sandstone. Such outliers are commonly intensely lateritized near the surface and the lateritized horizons are permeable relative to the shale bedrock beneath them. This forms an environment conducive for the development of perched aquifers.

The depth to the perched water table generally varies between 3–10 m below ground and depends on the relative elevation of the measurement point as well as on the season. Measured values range from 5–9 m at the beginning of the recharge season (April) and 3.6–6.5 m at the end of the recharge season (October). The saturated thickness of the perched aquifer is generally less than 5 m. Perched aquifers also occur within area underlain by the Mamu and Nkporo Formations and constitute the main drinking water supply source for the rural communities in this area especially those around Awgu and Okigwe. Generally the perched aquifer system is insignificant for large-scale groundwater development, but forms the source of many of the springs and streams dotting the severally shaped hills in the area.

(3) Upper Hydrogeological Group
The Imo Shale (Paleocene–Lower Eocene) predominantly comprises shales, claystones, calcareous mudstones, siltstones, ironstones and lenses of sandstone. The shales are fissile and occasionally are interbedded with sandstone intercalations giving rise to aquifer-aquitard systems of local significance.

The Bende-Ameki/Nanka Sand (Middle Eocene) dominantly consists of sands. Many lakes exist in the area due to discharge of groundwater from the flanks of depressions where clay units underlying thick unconsolidated sands create spring situation that feeds the lakes.

The Ogwashi-Asaba Formation which overlies the Ameki Formation is dominantly gritty clay. Lignite seams inter-bedded with clay occurrence at various horizons. Some limited sandy units are also present. The formation is not of much significance in terms of groundwater potential.

The youngest formation, the Benin Formation, overlies the Ogwashi-Asaba Formation and consists of friable sands that are pebbly in places. Intercalations of clay and sandy clay also occur.

The Ajali Sandstone, the Bende-Ameki/Nanka Sand and the Benin Formation/Alluvium are the most permeable formations in southeastern Nigeria and hold the greatest potential for groundwater storage and exploitation.

5 THE QUALITY OF GROUNDWATER IN NIGERIA

5.1 *Distribution of chemical quality according to the groundwater regions in Nigeria*

There is a growing concern at the moment of landfills, subsidence and moderate to severe pollution in both surface and groundwater in some parts of Nigeria. These have attracted more hydrochemical studies in relation to water quality and environmental impact assessment in the various regions in the last twenty years. Table 3 shows selected references related to groundwater quality assessment in Nigeria in the last 25 years.

Groundwater across the country is of variable but generally good quality. However, groundwater is exposed to active pollution in major cities and rural communities of Nigeria due to increased urbanization, indiscriminate waste disposals, intensive agricultural practices and industrial activities. Since the 1980's various chemicals and wastes have been found in groundwater. Nitrate, the most ubiquitous contaminant, has been found at levels in excess of drinking water standards in many regions of Nigeria (Adelana, 2006).

There are cities without organized waste disposal systems; equally where municipal landfills exist they are poorly managed. Groundwater in the coastal areas is often drawn from shallow, thin, and, sometimes, regolith aquifers that are also susceptible to contamination from septic tank seepage. The most significant groundwater contamination problems in southwestern and eastern Nigeria are nitrate and nitrite; total heterotrophic bacteria, THB; total faecal coliform, TFC (Adelana, 2004; Adelana & Olasehinde, 2004; Ugbaja & Edet, 2004). In the Niger Delta elevated nitrate concentrations exceeding 50 mg/l occur over a wide area (cutting across Delta, Anambra, Imo, Bayelsa, Rivers and Cross-River States).

Table 3. References related to groundwater quality case studies in Nigeria.

Author (s)	Year of publication	Comments
Adelana	2006	Nationwide investigation of nitrate in groundwater, based on analyses of samples from about 2, 200 water supply wells; 33% of the wells with nitrate values above 45 mg NO_3/L
Graham *et al.*	2006	Monitored groundwater quality in the Sokoto-Rima Basin, NW Nigeria; high NO_3 found
Ophori	2006	Groundwater quality in shallow domestic water wells, Ughelli, Nigeria; deteriorating quality
Adelana *et al.*	2005	The impact of anthropogenic activities on groundwater quality of a coastal aquifer in SW Nigeria; high NO_3, Cl, trace metals (Cu, Fe, Mn, Al, Zn, Pb, As, Cd, Cr) & H_2S of concern
Ofoma *et al.*	2005	Physico-chemical quality of groundwater in parts of Port Harcourt city, SE Nigeria
Adelana	2004	Water pollution by nitrate in weathered/fractured rocks in Offa, SW Nigeria
Adelana *et al.*	2005	Quality assessment and pollution vulnerability of groundwater in Lagos, SW Nigeria
Ugbaja & Edet	2004	High levels (NO_3, THB, TFC) in groundwater in Calabar, SE Nigeria; 58 and 62 % of THB and TFC counts respectively recorded exceed the WHO limits
Adelana & Olasehinde	2003	Report the occurrence and distribution of nitrate in over 1200 domestic wells in Nigeria
Adelana, *et al.*	2003	Deteriorating water quality with respect to chloride and nitrate in a coastal city, SW Nigeria
Adelana *et al.*	2003	Hydrochemical characteristics with comments on high Cl, NO_3 content in Offa, SW Nigeria
Ofoma *et al.*	2003	Hydrochemical investigation in Port Harcourt, low NO_3 content, high potential for metals
Abimbola *et al.*	2002	The environmental impact assessment of waste disposal site and groundwater quality in Oke-Ado, Lagos was evaluated; pollution traced to the waste disposal sites
Aremu *et al.*	2002	Heavy metals in groundwater from Warri, SE Nigeria
Awalla & Ezeigbo	2002	Appraisal of water quality in the Nturu-Okposi area, Ebonyi State, Southeastern Nigeria
Edmunds *et al.*	2002	High nitrate concentration in the unsaturated soil zone in northern Nigeria
Eniola	2002	Relationship of the hydrogeochemical characteristics of southwestern Nigeria Basement
Ikem *et al.*	2002	Summarizes the groundwater quality characteristics near two waste sites in Ibadan and Lagos, Nigeria emphasizing the extent of soil, water and air pollution
Adelana *et al.*	2001	Hydrochemical characteristics with comments on the elevated concentration of chloride, sulphate and nitrate in Sokoto sedimentary basin NW Nigeria

Table 3. (Continued)

Author (s)	Year of publication	Comments
Edet & Okereke,	2001	A regional study of saltwater intrusion in Southeastern Nigeria, high nitrate content
Edet *et al.*	2001	Chemical characteristics of Odukpani junction springs, Calabar Flank, SE Nigeria
Ntekim & Hussaini	2001	Heavy metal contents of soils & well water in Yola, NE Nigeria; 43% of wells >45 $mgNO_3/L$
Olatunji *et al.*	2001	Hydrochemical evaluation of the Water Resources of Oke – Agbe Akoko, SW Nigeria
Bassey & Opeloye	2000	Report of groundwater quality in the Yola arm sedimentary Basin, NE Nigeria
Edet	2000	Nitrate pollution in coastal city of Calabar (SE Nigeria); 80% of the wells exceeded 45 mg/l
Raji & Alagbe	2000	A topo-geochemical sequence of groundwater in Asa drainage basin, Kwara State, Nigeria
Ajayi & Umoh	1998	Groundwater quality in the Coastal Plain Sands Aquifer of the Akwa Ibom State, Nigeria
Goni & Agbo	1998	Geochemical Studies of Groundwater in the semi-arid Manga Grasslands, NE Nigeria.
Edmunds & Gaye	1997	High nitrate baseline concentrations in groundwaters from the Sahel
Mbonu & Ibrahim-Yusuf	1994	A preliminary survey of nitrate concentrations in groundwater in north central Nigeria
Okafor	1994	The physico-chemical qualities of waters of River Bakogi catchment area of Niger State
Shekwolo & Shoeneich	1994	Hydrochemistry of the Bida Basin and its influence on agricultural productivity
Edet	1993	Groundwater quality assessment in parts of Eastern Niger Delta, Nigeria
Uma	1993	Nitrates in shallow (regolith) aquifers around Sokoto Town, Nigeria
Egboka & Ezeonu	1990	Nitrate and nitrite pollution in groundwater in parts of SE Nigeria; 83% with NO_3 >50 mg/l
Amadi *et al.*	1989	Hydrogeochemical assessment of groundwater quality in parts of the Niger delta, Nigeria
Amajor	1986	Geochemical characteristics of groundwaters in Port Harcourt and environs, Niger Delta
Nwogute	1986	Highlight the influence of the geology of aquifers on groundwater quality of Kaduna State
Langenegger	1981	High nitrate concentrations in shallow aquifers in a rural area of Central Nigeria caused by random deposits of domestic refuse and excrement

In fact, this area with high nitrate in groundwater also extends several kilometers into Akwa-Ibom State where concentrations exceeded 200 mg/l (Egboka & Ezeonu, 1990). According to Egboka & Ezeonu (1990) high concentrations of nitrate (up to 636 mg/l) with an average of 110 mg/l were recorded in surface water of this area.

Although nitrate measurements are few and lacking in mostly reported hydrochemical investigations in the western part of the Niger Delta, indications are that nitrate concentrations are lower. In the vicinity of Port Harcourt, nitrate concentrations are generally low in groundwater (Egboka & Ezeonu, 1990; Ofoma *et al.*, 2003). About 87 wells (boreholes and dug wells) were analyzed chemically within Port Harcourt metropolis (Ofoma *et al.*, 2003) and all were found to be below 10 mg/l. This was in agreement with an earlier hydrogeochemical investigation of subsurface waters (Etu-Efeotor, 1981) where nitrate values were extremely low (<10 mg/l) within Port Harcourt and Patani areas. However, in Bonny, Omoku, and Akassa areas concentrations range between 30–60 mg/l. These isolated occurrences of high nitrate are observed in rural areas and are considered to be due to anthropogenic effects and indiscriminate waste disposal near unprotected wells.

In the past, elevated nitrates occur in shallow volcanic and crystalline basement aquifers of north-central Nigeria (Langenneger, 1981) as well as in the sedimentary basin of northwestern Nigeria (Uma, 1993). Recent groundwater quality assessment in the southwest Sokoto-Rima Basin (northwest Nigeria) still show that nitrate concentrations are elevated and attributable to excessive application and leaching of nitrogen fertilizers (Graham *et al.*, 2006). The high nitrate concentrations in shallow aquifers in the rural areas of Central Nigeria were reported as caused by random deposits of domestic refuse and excrement. However, intensive regulatory campaign and provision of basic sanitation in the rural settlements in Nigeria have helped to improve the situation over the years.

Some shallow groundwater has naturally high iron content (Etu-Efeotor, 1981; Ohagi & Akujieze, 1989; Oteze, 1991; Ophori, 2006). Nigeria has a southern coastline and high water demand in the densely populated coastal cities so there is potential for salt-water contamination of the coastal aquifers in the Lagos area (Adelana *et al.*, 2003b; Adelana *et al.*, 2005), Port Harcourt area (Etu-Efeotor & Odigi, 1983; Ofoma *et al.*, this volume) and Calabar area (Edet & Okereke, 2001). The extent of seawater intrusion in the coastal beach ridge of the Forcados, Niger Delta has been delineated using geoelectrical survey data (Oteri, 1990) and the pattern of migration discussed in Amadi & Amadi (1990). Hindrance to groundwater development in most of this area, especially within the first hydrogeological group in southeastern Nigeria, is the occurrence of saline groundwater. The salinity of groundwater in these areas varies from <5,000 mg/1 (of NaCl) to >30,000 mg/1 (Uma, 1998). The situation is similar to the Cretaceous sediments of the Benue Trough (Ogoja area) where groundwater exhibit >20,000 mg/l Na and >50,000 mg/l Cl (Uma, 1998; Tijani *et al.*, 1996).

6 MANAGEMENT AND PROTECTION OF AQUIFERS IN NIGERIA

Groundwater is a vital natural resource for the reliable and economic provision of safe water supplies in both the urban and rural environment. In spite of the fundamental role groundwater play in human well being, as well as that of many ecosystems, it is yet to be fully appreciated and adequately protected in Nigeria. Groundwater is generally taken as been the 'free gift from God'. Therefore, more than often, those exploiting this natural resource have taken no action to protect its quality as well as ensuring its sustainable management.

Groundwater contamination in rural areas has recently become the primary subject of groundwater investigations, particularly in developing Africa, partly as a result of the diversity, and increasing use of potential contaminants, and because most rural communities rely heavily on groundwater as their main source of drinking water. Even though the demand for groundwater is becoming higher in both rural and urban centres in Nigeria, yet the management of the aquifers or well fields is not closely monitored. Careful management is required to avoid further degradation of groundwater quality. There is a pressing need for a groundwater quality and monitoring network to investigate baseline character and trends. Point-source discharges and significant use need to be regulated through conditions for consents and regional rules to ensure sustainable resource management. National or regional codes of practice for managing non-point sources and commitment to environmental education as a means for managing groundwater quality are further required on the part of the government. For municipal water supply, high and stable raw-water quality is a prerequisite, and one best met by protected groundwater sources. Recourse to treatment processes (beyond precautionary disinfection) in the achievement of this end should be a last resort, because of their technical complexity and financial cost, and the operational burden they tend to impose.

6.1 *Defining strategy and setting priorities*

Vulnerability assessment is little known or researched in Nigeria. While regional plans and rules exist to protect surface waters, groundwater protection is in its infancy in Nigeria, and can be regarded as inadequate. Hanidu (1990) in his paper on 'National growth, water demand and supply strategies in Nigeria in the 1990's' emphasized the need to settle the ownership of water through legislation before any meaningful strategy could be adopted. The water resources legislation is already in existence but it's effect is yet to be felt in terms of true ownership and usage of the nation's water resources. The reasons for these are obvious. For example, the Decree 101 and water-related regulations are based solely on socio-economic factors and not necessarily on scientific research such as, groundwater vulnerability or aquifer sensitivity. Federal, States and Local Government programmes need to be criticized for not formulating or implementing policies on on-site effluent disposal.

The requirements of Decree 101 of 1990 have stimulated investigations such as impact assessments of waste disposal sites on groundwater quality. These investigations are aimed at existing or recognized potential problem sites but are often only 'token' investigations carried out by companies or corporations, to meet their obligations under the Federal, States and Local Governments conditions of award of contracts. Nigeria has standards for neither the evaluation of existing landfills nor for the investigation of potential sites. In some cases, it is also difficult to find the party responsible for degrading the resource or it is difficult to get such a party to assume liability. This alone emphasizes the need for a proactive approach to assess groundwater vulnerability under present conditions.

National plans and rules must be put in place to assist Local Government or Municipal Councils to carry out their functions relating to resource management. Unfortunately, this is still a game of chance under the present water management policies in Nigeria. There is need to restate and reform the law relating to the use of land, air, and water. This will help to promote the sustainable management of natural and physical resources, including groundwater. 'Sustainable management' must ensure that resources are sustained for present and future generations and adverse environmental effects are avoided, remedied, or mitigated. Therefore, it is the responsibility of the national and regional water councils to implement the water legislation or Decree 101 and incorporate aspects relating to management of groundwater or prevents groundwater contamination based on current scientific evidence.

Groundwater protection should be emphasized in order to prevent deteriorating conditions of the main aquifers in Nigeria. The assessment of aquifer vulnerability and sensitivity to pollution on a national scale is very necessary in Nigeria under the present conditions. The most important potential use of aquifer protection is in raising public awareness which, in turn, may result in positive reactions or more informed land use decisions. Groundwater vulnerability should be a dominant factor for analyzing alternatives to a particular groundwater quality plan and the response to a particular pollution event. Vulnerability can also aid the zoning, or assigning the range of acceptable activities, to the land surface. Aquifers with high sensitivity should be monitored closely while aquifers with low sensitivity may not require detail monitoring. In addition, permits or consents for environmental activities should have more demanding conditions imposed on them in areas of high as opposed to low sensitivity. Conditions may refer to quantities, treatment, and containment of contaminants or limitation of the activity to a certain time period.

6.2 *Recommendations*

This paper concludes with the following recommendations for managing groundwater and protecting aquifers in Nigeria.

To protect aquifers against pollution it is essential to control and manage land use, effluent discharge and waste disposal practices in Nigeria. However, in practice it is necessary to define groundwater protection strategies that accept trade-offs between competing interests. Thus, instead of applying universal controls over land use and effluent discharge, it is more cost-effective to utilize the natural contaminant attenuation capacity of the strata overlying the aquifer, when defining the level of control required to protect groundwater quality.

Aquifer pollution vulnerability assessment on a national scale is desirable and should be carried out. Local and municipal governments should also design or approve similar projects based on the resources available to them. Simple and robust zones (based on aquifer pollution vulnerability and source protection perimeters) need to be established, with matrices that indicate what activities are possible where at an acceptable risk to groundwater. Groundwater protection zoning also has a key role in setting priorities for groundwater quality monitoring, environmental audit of industrial premises, pollution control within the agricultural system, determining priorities for the cleanup of historically contaminated land, and in public education generally. All of these activities are essential components of a sustainable strategy for groundwater quality protection.

There is the need to strike a balance between the protection of groundwater resources (aquifers as a whole) and specific sources (boreholes, wells and springs). While both approaches to groundwater pollution control are complementary and have been applied in many countries, the emphasis in Nigeria has been misplaced and as such redirecting efforts to achieving aquifer protection will depend on the resource development situation and on knowledge of the prevailing hydrogeological conditions.

REFERENCES

Abimbola, A. F., Odukoya, A. M. & Adesanya, O. K. 2002. *The environmental impact assessment of waste disposal site on groundwater in Oke-Ado, Lagos, Southwestern Nigeria*. Proc. 15th Annual Conf. Nigerian Association Hydrogeologists, Kaduna, Nigeria, pp. 42.

Adegoke, O. S. 1969. *Eocene stratigraphy of southern Nigeria*. Bull. Bur. Rech. Geol. Mem. **69**, 23–48.

Adelana, S. M. A. 2006. *Nitrate pollution in Nigeria*. In: Xu & Usher (eds.) Groundwater Pollution in Africa. Taylor & Francis, London, 37–45.

Adelana, S. M. A. 2004. *Water pollution by nitrate in a weathered/fractured basement rock aquifer: The case of Offa area, West central Nigeria*. In: X. Ru-Ze, G. Wei-Zu, K. & P. Seiler (eds.) Research Basins & Hydrological Planning. A.A. Balkema, Amsterdam, 93–100.

Adelana, S. M. A. & Olasehinde, P. I. 2004. *Characterization of groundwater flow in fractured/weathered hard rock using hydrogeochemical and isotopic investigations*. Proc. XXXIII Congress of the International Association of Hydrogeologists (IAH), 11–15 October, Zacatecas, Mexico.

Adelana, S. M. A. & Olasehinde, P. I. 2003. *High nitrate in water supply in Nigeria: implications for human health*. Water Resources, **14**, 1–11.

Adelana, S. M. A., Olasehinde, P. I. & Vrbka, P. 2006. *A quantitative estimation of groundwater recharge in parts of Sokoto Basin, Nigeria*. Journal of Environmental Hydrology, **14**, 1–17.

Adelana, S. M. A., Bale, R. B., Olasehinde, P. I. & Wu, M. 2005. *The impact of anthropogenic activities over groundwater quality of a coastal aquifer in Southwestern Nigeria*. Proc. Aquifer Vulnerability & Risk, 2nd International Workshop & 4th Congress on the Protection and Management of Groundwater, 21–23 September 2005, Reggia di Colorno – Parma.

Adelana, S. M. A., Olasehinde, P. I. & Vrbka, P. 2003a. *Isotopes and geochemical characterization of surface and subsurface waters in the semi-arid Sokoto Basin, Nigeria*. African Journal of Science and Technology, **4**, 76–85.

Adelana, S. M. A., Bale, R. B. & Wu, M. 2003b. *Quality assessment and pollution vulnerability of groundwater in Lagos metropolis, SW Nigeria*. Aquifer Vulnerability at Risk (AVR03), Salamanco, Mexico, Vol. 2: 1–17.

Adelana, S. M. A., Olasehinde, P. I. & Vrbka, P. 2001. *Hydrochemical Characteristics of groundwater in the Sokoto Basin, NW Nigeria, West Africa*. Zentralblatt für Geologie und Paläontologie, Teil I, Heft 3: 365–374, Stuttgart, Germany.

Adelana, S. M. A., Olasehinde, P. I. & Vrbka, P. 2006 *Identification of groundwater recharge conditions in crystalline basement rock aquifers of the southwestern Nigeria*. In: Recharge systems for protecting and enhancing groundwater resources, IHP-VI, Series on Groundwater No. 13, 649-655, UNESCO, Paris.

Adeniran, B. V. 1991. *Maastrichtian tidal flat sequences from the northern Anambra Basin, southern Nigeria*. Nigerian Association of Petroleum Exploration Bulletin, **6**, 56–66.

Ajayi, O. & Umoh, O. A. 1998. *Quality of groundwater in the Coastal Plain Sands Aquifer of the Akwa Ibom State, Nigeria*. Journal of African Earth Sciences, **27**, 259–275.

Amadi, U. M. P. & Amadi, P. A. 1990. *Saltwater migration in the coastal aquifers of southern Nigeria*. Nigerian Journal of Geology and Mining, **26**, 35–44.

Amadi, P. A., Ofoegbu, C. O. & Morrison, T. 1989. *Hydrogeochemical assessment of groundwater quality in parts of the Niger delta, Nigeria*. Environmental Geology, **14**, 195–202.

Amajor, L. I. 1986. *Geochemical characteristics of groundwaters in Port Harcourt and environs*. Proc. NIWASA Symp., Ikeja, Nigeria, 366–374.

Anderson, H. R. & Ogilbee, W. 1973. *Aquifers in the Sokoto Basin, Northwestern Nigeria, with a description of the general hydrogeology of the region*. Geological Survey Nigeria, Water Supply Paper, **1757-L**.

Annor, A. E. & Olasehinde, P. I. 1996. *Vegetational niche as a remote sensor for subsurface aquifer: A geological-geophysical study in Jere area, Central Nigeria.* Water Resources, **7**, 26–30.

Aremu, D. A., Olawuyi, J. F., Meshitsuka, S., Sridhar, M. K. & Oluwande, P. A. 2002. *Heavy metal analysis of groundwater from Warri, Nigeria.* International Journal of Environmental Health Research, **12**, 261–267.

Asiwaju-Bello, Y. A. & Oladeji, O. S. 2001. *Numerical modelling of ground water flow patterns within Lagos metropolis, Nigeria.* Nigerian Journal of Mining and Geology, **37**, 185–194.

Awalla C. O. & Ezeigbo, H. I. 2002. *An appraisal of water quality in the Nturu-Okposi area, Ebonyi State, Southeastern Nigeria.* Water Resources, **13**, 33–40.

Barber, W. & Jones, D. G. 1960. *The geology and hydrology of Maiduguri, Bornu Province.* Record of the Geological Survey Nigeria.

Bassey, N. & Opeloye, S. A. 2000. *Groundwater quality data in the Yola arm sedimentary Basin of Adamawa State, NE Nigeria.* Water Resources, **11**, 26–30.

Bassey, J. O., Maduabuchi, C., Onugba, A., Verhagen, B. T. & Vrbka, P. 1999. *Preliminary results of hydrogeological and isotopic research in the Rima River basin, NW Nigeria.* Water Resources, **10**, 31–37.

Beacon Services Limited 1979. Maiduguri water supply, Hydrogeological report Vol. 1, Hydrogeology and future exploitation of groundwater. Consulint Int. S.r.l., 158p.

Carter, J. D., Barber, W. & Tait, E.A. 1963. *The geology of parts of Adamawa, Bauchi and Bornu Provinces in north-eastern Nigeria.* Bulletin of the Geological Survey Nigeria, **30**.

Du Preez, J. W. & Barber, W. 1965. *The distribution and chemical quality of groundwater in northern Nigeria.* Geological Survey of Nigeria Bulletin, **36**, 1–93.

Edet, A. E. 2000. *Water pollution by nitrate near some waste disposal sites in Calabar, Nigeria.* Groundwater 2000. Proc. Int. Conf. Groundwater research, Copenhagen, 6–8 June, 239–240.

Edet, A. E. 1993. *Groundwater quality assessment in parts of Eastern Niger Delta, Nigeria.* Environmental Geology, **22**, 41–46.

Edet, A. E. & Okereke, C. S. 2001. *A regional study of saltwater intrusion in Southeastern Nigeria based on analysis of geoelectrical and hydrochemical data.* Environmental Geology, **40**, 1278–1289.

Edet, A. E., Uka, N. K. & Offong, O. E. 2001. *Chemistry and discharge characteristics of Odukpani junction springs, Calabar Flank, Southeastern Nigeria.* Environmental Geology, **40**, 1214–1223.

Edmunds, W. M. & Gaye, C. B. 1997. *High nitrate baseline concentrations in groundwaters from the Sahel.* Journal of Environmental Quality, **26**, 1231–1239.

Edmunds, W. M., Fellman, E., Goni, I.B. & Prudhomme, C. 2002. *Spatial and temporal distribution of groundwater recharge in northern Nigeria.* Hydrogeology Journal, **10**, 205–215.

Egboka, B. C. E. & Ezeonu, F. C. 1990. *Nitrate and Nitrite pollution and contamination in parts of SE Nigeria.* Water Resources, **2**, 101–110.

Ekwueme, B. N. 1987. *Structural Orientations and Precambrian deformation episodes of Uwet area, Oban massif.* Precambrian Research, **34**, 269–289.

Ekwueme, B. N., Nyong, E. E. & Petters, S. W. 1995. *Geological excursion guidebook to Oban massif, Calabar Flank and Mamfe Embayment, southeastern Nigeria.* Decford Publishing, Calabar, Nigeria.

Eniola, O. A., Opoola, A. O. Adesokan, H. A. 2006. *Empirical analysis of electromagnetic profiles for groundwater prospecting in rural areas of Ibadan, southwestern Nigeria.* Hydrogeology Journal, **14**, 613–624.

Etu-Efeoter, J. O. 1981. *Preliminary hydrogeochemical investigations of sub-surface waters in parts of the Niger Delta.* Nigeria Journal of Mining and Geology, **18**, 103–105.

Etu-Efeotor, J. O. & Odigi, M. I. 1983. *Water supply problem in the Eastern Niger delta.* Nigerian Journal of Mining and Geology, **20**, 183–195.

FAO 2005. *AQUASTAT-FAO's information system on water and agriculture in Nigeria, Water Report no.29.* Food and Agricultural Organisation, Rome.

Farnbauer, B. & Tietz, G. 2000. *The individuality of laterites developed on the Jos-Plateau/Central-Nigeria (in Deutsch)*. Zbl. Geol. Palaeont. Tiel I, Heft 5/6, 509–525.

Gebhardt, H. 1998. *Benthic foraminifera from the Maastrichtian lower Mamu Formation near Leru (southern Nigeria): paleoecology and paleogeographic significance.* Journal of Foraminiferal Research, **28**, 76–89.

Goni, I. B. 2002. *Realimentation des eaux souterraines dan le secteur Nigerian du bassin du lac Tchad: approche hydrogeochimique.* These de Doctorat d'etat, Universite d'Avignon, France.

Goni, I. B. & Agbo, J. U. 1998. *Geochemical Studies of Groundwater in the semi-arid Manga Grasslands, NE Nigeria.* Water Resources, **8**, 34–40.

Graham, W. B. R., Pishiria, I. W. & Ojo, O. I. 2006. *Monitoring of groundwater quality for a small-scale irrigation: Case studies in the southwest Sokoto-Rima Basin, Nigeria.* Agricultural Engineering International: the CIGR Ejournal, Vol. VIII, LW 06 002.

Hanidu, J. A. 1990. *National growth, water demand and supply strategies in Nigeria in the 1990s.* Water Resources, **2**, 1–6.

Ikem, A., Osibanjo, O., Sridhar, M. K. C. & Sobande, A. 2002. *Evaluation of groundwater quality characteristics near two waste sites in Ibadan and Lagos, Nigeria.* Water, Air and Soil Pollution, **140**, 307–333.

Jones, B. 1948. *The Sedimentary Rocks of Sokoto Province.* Bulletin of the Geological Survey of Nigeria, **18**.

Jones, H. A. & Hockey, R. D. 1964. *The geology of part of southwestern Nigeria.* Bulletin of the Geological Survey of Nigeria, **31**, 101pp.

Kehinde, M. O. 1990 Die *Grundwasser-Ressourcen des Bida-Beckens, Zentral-Nigeria.* Ph.D, University of Münster, Germany.

Kogbe, C. A. 1989. *Geology of Nigeria.* Rock View (Nigeria) Limited. Jos, Nigeria.

Ladipo, K. O. 1988. *Paleogeography, sedimentation and tectonics of the Upper Cretaceous Anambra Basin, southeastern Nigeria.* Journal of African Earth Sciences, **7**, 865–871.

Langennegger, O. 1981. *High nitrate concentrations in shallow aquifers in a rural area of Central Nigeria caused by random deposits of domestic refuse and excrement.* In: W. van Duijvenbooden, P. Glasbergen, P. & H. van Lelyveld (eds.) Quality of Groundwater: Studies in Environmental Science, **17**, 135–140.

Longe, E. O., Malomo, S. & Olorunniwo, M. A. 1987. *Hydrogeology of Lagos metropolis.* Journal of African Earth Sciences, **6**, 163–174.

MacDonald, A. M., Davies, J. & Peart, R. J. 2001. *Geophysical methods for locating groundwater in low permeability sedimentary rocks: Examples from southeast Nigeria.* Journal of African Earth Sciences, **32**, 115–131.

MacDonald, A. M., Kemp, S. J. & Davies J. 2005. *Transmissivity variations in mudstones.* Ground Water, **43**, 259–269.

MacDonald, A. M., Cobbing, J. & Davies J. 2005. *Developing groundwater for rural water supply in Nigeria.* British Geological Survey Commissioned Report **CR/05/219N**.

Mands, E. 1992. *Kritische Betrachtung der Wasserbilanzparameter und hydrologische Untersuchungen der Einzugsgebiete River Gbako und River Gurara (Mittelnigeria).* Giessener Geologische Schriften, **47**, 1–174.

Mbonu, M. & Ibrahim-Yusuf, A. I. 1994. *Groundwater quality in the Basement Complex region of north-central Nigeria – A preliminary survey of nitrate concentration.* Water Resources, **5**, 16–21.

Murat, R. C. 1972. *Stratigraphy and Paleogeography of the Cretaceous and Lower Tertiary in Southern Nigeria.* In: T. F. L. Dessavuagie & A. J. Whiteman (eds,), African Geology, Ibadan, Geol. Dept. Univ. Ibadan, Nigeria, 251–266.

NEDECO, 1961. *River studies and recommendations on improvement on the Niger and Benue.* North Holland Publishing Company.

Ntekim, E. E. & Wandate, S. D. 2001. *Groundwater chemistry of the Basement Rock units of Adamawa area NE Nigeria: A reflection of the aquifer rock types.* Water Resources, **12**, 67–73.

Nwogute, N. S. 1986. *The influence of the geology of aquifers on groundwater quality of Kaduna State of Nigeria.* Proc. NIWASA Symp., Ikeja, Nigeria, pp. 287–304.

Offodile, M. E. 2002. *Groundwater study and development in Nigeria.* Mecon, Jos, Nigeria, 451 pp.

Offodile, M. E. 1992. *An approach to ground water study and development in Nigeria.* Mecon, Jos, Nigeria, 247 pp.

Offodile, M. E. 1972. *Groundwater level fluctuations in the East Chad basin of Nigeria.* Nigerian Journal of Mining and Geology, **7**, 19–34.

Ofoma, A. E., Ngah, S. A. & Onwuka, O. S. 2008. *Salinity problems in coastal aquifers: Case study from Port Harcourt City, Southern Nigeria.* This Volume.

Ofoma, A. E., Omologbe, D. & Aigberua, P. 2005. *Physico-chemical quality of groundwater in parts of Port Harcourt city, Eastern Niger Delta.* Water Resources, **16**, 18–24.

Ofoma, A. E., Omologbe, D. A. & Aigberua, P. 2003. *Hydrochemical investigation of groundwater samples from some parts of Port Harcourt City area east of Nigeria Delta.* Proc. 39th Annual Int. Conf. NMGS, 2–3 March 2003, Itakpe, Nigeria, 13–14.

Ohagi, S. M. O. & Akujieze, C. N. 1989. *Iron in borehole water sources in Bendel State Nigeria.* Water Resources, **1**, 192–196.

Ojo, S. B. & Ajakaiye, D. E. 1989. *Preliminary interpretation of gravity measurements in the Middle Niger Basin area, Nigeria.* In: Kogbe, C. A. (ed.) Geology of Nigeria. Rock View (Nigeria) Limited, Jos, Nigeria, 347–358.

Ojoh, K. A. 1990. *Cretaceous geodynamic evolution of the southern part of the Benue Trough (Nigeria) in the equatorial domain of the South Atlantic stratigraphy, basin analysis, and paleo-oceanography.* Centres Recherche' Exploration-Production Elf-Aquitaine, Bulletin, **14**, 419–442.

Okafor, D. U. 1994. *The physico-chemical qualities of waters of River Bakogi catchment area of Niger State, Nigeria.* Water Resources, **5**, 21–27.

Okosun, E. A. 1998. *Review of the early Tertiary stratigraphy of southwestern Nigeria.* Journal of Mining Geology, **34**, 27–35.

Olaniyan, O. & Olobaniyi, S. B. 1996. *Facies analysis of the Bida Sandstone Formation around Kajita, Nupe Basin, Nigeria.* Journal of African Earth Sciences, **23**, 253–256.

Olatunji, A. S., Tijani, M. N., Abimbola, A. F. & Oteri, A. U. 2001. *Hydrochemical evaluation of the Water Resources of Oke – Agbe Akoko, Southwestern Nigeria.* Water Resources, **12**, 81–87.

Oluyide, P. O., Nwajide, C. S. & Oni, A. O. 1998. *The geology of Ilorin area with explanations on the 1:250,000 series, sheet 50 (Ilorin).* Geological Survey of Nigeria Bulletin, **42**, 1–84.

Ophori, D. 2006. *Groundwater Quality in Shallow Domestic Water Wells, Ughelli, Nigeria.* Web Address: http://www.iseg.giees.uncc.edu/abuja2006/Abstracts/Abstract_ID_284

Oteri, A. U. 1990. *Delineation of saltwater intrusion in a coastal beach ridge of Forcados.* Nigerian Journal of Mining and Geology, **26**, 35–44.

Oteze, G. E. 1991. *Potability of groundwater from the Rima Group aquifer in the Sokoto Basin Nigeria.* Nigerian Journal of Mining and Geology. **27**, 12–23.

Oteze, G. E. 1989a. *Recharge characteristics of Rima aquifers, Sokoto Basin.* Water Resources, **1**, 154–160.

Oteze, G. E. 1989b. *The Hydrogeology of the North-Western Nigeria Basin.* In: Kogbe, C. A. (ed.) Geology of Nigeria. Rock View (Nigeria) Limited, Jos, Nigeria, 455–472.

Oteze, G. E. 1983. *Groundwater levels and ground movements.* In: Ola, S. A. (ed.) Tropical soils of Nigeria in engineering practise. Balkema, Rotterdam, The Netherlands, 39–58.

Oteze, G. E. 1981. *Water resources in Nigeria.* Environmental Geology, **3**, 177–184.

Oteze, G. E. & Fayose, S. A. 1988. *Regional development in the Hydrogeology of the Chad basin.* Water Resources, **1**, 9–29.

Oyawoye, M. O. 1972. *The basement complex of Nigeria.* In: TFJ Dessauvagie, T. F. J. & Whiteman, A. J. (eds.) African Geology. University of Ibadan Press, Ibadan, Nigeria, 66–102.

Petters, S. W. 1982. *Central West African Cretaceous – Tertiary benthic Foraminifera and stratigraphy*. Paloeontographica, A. 179, 1–104.

Reyment, R. A. 1965. *Aspects of the Geology of Nigeria*. University of Ibadan Press, Nigeria, 145pp.

Raji, B. A. & Alagbe, S. A. 2000. *A topo-geochemical sequence study of groundwater in Asa drainage basin, Kwara State, Nigeria*. Environmental Geology, **39**, 544–548.

Shekwolo, P. D. & Shoeneich, K. 1994. *Hydrochemistry of the Bida Basin and its influence on agricultural productivity*. Water Resources, **5** 28–33.

Tijani, M. N., Loehnert, E. P. & Uma, K. O. 1996 *Origin of saline groundwaters in the Ogoja area, Lower Benue Trough, Nigeria*. Journal of African Earth Sciences, **23**, 237–252.

Vrbka, P., Ojo, O. J. & Gebhardt, H. 1999. *Hydraulic characteristics of the Maastrichtian sedimentary rocks of the southeastern Bida basin, central Nigeria*. Journal of African Earth Sciences, **29**, 659–667.

Ugbaja, A. N. & Edet, A. E. 2004. *Groundwater pollution near shallow waste dumps in Southern Calabar, South-Eastern Nigeria*. Global Journal of Geological Sciences, **2**, 199–206.

Uma, K. O. 1998. *The brine fields of the Benue Trough, Nigeria: a comparative study of geomorphic, tectonic and hydrochemical properties*. Journal of African Earth Sciences, **26**, 261–275.

Uma, K. O. 1993. *Nitrates in shallow (regolith) aquifers around Sokoto Town, Nigeria*. Environmental Geology, **21**, 70–76.

Uma, K. O. & Onuoha, K. M. 1991. *Groundwater resources of the Lower Benue Trough, Southern Nigeria*. In: Offeagbu, C. O. (ed.) The Benue Trough, 77–91.

CHAPTER 12

The occurrence of groundwater in northeastern Ghana

W.A. Agyekum & S. Dapaah-Siakwan
CSIR Water Research Institute, Achimota-Accra, Ghana

ABSTRACT: The groundwater resources of northeastern Ghana have been assessed based on the interpretation of existing hydrogeological and water quality data from 2458 boreholes drilled throughout the region between 1954 and 2004 for domestic water supply. Single parameter groundwater and hydrochemical maps were developed to show the spatial distribution of groundwater and the geochemical characteristics for the region. Groundwater occurrence in the area is largely influenced by rainfall, topography, overburden thickness and geology. Aquifers are discrete, localised and discontinuous and groundwater flow is controlled by fracture intensity and the degree of interconnections. The recorded data indicate an average success rate of approximately 65%. Borehole depths range from 28 to 60 m. Groundwater was encountered at a mean depth of 30 m (range 13–42 m), and static water-levels stood at a mean depth of 8 m below ground level. Estimated yields of successful boreholes averaged 0.8 l/s and varied between 0.1 and 6 l/s, with higher yields occurring in the highly-fractured granitic and metamorphic rock areas. The total annual groundwater abstraction comprises approximately 74% of the total domestic water requirement of the region. Concentrations of nitrate, fluoride and manganese higher than WHO guideline levels were recorded in some boreholes in the Bongo, Bolgatanga and Kassena-Nankana districts.

The Upper East region of Ghana has a total land size of 8842 km^2, and forms only 4% of the total area of Ghana. By this size, it is considered one of the smallest regions in the country. In the 2000 population census, the population of the region was estimated to be 920,089 inhabitants, comprising 144,282 urban and 775,807 rural. The study area is located within the semi-arid region where the mean annual potential evapotranspiration exceeds the mean annual rainfall by about 49% (Gyau-Boakye, 1994). Due to this high potential evapotranspiration rate nearly all the surface water sources dry up completely during the dry season. Consequently, about 80 % of the entire population of the region relies on groundwater for water supply. Until the late sixties, most of the communities relied on shallow hand-dug-wells that were constructed by the inhabitants using simple technology and local tools. The hand-dug-wells were constructed into the overburden, and were limited to low-lying areas and valley bottoms. However, owing to the extreme seasonal variations of the water-levels, most of the hand-dug-wells experience decline in their water levels during the dry seasons, and consequently dry up completely at the peak of the dry season.

The inhabitants consequently have no alternative than to return to polluted surface water sources, consisting of pools of water collected along river channels, ponds and dug-outs, resulting in the outbreak of water-related and water-borne diseases in the region.

Consequently, the government of Ghana in cooperation with foreign donors embarked on comprehensive drilling of 2500 deep wells in the two northernmost regions (Upper East and Upper West) of the country between 1974 and 1979. About 1280 of these boreholes, representing 51.8%, were drilled in the Upper East region. Series of borehole drilling programmes continued to provide safe and reliable water supply to the rural communities throughout the country, and by the end of 2004, data on 2458 boreholes drilled in the study area were available. The basic hydrogeological and water quality data generated during the drilling of these wells were gathered, collated, analyzed and the results used to assess the potential of groundwater resources of the region. However, due to the fact that the boreholes were designed for hand-pump water supplies, no pumping tests were carried out on them.

In 1961, the Ghana Geological Survey Department produced a preliminary 'Hydrogeological map of Ghana' on the scale of 1:1,000,000. From the results of the study, the country was broadly classified into 10 hydrogeological provinces on the basis of their geology and groundwater potential. Based upon the classification, the study area was divided into two main hydrogeological provinces (Gill, 1969). The northern part of the region, which is underlain mainly by Precambrian granite, phyllite and quartzite, was classified as having a mean yielding capacity of about 1.25 l/s. Surface water supply during the dry season was noted to be poor in this region, and shallow wells rarely produced appreciable groundwater; therefore, boreholes fitted with hand-pumps were subsequently recommended for this region. The southern half of the region, which is underlain by Palaeozoic sedimentary mudstones and shale (often referred to as Voltaian), was considered neither suitable for the development of boreholes nor hand-dug-wells due to the rate of dry wells in the area; even though the study by Darko (2001) revealed the Voltaian as a promising formation with the highest prevailing transmissivity from all the regional units. It was, however, recommended that boreholes could be constructed at a few carefully selected localities using geophysical investigations to select the drilling sites.

1 BACKGROUND

1.1 *Location*

The study area is located at the northeastern corner of Ghana, and lies within longitudes 0°02′E and 1°32′W and latitudes 10°22′N and 11°11′N. It is bordered to the north by the Republic of Burkina Faso, to the east by the Republic of Togo and to the south and west by the Northern and Upper-West regions of Ghana respectively. The region has an estimated size of 8842 km^2, representing about 4% of the total area of Ghana. For administrative purposes, the region was initially divided into six districts, namely Bawku-East, Bawku-West, Bolgatanga, Bongo, Builsa and Kassena-Nankana. However, based upon the results of 2000 population census in Ghana, the study area was re-zoned to include two additional districts (Talensi-Nabdam and Garu-Tempane) in 2004. This groundwater assessment was, however, made based upon the original six districts. Figure 1 is a location map for the region, showing the six districts and their capitals.

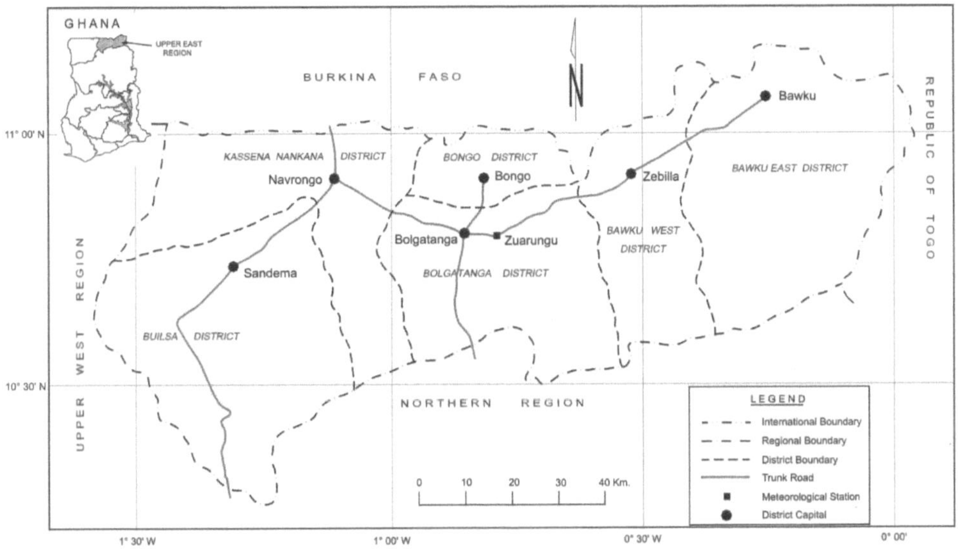

Figure 1. The location of Upper East region, Ghana.

1.2 *Climate and vegetation*

The study area falls within the semi-arid climatic region, otherwise referred to as the Tropical Continental or Interior Savannah climatic region, and is influenced by two main air masses, namely the Southwest Monsoon and the Northeast Trade Winds (Harmattan). Like other areas in West Africa, the climate of the region is highly influenced by the movement of the Inter-Tropical Convergence zone (ITCZ). The rainfall values are amongst the lowest in the country, and its distribution is markedly seasonal. The annual rainfall values, their intensity and duration, vary considerably from month to month. Available data on rainfall and temperature values over a 24-year period for Zuarungu (near Bolgatanga), which is centrally located within the study area, is shown in Figure 2. The graph shows that an estimated average of about 87% of the rainfall is recorded between May and October each year. The rain commences in April with low monthly mean rainfall value of about 30 mm and increases gradually to attain a peak of about 260 mm in August and then decreases sharply to about 50 mm in October. The rest of the year, spanning between November and March, constitutes the long dry season, with January being the driest month (Dickson & Benneh, 1974).

The study area is characterized by generally high temperatures with low yearly variations. The highest mean monthly and daily temperatures of 33°C and 42°C respectively are recorded in April, whilst the lowest mean monthly value of 26.5°C is registered during the peak of the dry season in December and January each year. Relative humidity varies from 70–90% during the rainy season and it falls to 20% during the dry season. Since the climatic conditions of an area have direct influence on plant life, the vegetation of the study area is the Wooded Savannah type. Consequently, Baobab, dawadawa, acacia and the shea-butter trees that are adapted to enduring long dry conditions are the major trees found in the area, while grasses that can reach as high as 3 m are the dominant vegetation type.

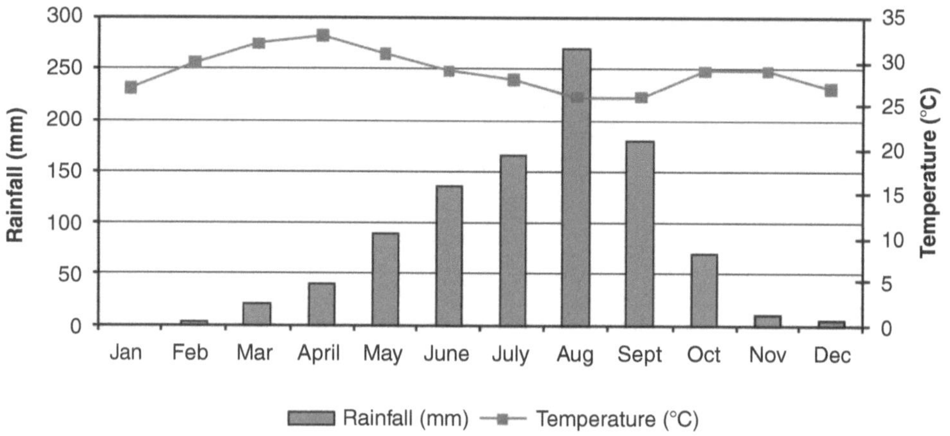

Figure 2. Mean Rainfall and Temperature Curves of Zuarungu, near Bolgtanga (1976–1999).

1.3 *Geomorphology, drainage and geology*

The topography of the Upper East region is fairly flat to slightly undulating with average topographic heights ranging between 180 and 300 m above mean sea level. Small rounded inselbergs of granitic rocks are found on high plains, especially in the Bongo district. Four perennial rivers (Kulpawn, Yaragatanga, Red and White Volta) and their ephemeral tributaries drain the entire region. The soils of the region are ferruginous, and have developed over both the parent sedimentary and igneous rocks. Texturally, the soils are silty or sandy-loam when developed over Voltaian rocks, and coarse-grained sandy-loam when developed on granite (Kesseh, 1985). The overburden is often shallow, and is poor in organic content.

The geology of the study area is shown in Figure 3. Two main geological formations underlie the study area. These are the Precambrian crystalline igneous and metamorphic rocks, which cover about 92% of the region; and the Palaeozoic Consolidated Sedimentary rocks (the Voltaian Series) occupying the remaining 8% of the project area.

Granites underlie almost 65% of the entire region, and are found in all the six districts. Three distinct granites (Bongo, Cape-Coast and Dixcove) are found in the region. The Bongo granite, which is mainly reddish microcline granites, is found in isolated patches in the central part of the study area, particularly in the Bongo district, and underlies just about 2% of the entire area. The Bongo granite is foliated, undifferentiated, and outcrops as boulders of varying sizes. They are often associated with gneisses and are highly migmatic. The Cape Coast granites are gneissic, migmatic and potassium-rich. The main rock types are biotite and muscovite-rich granites, grano-diorites, pegmatite, aplite and biotite-schist pendants. They are characterized by the presence of many enclaves of schist and gneiss. The Dixcove granite consists of highly altered feldspars, typically unfoliated and comprise hornblende granites, grano-diorites and hornblende-diorites. Porphyritic biotite and biotite-gneisses are also found in association with this rock type.

The Precambrian Birimian formation constitutes the second largest geological formation in the study area, and covers about 25% of the region. It trends in a NE-SW direction cutting diagonally across all the districts of the region from the southwestern to the northeastern corner. As a result of structural differences and lithological composition, the Birimian

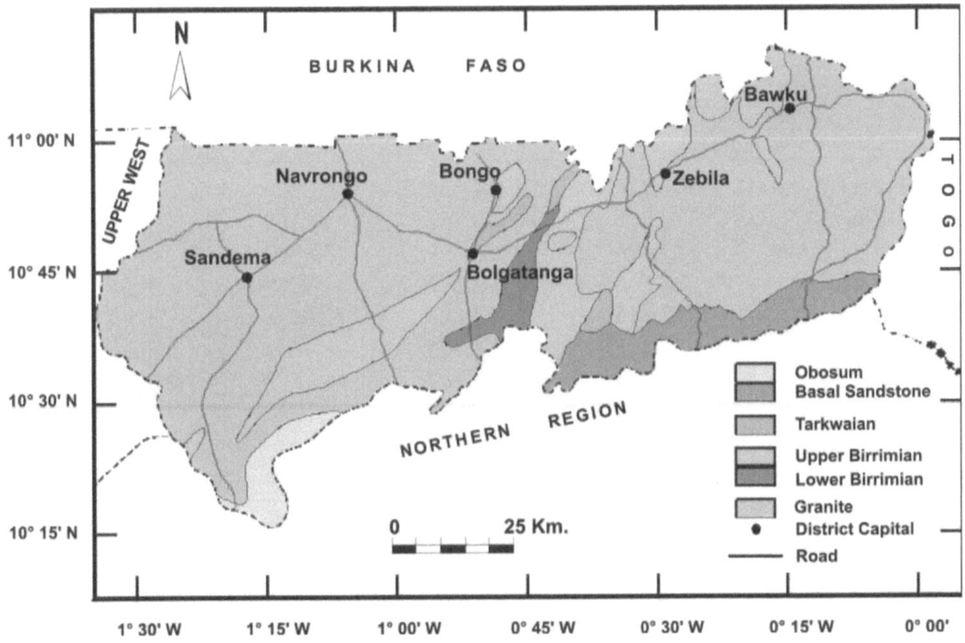

Figure 3. Geological map of Upper East, Ghana.

formation is sub-divided into two: the Lower and Upper Birimian series. However, many geologists such as Kesseh (1985) have the view that the two series can be considered as lateral facies equivalents due to the lack of sharp junctions between them. The Lower Birimian rocks are mainly of pelitic origin, and consist of strongly-foliated and jointed rocks. It comprises predominantly great thickness of alternating arenaceous and argillaceous meta-sediments, comprising shales, phyllites, siltstones, schists and greywacke. The Upper Birimian overlies the lower Birimian formation, and comprises great thickness of beds of agglomerates, and some pyroclastic and basic intrusive rocks. Generally, folding is intense in this rock formation.

The Tarkwaian rock formation underlies just about 2% of the area, and consists of thick series of argillaceous and arenaceous sediments. The rocks comprise coarse, poorly-sorted immature sediments with low roundness. It is considered to be of shallow water continental origin, derived from the Birimian granitic complexes and comprises mainly quartzites, phyllites, grits, conglomerates and schists. It rests unconformably on the Birimian rock formation, and has been subjected to low-grade metamorphism. Kesseh (1985) indicates that even though minor folds and flexures exist in this rock on the local scale, their relationship with the major folds have not been established.

The Voltaian Series, comprising Palaeozoic age consolidated sedimentary rocks, underlies about 8% of the study area. They are generally well-consolidated and gently-folded with an average dip of 5°, and are generally characterized by escarpments. Main rock types are quartz-sandstones, sandstones, pebbly grits and grits with ripple marks. Others include arkose, mudstone, conglomerate and limestone. These rocks are associated with cross-bedding, nodular structures and intense weathering. Groundwater occurs in joints and fractures as well as bedding and cleavage planes.

Table 1. Rural and Urban Population Distribution (Ghana Statistical Services, 2002).

District	Total population	Rural	Urban	District share of total population (%)
Bawku East	307,917	244,658	63,259	33.5
Bolgatanga	228,815	179,653	49,162	24.9
Kassena Nankana	149,491	125,689	23,802	16.2
Bawku West	80,606	72.547	8059	8.8
Bongo	77,885	77,885	–	8.5
Builsa	75,375	75,375	–	8.2
Total	920,089	775,807	144,282	100

Table 2. Population and their Sources of Water Supply (Ghana Statistical Services, 2002).

Water sources	Water type	Population	% Dependence
Surface water	Pipe-borne	124,415	13.5
	Dams/ponds	18,387	2
	Dug-outs	19,115	2.1
	Streams	57,297	6.2
Groundwater	Boreholes	588,136	63.9
	Hand-dug-wells	104,660	11.4
	Springs	8079	0.9
Total		920,089	100

1.4 *Population Distribution and Socio-economic Activities*

From the 2000 population census report of Ghana, it was revealed that the study area has a total population of 920,089 inhabitants, and represents 4.9% of the total population of Ghana. Out of this population, 144,282 representing approximately 15.7% live in the urban centres, whilst the remaining 84.3% (775,807) dwell in dispersed rural communities. The region has an inter-censual growth rate of 1.1%, considered to be the lowest in the country. The population density was computed to be 104.1 persons per square kilometre. Based upon the census report, as well as the population classification in Ghana, only seven towns (Bawku, Bolgatanga, Navrongo, Zebila, Paga, Pusiga and Garu), whose populations are higher than 5000 inhabitants, are considered urban. The distribution of the region's population showing both the urban and rural percentages on district basis is shown in Table 1.

The census report further revealed that about 23.8% of the entire population of the study area depends on surface water sources (including pipe-borne, dams/ponds, dug-outs and streams), whilst the bulk of the population (76.2%) rely on groundwater sources (boreholes, hand-dug-wells and springs). Table 2 shows the population and their source of water supply.

The inhabitants are mainly food and animal farmers. Since the vegetation of the area is dominantly grassland, large-scale crops such as rice, wheat, sorghum, millet, maize and guinea-corn are largely cultivated for both local and international markets. Other cash crops that thrive well in the area include yams, groundnuts, bambara & soya beans, cotton etc.

Table 3. Main borehole drilling programmes in Upper East.

Implementing agencies	Period	Number of bhs drilled	Percentage of total
Adventists Relief Agency (ADRA)	1954–1958	125	5.1
Upper Region Water Supply Project (URWSP)	1974–1981	1280	52.1
Ghana Water and Sewerage Corporation (GWCL)	1962–1985	162	6.6
Catholic Diocese of Bolgatanga	1993	50	2
District Assemblies and Community Water and Sanitation Agency (CWSA)	1999–2004	841	34.2
Total		2458	100

Table 4. Distribution of boreholes on District basis.

District	District size (%)	Number of bhs	Percentage of total
Bawku East	20	713	29%
Bawku West	14	203	8.2%
Bolgatanga	17	472	19.2%
Bongo	10	341	13.9%
Builsa	21	354	14.4%
Kassena-Nankana	18	375	15.3%
Total	100%	2458	100%

2 BOREHOLE SURVEY

The data from 2458 boreholes used for this assessment were drilled by five main implementing agencies. The list of the implementing agencies, indicating the number of boreholes and the period within which they were drilled, is shown in Table 3, and the distribution of these boreholes on the basis of the various districts of the region is presented in Table 4. The distribution of the 2458 boreholes in the study area is shown in Figure 4.

Analyses and interpretation of lithological logs of all 2458 boreholes drilled in the various geological formations in the region was made. Aquifer locations, static water levels, yields, drilled depths and overburden thickness were extracted from drilling and construction logs. In addition, any chemical analyese of water from the boreholes were retrieved from files. Simple statistical analyses were carried out on both the hydrogeological and hydrochemical data collected, using the SPSS statistical program. The means and ranges of each of the hydrogeological and water quality parameters were computed separately for each geological formation as well as for each district. The results obtained were used to prepare single parameter hydrogeological and hydrochemical maps such as borehole yield, depth-to-aquifers, static water-levels, fluoride and nitrate maps to portray the spatial distribution of these parameters in the study area.

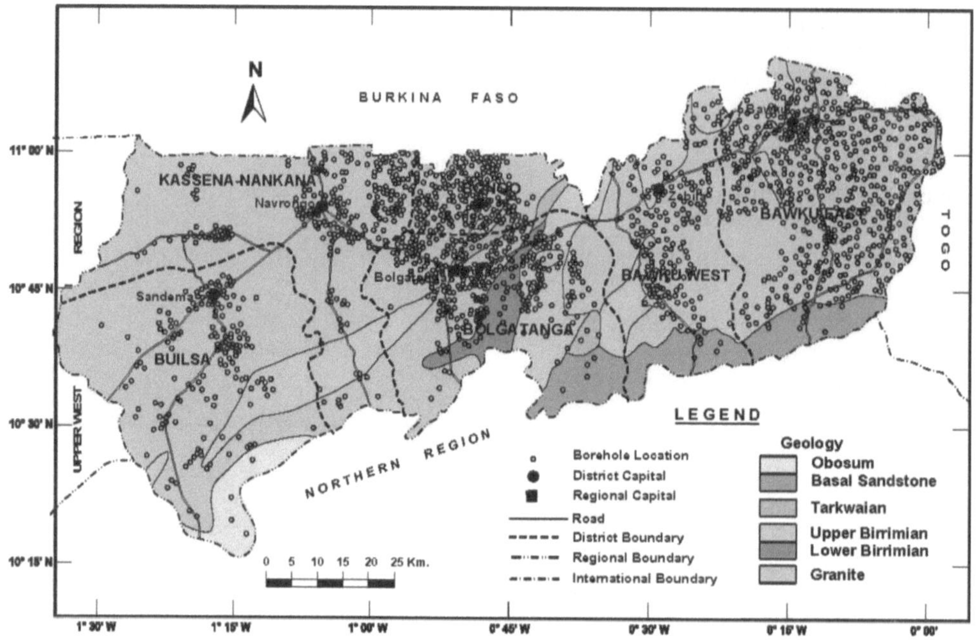

Figure 4. Borehole distribution across Upper East, Ghana.

Based upon the results obtained, the groundwater potential for the entire region was estimated. On the basis of the regional population figures, the domestic water requirements for both the urban and rural population were estimated. The computations were based on the mean yield of the existing groundwater resources, the daily per capita water consumption rates for both urban and rural population as well as the Ghana Water Company's per capita consumption rates of 75 litres per day and 150 litres per day for rural and urban population respectively. In estimating the future regional water demand, population projections were made using the 2000 rural and urban population figures, and a regional population growth rate of 1.1% (Ghana Statistical Service, 2002).

3 RESULTS AND DISCUSSIONS

3.1 *Borehole yield and success*

Groundwater occurrence in the Precambrian crystalline igneous and metamorphic rocks is generally controlled by the presence of fractures and degree of their inter-connectivity (Bates, 1995). Analyses and interpretation of lithological logs of boreholes drilled in the various formations in the region revealed that groundwater occurred at four main zones within the overburden and in the fresh rock itself. These are weathered zone, bedrock-weathered zone interface, fractured zones, and zones intruded by quartz-veins and pegmatites.

Weathered zone aquifers were observed to develop within the regolith, with their yield highly related to the thickness of the weathered mantle and the nature of the topography. Boreholes that were drilled through thick overburden (>25 m) yielded some appreciable quantities of groundwater. Similarly, regolith aquifers in the weathered zone overlying

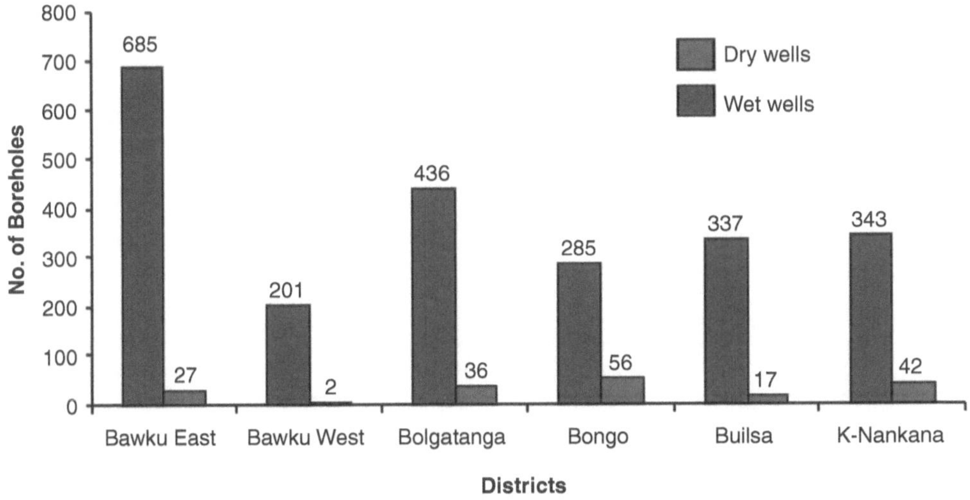

Figure 5. Distribution of borehole on the basis of the various districts.

the Precambrian basement rocks serve as the main source of groundwater in a large part of the Volta River basin (Martin, 2006). The spatial distribution of groundwater use and groundwater potential in the Volta River basin of Ghana has been discussed in Martin & de Giesen (2005). Moreover, flanks of valleys and proximity to streams also offered good points for higher yields. These weathered-zone/overburden aquifers (where most of the boreholes tap water from) are either phreatic or semi-confined, with high static water-levels. However, Agyekum *et al.* (2005) noted that such aquifers are susceptible to large groundwater level fluctuations due to the large temperature variations between the rainy and dry seasons.

An appreciably large volume of groundwater is also obtained between the regolith and hard rock interface. The thickness of the overburden in the region varied from place to place, but averaged approximately 7 m. Fractured and quartz-veined aquifers were mainly confined or semi-confined with relatively higher yielding capacities. Figure 5 shows the distribution of wet and dry boreholes in the respective districts of the region. The dry boreholes were considered as those with small yield, not even sufficient for the installation of a hand-pump. The success rate is that recorded by the programme – there may have been a larger number of dry boreholes that were unrecorded, or boreholes that have since failed.

Analyses of recorded successful boreholes in the study area have indicated that the rate of drilling successfully through the Birimian rocks averaged 96%. The rate was 82% in the granitic rock areas, and about 45% in the sedimentary rock areas. Out of the three different granitic rock types in the study area, the success rate was observed to be relatively lowest in the Bongo granitic rocks owing to the peculiarly impervious nature of this rock. The yielding capacity and groundwater potential in the study area was generally observed to depend upon an interplay of factors including type of geology, degree of structural deformation and the presence and nature of discontinuities in the underlying bedrock. Similarly, the degree of interconnectivity and intensity of fractures, nature of topography, as well as the extent of weathering and regolith development, play important roles.

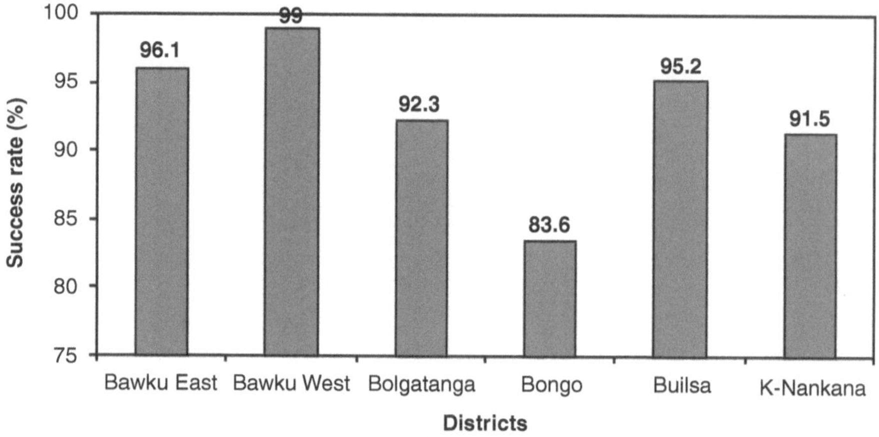

Figure 6. Borehole success rates in the various districts of the region.

Table 5. Percentage of recorded borehole yields in Upper East, Ghana.

District	No. of boreholes	Borehole yield values (m^3/h)				
		<1	1–3	3–5	5–7	>7
Bawku East	713	10%	70%	12%	5%	3%
Bawku West	203	20%	75%	–	–	5%
Bolgatanga	472	20%	57%	13%	7%	3%
Bongo	341	3%	80%	10%	5%	2%
Builsa	354	29%	39%	20%	8%	4%
Kassena-Nankana	375	25%	40%	21%	8%	6%
Mean		18%	60%	13%	5%	4%

On the basis of the individual districts, the success rate was found to be highest in the Bawku-East, Bawku-West and Builsa districts that are underlain mainly by granitic and Birimian rocks as shown in Figure 6. The distribution of boreholes with specific yields in the various districts of the study area is shown in Table 5. The borehole success rate is lowest (83.6%) in the Bongo district, which is predominantly underlain by Bongo granitic rock types.

It can be inferred from Table 5 that nearly 80% of the boreholes drilled in the region have a yielding capacity of greater than 1 m^3/h (0.28 l/s), while 9% of them yield higher than 5 m^3/h (1.4 l/s). The results indicate that there is very little chance of obtaining boreholes that would yield higher than 8 m^3/h (2.2 l/s) in the region, except at a few isolated locations that are underlain by fractured Precambrian phyllitic and quartzitic rocks of the Birimian formation. The map of borehole yields developed for the region is shown in Figure 7, and it indicates that yield values higher than 7 m^3/h (1.9 l/s) could be located in pockets in the Builsa, Kasena-Nankana. Bolgatanga and Bawku-East districts of the region.

Table 6 presents the summary of the results obtained from the statistical analyses of the hydrogeological data from boreholes in the various geological formations in the districts. A study of the available core logs showed that borehole yields were higher in the fractured

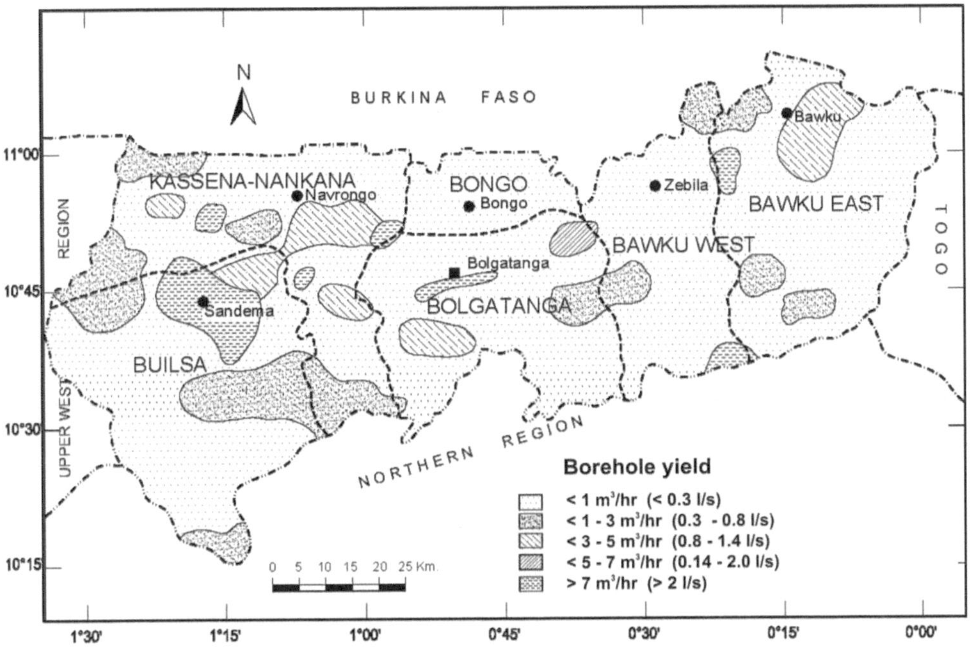

Figure 7. Yield map of boreholes in Upper East, Ghana.

Table 6. Summary of borehole information in the Upper East Region.

District	Geology	No. of boreholes	Yield (m^3/h)		Depth (m)		Water-level (m)		Water strikes	
			Mean	Range	Mean	Range	Mean	Range	Mean	Range
Bongo	Granite	300	2.4	0.78–7.2	38	31–49	8.7	3.9–17.4	33	16–47
	Birimian	41	6.7	1.4–12	39	31–46	10.8	5.8–16.6	31	19–45
Bawku East	Granite	513	4.5	0.78–21	38	28–76	6.3	0.59–18.4	30	15–55
	Birimian	132	3	0–6	35.5	31–46	7.9	2.4–18.3	30	16–45
	Voltaian	18	1.1	0.7–4.8	31	30–33	9.4	4.8–12.3	26	23–30
Bolgatanga	Granite	384	2.3	0.7–7.2	36	31–55	7.1	2.3–15.5	32	14–50
	Birimian	92	3.6	0.9–9.0	39.1	31–55	6.8	1.3–9.7	30	14–54
Builsa	Granite	300	4.4	0.5–10.8	47.8	31–60	9.4	0–9.2	32	18–60
	Birimian	54	4.4	1.8–7.2	53.3	43–62	8.5	0–13.1	30	22–55
Bawku West	Granite	153	10.8	3.6–18.0	39.2	31–46	8	7.5–8.5	37	18–45
	Birimian	50	3.2	0.9–5.6	34.4	31–42	8	7.2–9.6	26	13–29
K. Nankana	Granite	350	4.4	0.5–14.4	43.8	29–60	9.6	3.2–19.2	32	14–56
	Birimian	25	2.3	0.8–3.6	38.9	33–45	12	5.9–25.8	30	15–44

portions of the bedrock. Boreholes that intercepted quartz-veins and pegmatites registered significant yields. Depths to fracture zones were variable, ranging between 12–30 m.

The static water-levels recorded for the area show that the water-levels are generally less than 10 m below ground level. However, static water-levels deeper than 10 m were observed in the western part of the region and at some few isolated pockets in the Bawku, Bolgatanga

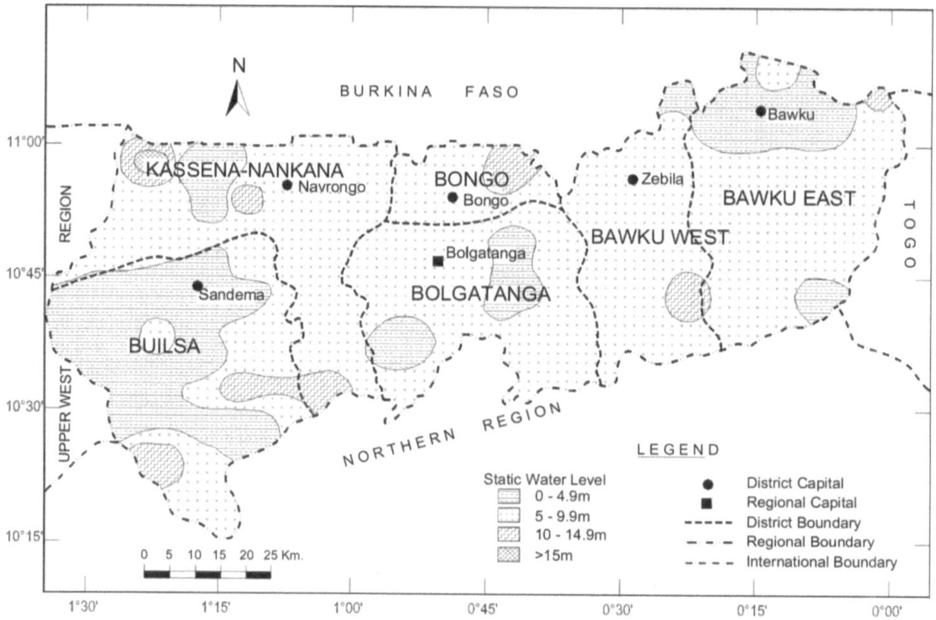

Figure 8. Groundwater level map for Upper East, Ghana.

and Kassena-Nankana districts of the region as shown in Figure 8. The lowest static water levels (> 15 m below ground level) were found to occur at the north-western corner of the region in granitic rocks of the Kassena-Nankana district.

3.2 *Groundwater recharge and abstraction*

Knowledge of recharge processes and recharge rates are basic to assessing and ascertaining the sustainability of a groundwater resource. In spite of the fact that evapotranspiration far exceeds precipitation in the study area, there is adequate recharge into the fractured bedrock aquifers to sustain the boreholes. On their separate works on the groundwater recharge of the eastern and western parts of northern Ghana, Bannerman & Ayibotele (1984), and Apambire (1996) estimated the groundwater recharge values of northeastern and northwestern parts of the country to range between 2.5% and 4% of the total annual rainfall. Recharge rate for the Atankwidi catchment, a 275 km^2 sub-catchment of the White Volta in northern Ghana, was estimated under the GLOWA-Volta project funded by GLOWA (Globaler Wandel des Wasserkreislaufs) programme of the German Ministry for Water Education and Research (BMBF). Using an integrated approach and a combination of several field methods with water budget modelling, recharge estimates ranging from 2% to 13% of annual rainfall were obtained for the catchment (Martin, 2006). These recharge rates vary considerably between wet and dry years and between locations. Inter-annual comparison of water-level fluctuations showed that a decrease in annual rainfall of 20% caused a reduction of groundwater recharge of 30–60% (Martin, 2006). The results further showed that a long-term average groundwater recharge of 60 mm/a compares to a total current groundwater abstraction of 4 mm/a in the area, which is one of the areas with the

Table 7. Domestic water demand.

Population class	Daily domestic water demand (m^3/d)	Population	Total domestic water demand (m^3/d)	Total annual domestic water demand (M m^3)
Rural	0.075	775,807	58,186	21.2
Urban	0.15	144,282	21,642	7.9
Total		920,089	79,828	29.1

Table 8. Statistical description of groundwater quality in the Upper East Region.

District	n		pH	Conductivity (μS/cm)	Chloride (mg/l)	Total iron (mg/l)	NO_3-N (mg/l)	F (mg/l)	Mn (mg/l)
Bawku East	80	range	6.9–7.6	167–589	0.1–8.0	0.01–0.06	0.4–51	0.2–1.7	0.1–0.4
		mean	7.3	280	2.0	0.03	7.8	0.6	0.2
Bawku West	8	range	6.9–7.4	210–871	0.1–0.3	0.01–0.03	1.9–15	0.1–0.4	0.1–0.2
		mean	7.2	400	0.2	0.02	3.2	0.3	0.15
Bolgatanga	44	range	6.9–7.6	260–748	0.3–3.0	0.01–0.03	0.3–15	0.4–2.8	0.1–0.7
		mean	7.2	320	0.5	0.02	7.5	0.8	0.3
Bongo	33	range	6.9–7.6	245–752	0.1–51	0.01–0.12	0.8–49	0.2–4.0	0.1–0.8
		mean	7.2	410	2.4	0.04	12.0	1.2	0.3
Builsa	17	range	6.9–8.4	147–621	1.4–13.0	0.01–0.11	0.06–24	0.2–2.1	0.1–1.0
		mean	8.0	320	3.1	0.06	6	0.7	0.2
Kassena-Nankana	32	range	7.1–8.3	279–660	0.1–21	0.01–0.13	0.7–20	0.2–2.1	0.1–0.6
		mean	7.2	400	8.4	0.05	3.8	0.5	0.15
WHO Guideline Levels (WHO, 1985)			6.5–8.5	–	250	0.3	11.4	1.5	0.1

highest groundwater use per square kilometre in the Volta River basin. Recharge is therefore currently not a limiting factor for groundwater resource development in this area.

Based upon the region's total urban and rural population figures, as well as Ghana Water Company's domestic daily per capita water consumption figures, the total yearly domestic water requirement of the region was estimated to be approximately 29.1 Mm^3. The results of the water demand computations are shown in Table 7.

Given that most of the 2458 successful boreholes in the region can sustain a yield of 0.3 l/s, then the total yearly groundwater abstraction in the region is estimated to be 21.53 M m^3. Given that water from the boreholes is utilized for domestic purposes, then water supply through boreholes alone represents 73.9% of the total domestic water demand.

3.3 *Water quality*

The water quality in the region based upon analyses of 214 borehole water samples is shown in Table 8. Amuzu (1974) indicated that groundwater quality data generally gives information about the geological history of the host rock and further provides some clue about the types of minerals, which the groundwater has come into contact with.

In the study area, manganese concentrations in excess of 0.1 mg/l were recorded in most boreholes in the Bolgatanga and Bongo districts. Manganese concentrations ranging between 0.3–0.8 mg/l were recorded in some boreholes in the Bongo district.

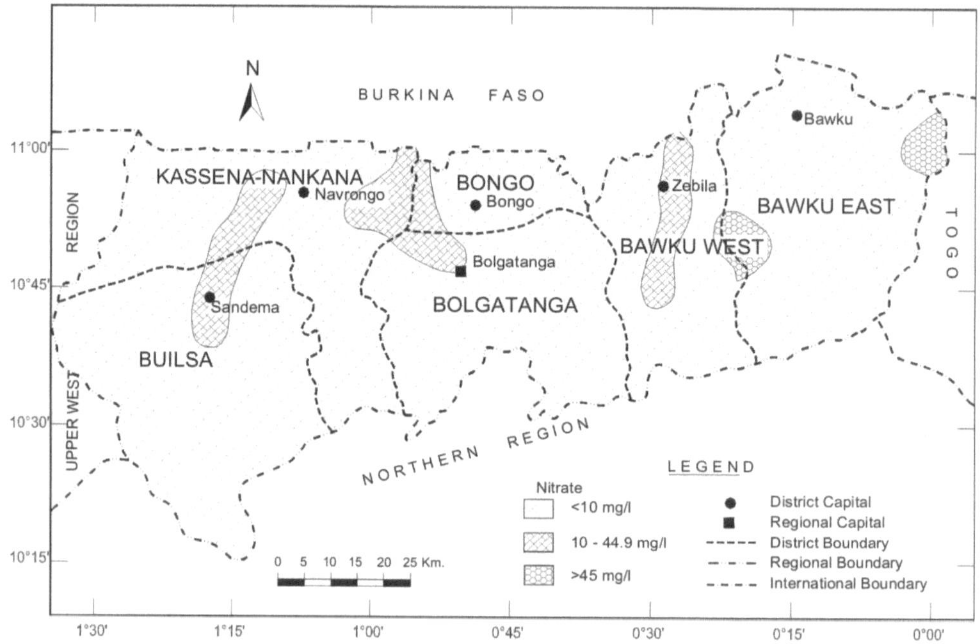

Figure 9. Concentrations of nitrate in sampled borehole water in the Upper East region of Ghana.

Even though the general nitrate levels of water from boreholes in the region were within the WHO recommended levels of 11.4 mg/l as NO_3-N, relatively higher concentrations were recorded in pockets in all the six districts of the region. The highest nitrate concentrations of 51 mg/l were observed particularly in boreholes located in the eastern and western parts of Bawku-East district as shown in Figure 9. Furthermore, analyses of nitrate concentrations show that about 38% of water from boreholes in the Bongo district showed concentrations in excess on 10 mg/l. The high concentrations of nitrate in groundwater in the district may be due to the excessive use of chemical fertilizers and manure for all-year-round farming activities arising out of the creation of the Vea irrigation dam, which is located near Bongo.

Elevated fluoride concentrations have been found throughout the area, but mostly in the Bongo and Bolgatanga districts, associated with the Bongo granite. Consequently, the Bongo district is severely affected by endemic tooth decay as a result of ingestion of groundwater containing excessive fluoride (Apambire *et al.* 1997).

4 CONCLUSIONS

The results of the general assessment of the potential of groundwater resources of North-Eastern part of Ghana have indicated the existence of high groundwater potential, which could be developed for various uses in the region. However, their availability is unevenly distributed throughout the study area. Despite this observation, the existing resource is capable of sufficiently meeting the total water requirement of the region through proper management and careful utilization. Geology, structural features, overburden thickness and the nature of topography largely control groundwater occurrence. Aquifers have been found to be discrete, isolated and laterally limited in extent. Borehole yield is highly variable, with successful boreholes having yields in the range 0.1 to 6 l/s, with a mean of 0.8 l/s.

Relatively higher yields occur in areas underlain by highly-fractured Birimian and granitic rock areas. The recorded data (which may be optimistic) indicate approximately 60% chance of obtaining successful boreholes that could yield higher that 0.3 l/s, and only 9% chance of obtaining 2 l/s or higher yields. Borehole depths were found to be generally shallow in the entire region, and ranged between 28–62 m. Aquifers were encountered at a mean depth of 30 m (range 13–60 m), while static water-levels occurred at a mean depth of 8 m below ground level. On average, the recorded information indicate the chance of obtaining a successful borehole in the study area is about 65%. The rate is higher in the Precambrian Birimian rock areas and those underlain by granites. The rate is however lower in areas underlain by Palaeozoic sedimentary mudstone and shale rocks as well as those underlain by Bongo granites. The total yearly groundwater abstraction in the region has been estimated to be about 74% of the total domestic water requirement. Most borehole water samples in the Bongo, Bolgatanga and Kassena-Nankana districts have problems of high nitrate, fluoride and manganese concentrations.

ACKNOWLEDGEMENT

The authors wish to express their profound appreciation to all the water implementing and delivery agencies such as Community Water and Sanitation Agency (CWSA), Adventists Relief Agency (ADRA), the District Assemblies, Catholic Dioceses of Bolgatanga and the CIDA-sponsored Upper Region Rural Water Supply project (URWSP) of the Upper East Region for making available the hydro-geological and hydro-chemical data for this study. The Groundwater Division of the CSIR Water Research Institute (WRI) also deserves appreciation for compiling all the borehole and water quality data.

REFERENCES

Agyekum, W. A., Dapaah-Siakwan, S., Darko, P. K., Peligba, K., Dankyi, N. O. & Asiamah, G. 2005. *An Assessment of Groundwater Resources of the Upper-East Region of Ghana.* WRI Technical Report, Accra, Ghana.

Apambire, W. B., Boyle, D. R. & Michel, F. A. 1997. *Geochemistry, genesis, and health implications of fluoriferous groundwaters in the upper regions of Ghana.* Journal Environmental Geology, **33**, 13–24.

Apambire, W. B. 1996. *Groundwater geochemistry and the genesis and distribution of groundwater fluoride in the Bolgatanga and Bongo districts, Ghana.* MSc, Carleton University, Ottawa, Canada.

Amuzu, A. T. 1974. A Study of the Drinking Water Quality of Boreholes in the Rural Areas of Ghana. WRRI Publication, Accra, Ghana.

Bannerman, R. R. & Ayibotele, N. B. 1984. *Some critical issues with monitoring crystalline rock aquifers for groundwater management in rural areas.* IAHS Publications, **144**, 47–57.

Bates, D. A. 1995. *Geological Map of Ghana.* Ghana Geological Survey, Accra, Ghana.

Darko, P. K. 2001. *Quantitative aspects of hard rock aquifers; regional evaluation of groundwater resources in Ghana.* PhD, Charles University, Prague.

Dickson, K. B. & Benneh, G. 1980. *A New Geography of Ghana.* Metricated Edition, Pearson Education Limited, Harlow, UK.

Ghana Statistical Service 2002. *Population and Housing Census Report of Ghana 2000.* Ghana Statistical Service, Accra, Ghana.

Gill, H. E. 1969. *A groundwater reconnaissance of the Republic of Ghana, with a description of geohydrologic provinces*. USGS Water Supply Paper, **1757-K**.

Gyau-Boakye, P. 1993. *Filling Gaps in Hydrological Run-off Data Series in West Africa.* PhD, University of Bonn, Germany.

Kesseh, G. O. 1985. *The Mineral and Rock Resources of Ghana*. AA Balkema, Rotterdam, The Netherlands.

Martin, N. 2006. *Development of a water balance for the Atankwidi catchment, West Africa – a case study of groundwater recharge in a semi-arid climate*. In: Vlek, P. L.G., Denich, M., Martius, C. & Rodgers, C. (eds.) Ecology and Development Series No. 4.

Martin, N. & de Giesen, N. 2005. *Spatial distribution of groundwater use and groundwater potential in the Volta River basin of Ghana and Burkina Faso.* Water International, **30**, 239–249.

World Health Organisation WHO, 1985. Guidelines for Drinking Water Quality, Geneva.

CHAPTER 13

Application of remote-sensing and surface geophysics for groundwater prospecting in a hard rock terrain, Morocco

S. Boutaleb
Department of Geology, Polydisciplinare Faculty of Taza, Sidi Mohamed Ben Abdallah University, Morocco

M. Boualoul
Department of Geology, Faculty of sciences, Moulay Ismail University, Meknes, Morocco

L. Bouchaou & M. Oudra
Department of Geology, Faculty of sciences, Ibnou Zohr University, Agadir, Morocco

ABSTRACT: An integrated study was carried out to investigate the subsurface geological conditions in a hard rock environment, with the aim to identify groundwater potential zones. The study considered the use of remote sensing, geo-electrical methods and nuclear magnetic resonance sounding techniques (NMRS). Remote sensing was used to identify lineaments which were extracted from the satellite images by directional filtering of the image. Resistivity profiles and NMRS were used to locate the lineaments more precisely on the ground. The identification of geomorphological significance of each mapped lineament has enabled the explanation of the relationship between lineament distribution and groundwater in fractured rocks. The study made it possible to select localised sites for the drilling of successful boreholes, after taking into account the hydrogeological and climatic conditions of the area. The results of this study, as tested in the Moroccan Anti-Atlas chain, show the importance of using an integrated approach for siting boreholes in hard rock terrain in a semi-arid climate.

1 INTRODUCTION

Groundwater is the major source of water supply needed for industrial, agricultural and domestic purposes in many semi-arid regions of the world. In some cases, over-exploitation has resulted in declining groundwater levels and has consequently confined groundwater flow to deeper weathered/fractured zones. Therefore, it becomes essential to study these zones to determine their potential for supporting groundwater resources in order to meet demand.

Lineaments (large-scale linear features expressed in topography including valleys and underlying structural features controlled by faults or joints) are often good water potential zones (Cazabat, 1975; Boehmer & Boonstra, 1987; Ozer *et al.*, 1987; Lloyd, 1999; Neuman, 2005). However, mapping these lineaments is not sufficient to locate groundwater potential zones without considering other parameters like geology, topography and proximity to rivers and streams.

Geology, geomorphology and structure are the controlling factors for groundwater storage, occurrence and movement in hard rock terrain. These features can often be identified through satellite remote sensing. Geophysical methods can also play a major role in defining these parameters. Nevertheless, the traditional application of geophysical prospecting in Morocco has often failed for variety of reason such as: ambiguities in identifying geophysical anomalies; and the lack of adequate understanding of the occurrence of groundwater.

To face these problems, an integrated study was commissioned, in which the results of remote sensing, geophysical methods and geomorphological studies were combined to optimize the choice of potential groundwater sites in the Anti-Atlas chain, Morocco. This study consists of two stages:

1) Low level processing module: the aim of this first level was to determine and extract lineaments from remote sensing by means of classical contour detectors (Sobel and Laplacian). The obtained results from mathematical processing were compared and a map of major lineaments established.
2) High level processing module: this second level consisted of analysis (by a structural geologist) of the above low level results incorporating topography, geology, aerial photos, and ground checking. This analysis permitted the determination of orientation and coordinates of most important lineaments. This step is followed by determining the exact position of the various lineaments, using electrical prospecting method. The NMRS method is used finally to verify the water presence and to estimate the hydraulic permeability of the fractured zone (Legchenko *et al.*, 1997).

2 GEOLOGY, GEOMORPHOLOGY AND HYDROGEOLOGY OF THE STUDY AREA

The study area, situated at a distance of 180 km south of Agadir city in Morocco, covers 170 km^2 (Figure 1) and lies between 29°26′ to 29°33′N latitude and 9°53′ to 9°46′E longitude. This zone is located in the western Anti-Atlas Mountain at an elevation of 70 to 800 m. The climate of this zone is continental and semi-arid, with low rainfall (<200 mm/a) and high average summer temperatures (>32°C). The range of daily and seasonal temperatures is also high (18°C in winter and 27°C in summer). The wet period is between November and March, and the dry period can extend from April to October.

Geologically, most of the formations underlying the study area consist of a complex assemblage of crystalline, metamorphic and volcanic Precambrian rocks (Figure 1). These formations crop out at the western part of the study area. Palaeozoic aged rocks crop out along the eastern and the central part of the study area (Soulaimni & Piqué, 2004). These Infra-cambrien and Georgian formations are characterized by alluvial fans and fluvial deposits, and by a large development of limestone formations, known as the Adoudounian limestone. Some conglomeratic horizons, belonging to this zone, comprise palaeo-valley molasse fill.

Within this study area, three fault systems exist: (1) a NW-SE striking system due to the Eburnian deformation (this phase affects the Precambrian and Palaeoproterozoic basement rocks); (2) a NE-SW striking system due to the Pan-African deformation; and (3) a sub-longitudinal striking system including the youngest faults (Hassenforder, 1987).

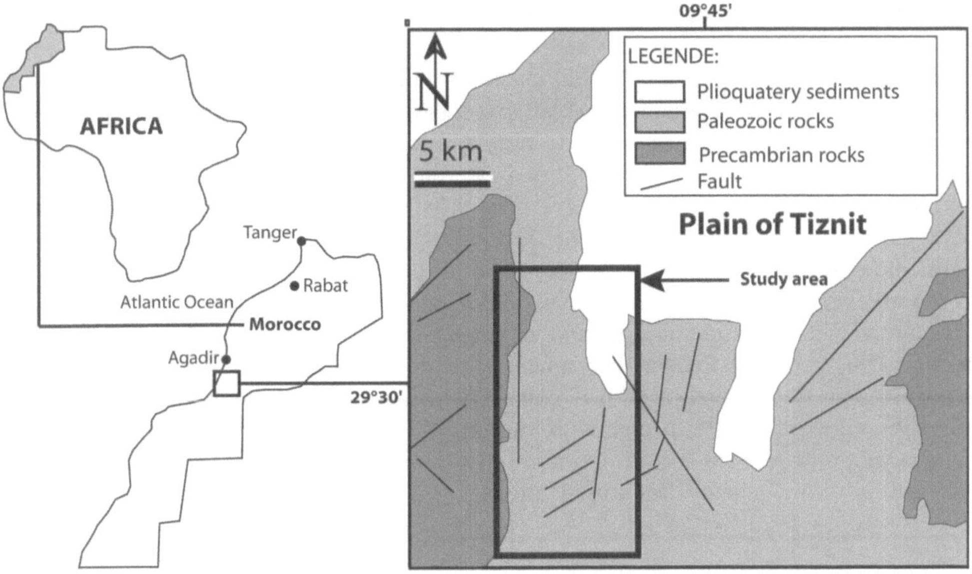

Figure 1. Location of the study area (Western Anti-Atlas), Morocco.

Geomorphologically, the main study area is further divided into several small catchments with succession of three types of geomorphic units: plateau, escarpment and valley.

From a hydrogeological point of view, all the rock formations of this zone present almost the same hydrogeological parameters except for the Adoudounian limestone. The groundwater storage of these formations is insignificant in comparison with the storage of the Adoudounian limestone. In the latter formation, groundwater is mainly stored in fractures and crushed zones and circulates within a network of high hydraulic conductivity zones. It receives water mainly by regional groundwater flow, with additional, important in-situ recharge by rainfall, surface water, and water from higher lying non-karstic denuded areas. Discharge of the groundwater takes place in river valleys and depressions. Many small rivers constitute the drainage system in the study area. They are characterised by a low mean discharge not exceeding 1 m^3/s and can be up to 15 m^3/s during flood periods.

An analysis of borehole data within the study area indicates a complex interaction between surface water and groundwater. This current study will focus on explaining the relationship existing between the rock formations and the potential groundwater as well as the interaction between these, including the geomorphological characteristics of the study area.

3 PROSPECTING METHODS USED IN THE STUDY

3.1 *Remote sensing*

The mapping of lineament features on satellite images has been an integral part of many groundwater exploration programmes in hard rock terrain. In these terrains streams, escarpments, mountain ranges, and variations in soil type are often visible on satellite images as linear features (O' Leary *et al.*, 1974; Siegel & Abram, 1976; Condit & Chaves, 1979; Rothery, 1985; Wright & Burgess, 1992).

The lineaments in the study area were extracted from a Panchromatic Spot image which was pre-processed, including radiometric correction and noise reduction, to remove systematic defects and undesirable sensor effects. For lineament extraction, a Matlab program including line algorithm was used. Sobel and Laplacian kernels were used because they generate an effective and faster way to evaluate lineaments in four principal directions (Suzen & Toprak, 1998). A lineament in a satellite image can show up either as darker pixels in the middle and lighter on both sides or is lighter on one side and darker on the other side. When directional filtering is applied to the image, the lineaments show up in light colour, which is surrounded by dark coloured pixel.

Lineaments of less than 1 km in length and those corresponding to the roads, limits of agricultural fields, mountain ridges and village houses, were discarded, as they distort the results of this study. The extracted lineaments were done manually and were geo-referenced using an affine transformation with 12 control points, of which the coordinates were determined by differential GPS. A map with lineaments longer than 1 km was established and the junction points of the most important lineaments (JPL), near the main villages were selected for further geophysical investigations.

3.2 *Surface geophysical methods*

Two geophysical methods were used in this study: surface electrical prospecting to confirm and accurately locate lineaments, and magnetic resonance sounding (NMRS) to verify and estimate the water content of the aquifer.

3.2.1 *Electrical prospecting methods*

The junction points of lineaments were investigated by profiling using a Schlumberger array with a maximum half electrode (AB/2) distance of 100 m. The array consists of four electrodes arranged symmetrically along a straight line with the outer two for current injection and the inner two for potential measurement. Spacing between the current electrodes is greater than four times the spacing between the potential electrodes (Kunetz, 1966; Durand, 1986).

The use of simple electrical resistivity profiles allowed for the confirmation of fracture locations, anomalous contacts or boundaries. A repeated resistivity profile with different lengths permit the estimation of dip of these anomalies. The electrode spacing is determined beforehand from a vertical electrical sounding (VES) which is also used to calculate the NMRS matrix. In each VES, current electrodes were spread out step by step. Apparent resistivity was calculated in the field for inspecting the data quality. If a distortion in data appeared, measurement was repeated or the current electrode position was changed to improve the quality of data. All the sounding curves were interpreted using automatic WinSev 6.0 program.

3.2.2 *Nuclear Magnetic Resonance Sounding (NMRS)*

At each junction point of lineaments (JPL), checked and specified by the horizontal resistivity profiles, one NMRS was undertaken to verify the presence of water content (Legchenko *et al.*, 1996). NMRS consists of exciting the protons of the water molecules with a magnetic field at a specific frequency (the Larmor frequency, depending on the amplitude of the Earth field) and measuring the magnetic field produced in return by these protons (Legchenko *et al.*, 1997; Yaramanci, 2002; Legchenko, 2004).

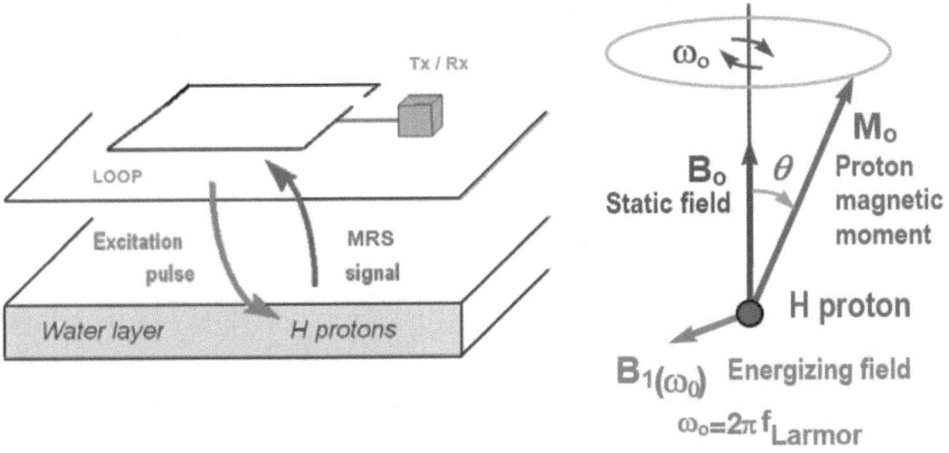

Figure 2. The principles of the magnetic resonance sounding technique.

The same loop is used for transmitting the excitation pulse and for analysing the relaxation signal (Figure 2). The initial amplitude E_0 of the relaxation, measured just after the excitation current has been switched off, is directly proportional to the number of protons which have been reacting, namely the water content.

The decay time (T2) of the relaxation curve is linked to the mean pore size of the material (Lubzcynski & Roy, 2003), fine grain sediments giving short decays (a few tens of milliseconds) while coarser grain sediments lead to longer decays (Shirov *et al.*, 1991).

The time constant is thus related to the permeability of the layer. The intensity of the excitation pulse (its moment $I.\Delta t$, where I is the current and Δt the pulse duration) controls the depth of investigation, small pulses for shallow, high pulses for deeper. The size of the loop used on the surface controls the maximum depth reachable.

This property of the pulse moment makes it possible to sound the ground at a given position of the transmitting/ receiving loop laid on the surface of the ground (Israel *et al.*, 1996). A sounding curve represents the variations of the signal initial amplitude versus the pulse moment, which gives, after inversion, the porosity versus the depth. The determination of the porosity is submitted to equivalence laws, the invariant parameter being the product of the porosity by the thickness, that is to say the total quantity of water located in a given layer.

The NUMIS proton magnetic resonance (PMR) system, developed by IRIS Instruments, was used in this study with a 75-m sided square loop (for which the maximum depth of exploration is about 80–100 m). The inversion of the MRS sounding data was done using the Numis-Plus software with Samovar Program of Automatic Inversion of the Surface Nuclear Resonance Sounding, version 4.04.

4 RESULTS

4.1 *Remote sensing*

Figure 3 show the results of the application of directional filtering to the study area satellite image. Only lineaments longer than 1 km are included in this map because of the small scale of presentation. The map shows that the majority of lineaments develop in NE-SW (which is also the principal regional structure), and N-S directions. Minor lineaments trend

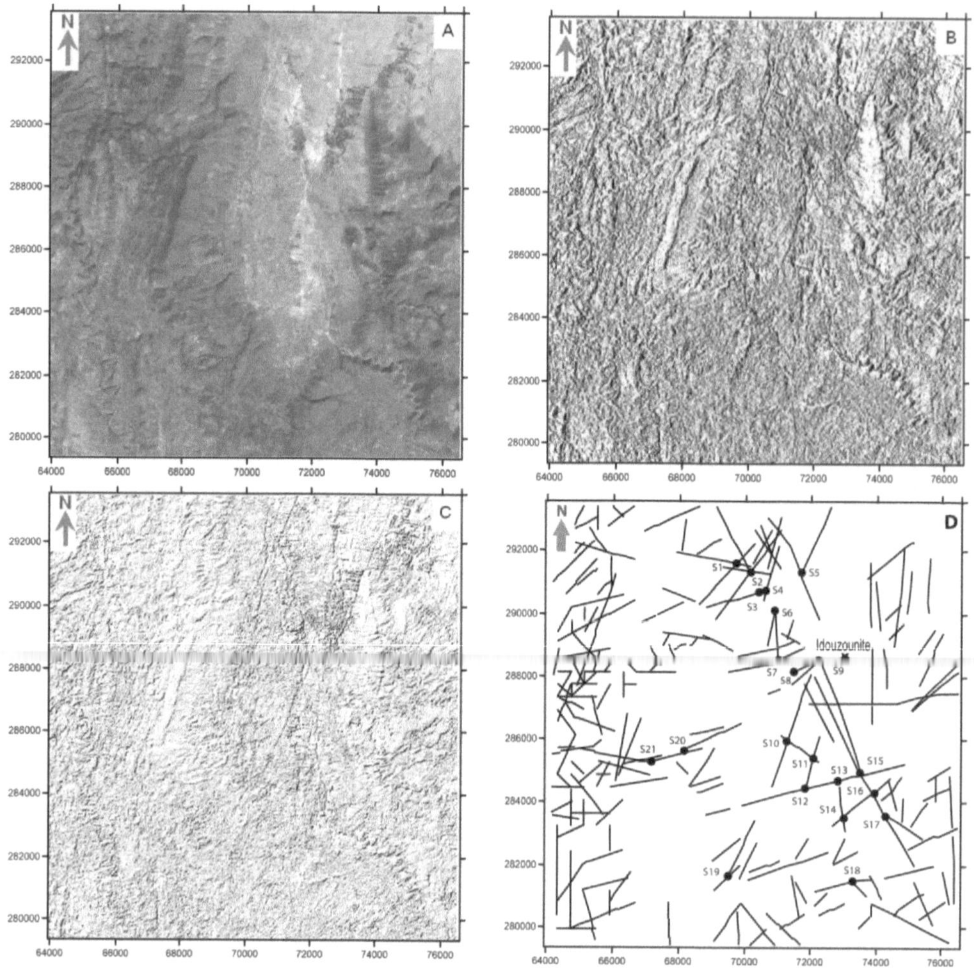

Figure 3. A) Panchromatic Spot Image of the study area (Western Anti-Atlas), B) Results of the Sobel filtering treatment of the image, C) Results of the Laplacien filtering treatment of the image, D) Manual digitized Lineaments.

generally E-W, NW-SE and ENE-WSW and cut across the major regional lineaments. Based on statistical work of distribution springs and their discharge values, the most prominent groundwater occurrences can be found along the NE-SW striking faults. These can be an indicator of the direction of groundwater movement in the studied area.

4.2 *Resisitivity*

These first results oriented the study and allowed the selection of 21 junction points of major lineaments and 4 boreholes sites to carry out the geophysical characterisation. The analysis of all the geophysical results shows one single point (S12) which presents an interesting potential groundwater. We will try, after analyzing its geophysical results, to explain the controlling factors which must be taken into account for determining the groundwater potential lineament zones.

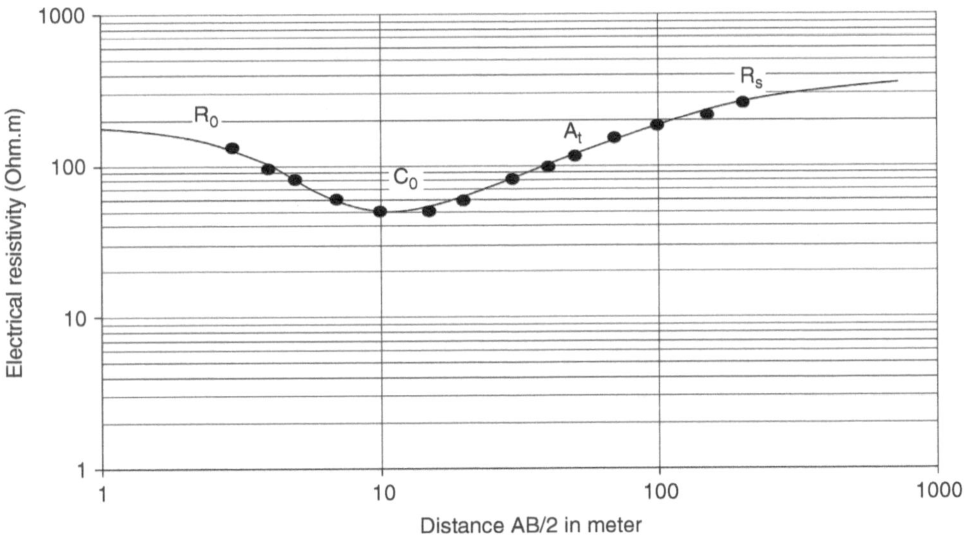

Figure 4. Representative apparent resistivity curve of the vertical sounding S12.

Table 1. Summary of the interpretation of VES data for the study area.

Layer N°	Resistivity Ohm-m	Thickness (m)	Depth to top (m)	Lithology
1	170	2	0	Dry alluvium R0
2	37	14	−2	Humid alluvium C0
3	100	20	−16	Fractured limestone At
4	350	Inf.	−34	Conglomerate Rs

All the VES curves collected in the area are in general of S12 type (Figure 4). The result of the mathematical inversion of this VES (summarized in Table 1) shows:

- upper layer (R0) with a maximum thickness of 2 m and apparent resistivity of 170 Ohm-m corresponding probably to dry alluvium.
- a second layer (C0) is characterized by lower resistivity value (37 ohm-m) corresponding to humid alluvium.
- the third layer (At), the most important one, corresponds to the fractured limestone which is characterised by an average resistivity of 100 Ohm-m and its thickness is of 10 m.
- the last layer (Rs) corresponds to the dry conglomerates.

Resisitivity profiles (AB/2 = 100 m) were carried out across the lineament junction points (JPL). In 90% of the cases, negative anomalies (low resistivity) corresponding to the fractured rock zones. In the S12 JLP taking as an example in this study (Figure 5), the lower anomaly is shown at station n° 13 of the profile which corresponds to a fractured limestone. This anomaly is surrounded by materials of 300 ohm-m which probably corresponds to the Precambrian conglomerates. Micro-fractures measured in this study at this location show a N05° and N20° direction (Figure 6). The second electrical lateral profile, executed in

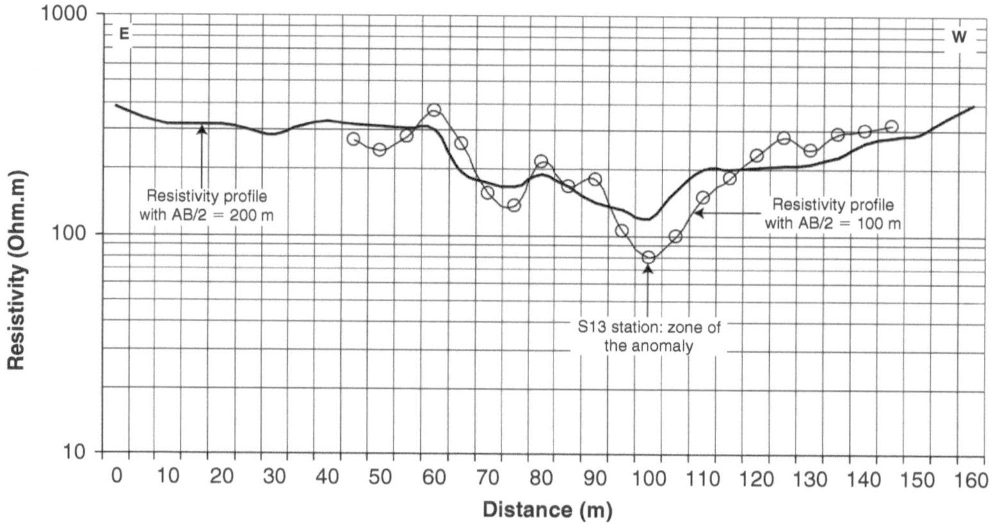

Figure 5. Horizontal resistivity profile of the S12 joint point of lineaments.

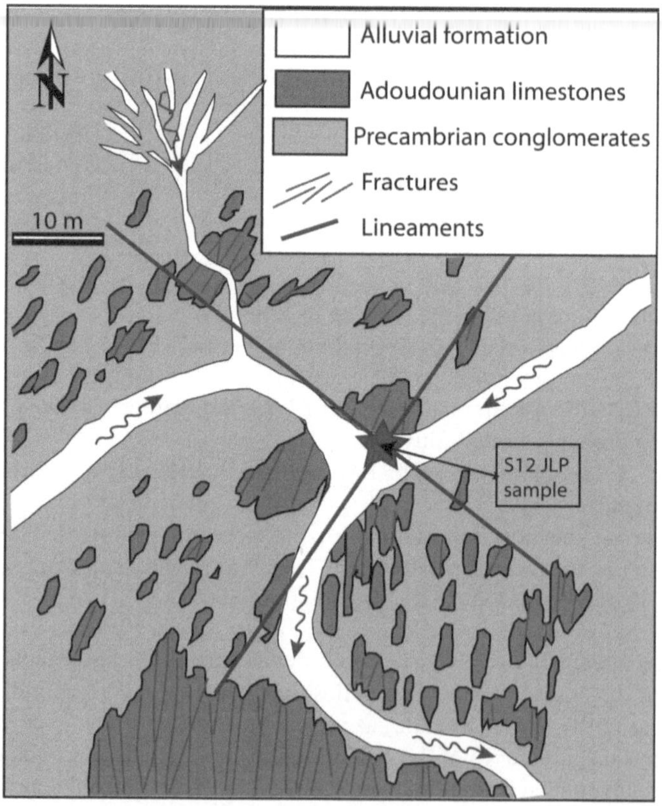

Figure 6. Geological map located in S12 joint point of lineaments.

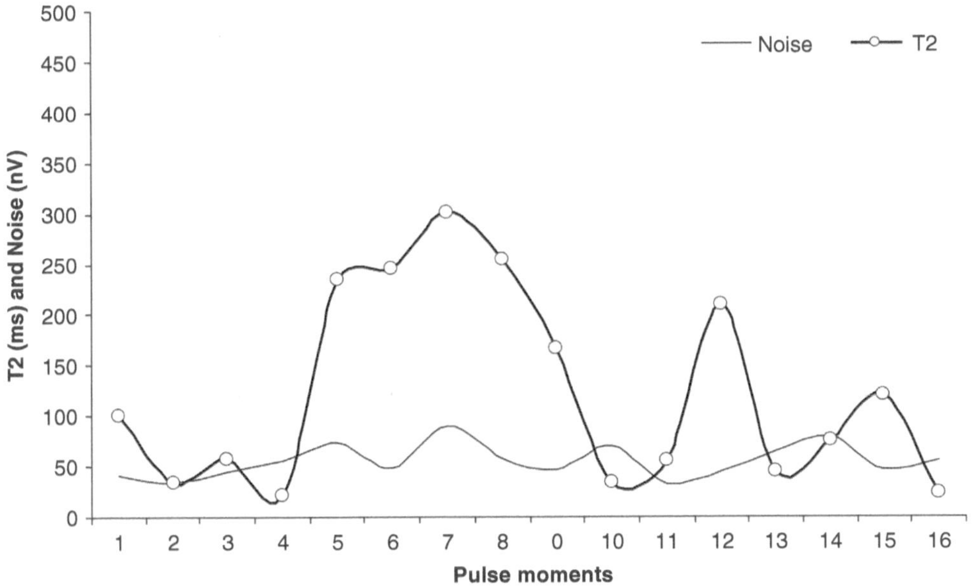

Figure 7. Variation of ambient noise and t decay time values during the acquisition of S12NMRS.

the same station with a different length (AB/2 = 200 m) confirmed the first conducting anomaly at the same station. This implies that the lineament is sub-vertical in depth.

4.3 *Nuclear magnetic resonance*

Four NMRS were performed near existing boreholes (whose yields vary from 0 to 3 l/s) as a control, before applying NMRS in the electrical anomalous zones. The raw data and the inversion results of all the NMRS do not show any significant response (signal amplitude) except for the S12 JPL which registered a good signal with maximum amplitude of 90 nV observed in the 6th impulsion. The conditions for applying NMRS were adequate in all cases, since the noise did not surpass 100 nV (Figure 7).

For S12, a high decay time (>200 ms) registered between 6th and 8th impulsions implies a high permeability at depth (Figure 7). Measurements over fractured limestone aquifers at the four control boreholes also indicated a decay time of 180–250 ms. With some degree of reliability this decay time value can be used as a specific characteristic of the water-saturated fractured limestone in this study area. The repeated values of both excitation frequency and phase values, surrounding the 7th impulsion confirms the anomaly and the possibility of a water bearing zone (Figure 8). The water-level, calculated after the mathematical inversion for the S12 NMRS, is estimated at 9 m in depth (Figure 9).

The results for S12 were confirmed by drilling. A borehole encountered fractured limestone. The water-level was 12 m deep and the yield of the borehole 2 l/s.

4.4 Geomorphology

The analysis of all results of this study demonstrates that mapping of lineaments alone cannot give very good results for borehole siting. The contribution of geophysical methods

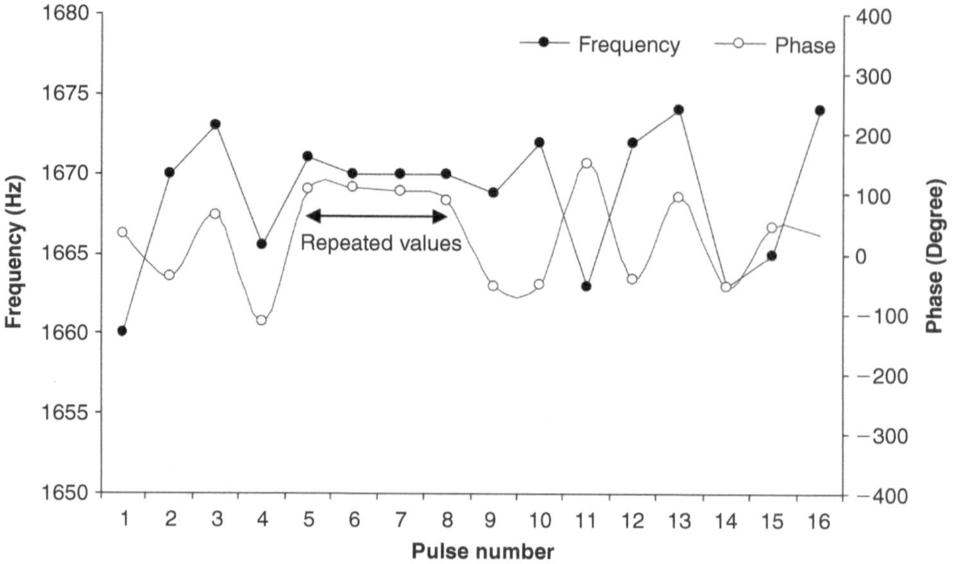

Figure 8. Variation of decay time T2 and the ambient noise during the acquisition of S12MRS sounding.

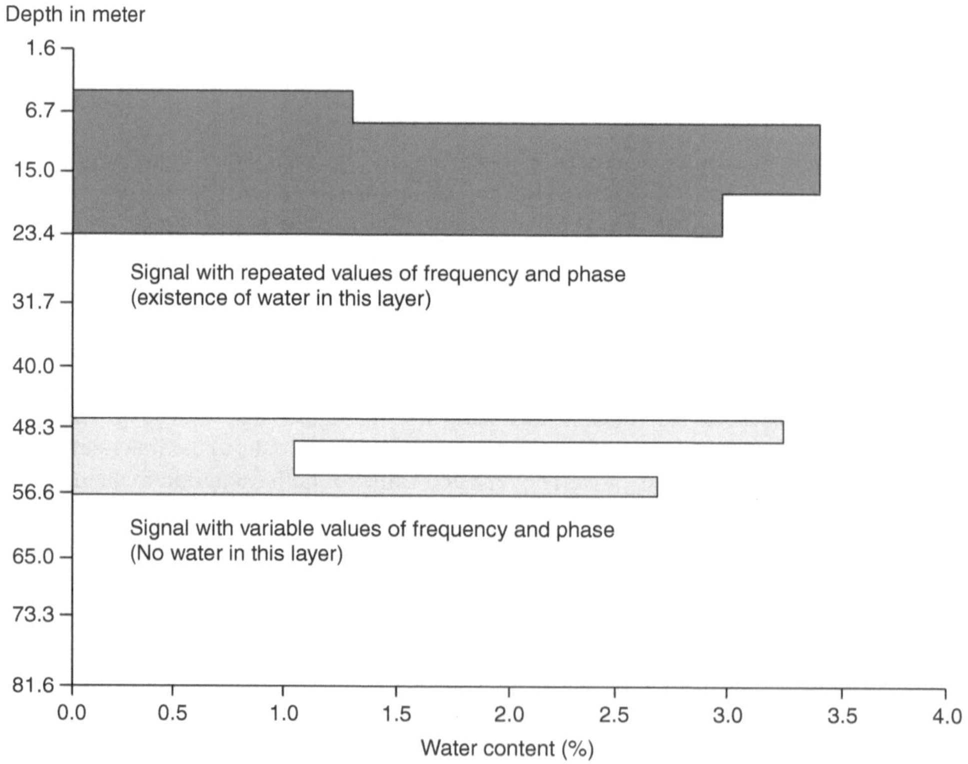

Figure 9. Water content in the S12 NMRS.

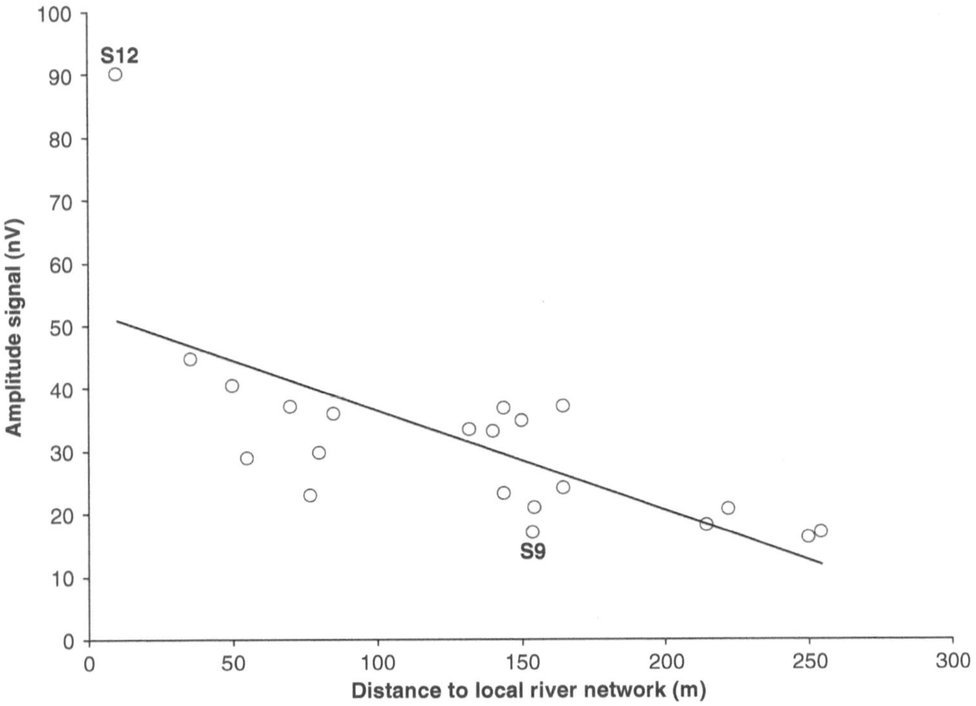

Figure 10. Variation of amplitude signal versus distance to local river network.

is fundamental for the success of the methodology. Nevertheless, available data on geomorphological features, such as elevations of local river courses, drainage density and distance from the JPL to local rivers, must be also be taken into account to explain the relationship between these MRS results and the lineaments.

It is important to indicate that the majority of the boreholes of the study area are located along river segments, where interaction between groundwater and surface water occurs. It is also observed that the water-levels in existing boreholes are similar to the level in the rivers. Figure 6 shows that limestone aquifer at S12 JPL is situated in hydraulic contact with the river. For the majority of other JPL, the fractured limestone is confined with a thick layer of impermeable materials, and interaction is unlikely to occur. Figure 10 shows that the higher NMRS signals are found closest to the local river network. In fact, the S12 JLP presents the shortest distance to the river in comparison with the other points. Therefore, the identification of geomorphological significance of the mapped lineament is crucial to explain the relationship between lineament distribution and groundwater of fractured limestone.

5 CONCLUSIONS

Remote sensing and geophysical investigations were carried out in the western Anti-Atlas chain to delineate potential zones for future groundwater exploration. The study area is underlain by the granitoids, gneisses and a complex series of metamorphic rocks covered by Adoudounian limestone. A realised lineaments map shows that the majority of lineaments develop in NE-SW and N-S directions. Lateral resistivity profiles were used to confirm the presence of these lineaments in 90% of the JLP points. The application of nuclear magnetic resonance techniques indicated a discernable anomaly at the S12 JPL point. The analysis of the different physical parameters indicates the occurrence of groundwater between the 6th and the 8th impulse. This result was confirmed by drilling.

Finally the identification of geomorphological significance of each mapped lineament gave an excellent explanation of the relationship between lineament distribution and groundwater in fractured limestone. The distance between the junction lineaments points and the river network, which play an important role in the recharge of the fractured formations, was inversely related to the likely presence of groundwater as indicated by NMRS. The results of this study show the importance of using an integrated methodology for groundwater prospecting in the hard rock terrains, and not relying on one technique alone.

ACKNOWLEDGEMENTS

The authors wish to thank the Hydraulic department of Souss-Massa-Drâa Basin Agency for there help during different stages of this study and the assistance that it gave for the scientific research in the framework of the convention between University and the hydraulic Agency.

REFERENCES

Boehmer, W. K. & Boonstra, J. 1987. *Flow to wells in intrusive dikes*. PhD, Vrije Universitiet Amsterdam, The Netherlands.

Cazabat, C. 1975. *Topologie ertsienne de la France*. Bulletin dc la Société Française de Photogrammétrie, **60**, 21–36.

Condid, C. D. & Chavez, P. S. 1979. *Basic concepts of computerized digital image processing for geologists*. USGS Bulletin, **1462**.

Durand, A. 1986. *Réflexions sur les principes. Domaines de validité et applications des méthodes de prospection géophysique de surface*. Rapport 86 SGN 247 EAU.

Hassenforder, B. 1987. *La tectonique panafricaine et varisque de l'Anti-Atlas dans le massif de Kerdous (Maroc)*. Thése Doctorat es-Sciences, Université Louis-Pasteur.

Israel, G., Goldman, M., Rabinovich, B., Rabinovich, M. & Issar, A. 1996. *Detection of the water level in fractured phreatic aquifers using nuclear magnetic resonance (NMR) geophysical measurements*. Journal of Applied Geophysics, **34**, 277–282.

Kunetz, G. 1966. *Principles of Direct Current Resistivity Prospecting*. Gebrülder Borntraeger, Berlin.

Legchenko, A. 2004. *Magnetic resonance sounding: enhanced modeling of a phase shift*. Applied Magnetic Resonance, **25**, 621–636.

Legchenko, A. 1996. *Some aspects of the performance of the surface NMR method*. Extended Abstracts of the International Exposition and SEG 66th Annual Meeting, November 10–15, Denver, USA, 1–4.

Legchenko, A., Baltassat, J. M. & Beauce, A. 1997. *Application of proton magnetic resonance for detection of fractured chalk aquifers from surface*. Proc. EEGS. 3rd Meeting on Environmental and Engineering Geophysics, Aarhus, Denmark, 8–11 September 1997. Environmental and Engineering Geophysics Society, Englewood, UK, 115–118.

Lloyd, J. W. 1999. *Water resources of hard rock aquifers in arid and semi-arid zones*. UNESCO Publication, **58**, Paris.

Lubczynski, M. W. & Roy, J. 2003. *Hydrogeological interpretation and potential of the new Magnetic Resonance Sounding (MRS) method*. Journal of Hydrology, **283**, 19–40.

Neuman, S. P. 2005. *Trends, prospects and challenges in quantifying flow and transport through fractured rocks*. Hydrogeology Journal, **13**, 124–147.

O'Leary, D. W., Friedman, J. D. & Pohn, H. A. 1974. *Lineament, linear lineation some proposed new standards for old terms*. Geological Society of America Bulletin, **87**, 1463–1469.

Ozer, A., Marion, J. M., Roland, C. & Trefois, P. 1987. *Signification de linéaments sur une image SPOT dans la région liégeoise*. Bulletin de la Société Belge de Géologie, **97**, 153–172.

Rothery, D. A. 1985. *Interactive processing of satellite images for geological interpretation – a case study*. Geological Magazine, **122**, 57–63.

Siegal, B. S. & Abrams, M. J. 1976. *Geological mapping using Landsat data*. Photogrammetric Engineering and Remote Sensing, **42**, 325–337.

Schirov, M., Legchenko, A. & Creer, G. 1991. *A new direct non-invasive groundwater detection technology for Australia*. Exploration Geophysics, **22**, 333–338.

Soulaimani, A. & Pique, A. 2004. *The Tasrirt structure (Kerdous inlier, Western Anti-Atlas, Morocco): a late Pan-African transtensive dome*. Journal of African Earth Science, **39**, 247–255.

Suzen, M. & Toprak, V. 1998. *Filtering of satellite images in geological lineament analyses: an application to a fault zone in central Turkey*. International Journal of Remote sensing, **19**, 1104–1114.

Wright, E. P. & Burgess, W. G. (eds.) 1992. *The hydrogeology of crystalline basement aquifers in Africa*. Geological Society London Special Publications, **66**.

Yaramanci, U., Lange, G. & Hertrich, M. 2002. *Aquifer characterization using Surface NMR jointly with other geophysical techniques at the Nauen/Berlin test site*. Journal of Applied Geophysics, **50**, 47–65.

Urban Groundwater

CHAPTER 14

Urban groundwater management and protection in Sub-Saharan Africa

S.M.A. Adelana
Geology Department, University of Ilorin, Ilorin, Nigeria

T.A. Abiye
School of Geosciences, University of the Witwatersrand, Private Bag 3, South Africa

D.C.W. Nkhuwa
The University of Zambia, School of Mines, Geology Department, Lusaka, Zambia

C. Tindimugaya
Ministry of Water and Environment, Water Resources Management Department, Entebbe, Uganda

M.S. Oga
Université de Cocody, UFR des Sciences de la Terre et des Ressources Minières, Abidjan, Côte d'Ivoire

ABSTRACT: Groundwater is the preferred source for piped water supplies in many urban areas across Sub-Saharan Africa and its development is forecast to increase dramatically in an attempt to improve urban water supply coverage. The provision of clean drinking water while providing adequate sanitation and storm-water disposal has become a major challenge for many cities. The hydrogeology and groundwater situation of Addis Ababa, Abidjan, Cape Town, Dakar, Lagos, and Lusaka are highlighted as examples that illustrate the status of urban groundwater in Sub-Saharan Africa. The history of urban development and the current groundwater management practices under each case example is also discussed. The main man-made impacts on groundwater in the various cities under consideration are rapid urbanisation and changes in land use surrounding cities. The impact of urbanisation is not only viewed in terms of groundwater quality but as it affects recharge.

1 INTRODUCTION

In urban areas across Sub-Saharan Africa (SSA), the density of population and industry often results in serious problems of groundwater quantity and quality. The United Nations Population Division (UNPD, 2005) predict that global urban population will increase from 3 billion in 2000 to nearly 8 billion in 2030. Nearly all of this increase is expected to take place in developing countries as urban population will grow at an annual average rate of 1.9% compared to global population growth rate of 1%. The rapid rate of urbanisation in Africa has far exceeded the management and financial capabilities of all the levels of governments since the 1960s. This is a crucial fact that poses challenge to water resources management in SSA.

African cities have a long history of water supply both from surface water and groundwater. However, due to the deteriorating quality and quantity of surface water through increased urbanisation and industrialisation, and the high costs of developing new dams, urban groundwater is being seen as an increasingly valuable resource. However, recent investigations reveal that key groundwater resources are either polluted or at risk of pollution.

The occurrence and use of groundwater within the general area of any city depends not only on geology, climate, and the availability of other sources of water, but also on the history of development. Investigations in African capital cities show similar problems across the continent, but with some key differences due to the diverse political, cultural and economic conditions across SSA. Yet a unifying theme for all is the need for freshwater and its sustainability. The provision of clean drinking water, adequate sanitation and stormwater disposal has become a major challenge for most African cities.

Groundwater resources are used to supply many urban centres across Sub-Saharan Africa. The large cities that are groundwater-dependent are shown in Figure 1 (Morris *et al.*, 2003). Recent development and increased groundwater demand would have added more cities to this list in the last few years. In cities where groundwater is a small fraction of total water use, it still represents a stable source of water, when surface water resources may fluctuate.

This paper reviews the current status of urban groundwater in Sub Saharan Africa by means of some selected case histories. The objective is to describe the importance of groundwater in water supply to major cities in SSA and examine the impact of urbanisation on quality and quantity of groundwater. The history of urban development, current groundwater management practises, policy implications and future challenges are also discussed.

2 INTRODUCTION TO THE CASE STUDIES

2.1 *Addis Ababa*

The city of Addis Ababa has a population of 4 million of which 50% have no direct access to the Municipal water supply system (derived from both surface water and groundwater). The city was established as a capital of Ethiopia in 1886 and has since grown to become the largest urban and commercial centre in the country. It is located in the central part of the country at the edge of the western escarpment of the Ethiopian Rift (Figure 1). In the 1970's the size of the city was 37 km^2, while a satellite image in 1999 revealed that the size increased 230 km^2 (Figure 2). Within 29 years the city expanded at an average rate of 6.7 km^2/a. According to projection of the United Nations, Addis Ababa will become the fourth largest city in Africa by 2015 (UNPD, 2005).

For the first 58 years (1886–1944), the water supply for the city was derived from groundwater in the form of springs located at the foot of the Entoto ridge (northern part of the city) and also from dug wells located in the central and southern lower part of the city (Alemayehu, 2006). Additional demand necessitated treatment of surface water derived from three surface dams (Gefersa, Legedadi and Dire). However, the exponential population growth in the city demanded further water resources, and the Akaki well field in the southern parts of the city was developed to supplement the growing water demand. Individual boreholes at the Akaki well field can yield as much as 80 l/s.

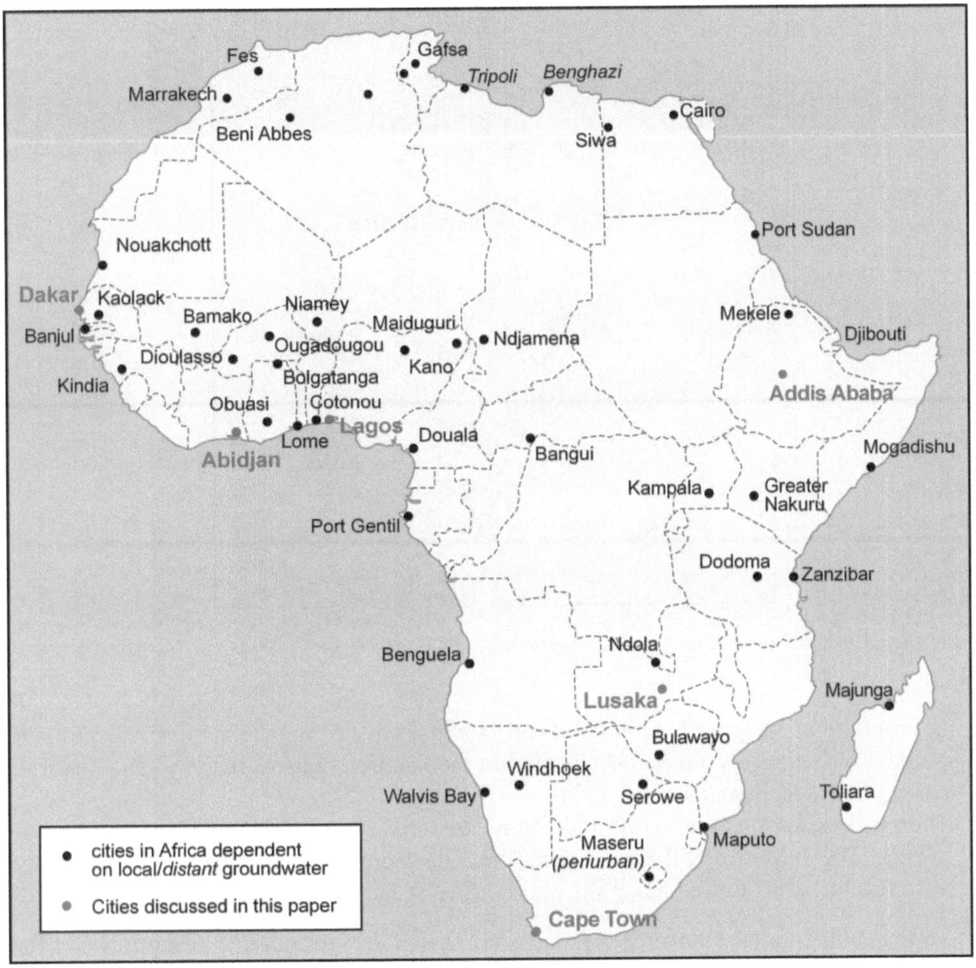

Figure 1. Examples of cities heavily dependent on groundwater in Africa (adapted from Morris *et al.*, 2003).

The current demand has created a supply shortfall of about 50%. Large numbers of private boreholes abstract water from the volcanic aquifer with yields as much as 15 l/s. The large number of private boreholes has caused well interference and a lowering of the water-table. The water quality from private boreholes within the city can also be poor.

Geologically, the city is dominated by volcanic material of different ages and compositions (Alemayehu *et al.*, 2005). The Miocene-Pleistocene volcanic succession in the Addis Ababa area from bottom to top are: Alaji basalts and rhyolites, Entoto silicics, Addis Ababa basalts, Nazareth group, and Bofa basalts. The Alaji group volcanic rocks (rhyolites and basalts) show variation in texture from highly porphyritic to aphyric basalts and there is an intercalation of grey and glassy welded tuff.

The main aquifers in Addis Ababa are:

1. Shallow aquifer: composed of slightly weathered volcanic rocks and alluvial sediments. Depth to groundwater in this aquifer reaches up to 50 m.

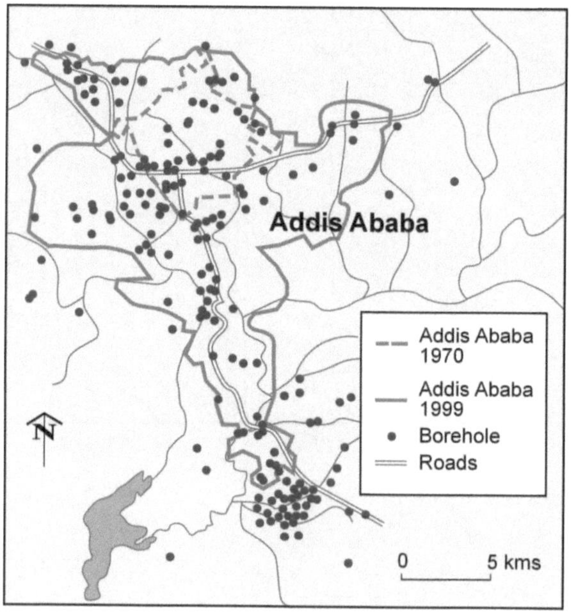

Figure 2. Borehole locations in Addis Ababa (after Alemayehu *et al.*, 2005).

2. Deep aquifers: composed of fractured volcanic rocks that contain relatively fresh groundwater. These aquifers are mainly located in the southern part of the city. The depth in some places reaches as high as 180 m.
3. Thermal aquifer: that is situated at depth greater than 300 m and located in the centre of the city. The existence of these aquifers is manifested by deep circulating thermal water with few hot wells drilled along the major Filwoha fault.

More than 500 boreholes have been recorded in the city (Figure 2) that provides water for domestic and industrial uses. These boreholes are currently tapping water from the volcanic rock reservoir at a variable rate. The water abstracted from the volcanic aquifers by the municipality is around 40,000 m^3/d (out of which 30,000 m^3/d is from Akaki well field). Other private and governmental institutions abstract as much as 50,000 m^3/d with overall total abstraction of 90,000 m^3/d. Both in central and eastern part of the city, groundwater occurs within the confined aquifer. Therefore, the main groundwater potential areas are the eastern and southern part of the city (Alemayehu, 2006).

2.2 *Abidjan*

Abidjan is the most populous city of Cote D'Ivoire (formerly Ivory Coast) with 4 million people, more than a fourth of the total country's population estimated at 15 millions (INS, 2001) It is an industrial city situated along the Atlantic Ocean coastline (Figure 1). The great region of Abidjan constitutes the central part of a coastal sedimentary basin which covers a surface of 16,000 km^2 between the latitudes of 5°00 and 5°30 N and the longitudes of 3°00 and 6°00 W (Figure 3). The climate of the area is sub-equatorial, characterized by two rainy seasons (March–July and September–November) separated by two relatively

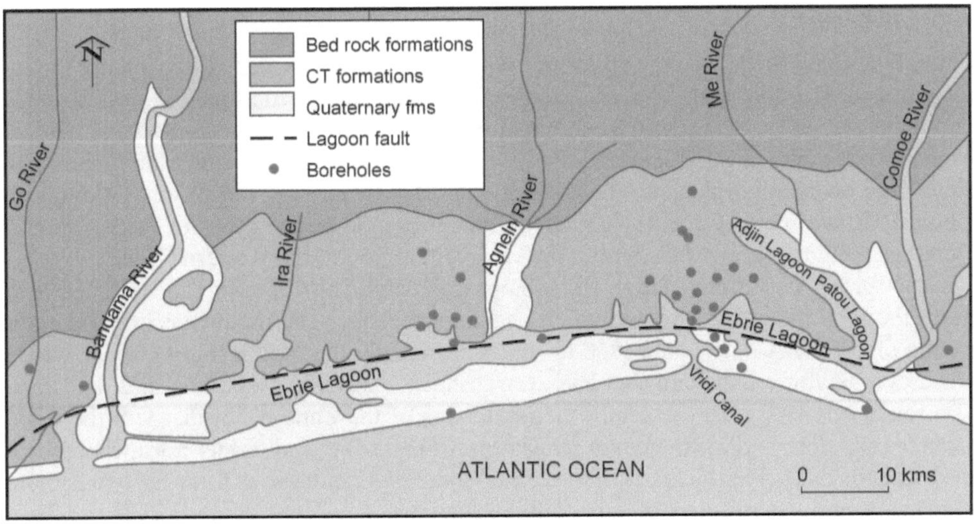

Figure 3. Geology of Abidjan and location of main boreholes.

dry periods. The annual rainfall is between 1500 and 2000 mm for an average temperature of 27°C (monthly temperature range 24–30°C). The vegetation is a clear forest near the coastline and becomes dense further inland.

Like other coastal cities of Western Africa, the recent increase of population has resulted in environmental problems such as the deterioration of drinking water quality and quantity (Kouadio *et al.*, 1998; Jourda, 2003, 2004; Soro, 2003). Although there are rivers in this area, the water supply of the Abidjan area is mainly from two shallow unconfined sedimentary aquifers: the Continental Terminal aquifer (CT), the most important aquifer; and the Quaternary aquifer along the coast.

The Quaternary deposits are marine sands (Nouakchottien) and fine sands (Oogolien). This aquifer is highly vulnerable to pollution because its potentiometric surface is very close to the ground surface. In the Continental Terminal aquifer, both fluviatile (CT3) and clayey sands (CT4) are hydrologically important. The aquifer (CT3) may be confined when the upper parts of CT3 are clay rich; where there is no clay between CT3 and CT4 deposits, the CT3 and the CT4 are interconnected.

The hydraulic conductivity of the CT aquifer is variable due to lateral changes in the grain size of water-bearing sediments (100 m/d in the sands and the sandstones and 0.1 m/d in the clayey sands). The transmissivity values can be as high as 10,000 m^2/d while porosity values range from 0.05 to 0.20. The regional groundwater flow occurs from north to south, i.e. towards the lagoon. The hydraulic conductivity of the Quaternary aquifer ranges between 3.5 and 100 m/d. The hydraulic gradient increases up to 3‰ close to the lagoon. The flow rates are low compare to that of the CT aquifer: 0.6 to 6 l/s for the Quaternary aquifer and 2 to 90 l/s for the CT aquifer (Oga *et al.,* this volume).

Deeper in the basin, around 200 m below the ground surface, the Maestrichtian carbonates and sandstones constitute a confined aquifer. This aquifer is artesian with a potentiometric surface at +27 m above the sea level. Only one borehole of depth 190 m (from the SADEM Company) draws its water from this aquifer.

2.3 *Cape Town*

The City of Cape Town is a large urban area with a high population density, an intense movement of people, goods and services, extensive development and multiple business districts and industrial areas. It represents centres of economic activity with complex and diverse economies, a single area with integrated development planning and strong interdependent social and economic linkages. The City of Cape Town includes the Cape Metropolitan Council, Blaauwberg, Cape Town CBD, Helderberg, Oostenberg, South Peninsula and Tygerberg (Figure 4); and falls within the semi-arid region of the Western Cape.

Cape Town in its eastern edge and its suburbs is underlain by Cenozoic sands (Cape Flats) up to the foothills of the first mountain chains to the east. The mountains are underlain by the oldest rocks in the area (830 to 980 Ma), namely the Malmesbury Group, which consists of phyllitic shale and siltstone, quartz schist, quartzitic greywacke and sandstone (Theron *et al.*, 1992). To the south, the mountains of the Cape Peninsula with the Cape Granite intrusions and the mountain ranges east of False Bay and farther northwards form the higher terrains. The thickness of the Table Mountain Group varies from 1200 to 2100 m in the area, and it is sub-divided into eight Formations.

The Cape Flats is within the sphere of the major catchments within greater Cape Town area, where runoff is mostly generated in the mountain ranges in the southeast and the Table Mountain and Cape Peninsula Mountains in the southwest. The Berg, Diep, Mosselbank and Eerste Rivers constitute the most important drainage systems. Sandy lowlands with minimal runoff and a high water-table extend over the central area. The greater Cape Town Metropolitan Area lies on one of the most extensive sand aquifers in South Africa, and the supply potential of groundwater from this aquifer is highly significant. This extensive sand aquifer, called the Sandveld Group, is hydrogeologically divided into four main units: the Cape Flats unit, the Silwerstroom-Witzand unit in the Atlantis area, the Grootwater unit in the Yzerfontein region and the Berg River unit in the Saldanha area. Yield analysis of about 497 boreholes in the Sandveld Group indicates that 41% of boreholes yield 0.5 l/s and less while 30% yields 2 l/s and more (Meyer, 2001).

2.4 *Dakar*

Dakar is the capital city of Senegal (Figure 1), and one of the cities in SSA that relies heavily on groundwater for potable supply, both from shallow private hand-dug-wells and from deeper public water boreholes. The city has witnessed high population growth (2 million people) during the last three decades (Cissé *et al.*, 2000). It is not only a fast growing urban centre but the high density of population (much of which are in unplanned substandard housing with no services) has resulted in serious degradation in groundwater quality. In Dakar and its suburbs, on-site sanitation, lack of organized domestic waste disposal and pollution from industries pose a serious threat to water supply (Cissé *et al.*, 2004).

There are two aquifer systems in the area around Dakar (Figure 5): a semi-confined basaltic aquifer in the western part and the unconfined Thiaroye aquifer in the eastern part (Tandia *et al.*, 1998). Sediments of the Thiaroye aquifer belong to the Senegalese superficial aquifer, which consists largely of fine- and medium-grained sands with porosity of 20% and hydraulic conductivity between 10 and 5000 m^2/d (Cisse *et al.*, 2000). With population growth, water demand has drastically increased, inducing saltwater intrusion in many dugwells and piezometers in the local coastal aquifers (Tandia *et al.*, 1998).

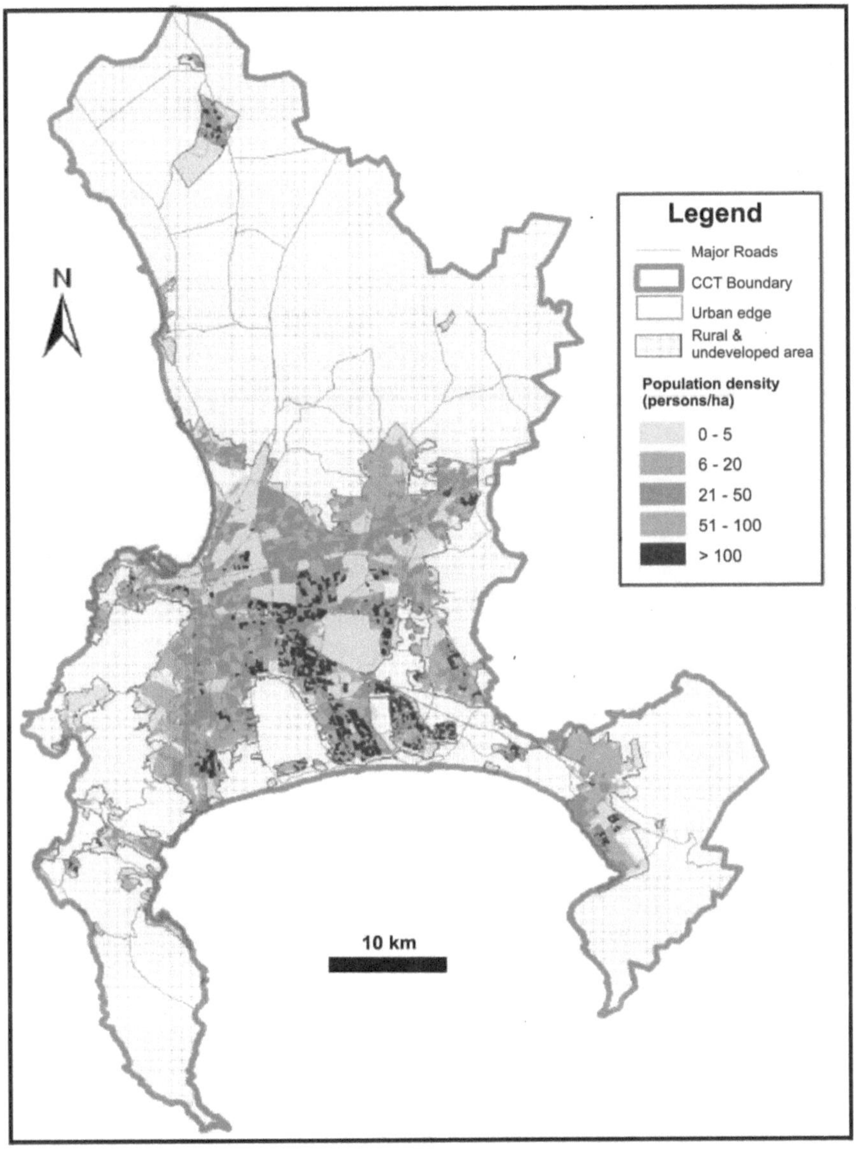

Figure 4. The city of Cape Town Metropolitan Municipality with population density.

2.5 *Lagos*

Lagos is located within the Western Nigeria Atlantic coastal zone consisting largely of coastal creeks and lagoons developed by barrier beaches and situated on stratified series of sedimentary rocks made up of silt, clay, peat or coal associated with sand deposition. Lagos is located on latitude 6° 34′50″ North, and 3° 19′59″ East of the Greenwich Meridian (see Figure 1).

Lagos, which was until 1991 the federal capital city of Nigeria, continued to witness a high increase in population growth and presently has the highest population density in

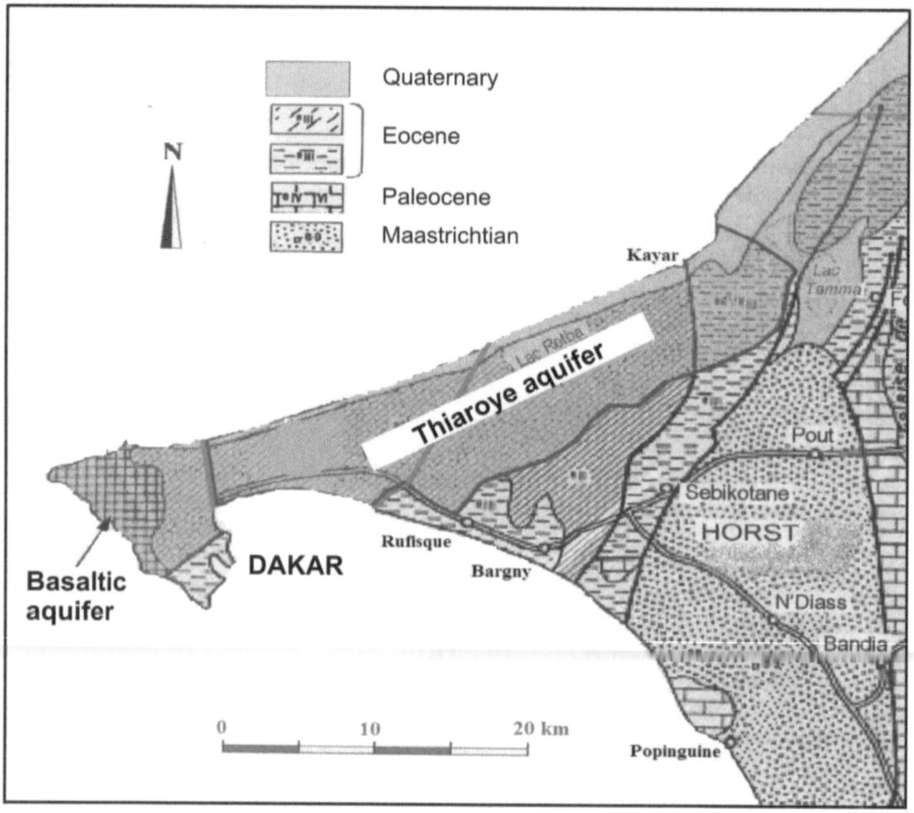

Figure 5. Location of Dakar city in the western part of Senegal showing the Thiaroye and basaltic aquifers.

Nigeria. It is one of the largest cities in Africa (second only to Cairo in Egypt) and remains the commercial capital of Nigeria even after the seat of government was moved to Abuja. The rate of population growth is about 300,000 persons per annum with an average density of 20,000 persons/km^2. With a total land area of about 3600 km^2 and current annual growth rate of 4%, Lagos is expected to be one of the world's five megacities in 2015. This demographic expansion had great consequence on the municipal water supply system and increased the problem of waste management within the city. The area is generally low-lying with several points virtually close to sea-level. Lagos is built on the mainland and the series of islands surrounding Lagos Lagoon (Figure 6).

In the city of Lagos, there is pressure on groundwater resources; more problems arise because of the potentially contaminating human activities developed above the aquifers. Therefore, groundwater contamination is a major public health and environmental concern in the coastal city of Lagos, partly because the majority of the population uses wells (either boreholes or hand-dug) for drinking and domestic purposes.

Quaternary geology of the area comprises the Benin Formation (Miocene to Recent), recent littoral alluvium and lagoon/coastal plain sands (Jones & Hockey, 1964; Durotoye, 1989; Longe *et al.*, 1987). Lagos metropolis is underlain by a 3-layer aquifer system

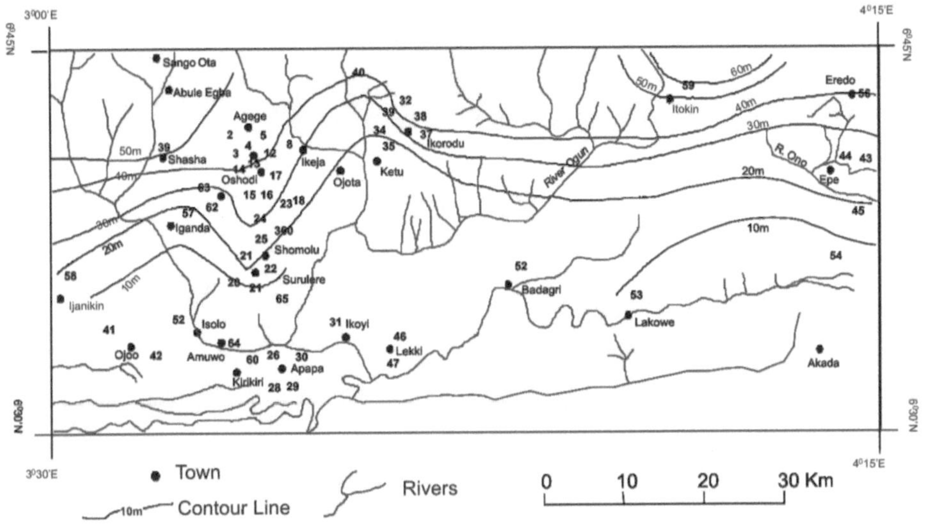

Figure 6. Map of densely populated Lagos (the names represent the densely populated areas and numbers indicate wells sampled for water quality see later).

with varying hydrogeological properties and homogeneities. The aquifers have variable thickness with the first and third horizons attaining a thickness of approximately 200 m and 250 m respectively at Lekki peninsula while the second horizon attains a thickness of about 100 m at Ijanikin (Asiwaju-Bello & Oladeji, 2001). About 75% of groundwater abstractions for domestic and industrial purposes in Lagos are obtained from the second aquifer. Transmissivities range from 100 to 500 m^2/d and an average storage coefficient of approximately 0.003 has been estimated for the first aquifer horizon (Asiwaju-Bello & Oladeji, 2001).

2.6 *Lusaka*

Lusaka, the capital city of Zambia, is located on a wooded ridge, which runs from south-east to the north-west. The location of Lusaka for a new capital city of the then Northern Rhodesia was chosen after many years of debate and after examination of a number of possible locations, including Chilanga, Broken Hill (now Kabwe), and four Copperbelt towns. After ascertaining that there were adequate groundwater supplies, Lusaka was chosen partly to avoid domination by the mining companies on the Copperbelt and partly because of its location at the intersection of the main route network. It was inaugurated the new capital city of Northern Rhodesia on May 30, 1935.

Settlement and development patterns in Lusaka have been greatly influenced by the growth of population. With a population of only 195,753 at independence in 1964, there have been progressive increases in population over the years, rising to 535,850 in 1980 and 769,353 in 1990 (CSO, 1990). In 2000, the population of Lusaka reached 1.2 million, and is estimated at 2 million in 2007. Other than the high birth rate (3%), the drive for most of this population has been the rural-urban migration in search of employment and better livelihoods. Consequently, Lusaka has experienced rapid population growth and

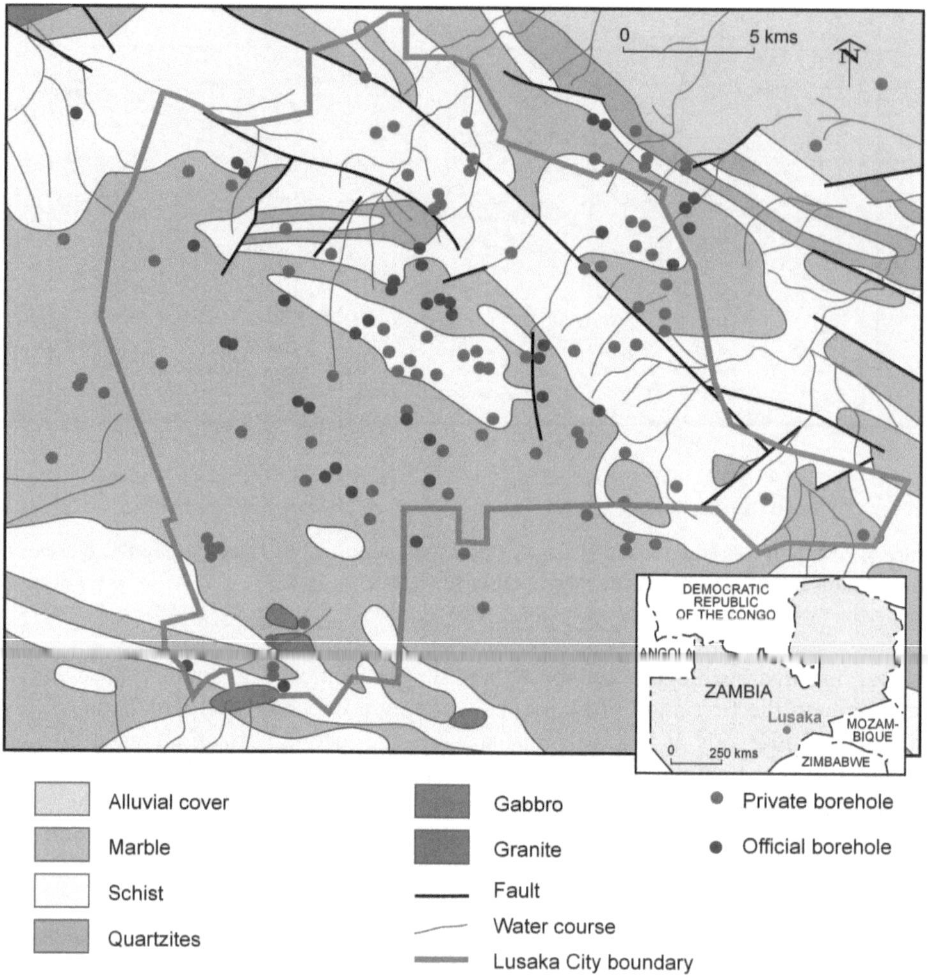

Figure 7. Geology and location of private and public water supply boreholes in the Lusaka aquifer.

uncontrolled rates of urbanisation (Figure 7), which has been a recipe for the mushrooming of informal high-density settlements and the sprawling of unplanned low-density residential settlements. The rapid growth of population in Lusaka has mismatched the development of infrastructure.

Rocks underlying the city of Lusaka consist of schists interbedded with quartzites and dominated by thick and extensive sequences of marbles (Figure 7). Available borehole drilling data indicates carbonate rocks extending to depths in excess of 100 m with variations in the fracturing intensities (Nkhuwa, 1996). While some of the solution features may not show any evidence of occurrence at the surface, they have great lateral extents in the subsurface, with some of them having been intersected in boreholes at depths in excess of 60 m below ground surface. The presence of these features has transformed these rocks into a favourable and comparatively cheap source of water supply to the city and they appear to have exerted a lot of control on groundwater flow in the aquifer. Further, the occurrence and orientation of these fractures have dictated the general course of groundwater flow in the aquifer. And an evaluation of water strikes in boreholes is indicative of their close association with discontinuities in the rock mass. The marbles have an average transmissivity of approximately 600 m^2/d. Thus, the hydrogeology of Lusaka indicates that the aquifer has the best groundwater potential to support the city water supply for large-scale exploitation if well managed.

Currently, rapid growth of population in Lusaka has outstripped the rate of provision of basic social goods as well as adequate sanitation services. Peak water demand to cope with current requirements (estimated at an average per capita consumption of about 200 litres per day) stands at about 400,000 m^3 per day for a population that was estimated at two million in 2007. Measured against actual daily production of about 200,000 m^3 per day currently supplied by the water authority, the deficit in supply raises concern. The current unsatisfied water demand has triggered a process of indiscriminate borehole drilling (Figure 7) and *excessive* abstractions of groundwater from the aquifer.

3 STATUS OF URBAN GROUNDWATER IN SUB-SAHARAN AFRICA

Urban groundwater development in Sub-Saharan Africa (according to Taylor *et al.*, 2004) can be categorised into (i) low-intensity abstraction ($<$0.2 l/s) and high-intensity abstraction ($>$2 l/s). The high-intensity groundwater abstraction category is less common and is for piped supplies, usually achieved using wellfields consisting of one or more high-yielding boreholes equipped with motorised pumps. Low-intensity groundwater use occurs via manually pumped boreholes and shallow wells, and is often private and unregulated.

The examples introduced above are used to discuss the impact of urbanisation. A summary of the case studies is given in Table 1.

3.1 *Recharge*

The impact of urbanisation is not only viewed in terms of groundwater abstraction but as it affects recharge. The general concept of urbanisation in this regard is that it reduces recharge by water-proofing surfaces (Lerner & Barret, 1996; Barret, 2004), but the network of water-carrying pipes under most of African cities are leaky, sometimes old and rusty. This in addition to leaking sewers, septic tanks and storm drains constitutes high potential for urban recharge (Table 2). According to Krothe *et al.* (2002), sewer lines are designed for leakage (typically about 10%). In Cape Town, it is said that nearly 40% of the water from the supply dams to consumers are lost through pipe bursts and leakages (comparable to 30% of recharge from utility system leakage in San Antonio (Sharp *et al.*, 2000); 12% in Austin, Texas (Lorenzo-Rigney & Sharp, 1999)). Although few quantitative data are available, the general belief is that much water is lost through the supply mains and distribution channels. Estimates of water main leakage in urban areas are presented in Table 3. This loss, coupled irrigation of farmlands and gardens in most African cities, contributes significantly to recharge.

Generally, the quantification of natural recharge can be subject to a whole range of difficulties; no clear methodologies; data deficiencies, and the resultant uncertainties that may be relate to: (i) wide spatial and temporal variability in rainfall and run-off events; (ii) lateral variation in soil profiles and hydrogeological conditions. These constraints notwithstanding, estimates based on available data are useful for initial management

Table 1. Summary of groundwater status of selected cities in SSA.

Urban area	Addis Ababa	Abidjan	Cape Town	Dakar	Lagos	Lusaka
Main aquifer type	Multilayer volcanic rocks	Unconfined Quaternary deposits	Unconfined sand (semi-confined in places), Fractured sandstone	Semi-confined basalts and unconfined sands and gravels	Thick sands and gravel; deeper confined sands	Unconfined sands and gravels; Karstified marble aquifer.
Aquifer storage	Low in N and central part of the city and high in E and S	Moderate, but progressive degradation of the quality from S	Moderate but towards margins increases salinity	Low with increasing salinity from the NE towards the NW.	Moderate with increasing salinity towards the west margins	n.a.
Average rainfall	1150 mm/a	1600 mm/a	600 mm/a	485 mm/a	1700 mm/a	900 mm/a
Source of primary recharge	Rainfall and dams	Mainly rainfall	Mainly rainfall, periodic infiltration of runoff	Mainly rainfall	Excess of rainfall, canal and riverbed infiltration	Mainly rainfall, but with a substantial contribution from water utility supply and sewer leakage
Aquifer depletion	There is potential depletion in the southern part of the city	Not currently quantified but modelling projection shows 0.33 m/a	–	0.133 m/a (from model calculation)	Few data, probably averaging 0.5 m/a	Currently, not quantified
Sanitation risk	Extremely high	Extremely high	Very high in shallow aquifers; threatens deeper fractured aquifers	Very high	Very high in both shallow and deep aquifers	Very high arising from use of onsite sanitation; indiscriminate disposal of various forms of solid wastes
Land subsidence	Very low	Low	Generally low hazard	Low	Fairly low hazard but growing concerns	Would have been quite high, save for the *consolidation* of soils filling solution channels

Table 2. Sources of aquifer recharge in urban areas with implications for groundwater quality (Foster *et al.*, 1996).

Recharge source	Importance	Water quality	Pollution indicators
Leaking water mains	Major	Good	Generally no obvious indicators
On-site sanitation systems	Major	Poor	N, B, Cl, FC
Leaking sewers	Minor	Poor	N, B, Cl, FC, SO_4 (industrial chemicals)
Surface soakaway drainage	Minor to major	Good to poor	N, Cl, FC, HC, DOC (industrial chemicals)
Seepage from canals & rivers	Minor to major	Moderate to poor	N, B, Cl, SO_4, FC, DOC (industrial chemicals)

B – Boron
DOC – Dissolved organic carbon (organic load)
N – Nitrogen compounds (nitrate or ammonium)
Cl – Chloride and salinity in general
FC – Faecal coliforms
SO_4 – Sulphate

Table 3. Estimated leakage rates for municipal water in selected cities.

Urban area	Estimated water main losses (%)	Estimated increase in recharge (mm/a)
Addis Ababa	55	150
Abidján	40	100
Cape Town	40	30
Dakar	–	34
Lagos	45	240
Lusaka	55	140

decisions. Sufficient efforts are, however, needed to monitor and analyse aquifer response to medium-term abstraction in order to be able to refine the initial estimates of recharge.

3.2 *Impact of urbanisation on groundwater quality and quantity*

The main man-made impacts on groundwater in the various cities under consideration result from urbanisation and from changes in land use. Some of the changes over the years include: (i) clearing of bushland and shifting of vegetated areas to residential use; (ii) developing agricultural, industrial and residential uses in undeveloped sections of the coastal cities; (iii) increasing and high growth rate of urban population; (iv) high rural-urban migration. The ever-increasing growth in urban population in Sub-Saharan Africa since 1950s has been analysed by Harris (1990) and illustrated in Taylor *et al.*, (2004). Providing safe drinking water supply; treatment of domestic and industrial wastewater; and management of solid waste generated in the cities of Africa, will be some of the major challenges for urban planners, policy makers, politicians and social workers.

Typical situations of water pollution in the urban settlements of SSA arise due to:

- pollution of surface water by industrial effluents discharged into rivers or streams and pollution of groundwater by leakages and spills from industries;
- pollution of surface water by storm water, untreated sewage and wastewater, flowing from an urban agglomerate, into a river or stream;
- pollution of surface water by residential and industrial development in the vicinity of a reservoir, river or canal supplying water to an urban area; and
- pollution of groundwater in shallow aquifers due to leakage from septic tanks, soak pits and sewage lines of urban areas, and also from influent seepage from rivers/streams carrying polluted surface water.

Often urban groundwater use can lead to over-exploitation and consequent problems such as saline water intrusion. Examples include the south coast of Lagos, Dakar, and Cape Town. Groundwater quality is usually worse beneath cities than beneath nearby rural areas.

3.2.1 *Lagos*

During a comprehensive groundwater-quality study for the south-eastern part of the city of Lagos (1999–2001) groundwater samples from existing wells were analysed for many parameters (Adelana *et al.*, 2003, 2004, 2005). Urban impact is indicated for more than 60% of all samples by high concentrations of sulphate, nitrate (with NH_4-N in places) and chloride (Figure 8).

The mean nitrate content in groundwater in the Lagos area is 70.3 mg/l; 60% have NO_3 content above the WHO guideline of 50 mg/l. The groundwater in Lagos is particularly vulnerable to contamination since groundwater is shallow, and the aquifer comprises unconsolidated permeable sand and gravel. The concentration of nitrate measured in rainwater shows clearly that NO_3 is introduced into groundwater through urban activity, and most likely on site sanitation, rather than through a natural source. The fast rate of urbanisation in Lagos has brought most of the industrial layout, farmlands and swamps within residential areas.

3.2.2 *Addis Ababa*

A study of groundwater-quality for Addis Ababa included sampling of several observation and production boreholes and springs (Alemayehu *et al.*, 2005). The results revealed that the nitrate concentration ranged from 0.72 mg/l to 728 mg/l in groundwater and springs (Alemayehu, 2001). Other indicators of pollution are the occurrence of nitrogen as NH_4 and/or NO_2 and coliform bacteria. The concentration of total coliforms and *E.Coli* in the groundwater system showed large seasonal variations (Alemayehu *et al.*, 2005). Groundwater contamination problems in Addis Ababa are mainly related to poor borehole construction and leakage from defective sewerage lines and septic tanks.

3.2.3 *Abidjan*

During a 3-year assessment project on the pollution status and vulnerability of the Abidjan aquifer, there was a clear indication that the groundwater of Abidjan is affected by the progression of pollution from the south towards the north and west (Jourda *et al.*, 2006). The major pollution threat to the Abidjan aquifer is sewage: there are no developed network

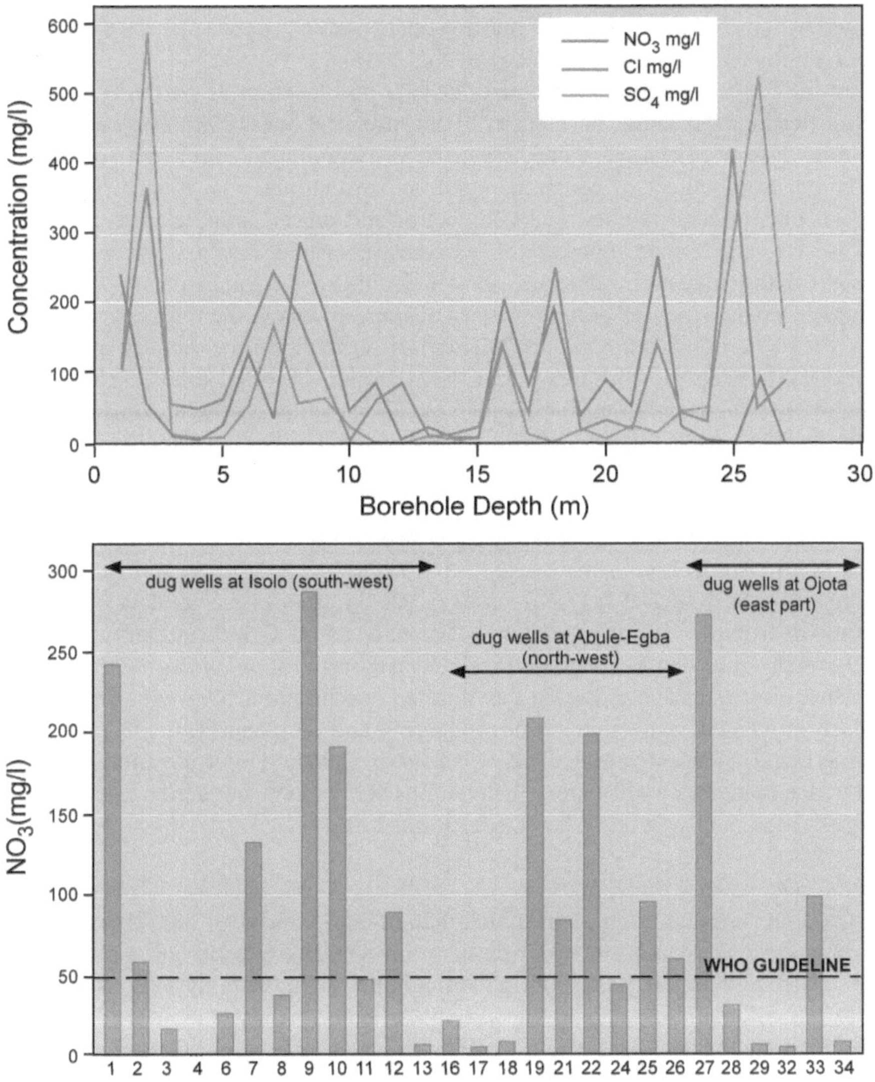

Figure 8. Water quality in shallow wells in Lagos (a) plot of Cl, SO_4 and NO_3 in groundwater versus the depth of borehole; (b) Nitrate concentrations of groundwater in hand-dug-wells.

of systems for the collection of waste and wastewater. This led to high nitrate concentrations (commonly 100–300 mg/l, maximum 500 mg/l as NO_3) in water supply boreholes (1994–2004) with cholera cases between 2001 and 2003 being attributed to contaminated groundwater (Jourda *et al.*, 2006).

3.2.4 *Cape Town*

In Cape Town the need to augment the present water supply through additional borehole drilling is expected to put more pressure on the groundwater resources in the area. Data from 20 boreholes located in the area show elevated Cl, Na, and total dissolved solids (TDS) associated with proximity to the sea. Chloride concentrations between 1500 and 2000 mg/l are recorded in some boreholes (Adelana & Xu, 2006).

Within the Cape Municipality, the size, density and location of settlements have been found to determine the degree or intensity of pollution (DWAF, 1999). The various sources of pollution have been classed according to the varying human activities (Adelana & Xu, 2006). Significant sources of contaminants in and around the city of Cape Town is from leakage of underground storage tanks for petrol and diesel, nutrients and pathogens in human wastes (e.g. nitrate, phosphate, potassium), cyanide and trichloroethene (TCE) from metal plating industry, chemicals used for cleaning and agrochemicals (fertilizers and pesticides). General findings around the city of Cape Town are the high concentrations of phosphorus (>7.5 mg/l), ammonium (200 mg/l as N) found in the vicinity of the unlined sludge ponds (Usher *et al.*, 2004), cyanide (15–210 mg/kg in soils), and TCE (6–4089 μ g/l) from unlined sewage sludge drying ponds and sludge spills on the unconfined, primary sand aquifer (Parsons & Taljard, 2000).

3.2.5 *Dakar*

The urban aquifer of Dakar contains nitrate concentrations that exceeds the standards of 50 mg/l. Monitoring boreholes revealed up to 300 mg/l as NO_3 in the unconfined aquifer to the eastern suburb of Dakar (Deme *et al.*, 2006). Earlier chemical data of the aquifer and spatial distribution of nitrate contamination showed up to 540 mg/l NO_3 (Cisse *et al.*, 2000). The high nitrate concentrations are related to the rapid population growth in this urban setting. Investigations have indicated that nitrate concentrations are increasing steadily in the densely populated zone (from Pikine to Thiaroye and Yeumbeul) from 26.7 mg/l in 1967 to 804 mg/l in 2004 (Deme *et al.*, 2006; Cisse *et al.*, 2004). This widespread pollution is caused by the inadequacy of sanitary installations and garbage collection infrastructure in the densely populated area of Dakar (Deme *et al.*, 2006).

3.2.6 *Lusaka*

In Lusaka, four sampling campaigns undertaken in mid-November 2003 (before the onset of the rainy season), March 2004 (during the rainy season), October 2004 (at the peak of the dry season) and March 2005 (during a drought period in the rainy season) revealed the variability of pollutant loading with the varying levels of the water-table (Nkhuwa, 2006). Faecal contamination increases with the rise in the water-table levels. Consumption of such water, which is usually of unfavourable quality, has heightened outbreaks of waterborne diseases, such as cholera and dysentery that have been experienced in many areas of the city.

In Lusaka groundwater mining and drought episodes appear to have resulted in a progressive decline in the aquifer water-table (Figure 9) because groundwater withdrawals are far in excess of the average rates of annual recharge. Such declines in the water-table lead to reduced borehole yields, which provoke an expensive and inefficient cycle of borehole deepening to regain productivity, or even premature loss of investment due to forced abandonment of boreholes. Progressive lowering of the water-table will undoubtedly increase production costs, thereby imposing restricted access to the resource by the low-income group of the city population, whose shallow wells may dry up.

High-density settlements have the highest growth of population in the city. Unfortunately, these areas rely solely on pit latrines to dispose of their excreta, while shallow wells have provided the most common sources of water supply. Most of the settlements have flourished over the aquifer recharge areas where there is indiscriminate disposal of different forms

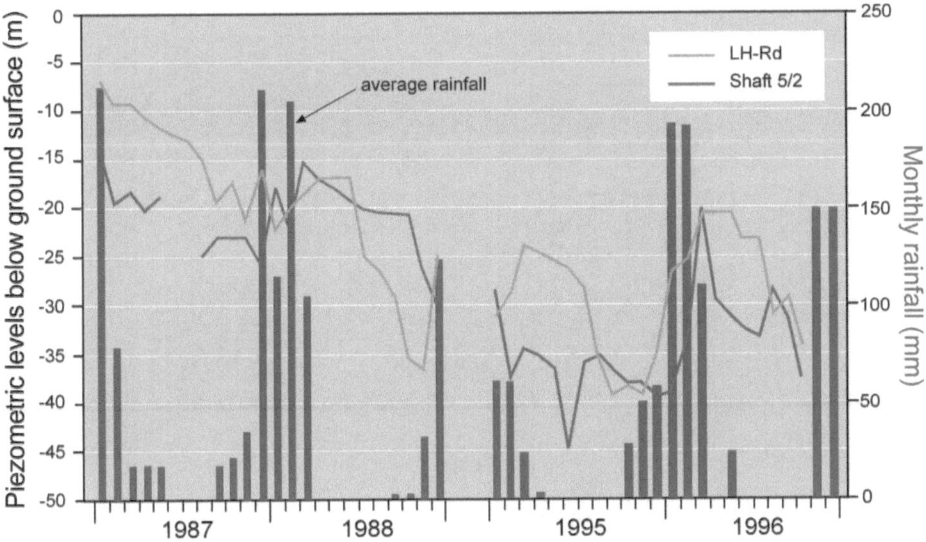

Figure 9. Fluctuations of the water-level in response to pumping and drought in Lusaka (After Nkhuwa, 1999). Annual fluctuations are superimposed on a long-term decline.

Figure 10. Photograph showing different forms of waste dumped in informal waste sites around Lusaka.

of solid wastes in solution features in karstified marbles underlying the city (Figure 10). Consequently, these practices pose serious risks to the potability of water in sufficient quantities to support socio-economic development of Lusaka's residents.

Another major consequence of the limited planning capacity and the ensuing unplanned settlements in Lusaka has been the settlement into areas that were previously considered unsuitable for habitation because of the high water-table. Most of the settlements use either pit latrines or septic tanks to dispose of excreta, while also being heavily dependent

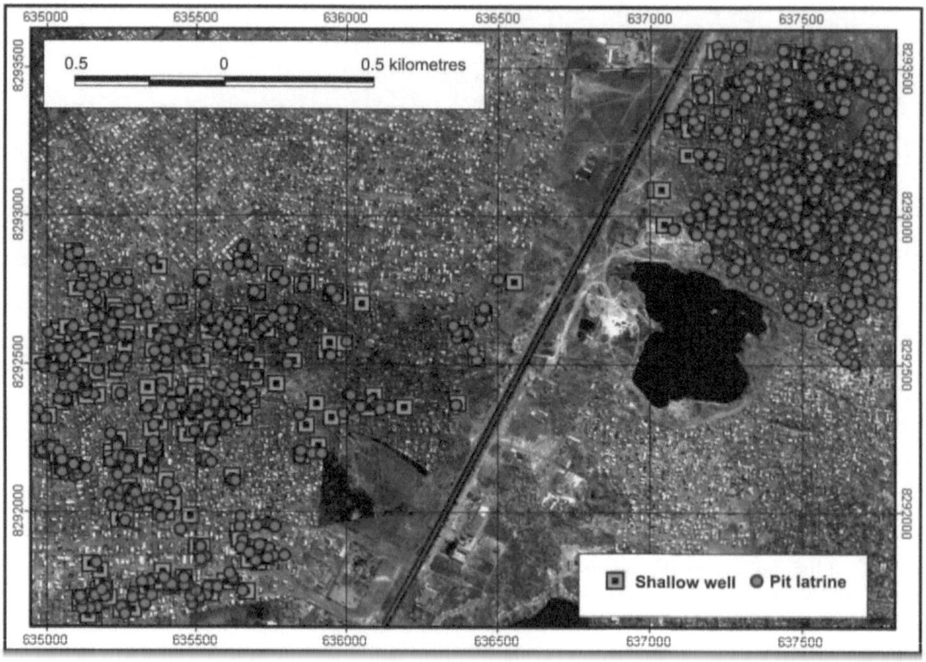

Figure 11. The distribution of excreta disposal points and shallow wells in some high-density townships of Lusaka (after Nkhuwa 2006).

on groundwater obtained from private sources lying in close proximity to the pit latrines (Figure 11).

4 HISTORY AND CURRENT GROUNDWATER MANAGEMENT PRACTICES

4.1 *Addis Ababa/Ethiopia*

The water management practice in the city of Addis Ababa is handled by the Addis Ababa Water and Sewerage Authority. The Authority provides clean water both from surface water and groundwater sources. In the outskirts of the city four water reservoirs were built for water supply purposes. Gefersa, Legadadi, and Dire reservoirs are the most important surface sources. Gefersa was the first dam built in 1944 about 18 km west of Addis Ababa. At present the dam has a reservoir capacity of 6.5 Mm^3 and the maximum capacity of the treatment plant is 30,000 m^3/d. Due to rapid growth of the population and expansion of the city from year to year, there is a serious shortage of water in different parts of Addis Ababa. To alleviate the problem Legedadi and Dire dams were built in 1970 and 1999, 33 km east of Addis Ababa. The treatment capacity of the Legedadi plant was upgraded from 50,000 m^3 to 150,000 m^3 per day. The Dire dam supplies 42,000 m^3 per day for the Legedadi plant, since 1999.

Groundwater abstraction from the Akaki well field was fixed (based on modelling results) at 35,000 m^3/d for a 20-year period; other municipal boreholes account for 10,000 m^3/d. Other private and governmental institutions tap as much as 50,000 m^3/d from groundwater.

Water quality monitoring is carried out regularly before supply. However, there is no adequate legislation to guide management practice in the city.

4.2 *Abidjan/Cote D'Ivoire*

In Cote D'Ivoire there is no overarching institutional framework governing water resources management. Water resources are managed by several institutions ranging from state ministries, private companies/organisations, universities and research institutes. This conforms to a partitioned organisational model (Jourda *et al.* 2006), with three groups (Ministry A, B, C), each with specialised structures. Interestingly, these structures work without systematic coordination with no joint actions in the management of water supply aquifers, particularly in Abidjan, and no synergy between the structures. These pose serious management problems for the Abidjan aquifers with consequent deterioration of water quantity and quality. According to Jourda *et al.* (2006) reforms have now been undertaken which have led to the establishment of the Ministry of Water and Forestry, with responsibility for the supervision and implementation of national policies on water.

4.3 *Cape Town/South Africa*

South Africa now has progressive water legislation and sophisticated water resources management adequately catered for under the National Water Acts of 1998. Under the Act, water is recognised as a national asset, and this permits its transfer from where it is available to where it is inadequate for the benefits of the populace. The previous Water Act linked access to water to land ownership. In both Act No. 108 of 1996 and Act No. 36 of 1998, the management of water resources is an exclusively national issue. The National Water Act 1998 mandates the Minister of Water Affairs and Forestry to ensure that water is protected, used, developed, conserved, managed and controlled in a sustainable and equitable manner for the benefit of all persons. The National Water Resource Strategy (NWRS) is focused on the sustainable use of both surface and groundwater resources. This includes the following:

- develop appropriate, achievable and easily implementable policies and systems;
- provide simple, efficient and effective procedures and guidelines;
- adopt a total systems approach to water resource management in the development of policy;
- effective and focused protection interventions will be facilitated by a differentiated approach, based on a system of resource classification designed specifically for groundwater resources.

Determining the practicable level of protection necessary for individual groundwater resources will take account of a number of factors as described in the Act. Aquifers which represent the sole source of water for communities will be afforded special status, and will enjoy the highest level of protection.

In Cape Town, as in most South African cities where there is a growing water requirement, demands are met by the development and utilisation of surface water (through construction of more dams and weirs); water resource management interventions such as diversions, storage and inter-catchment transfer of water is common practise. These impact on water quality and also can leave groundwater under-utilised. Optimal utilisation and management of groundwater resources will require improved capacity to assess groundwater potential and monitor trends, and a better understanding of aquifer functioning.

4.4 *Dakar/Senegal*

Water resources in Senegal are regulated by Law No. 81-13 of March 4, 1981, which gives ownership of the resource to the state (Law 81-13, art 8, Senegal Constitution adopted 2001). The preamble of the statutory framework defines water rights and the management of water resources (Salman & Bradlow, 2006). After the Sahelian drought of the 1970s the 'Organisation pour la Mise en Valeur du Fleuve Sénégal' (OMVS) was created. A pilot study funded by the Japan International Cooperation Agency (JICA, 1999) examined the potential for self reliance and community management. Rural residents suffered from drying wells in Senegal during drought. JICA (with the local authorities) has constructed deep boreholes and water supply facilities at 109 locations in rural areas. However, problems arose afterwards due to the lack of appropriate maintenance and management. This new assistance aims to improve the management capacity of the community water management cooperative including financial independence. The cooperative, as an organization run with the participation of local people, strives to select their members democratically and to increase the transparency of their operations through information disclosure.

In Dakar, new strategies are in place to develop rainwater harvesting. The new approach preserves rainwater by increasing the rainwater infiltrating rate in the dune sands through artificial recharge (Dasylva *et al.*, 2004). This management measure allows the lowering of the run-off volume and favours the increase of groundwater replenishment.

4.5 *Lagos/Nigeria*

Management of water resources in Nigeria has led to the establishment of several bodies/organisations in the past. Some of these were short-lived and closed abruptly without achieving the desired goal (usually as a result of the nation's political instability) while others were crippled by a lack of organisation and vision-oriented leadership or coordination. According to Akujieze *et al.* (2003) there are about 11 independent bodies charged with water resources management issues. The River Basin Development Authorities, for example, were established by the Federal Government of Nigeria in 1977 with emphasis on the concept of River catchment as a water resource management unit.

Nigeria's water legislation came in a decree referred to as Decree 101 of 1993. Although this was set out as a water resources development and management law it has little reference to groundwater. The requirements of Decree 101 of 1993 have stimulated investigations such as impact assessments of waste disposal sites on groundwater quality. These investigations are aimed at existing, or recognised potential problem sites but are often only 'token' investigations carried out by companies or corporations, to meet their obligations under the Federal, States and Local Governments conditions of award of contracts. Nigeria has no standards for the evaluation of existing landfills or the investigation of potential new sites. In some cases, it is also difficult to find the party responsible for degrading the resource or it is difficult to get such a party to assume liability. This alone emphasizes the need for a proactive approach to assess groundwater vulnerability under present conditions.

National plans and rules are to assist Local Government or Municipal Councils to carry out their functions relating to resource management. Unfortunately, this is still a game of chance under the present water management policies in Nigeria. There is need to restate and reform the law relating to the use of land, air, and water. This will help to promote the sustainable management of natural and physical resources, including groundwater. 'Sustainable management' must ensure that resources are sustained for future generations and adverse environmental effects are avoided, remedied, or mitigated. Therefore, it is the responsibility of the national and regional water councils to implement the water legislation or Decree 101 and incorporate aspects relating to management of groundwater or prevent groundwater contamination based on current scientific evidence.

4.6 *Lusaka/Zambia*

The Lusaka aquifer is generally characterized by shallow water-tables, a thin cover of coarse soils with low clay contents, unconfined conditions, and a flat topography. These factors generally facilitate increased recharge for an aquifer littered with pollution sources over its recharge areas. In this context, aquifer protection becomes very difficult. Consequently, aquifer protection in Lusaka has become very reactive as it continues to emerge in response to aquifer contamination resulting from human activities in the recharge zones. In other words, efforts in the protection of the aquifer have not attempted to prohibit any potentially contaminating development within the cone of depression of each borehole. This occurred because, in its current state, the Zambian Water Act neither regulates drilling of boreholes nor pollution of groundwater resources.

The Water Act – Cap 198 of the Laws of Zambia was passed by the Legislative Council of Northern Rhodesia in 1948 as the Water Ordinance, and came into force in 1949. After independence in 1964, the Water Ordinance was revised and transformed into an Act of Parliament – the Water Act. The Water Act has been amended 10 times, the last being Act. N°· 13 of 1994. Recognising these deficiencies, the Zambian Government embarked on the formulation of the new Water Bill, which is currently in its draft form. The Draft Water Bill seeks to address the deficiencies of the Water Act – Cap 198 of the Laws of Zambia, with a view to providing a framework that promotes Integrated Water Resources Management through: (a) Establishment of a national monitoring and information system, (b) Establishment of hydrological stations.

Parts VII and VIII of the Water Bill deal with regulation on the drilling of boreholes and preservation of good quality national water resources, respectively. However, any groundwater protection measures to be adopted in Lusaka must accept the fact that existing infrastructure and anthropomorphic activities cannot be moved, although they cannot be ignored. In this regard, the main route of aquifer protection in Lusaka must be proactive/preventive. Clearly, this route would be preferable as it tends to promote contaminant-free water sources. The latter represents the cheaper option since treatment costs becomes lower.

Further, due to reasons of economics and inadequate public awareness, it may be very difficult, if not impossible, to force changes in landuse practices in pursuit of aquifer-wide protection strategies. In order to assess the effectiveness of any protection measures instituted in Lusaka, it is strongly recommended that a monitoring system for a range of physico-chemical and microbial parameters be instituted. This will give an early warning of any contaminant event likely to occur in the aquifer or part thereof.

Table 4. Groundwater condition in selected Sub-Saharan Cities.

City	Country	Information status	Role of groundwater	Groundwater problems
Addis Ababa	Ethiopia	1	Major	urb poll
Abidjan	Cote D'Ivoire	2	Major	ss, urb poll, gwl
Cape Town	South Africa	1	Minor	urb poll, d-s poll
Dakar	Senegal	2	Major	ss, urb poll, sal int
Lagos	Nigeria	2	Major*	urb poll, sal int
Lusaka	Zambia	3	Major	urb poll, gwl

1 Full survey data
2 Useful summary document
3 General background only
major – Major source of public water supply
minor – Minor source of public water supply but high potential for future supply
major* – Major private domestic/industrial use
ss – sole source of water supply
urb poll – Groundwater pollution within urban area
sal int – Aquifer saline intrusion
d-s poll – downstream groundwater pollution
gwl – falling groundwater level

5 GROUNDWATER MANAGEMENT CHALLENGES AND PROTECTION STRATEGIES

Unrestricted use of groundwater through boreholes will result in a lowering of the water-table in many cities/urban centres and intrusion of saline water in coastal areas of SSA. The crucial issues of management of groundwater requires adequate planning and direction from hydrogeologists and a reasonable level of hydrogeological awareness amongst water managers, supply engineers, farmers and local villagers who may be responsible for managing abstraction and monitoring. It has been suggested that part of the management cycle in SSA should incorporate monitoring data, learning and feedback to improve the understanding of the resource under extreme conditions (Colvin & Chipimpi. 2004). The key challenge is developing capacity to manage and monitor intensive groundwater use in our urban centres. Appropriate technology and sustainable financing are also critical management constraints.

Furthermore, the present legal and jurisdiction framework for groundwater management is fragmented, inconsistent and incomplete. Groundwater management practises vary from region to region, and in some cases, do not exist at all. This is a long-standing problem in SSA, and as such requires a framework of collaboration on groundwater studies. Such collaborative efforts should focus on providing basic geological and groundwater data essential to manage groundwater resources in SSA.

Part of the challenge will be to develop locally appropriate groundwater protection plans for the cities; even with the limited resource and knowledge base in SSA. There is also a need for stakeholder consultation to contribute to the development of policy options for aquifer protection in respective cities of SSA. Table 4 shows the groundwater-dependent cities considered in this paper, and whose groundwater information/condition could be useful in formulating policy strategies.

In the present condition, any set of aquifer protection policies to be applied to an already-existing urban area will need to evolve strategies which, while they constrain land-use, accept trade-offs between competing interests and utilise the natural contaminant attenuation capacity of the strata overlying aquifers (Matthess *et al.*, 1985). In any case, to successfully implement such strategies hydrogeological understanding needs to inform land-use policies and provide simple robust matrices that indicate what activities are possible where, at an acceptable risk to groundwater. In turn, construction of such matrices requires pragmatic design criteria if planning is not to be so delayed as to irretrievably prejudice resource sustainability. In many cases important and influential stakeholders involved in urban water management decisions do not have a technical background either in engineering or in resource planning. Professional hydrogeological expertise in city water management is generally absent, and municipal water supply utilities may be more focussed on day-to-day operational needs of the present system, even where groundwater is a major urban resource.

5.1 *Addis Ababa, Ethiopia*

The main problem is lack of implementation of the current Environmental management plan to control surface water and groundwater from pollution. The concerned bodies are striving to implement the plan through community participation and by imposing laws. In an effort to fulfil the growing water demand the city is increasingly relying on supplies from groundwater sources such as the Akaki well field, which requires a delicate aquifer management strategy to avoid over-abstraction. There is also continuous effort by the municipality to establish new well fields to fill shortfalls in supply.

The growths of towns have had profound effects on the groundwater resources due to poor land use and waste disposal and sanitation practices. Population growth rate in Addis Ababa, estimated at 4% has moved at a higher pace than the ability of the government to provide the necessary infrastructure to handle domestic and industrial effluents and this has lead to widespread contamination especially of shallow groundwater. Due to the above challenges protection of groundwater in terms of quality and quantity is needed in order to avoid reduction in available groundwater resources, escalating water supply costs and potential impacts on human health.

5.2 *Abidjan, Cote D'Ivoire*

The problem of groundwater management in Cote D'Ivoire begins with poor organisational structure, where water resources are managed in a sectarian manner and a single policy of resource management. In reality the decision-making centres are varied and often antagonistic one to another (Jourda *et al.*, 2006). Factors contributing to poor management are: absence of a well-coordinated institutional framework governing water resources in Cote D'Ivoire; lack of proper dissemination of information to users of the resource; the lack of respect to public property; poor sanitation and waste collection network; increasing water demand (resulting from high population growth); and decrease of recharge (due to climatic variations). The growth of population (up to 3.7% in 10 years), increased agricultural and industrial practices of this area have resulted in an increasingly demand for water from the Abidjan aquifer and posing a greater challenge (Issiaka *et al.*, 2006; Jourda *et al.*, 2006). Management policies are rare and the lack of cooperation among various institutions responsible for resource management further heightens the problem.

New groundwater management and aquifer protection strategies are proposed by Jourda *et al.* (2006) and include: (i) creation of an institutional framework for integrated water resources management in order to establish lawful measurements in the various application of water protection; (ii) creation of synergy among the scientific community, the NGOs, stakeholders and authorities charged with aquifer management in order to collect and disseminate relevant information; (iii) identification of major pollution areas and threats to the water supply aquifer with the aim to draw attention of stakeholders and the general public to mitigate such threats; and (iv) establish urban water quality monitoring in order to prevent further pollution of the Abidjan aquifer.

5.3 *Cape Town, South Africa*

South Africa has in place a good management programme for its groundwater resources. However, the issue of protection of aquifers for the supply of good quality water require a more pragmatic approach. A systematic approach was neglected in the past as a result of its "private water" status under the previous legislation, and relatively little was invested in comprehensive resource assessment (DWAF, 1999).

The overall quality of groundwater in the greater Cape Town area is good enough to warrant its full development and utilisation in the water supply augmentation scheme for the City of Cape Town Municipality. However, occurrence of pollution seen in present and previous studies as highlighted earlier in this paper and the identification of point and non point sources calls for groundwater protection. The establishment of site specific protection zones, with regulation of land use within them and the Resource Directed Measures (RDM) under the South African Water Act of 1998 has been seen as a possible way forward for future groundwater management in South Africa. Through research and development investment in the past five to ten years it has become clear that groundwater in usable, quantities can be found in many places with the appropriate expertise. Deep drilling has shown the potential for large-scale development of groundwater in some areas such as those underlain by the Table Mountain Group geological formations. With a focus on the development of local resources groundwater's role in reconciling future demand and supply could rise significantly, and meeting relatively small water requirements from groundwater would be especially attractive.

5.4 *Dakar, Senegal*

Water resources management in Senegal is a crucial issue; explosive population growth and economic development have exacerbated and expanded the range of water-related problems, such as shortages of supply and pollution. Groundwater is abundant in Senegal except for Tambacounda region. Major constraints for the use of the groundwater are the lack of precise data on aquifers such as capacity, depth and contamination. Groundwater is a crucial resource for future development in Senegal. Therefore, certain policies and strategies to sustain the development and management of groundwater resources will have to be elaborated through a series of studies to secure safe water supply.

Future challenges to water resources management in Dakar include pollution and management of wastes. The amount of solid waste generated annually throughout the country is estimated to be 744,250 tons of which 280,000 tons are generated in Dakar (JICA, 1999). 5% of Dakar's population has no access to any sanitation facilities. 40% of households in urban area have water in their houses, 50% have access to a public water tap, and the remainder depend on wells.

At present, household waste is collected in metal containers and transported to the dumping site. The number of containers has increased in major cities but the waste collection service is not yet regularized because of the lack of finance and capacity of the organization. According to JICA report, collection vehicles cannot reach all the collection sites because of narrow streets or unpaved road. A significant proportion of household waste in urban area is still uncollected and remains in the streets, open canals or illicit dump sites. Untreated wastewater flows directly into rivers and subsequently into the aquifer systems (JICA 1999).

5.5 *Lagos, Nigeria*

Even though the demand for groundwater is becoming higher in both rural and urban centres in Nigeria, the management of the aquifers or wellfields is not closely monitored. In Lagos metropolis, in particular, careful management is required to avoid further degradation of groundwater quality. There is need to maintain a groundwater quality monitoring network to characterize groundwater quality and investigate trends. Point-source discharges and significant use need to be regulated through conditions for consents and regional rules to ensure sustainable resource management. National or regional codes of practice for managing non-point sources and commitment to environmental education as a means for managing groundwater quality are further required on the part of the government. For municipal water supply, high and stable raw-water quality is a prerequisite, and one best met by protected groundwater sources.

Vulnerability assessment is little known or researched in Nigeria. With many surface waters now polluted, the importance of groundwater as a source of drinking water has to increase. While regional plans and rules exist to protect surface waters, groundwater protection is in its infancy in Nigeria, and is regarded as inadequate. Water resources legislation is in existence but its impact is yet to be felt in terms of true ownership and usage of the nation's water resources. The reasons for these are obvious. For example, the Decree 101 and water-related regulations are based mainly on socio-economic factors and not necessarily on scientific research such as, groundwater vulnerability or aquifer sensitivity. Federal, States and Local Government programmes need to be criticized for not formulating or implementing policies on on-site effluent disposal.

Emphasis is here placed on groundwater protection in order to prevent deteriorating conditions of the main aquifers in Nigeria. The assessment of aquifer vulnerability and sensitivity to pollution on a national scale is very necessary in Nigerian urban centres under the present conditions. The most important potential use of aquifer protection is in raising public awareness which, in turn, may result in positive reactions or more informed land use decisions. Aquifers with high sensitivity should be monitored closely while aquifers with low sensitivity may not require detailed monitoring. In addition, permits or consents for environmental activities should have more demanding conditions imposed on them in areas of high as opposed to low sensitivity.

5.6 *Lusaka, Zambia*

The presence of a well developed system of conduits, solution channels and subterranean cavities in the Lusaka aquifer(s) reduces and/or completely eliminates the natural attenuation of pollutants through dilution and natural filtration. Therefore, the absence of legislation to limit or inhibit human activities on the aquifer recharge areas pose a great risk of groundwater pollution and rendering it unusable for any purpose.

The quality of most physico-chemical and bacteriological parameters over much of the aquifer show a general decrease from the dry season into the wet season. This is probably resulting from dilution arising from increased saturation in the aquifer. Therefore, regular water quality monitoring from supply points for physico-chemical and microbial parameters will give an early warning of any contaminant event likely to occur in any part of the aquifer. This will allow timely action by concerned authorities to avert the possibility of the population consuming water of insufficient quality.

The long-term deterioration of water quality, leading to progressively more costly water treatment, is the inevitable result of current ad-hoc development reminiscent of a thriving city of Lusaka located largely on a karstic aquifer. In the long-term, groundwater beneath Lusaka is likely to become unfit for human consumption even with expensive treatment. In which case, there will be need to look for new sources of water supply away from current sources or a reconsideration of whether or not a new site should be sought for the city (Mpamba *et al.*, this volume).

6 FUTURE FOCUS OF URBAN GROUNDWATER MANAGEMENT AND PROTECTION IN SUB-SAHARAN AFRICA

Groundwater is the preferred source for piped water supplies in many urban areas across Sub-Saharan Africa and its development is forecast to increase dramatically in an attempt to improve urban water supply coverage. Heavy groundwater abstraction in some urban areas has already resulted in lowering of groundwater levels and competitive pumping between water sources. Similarly, poor landuse practices and onsite sanitation systems in the form of septic tanks and pit latrines have caused contamination of groundwater resources in many urban areas especially where the groundwater table is shallow. Concrete action in the form of improved management therefore needs to be urgently taken in a number of areas to mitigate actual and potential derogation caused by excessive exploitation and inadequate pollution control. Key areas where future action is required include: (a) further research on groundwater occurrence and movement, (b) development of institutional frameworks for groundwater management, (c) advocacy and raising awareness, (d) stakeholder involvement.

6.1 *Research in groundwater occurrence and movement*

Understanding the occurrence and movement of groundwater in urban areas is key to its sustainable management and protection. Key information for guiding groundwater management and protection decisions is however unavailable in most places. The continued lack of this information implies that decisions are either not made or have no good basis. It is for example, not known how far water sources should be from each other and from sources of pollution to avoid competitive pumping or pollution and this constrains decision making regarding siting of various facilities. It is therefore necessary to carry out research to resolve key hydrogeological questions related to the protection of boreholes from competitive pumping and siting, in relation to onsite sanitation systems, in various geological environments. Availability of this information will make it possible to develop strategies and guidelines for optimal groundwater development, management and protection in various settings.

6.2 *Institutional frameworks for groundwater management*

While institutional frameworks for groundwater management exist in many national and local level government institutions, these are not normally replicated in urban areas. Thus, groundwater management and protection continues to be done in an adhoc manner, if done at all. To improve this situation, institutional frameworks for groundwater resources management in urban areas need to be developed.

6.3 *Advocacy and raising awareness*

Groundwater, in many places, is considered to be abundant due to its concealed nature, and its susceptibility to pollution is not apparent. Thus, its management and protection often do not attract attention from policy and decision makers, funding organisations, users and other stakeholders. In view of the current and planned heavy abstraction of groundwater in many urban areas, and the increase in pollution from various sources, it is essential that advocacy and awareness raising programmes are instituted to ensure that the susceptibility of groundwater to overexploitation and pollution is appreciated by all the stakeholders. This will enable them to appreciate the benefits of groundwater management and protection and ensure that appropriate actions are taken.

6.4 *Stakeholder participation*

Sustainable groundwater management and protection requires the active participation of all stakeholders. Stakeholder participation is the process of involving those who are affected by and thus have an interest in the management and protection of groundwater resources. Groundwater stakeholders may be users of groundwater, or those who carry out activities that could pollute groundwater, or those who are concerned with groundwater resources and general environmental management. Participation of stakeholders in groundwater management and protection is important for a number of reasons: it ensures that decisions regarding groundwater resources exploitation are integrated and coordinated with landuse and environmental management; it ensures that there is equity in allocation of groundwater resources to various users; it enables better estimation of current and future demands for groundwater resources; it can facilitate the optimization of groundwater use by competing users; it can facilitate the implementation of strategies and decisions regarding sustainable groundwater abstraction and protection; and it can enable active involvement of the stakeholders in data collection and follow up monitoring and inspection of groundwater use and pollution.

REFERENCES

Adelana, S. M. A., Bale, R. B. & Wu, M. 2003. *Quality assessment and pollution vulnerability of groundwater in Lagos metropolis, SW Nigeria.* In: Proceedings of the Aquifer Vulnerability Risk Conference AVR03, Salamanca, Mexico, **2**, 1–17.

Adelana, S. M. A., Bale, R. B. & Wu, M. 2004. *Water quality in a growing urban centre along the coast of southwestern Nigeria.* In: Seiler, K. P., Wu, C. & Xi, R. (eds.) Research Basins and Hydrological Planning. Balkema, The Netherlands, 83–92.

Adelana, S. M. A., Bale, R. B., Olasehinde, P. I. & Wu, M. 2005. *The impact of anthropogenic activities over groundwater quality of a coastal aquifer in Southwestern Nigeria.* In: Proceedings of Aquifer Vulnerability & Risk, 2nd International Workshop & 4th Congress on the Protection and Management of Groundwater, 21–23 September 2005, Reggia di Colorno – Parma.

Adelana, S. M. A. & Xu, Y. 2006. *Groundwater contamination and protection of the Cape Flats aquifer, South Africa.* In: Xu, Y. & Usher, B. (eds.) Groundwater pollution in Africa, Taylor & Francis, London, 265–277.

Akujieze, C. N., Coker, S. J. L. & Oteze, G. E. 2003. *Groundwater in Nigeria – a millennium experience – distribution, practice, problems and solutions.* Hydrogeology Journal, **11**, 259–274.

Alemayehu, T. 2001. *The impact of uncontrolled waste disposal on surface water quality in Addis Ababa.* SINET: Ethiopian Journal of Science, **24**, 93–104.

Alemayehu, T. 2006. *Groundwater occurrence in Ethiopia.* Addis Ababa University Press, Addis Ababa, Ethiopia.

Alemayehu, T., Dagnachew, L. & Tenalem, A. 2005. *Hydrogeology, water quality and the degree of groundwater vulnerability to pollution in Addis Ababa, Ethiopia.* UNEP/UNESCO/UN-HABITAT/ECA publication, Nairobi, Kenya.

Asiwaju-Bello, Y. A. & Akande, O. O. 2001. *Urban groundwater pollution: case study of a refuse disposal site in Lagos metropolis.* Water Resources, **12**, 22–26.

Barret, M. H. 2004. *Characteristics of urban groundwater.* In: Lerner, D. N. (ed.) Urban groundwater pollution. IAH International Contributions to Hydrogeology, **24**, Balkema, The Netherlands, 29–51.

CCT 2006. *Annual Report 2005/2006.* City of Cape Town.

Cissé, S. Faye, S., Wohnlich, S. & Gaye, C. B. 2004. *An assessment of the risk associated with urban development in the Thiaroye area (Senegal).* Environmental Geology, **45**, 312–322.

Cisse, S., Faye, S., Gaye, C. B. & Faye, A. 2000. *Effect of rapidly urbanizing environment on the Thiaroye, Senegal unconfined sandy aquifer.* In: Silolo (ed.) Groundwater: past achievements and future challenges, Balkema, The Netherlands, 719–723.

Colvin, C. & Chipimpi, B. 2005. *Opportunities and challenges of intensive use of groundwater in Sub-Saharan Africa.* In: Sahuquillo, A., Capilla, J., Martinez-Cortina, L. & Sanchez-Vila, X. (eds.) Groundwater intensive use. IAH Selected papers, **7**, Balkema, The Netherlands, 147–156.

CSO 1990. *Census of population, housing and agriculture 1990 – Prelim report.* Central Statistical Office, Lusaka, Zambia.

Dasylvia, S., Cosandey, C., Orange, D. & Sambou, S. 2004. *Rainwater infiltration rate and groundwater sustainable management in the Dakar region.* Agric Eng International: CIGR Journal of Scientific Research & Development, **6**, 1–11.

Deme, I., Tandia, A. A., Faye, A., Malou, R., Dia, I. & Diallo, M. S. 2006. *Management of nitrate pollution of groundwater in African cities: The case of Dakar, Senegal.* In: Xu Y & Usher B (eds.) Groundwater pollution in Africa. Taylor & Francis, London, 181–192.

Durotoye, A. B. 1989. *Quaternary sediments in Nigeria.* In: Kogbe, C. A. (ed.) Geology of Nigeria, Elizabeth Press, Lagos, 431–451.

DWAF 1999. *Water resources protection policy implementation, resource directed measures for protection of water resources*, Integrated Manual, Department of Water Affairs and Forestry Report, **N/28/99**.

Foster, S. S. D., Lawrence, A. R. & Morris, B. L. 1996. *Groundwater resources beneath rapidly urbanizing cities-implications and priorities for water supply management.* In: Report of the Habitat II Conference, Beijing, China, March 1996, 356–365.

Harris, N. 1990. *Urbanization, economic development and policy in developing countries.* Habitat International, **14**, 3–42.

INS 2001. *Recensement General de la Population et de l'Habitation (RGPH) 1998.* Domees socio-demographiques et economiques des localites, resultats definifs par localites, region des lagunes, **3**, National Statistics Institute of Ivory Coast, 43p.

Issiaka, S. Albert, Bi T, Ariiistide D.G. & Innocent K.K. 2006. *Vulnerability assessment of the Abidjan Quaternary aquifer using DRASTIC method.* In: Xu, Y. & Usher, B. (eds.) Groundwater pollution in Africa, Taylor & Francis, London, 115–124.

JICA 1999. *Country profile on environment and water development projects – Senegal.* Japan International Cooperation Agency Annual Report, November 1999.

Jones, H. A. & Hockey, R. D. 1964. *The geology of part of southwestern Nigeria.* Geological Survey of Nigeria, Bulletin, **31**.

Jourda, J. R. P. 2003. *Evaluation de la pollution et de la vulnérabilité des aquifères des grandes cités urbaines d'Afrique.* Project UNEP/UNESCO/UN-HABITAT/ECA. Rapport sur les activités en Côte d'Ivoire. Rapport à mi-parcours de la 2 ème phase. Décembre 2003.

Jourda, J. R. P. 2004. *Qualité des eaux souterraines de la nappe d'Abidjan.* Bulletin N° 4. Juin 2004.

Jourda, J.P., Kouame, K.J., Saley M.B., Kouame, K.F., Kouadio, B.H. & Kouame, K. 2006. *A new cartographic approach to determine the groundwater vulnerability of the Abidjan aquifer.* In: Xu, Y. & Usher, B. (eds.) Groundwater pollution in Africa, Taylor & Francis, London, 103–114.

Kouadio, L. P., Abdoulaye, S., Jourda, P., Loba, M. & Rambaud, A. 1998. *Conséquences de la pollution urbaine sur la distribution d'eau d'alimentation publique à Abidjan.* Cahier de l'Association Scientifique Européenne pour l'Eau et la Santé, **3**, 61–75.

Krothe, J. N., Garcia-Fresca, B. & Sharp, J. M. 2002. *Effects of urbanisation on groundwater systems.* In: Proceedings of the XXXII IAH Congress on groundwater and human development, 21–25 October 2002, Mar del Plata, Argentina.

Lerner, D. N. & Barret, M. H. 1996. *Urban groundwater issues in the United Kingdom.* Hydrogeology Journal, **4**, 80–89.

Longe, E. O., Malomo, S. & Olorunniwo, M. A. 1987. *Hydrogeology of Lagos metropolis.* Journal of African Earth Sciences, **6**, 163–174.

Lorenzo-Rigney, B. & Sharp, J. M. Jr. 1999. *Urban recharge in the Edwards aquifer.* Geological Society of America Conference, **31**, A-12.

Matthess, G. Foster, S. S. D. & Skinner, A. C. 1985. *Theoretical background, hydrogeology and practice of groundwater protection zones*, IAH International Contributions to Hydrogeology, **6**, Verlag Heinz Heise, Hannover,167–200.

Meyer, P. S. 2001. *An explanation of the 1:500 000 hydrogeological map of Cape Town 3317.* Department of Water Affairs & Forestry, South Africa, 59p.

Morris, B. L., Lawrence, A. R., Chilton, P. J., Adams, B., Calow, R. C. & Klinck, B. A. 2003. *Groundwater and its susceptibility to degradation: A global assessment of the problem and options for management.* Early Warning and Assessment Report Series, RS. 03-3. UNEP, Nairobi, Kenya.

Mpamba, N. H, Nkhuwa, D. C. W., Nyambe, I. A., Mdala, C. & Wohnlich, S. 2008. *Groundwater mining: A reality for the Lusaka urban aquifer?* This volume.

Nkhuwa, D. C. W. 1999. *Is groundwater management still an achievable task in the Lusaka aquifer?* Proceedings of the XXIX Conference of IAH, Bratislava, Slovak Republic. 209–213.

Nkhuwa, D. C. W. 2006. *Groundwater quality assessments in the John Laing and Misisi areas of Lusaka.* In: Xu, Y. & Usher, B. (eds.) Groundwater pollution in Africa, Taylor & Francis, London, 239–251.

Oga, M. S., Marlin, C., Njitchoua, R., Dever, L. & Filly, A. 2008. Hydrochemical and isotopic characteristics of groundwater from coastal sedimentary basin in Abidjan. In: Adelana, S. M. A. & MacDonald, A. M. Applied Groundwater Studies in Africa. IAH Selected Papers on Hydrogeology, Volume 13, CRCPress/Balkema, Leiden, The Netherlands.

Parsons, R. & Taljard, M. 2000. *Assessment of the impact of the Zandvliet Wastewater Treatment Works on groundwater.* Biennial Conference, Sun City, 28 May – 1 June, 2000. Water Institute of South Africa.

Sharp, J. M. Jr. & Banner, J. L. 2000. *The Edwards aquifer: Water for thirsty Texans.* In: Schneiderman, J. S. (ed.) The Earth Around Us: Maintaining a Livable Planet. Freeman, New York, 154–165.

Soro, N. 2003. *Gestion des eaux pour les villes africaines : évaluation rapide des ressources en eau souterraine/occupation des sols.* UN-HABITAT. Avril 2003 (Rapport final).

Tandia, A. A., Gaye, C. B. & Faye, A. 1998. *Origin, process and migration of nitrate compounds in the aquifers of Dakar region, Senegal.* In: Application of isotope techniques to investigate groundwater pollution, IAEATECDOC-1046, 67–80.

Taylor, R. G., Barret, M. H. & Tindimugaya, C. 2004. *Urban areas of Sub-Saharan Africa: Weathered crystalline aquifer systems.* In: Lerner, D. N. (ed.) Urban groundwater pollution, IAH International Contributions to Hydrogeology, **24**, Balkema, The Netherlands, 155–179.

Theron, J. N., Gresse, P. G., Siegfried, H. P. & Rogers, J. 1992. *The geology of the Cape Town area.* Explanation on Sheet 3318, Geological Survey, South Africa.

UNPD 2005. World Population Prospects: The 2004 Revision and World Urbanization Prospects. Available on the web: http://esa.un.org/unpp.

Usher, B. H., Pretorius, J. A., Dennis, I., Jovanovic, N., Clarke, S., Titus, R. & Xu, Y. 2004. *Identification and prioritisation of groundwater contaminants and sources in South Africa's urban catchments.* WRC Report, **1326/1/04**.

CHAPTER 15

Urban groundwater and pollution in Addis Ababa, Ethiopia

T.A. Abiye
School of Geosciences, University of the Witwatersrand, Private Bag 3, South Africa

ABSTRACT: The city of Addis Ababa has a population of over 4 million of which 50% have no direct access to the Municipal water supply system – a combination of surface water and groundwater. Pollution of fresh water resources is currently one of the major problems faced by the Addis Ababa residents and Municipality decision makers. Fast population growth and rapid industrialization, on one hand, and lack of sewerage network and poor living condition, on the other, have led to the progressive deterioration of surface and groundwater quality across the city. Of particular concern is the widespread use of septic tanks, open dumps, land applications and surface impoundments to dispose of waste. Direct discharge of domestic and industrial wastes into water bodies have increased the level of undesirable constituents in rivers, streams, and groundwater reservoirs. Chemical and bacteriological analyses reveal that constituents such as ammonia, nitrate, chloride, and extremely high total coliform concentrations in the surface water have been introduced through anthropogenic activities. The high degree of hydraulic connection between surface water and groundwater means that the volcanic aquifer is vulnerable to pollution. Therefore, controlling the quality of surface water is important to reduce long lasting impact on groundwater quality and maintain good quality water supply.

1 INTRODUCTION

The city of Addis Ababa (*New Flower* in Amharic) was established as the capital city of Ethiopia in 1886 and has grown to become the largest urban and commercial centre in the country, hosting over 4 million residents. It is located in the central part of the country at the edge of the western escarpment of the main Ethiopian Rift (Figure 1). The city is the seat for the African Union and many other international organizations.

For the first 58 years (1886–1944), the water supply for the city was derived from groundwater in the form of springs located at the foot of the Entoto Ridge (northern part of the city) and also from dug wells located in the central and southern part of the city. Additional demand necessitated treatment of surface water derived from three surface dams namely: Gefersa, Legedadi and Dire. However, the exponential population growth in the city has led to further groundwater exploration in different localities resulting in the development of the Akaki well field in the southern part of the city.

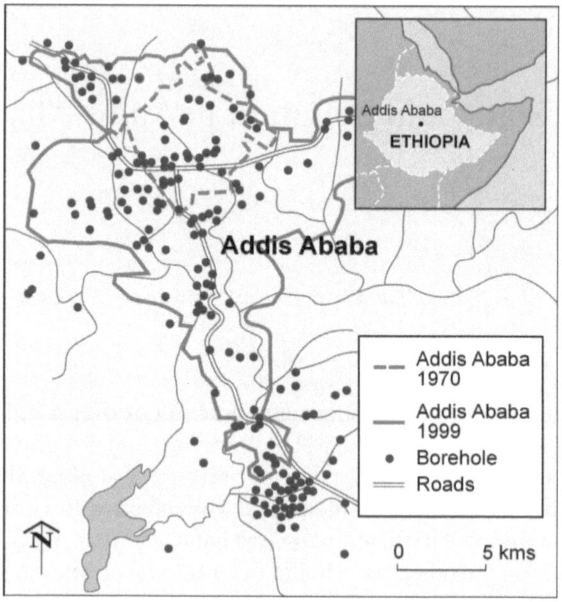

Figure 1. Boreholes in Addis Ababa.

According to projection of the United Nations, Addis Ababa will be the fourth largest city in Africa by 2015. In an effort to fulfill the growing water demand the city is increasingly relying on supplies from groundwater sources such as the Akaki Well field, which requires a robust aquifer management strategy to avoid over-exploitation and contamination. There is also a continual effort by the municipality to establish new well fields to fill the existing supply gap. It has also been estimated that per capita consumption of water will approximately double over the next 10 years as a result of several factors: improved living conditions; increased backyard plantation and sanitary facilities; good personal hygiene; and increase in commercial and industrial demand (Tamiru, 2005).

The current demand has created a supply shortfall of about 50%. Large numbers of private boreholes tap water from the same volcanic aquifer with yields of up to 15 l/s. Many of the private wells have excessively abstracted groundwater causing well interference and declining water-levels in the southeastern part of the city. Many of the boreholes tap shallow contaminated groundwater.

Aside from the water demand problems, the city is facing growing water quality problems due to pollution from point and diffuse sources. The foundation and expansion of the city was associated with the rapid conversion of land from rural to urban uses more than anywhere else in the country. Aside from the expansion of the city, the impact of urbanization on surface and groundwater is increasing with the development of industry and population size in the city (Adane, 1999; Tamiru, 2001). In many parts of the world, urbanization has been shown to have a direct impact on groundwater quality deterioration (Lerner & Barrett, 1996; Grischek & Nestler, 1996; Malcolm *et al.*, 1996; Chilton 1999).

The fast population growth, uncontrolled urbanization and industrialization, poor sanitation situation and uncontrolled waste disposal have resulted in serious quality degradation of both surface water and groundwater in the city and surrounding environment. Currently water quality degradation in Addis Ababa has become a major threat to the health of the population especially for those living downstream of the recharge areas and along the main rivers draining through the city. Recently there are signs of public awareness about the growing water quality problem; the city council has also currently implemented a policy that integrates economic development with environmental protection in the city (Tamiru *et al.*, 2005).

In the 1970's the size of the city was about 37 km^2, while a satellite image of 1999 revealed that the size had increased to 230 km^2. In other words, the city had expanded by 193 km^2 within a period of 29 years. The rate of expansion is, therefore, about 6.7 km^2/a, which demands huge additional volumes of water supply for the various urban uses. Urbanization may lead to reduction in the permeability of land surface, depending on land use changes that can modify the groundwater recharge (Lerner & Barrett, 1996; Barrett, 2004). In developing cities like Addis Ababa, economic and social developments do not take into account the long-term impact on the environment (including water) partly because the massive population growth takes place without a corresponding increase in wealth (Barrett, 2004). In addition, the unbalanced waste collection system in the city aggravates the deterioration of the surface and groundwater by reducing the volume of available good quality water. Therefore, the main concern of this paper is to provide some highlights regarding water quality problems in Addis Ababa through assessment of the chemical and bacteriological composition of surface and groundwater systems within the city.

2 BACKGROUND TO ADDIS ABABA

2.1 *Physiography, climate and landuse settings*

The urban topography of the study area reflects the presence of different volcanic stratigraphic successions, tectonic activity and the action of erosion between successive lava flows. The city is located at the southern flank of Entoto Ridge with an average elevation of 3200 m above sea level (asl). This ridge marks the northern boundary of the city following the east-west trending major fault (Ambo-Kassam). Other prominent volcanic features surrounding the city are Mt. Wochacha in the west (3385 m asl), Mt. Furi (2839 m asl) in the southwest and Mt. Yerer (3100 asl) in the east. These typical volcanic features are made up of mainly acidic and intermediate lava flows. Thus, they are characterized by rugged landscapes and steeper slopes. The general inclination of the slope becomes lower towards the southern part of Addis Ababa (a difference of about 900 m). The centre of the city lies on rough topography with some flat land areas and a graben structure. There are many streams throughout the area.

The seasonal distribution of rainfall in Addis Ababa can be attributed to the position of the Inter-Tropical Convergence Zone (ITCZ). There are seven rainy months from March to September, with most of the rainfall from June to September; dry months extend from October to February. The mean annual rainfall is 1150 mm. The monthly mean maximum and minimum temperatures are 22.7 and 9.9°C with the mean values of 16°C. Generally, the climate of the city is semi-humid.

The climatic condition and topography of the area favours the development of thick soil profiles through weathering of the parent rock. Thus, residual soils are commonly seen in most parts of the city with varying thickness. The dominant type of soil in the southern parts of the city is vertisol while in the northern sector is cambisol. However, due to intensive

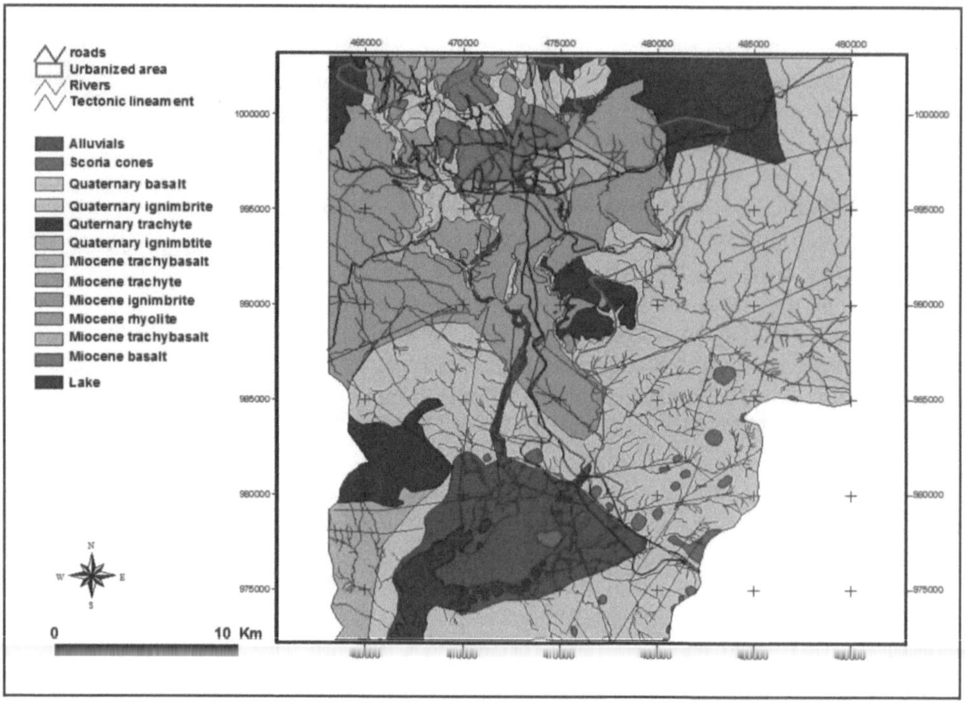

Figure 2. Geology of Addis Ababa.

erosional activities, most parts of the slope area are characterized by poor soil development with shallow soil profiles occurring as patches in some places.

2.2 *Geological and hydrogeological setting*

Geologically, the city is dominated by volcanic materials of different ages and compositions (Figure 2). The Miocene-Pleistocene volcanic succession in the Addis Ababa area from bottom to top is: Alaji Basalts and Rhyolites, Entoto Siliciclastics, Addis Ababa Basalts, Nazareth Group, and Bofa Basalts.

The Alaji volcanic rocks (rhyolites and basalts) show variation in texture from highly porphyritic to aphyric basalts and there is an intercalation of grey and glassy welded tuff. The outcrop of Alaji basalt extends from the crest of Entoto (ridge bordering the northern parts of Addis Ababa) towards the north. The age of this rock is about 22.8 Ma (Morton *et al.*, 1979). However, the Entoto Siliciclatics are composed of rhyolite and trachyte with minor amount of welded tuff and obsidian (Haileselassie & Getaneh, 1989). The rhyolitic lava flows outcrop on the top and the foothills of the Entoto ridge and in the central sector of the city. The rhyolites are overlain by feldspar porphyritic trachyte and underlain by a sequence of tuffs and ignimbrites. Tuffs and ignimbrites are welded and characterized by columnar jointing. The Addis Ababa basalt, mainly present in the central part of the city, are underlain by Entoto siliciclastics and overlain by welded tuff of the Nazareth Group. The Nazareth units are represented by welded tuff and aphanitic basalt. Bofa basalts, with the age of 2.8 Ma, outcrop southward from Akaki River with the thickness of about 50 m. They are restricted and dominant in the southeastern part of the city especially in the Akaki well field area and represent the major source of groundwater supply for large parts of Addis Ababa.

Tectonically, an elongated east-west trending fault line cuts across the western rift escarpment and uplifted the northern block about 8 Ma ago (Zanettin *et al.*, 1978). The fault marks the western margin of the Main Ethiopia Rift north of Addis Ababa and is downthrown to the south towards the city. Another prominent normal fault in the city is the Filowha Fault that trends in the NE-SW direction located within the city centre. Also a fault from the railway station crosses Bole Road and forms a prominent graben with the Filwoha fault at Meskel square, descibed in this study. The graben is dominantly filled with transported soil (1–8 m thick) while the faults act as a major conduit for deep heated water to the surface and horizontal diversion.

2.3 *Rivers*

The city of Addis Ababa lies within the upper Awash River basin and is drained by two Akaki Rivers that are the tributary of Awash River. The watershed between Awash Basin and the Blue Nile Basin lies on the top of the Entoto Ridge immediately north of the city. The catchment area of the Akaki River Basin which mainly includes Addis Ababa is divided into two sub-catchments by approximately a north-south running surface water divide. These are the Big Akaki River (eastern) sub-catchment and the Little Akaki River (western) sub-catchment. In Addis Ababa the streams drain south from the Entoto ridge; towards southeast direction from Mt. Wechecha and Mt. Furi; and towards southwest direction from Mt. Yerer and other elevated areas of the eastern outskirts of the city. The baseflow component of Big Akaki River from the data measured at Akaki Bridge was estimated as 1 m^3/s during the dry season (Tahal, 1992). However, recent work indicates baseflow as 8 m^3/s and the estimated wastewater discharge into the river as 2.5 m^3/s.

In the southern parts of the city, the drainage density is reduced with the main rivers showing a meandering drainage pattern due to reduced gradient of the valley floor and wide alluvial cover. However, in most parts of the city, the width of the channel has been controlled by the construction of man made structures like retaining walls on the bank of the streams, and the natural path of the flow changes accordingly. Moreover, significant decreases in the gradient of the topography, reduction in the eroding activity of the rivers and minimum flow velocity and transporting capacity towards the south led to the formation of alluvial deposits.

Like other large cities of the world, Addis Ababa is characterized by impervious features like asphalt or compacted gravel roads, concreted drainage channels, airfields, car parks, recreational areas and other man made impermeable structures. These human induced features significantly increase the amount of run-off water in the streams across the city.

2.4 *Groundwater*

To evaluate the groundwater reserve in the Akaki River Basin from where the water supply for the city is abstracted, a conventional water balance method was applied to give an estimated groundwater reserve in Addis Ababa of 140 Mm^3. More than 500 boreholes are currently abstracting water from the volcanic aquifer for the city supply. The water abstracted from volcanic aquifers by the Addis Ababa water and sewerage authority is around 40,000 m^3/d (out of which 30,000 m^3/d is from Akaki well field). Other private and governmental institutions also extract as much as 50,000 m^3/d with overall total abstraction of 90,000 m^3/d. In both the central and eastern part of the city, groundwater occurs within the confined aquifer while the main groundwater potential areas are the eastern and southern part of the city.

The main aquifers in the city and its surroundings are located at different depths and are made up of fractured basalts and ignimbrites of different types with confined or semi-confined conditions. In addition, some phreatic aquifers are formed by the alluvial sediments and residual materials that occupy valley sides. These aquifers supply groundwater to springs, boreholes and dug wells for domestic, agricultural and industrial supplies in the city. Generally, the main aquifers in Addis Ababa can be grouped as:

1. Shallow aquifer: composed of slightly weathered volcanic rocks and alluvial sediments with depths of the aquifer that reaches up to 50 m.
2. Deep aquifers: composed of fractured volcanic rocks that contain relatively fresh groundwater. These aquifers are mainly located in the southern part of Addis Ababa. The depth of aquifer in some places reaches 280 m.
3. Thermal aquifers: situated at depth greater than 300 m and located in the central portion of the city. The existence of these aquifers is manifested by deep circulating thermal water with a number of thermal wells completed along the major Entoto fault.

The 900 m elevation difference between the northern and southern part of the city could drive groundwater southward along with contaminants. The groundwater movement direction is dominated by north-south and northwest-southeast flows and the flow lines converge towards the southern parts of the Addis Ababa, around Akaki well field (Tamiru *et al.*, 2005). In some localities, however, the groundwater flow direction changes due to the occurrence of fault lines. In general, the groundwater movement is sub-parallel to the surface water flow direction and more or less controlled by the topography of the area.

3 METHODOLOGY

To prepare this paper, previous research experience of the author in the city has been used along with data and information collected over several years. The main approach is to integrate different techniques and data to draw conclusions. An attempt to assess the quality of surface water and groundwater in the city has been made by collecting water samples from streams, springs, boreholes and storm drains located in different parts of the city. Representative water samples for chemical analyses were collected (more than 20 samples) in polyethylene bottles, while sterilized glass bottles were used to collect water for bacteriological analyses (about 21 samples). The major ion compositions were determined by atomic absorption spectrophotometer and UV-visible spectrophotometer and for completeness the data on heavy metal from Tamiru (2006) were also incorporated into this study. Bacteriological sampling and analysis was performed for rivers, springs, boreholes and storm drains. The monitoring data BOD and DO were used to characterize the organic matter load of the rivers.

4 RESULTS AND DISCUSSION

4.1 *Surface water quality*

The analytical results for surface water in the city are given in Table 1. The rivers act as direct sources for water supply and indirectly recharge the groundwater. Therefore, controlling

Table 1. Chemical composition of surface water in Addis Ababa (data from March 2004).

	Kolfe Upstream (01)	Kolfe Down Stream (02)	Kebena Upstream (03)	Kebena Down Stream (04)	Big Akaki river (Lake inlet) (05)	Big Akaki river (Lake outlet) (06)	Little Akaki River (07)
pH	6.1	7.51	7.22	7.48	7.3	7.05	7.56
Na mg/l	31	90	74	28	14	12	147
K mg/l	3	24	21	17	13	11	15
Ca mg/l	25	50	67	36	54	49	48
Mg mg/l	12	23	35	15	8	29	18
HCO_3mg/l	58	176	220	122	146	134	244
COD mg/l	8	24	27.2	20	25.6	17.6	35.2
NO_2mg/l	0.046	0.05	0.053	0.35	0.75	<0.01	0.15
NO_3mg/l	86	9.7	188	9.0	13	7.0	5.4
Cl mg/l	49.5	103	106	34.2	30.7	75.3	69.4
F mg/l	<0.2	0.74	0.78	0.24	1.08	0.92	0.64
SO_4 mg/l	<1	0.44	19.7	9.26	2.2	12.6	64.4
PO_4 mg/l	0.13	3.03	<0.03	2.09	1.32	0.86	0.21
S mg/l	<0.1	<0.1	<0.1	<0.1	0.75	<0.1	0.1
SiO_2 mg/l	55.5	16.2	34.1	33.6	39.5	23.5	43.9
Mn μg/l	34.1	1760	6530	1220	1190	2040	2540
Cr μg/l	<0.1	14.12	<0.1	<0.1	2.28	<0.1	13.31
Ni μg/l	44.45	4.8	<0.1	<0.1	5.05	8.9	<0.1
As μg/l	<0.1	2.3	<0.1	0.6	0.42	2.2	2.88
Co μg/l	2.47	4.01	10.7	2.41	2.7	2.4	3.84
Ba μg/l	80.1	132	409	70.1	107	153	194
Cd μg/l	<0.1	<0.1	<0.1	<0.1	<0.1	<0.1	<0.1
Pb μg/l	<0.1	<0.1	<0.1	<0.1	<0.1	<0.1	<0.1
Zn μg/l	<0.1	<0.1	<0.1	<0.1	<0.1	<0.1	<0.1
Li μg/l	2.08	1.13	<0.1	0.52	4.93	1.26	<0.1

the quality of surface water could improve the quality of groundwater. Table 1 indicates that the major ion chemistry is dominated by sodium, bicarbonate, chloride and nitrate. Sodium, chloride and nitrate are likely to be related to the anthropogenic sources mainly from domestic activities. The impact of urban activities is indicated by high concentration of undesirable substances in all the analyzed samples.

At least three samples collected from each stream draining the city show that nitrate concentration decreases markedly downstream (Table 1). This could be explained by denitrification in the water. There is much algae development in the streams in the lower part of the city, the flow velocities decrease and allow for the development of algae that consumes nitrate from the water.

Nitrate contamination in the groundwater is a common problem in urban parts besides farming areas. Samples from springs contain extremely high nitrate concentration as much as 728 mg/l compared to the WHO, (1984) guideline value of 50 mg/l which can be attributed to leaking foul water systems. Nitrate pollution and higher concentration of nitrate reduces

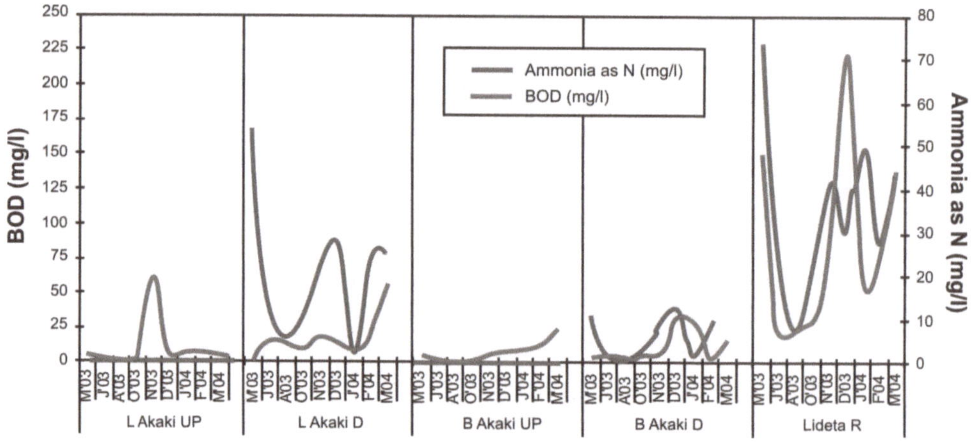

Figure 3. Monitoring results of BOD and Ammonia within the dominant rivers.

oxygen carrying capacity of bloods in infants (Hamill & Bell, 1987). It has also been linked with gastric and oesophagal cancer because of reaction of nitrate with amines in the diet forming carcinogenic nitrosamines and implicated diabetes (Chettri & Smith 1995). According to WHO (1984) for older age groups, certain forms of cancer might be associated with very high nitrate concentration.

Furthermore, the presence of large amounts of organic wastes in the streams of the city is indicated by a high Biological Oxygen Demand (BOD) derived from the decomposition of organic wastes. In most of the streams in the city, BOD increases downstream due to continuous input of nutrient rich liquid wastes from domestic and industrial activities (Tamiru, 2001). Hence, some of the rivers could be considered as dead rivers. The Lideta River is one of the most polluted rivers with respect to organic waste that has very high BOD, low dissolved oxygen and elevated coliforms (Figure 2, Table 2). The BOD in the Lideta River reaches as much as 150 mg/l while the maximum value in Little Akaki and Big Akaki are 18 mg/l and 33 mg/l respectively. Coliforms in the rivers are high, the Lideta River receives a large part of the waste generated from the central highly populated part of the city including the big market centre "Merkato" and from industries like winery, brewery, soft drink factory, tanneries, slaughterhouse that increase pollutant load (see photographs in Figure 4). Consequently, the concentration of some chemical constituents like Cl, Cr, etc. increase down stream mainly due to a progressive contribution of latrine and industrial effluents (Table 1). The observed increase in concentration of chromium downstream within the Lideta River could be attributed to the waste discharge from the leather processing industries (Tamiru, 2001). Details of sources and characteristics of urban groundwater pollutants in general are described by Zhang *et al.*, (2004).

From the factories located south of the city (around Kaliti), the Big Akaki River receives wastes rich in chromium while the main concerns from tanning industries are high levels of sulphide and chromium. Conventionally, chromium can be precipitated by chrome precipitating agents such as sodium hydroxide, magnesium oxide and lime (Mbuthia, 1989). In addition, the Aba Samuel Reservoir, located in the southern extreme part of the city, receives all type of wastes generated from the city and acts as a sink for most toxic substances.

Table 2. Total coliform count of different water points.

Water point	Sample code	Total Coliform
Rivers	01	560/ml
	02	TMC*
	03	660/ml
	04	TMC
	05	1010/ml
	07	TMC
Springs	Megenagna	290/ml
	R. Mekonen	350/ml
	Abo	Nil
Boreholes	14	10/ml
	15	160/ml
	16	24/ml
	17	16/ml
	18	340/ml
	19	24/ml
	20	8/ml
Storm drains	Arat kilo	3,000,000/ml
	Piassa	3,50,000/ml
	Merkato	6,000,000/ml
	Bole	20,000/ml
	Shola	1,00,000/ml

*TMC is too many to count

Figure 4. Photos of industrial effluent and algal development in the rivers in Addis Ababa.

The observed higher concentrations of sodium in the surface water are probably due to leakage from sewers, which drain directly into the rivers in many parts of the city. Other locally significant sources of sodium are washing powders, soaps and chemical detergents. A report of the Central Statistical Authority (1995) showed that 30% of the housing units in the city have no toilet facilities which is likely to contribute to much of the observed elevated nitrate concentrations. Measurements of electrical conductivity show

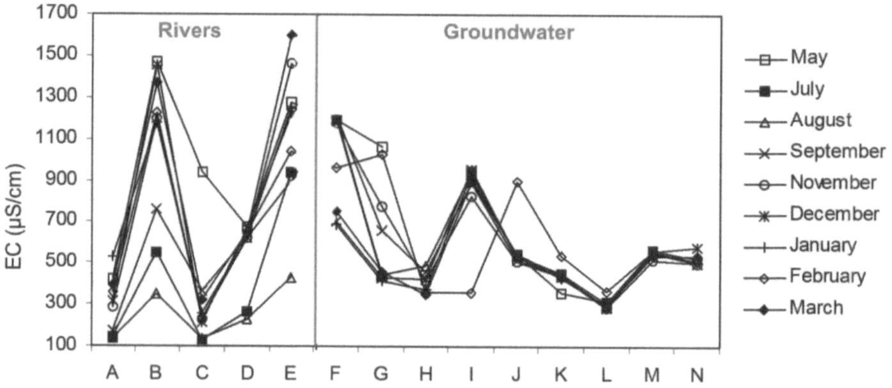

Figure 5. Rapid fluctuation of EC in surface water and shallow groundwater in the city (From May 2003 to March 2004).
(A = Little Akaki Upstream, B = Little Akaki downstream, C = Big Akaki upstream, D = Big Akaki downstream, E = Lideta river, F = Lideta spring, G = Building College BH, H = Legehar DW, I = Textile BH, J = EP-8 BH, K = Addis tyre BH, L = Kara BH, M = BH22, N = Alert BH)

more monthly variability than the groundwater sources (Figure 5), which reflects runoff and contamination.

The heavy metal concentration in the Addis Ababa surface waters is extremely variable and mainly depending on the location of sources of pollution (Tamiru, 2006). However, Mn concentrations in groundwater increases from shallow to deep aquifers detected in the boreholes that could be attributed to a longer residence time. Moreover, surface waters that have undergone anaerobic decomposition and reducing groundwater often contain high levels of dissolved manganese (Hem, 1989). Therefore anaerobic conditions that exist in most of the streams and rivers of the city may account for the elevated concentrations of manganese.

The chromium concentrations in water in the northern part of the city are low compared to the southern part. As the streams in the Little Akaki Basin approach the southern part of the city, where several tanneries are located, the chromium concentration reaches 14 µg/l (the sample was taken at 1 km downstream of the known point source: tannery). It is also thought that chromium rich water infiltrates into the shallow groundwater system through fractures. According to Tamiru (2006), the heavy metal concentration in soils of Addis Ababa is mainly derived from a natural source where anthropogenic sources are minor.

Biological indicators of pollution such as algae and bacteria could be used to show the extent of surface water and groundwater pollution in the city. From field survey it is found that green algae like *Cladophora* and *Spirogyra*, and blue green algae such as *Oscillatoria*, *Ulothrix* and *Rivularia* are the dominant species in the nutrient rich streams in the city. The total coliform count in the streams varies between 560/ml and several millions. In the springs, the count varies between 0–350/ml, while in the boreholes it ranges between 8/ml and 340/ml (Table 2). The maximum limit for potable water is 10 total coliform/ml (WHO, 1984) but in recent EU directive this value falls to zero. The presence of bacteria in groundwater indicates that the shallow unconfined aquifers in the central and southern part of the city are contaminated. As indicated in Table 2, the primary bacterial sources are storm drains that extensively pollute the rivers and the streams. Consequently, the shallow groundwater system is also affected through infiltration process within the fractured rocks. The bacterial population in the storm drains is in the range of millions. Hence, the connection of large numbers of drain lines with streams is thought to have increased the number of bacteria. It can be concluded that the streams are bacterially polluted and are also characterized by offensive odours.

4.2 *Groundwater quality*

The physical and chemical properties of groundwater are most importantly related to its relationship with the media which the water encounters and its residence time (Appelo & Postma, 2005). In addition to the natural factors major changes in the constituents of groundwater in the study area are a result of human activities.

The chemical analysis of major ions in groundwater shows sodium and calcium as the dominant cation (Table 3) and they are more or less in equal abundance. Groundwaters in Addis Ababa fall into two categories: Na-Ca-HCO_3, and Ca-Cl (Tamiru *et al.*, 2005). The variation of the type could be attributed to different host rocks and depth of circulation. High concentration of chloride in groundwater may be attributed to seepage from contaminated river water or surface runoff. This is due to the high mobility conservative nature of chloride.

Changes in recharge in urban areas could have an impact on the groundwater quality, either directly by allowing chemicals used in urban areas to be leached into a groundwater system, or indirectly by changing chemical conditions within an aquifer (Appleyard, 1995). On the other hand, urban wastewater could be considered useful for groundwater recharge if the media is capable of filtering the pollutant. The impact of wastewater infiltration on specific groundwater supply sources will depend not only on its impact on the shallow aquifer system, but also on their siting relative to wastewater infiltration area, the depth of water intake and the integrity of well construction (Foster *et al.*, 2004). The groundwater contamination problems in Addis Ababa are related to the poor well construction and defective sewerage line or septic tanks located close to wells. Moreover, groundwater becomes polluted as consequences of its interaction with the contaminated surface water that carries industrial effluent. The rocks have poor attenuation capacity. Widespread unplanned interaction of human beings with the physical environment especially in developing countries has a potential to reduce the quality of water through biological and chemical wastes. Such reduction in quality will bring water shortage and hence, enhance poverty.

The World Health Organization (WHO, 1984) suggested three modes of transmission of bacterial pathogens through ingestion of contaminated water and food; contact with infected persons or animals and exposure to aerosols. The presence of pathogen bacteria in natural water is mostly related to human or animal excrement. If the contributors are carriers of communicable decreases like cholera, the contamination of water becomes more significant. Therefore, besides the chemical constituents, it is necessary to determine the bacteriological quality of water for safe intended use. In the assessments of the bacteria for water quality, the determination of coliform concentration is universally accepted. Commonly it includes the detection and enumeration of coliform organisms (total coliform) and fecal (thermo tolerant) coliform organisms.

In general, the presence of faceal coliforms in groundwater indicates the infiltration of water contaminated by wastes from animals and man. Particularly the presence of *E.Coli* in the boreholes, where water is found at a depth within porous aquifer and the area is covered with thick black cotton soil, needs further investigation to infer the possible sources of

Table 3. Major ion and trace metal concentration in groundwater (March 2004).

Parameter	AAU (14)	Repi soap Factory (15)	Addis cement Factory (16)	Mekanissa (17)	Shola Dairy (18)	Building (19)	Dewero (20)
pH	7.62	7.44	6.98	7.15	6.42	8.03	7.68
Na mg/l	27	43	31	28	21	22	10
K mg/l	6	14	11	53.5	6	8	15.1
Ca mg/l	80	185	121	25.1	40	140	40
Mg mg/l	8	10	16	4.4	3	15	8.3
HCO_3mg/l	232	171	152	73.2	110	153	256.3
COD mg/l	6.8	7.6	9.2	8.8	7.2	8.1	40.8
NO_2mg/l	<0.01	<0.01	<0.01	<0.01	<0.01	<0.01	<0.01
NO_3mg/l	0.72	10.2	11.3	16.5	8.8	35	8.6
Cl mg/l	6.8	4.82	11.9	21.2	1.7	16.8	28.2
F mg/l	0.21	0.4	0.72	0.22	0.27	0.22	0.6
SO_4 mg/l	6.4	<1	12.7	2.9	2.5	4.8	82.4
PO_4 mg/l	<0.03	0.1	0.1	0.08	<0.03	<0.03	<0.03
S mg/l	<0.1	<0.1	<0.1	<0.1	<0.1	<0.1	<0.1
SiO_2 mg/l	23.5	21.4	20.4	17.3	20.6	22.9	24.2
Mn µg/l	0.86	0.64	5.25	3.48	1.97	16.9	1.75
Cr µg/l	0.69	0.86	3.29	0.22	0.72	0.24	1.8
Ni µg/l	0.57	0.31	0.33	0.69	0.63	<0.1	0.51
As µg/l	0.37	0.3	0.3	0.32	0.2	0.63	0.5
Co µg/l	<0.1	<0.1	<0.1	<0.1	<0.1	<0.1	<0.1
Ba µg/l	<0.1	<0.1	<0.1	<0.1	<0.1	<0.1	<0.1
Cd µg/l	<0.1	<0.1	<0.1	<0.1	<0.1	<0.1	<0.1
Pb µg/l	25.1	8.98	10	25.3	15.7	9.75	4.66
Zn µg/l	23.3	85.8	3.9	34.2	33.5	10.3	20.5
Li µg/l	0.77	0.49	0.44	<0.1	<0.1	0.39	2.58

contaminants. On the other hand faceal pollution on water bodies can be clearly attributed to direct or indirect discharge of wastes into the watercourse.

4.3 *Rating of groundwater contamination sources*

Once the major sources of water pollutants are identified, it is useful to rate the groundwater contamination sources. Rating is used where contamination sources are given quantitative or qualitative measures of the potential hazard they pose to groundwater (Johansson & Hirata, 2004). Rating of sources can be very useful to raise awareness and to put groundwater protection on the agenda of decision and policy makers. The identification of potential contamination sources is based on the existing information available from various institutions and ground survey. The main potential sources in Addis Ababa were classified into 9 classes. These are residences, food and beverages, printing, textile, chemical, tannery,

Table 4. Ratings of the groundwater contamination sources, higher scores pose the greatest risk.

	Residence	Food and drink	Printing	Textile	Chemical	Tannery	Metal	Garages Petrol stations	Hospitals
Contamination class	0.6	0.4	0.6	1	1	1	0.8	0.8	0.8
Relative concentration	1	0.8	0.4	1	1	1	0.8	0.6	1
Mode of disposition	1	0.6	0.6	0.8	0.8	0.6	0.4	1	0.4
Duration of load	1	0.4	0.2	0.6	0.6	0.6	1	1	0.4
Potential for remediation	0.6	0.6	0.8	0.8	0.8	0.8	0.4	0.8	0.4
Average	*0.84*	*0.56*	*0.52*	*0.84*	*0.84*	*0.84*	*0.68*	*0.84*	*0.6*

metal, garages and hospitals. The sources of pollution were weighted according to five categories (Table 4).

As shown in Table 4, wastes from residence, textile, chemical and garages and petrol stations pose the highest risks as a source for groundwater pollution and metal industries and hospitals are grouped as medium source while food and beverages and printing centres are categorized as low.

5 CONCLUSIONS

The deterioration of the quality of surface water resources in Addis Ababa is manifested mainly though high bacterial concentration derived from storm drains that receive wastes from domestic sources. All streams and rivers in the city are characterized by offensive odour due to disposal of organic wastes. The main surface water pollutant input comes from the direct disposal of wastes, and the linkage of storm drains with the rivers. This is attributed to the absence of a properly distributed sewerage system. Wastes generated from domestic activities, textile factories, chemical industries, garages and petrol stations are the main source for groundwater pollution in the city of Addis Ababa.

Generally, the quality of surface water and shallow groundwater found in different parts of the city is not suitable for human consumption. It is, therefore, absolutely necessary that the population should only utilize treated water provided by Addis Ababa Water and Sewerage Authority for cooking and drinking. This work indicated that the quality of groundwater in the city is dependent on the quality of surface water. Hence, for the quality management of groundwater, it is crucial to control the quality of surface water.

Due to increasing severity of the effect of pollution, municipal decision makers should:

- coordinate all researchers that are involved in the surface and groundwater pollution study to monitor environmental changes;
- encourage industries that are located along vulnerable aquifers and surface water bodies to build waste treatment facilities before discharging into the aquatic environment;

- encourage residents to implement integrated domestic waste management techniques;
- advise communities and the population to construct properly lined septic tanks.
- empower local communities to take an active participation in waste disposal and environmental control activities.

ACKNOWLEDGMENTS

The author is highly grateful to UNEP/UNESCO Nairobi office for sponsoring this work. Special thanks go to Prof. Salif Diop, Prof. Emmanuel Naah, Prof. Yongxin Xu, and Mr. Patrick M'mayi. I thank also Mr. Nuri Mohammed for handling the major element and BOD-DO analyses and Dr. Dagnachew Legesse for the GIS part. I am also grateful to Dr. Tenalem Ayenew, Mr. Yirga Tadesse and Mr. Solomon Waltenegus for data collection and constructive discussions.

REFERENCES

Adane, B. 1999. *Surface water and groundwater pollution problems in the upper Awash River basins.* MSc, University of Turku, Finland.

Appelo, C. A. J. & Postma, D. 2005. *Geochemistry, groundwater and pollution.* 2nd edition. Balkema. The Netherlands.

Appleyard, S. 1995. *The impact of urban development on recharge and groundwater quality in a coastal aquifer near Perth, Western Australia.* Hydrogeology Journal, **3**, 65–75.

Barrett, M. H. 2004. *Characterization of urban groundwater.* In: Lerner, D. N. (ed.) Urban groundwater pollution. IAH International Contributions to Hydrogeology, **24**.

Central Statistical Authority 1995. *The 1994 population and housing census of Ethiopia.* Volume 1. Statistical report, Addis Ababa, Ethiopia.

Chettri, M. & Smith, G. 1995. *Nitrate pollution in groundwater in selected districts of Nepal.* Hydrogeology Journal, **3**, 71–76.

Chilton, P. J. (ed.) 1999. *Groundwater in the urban environment – Selected city profiles.* IAH International Contribution to Hydrogeology, **21**, Balkema, The Netherlands.

Foster, S. S. D., Garduno, H., Tuinhof, A., Kemper, K. & Nanni, M. 2004. *Urban wastewater as groundwater recharge.* GW-MATE Briefing Note, **12**.

Grischek, T., Nestler, W., Piechniczek, D. & Fisher, T. 1996. *Urban groundwater in Dresden, Germany.* Hydrogeology Journal **4**, 48–63.

Hamill, L. & Bell, F. G. 1987. *Groundwater pollution and public health in Great Britain.* Bulletin of the International Association of Engineering Geology. **35**, 71–78.

Hem, J. D. 1989. *Study and interpretation of the chemical characteristics of natural water.* USGS Water Supply Paper, **2254**.

HaileSelassie, G. & Getaneh, A. 1989. *The Addis Ababa-Nazaret Volcanics: A Miocene-Pleistocene volcanic succession in the Ethiopian Rift.* SINET: Ethiopian Journal of Science, **12**, 1–24.

Johansson, P. O. & Hirata, R. 2004. *Rating of groundwater contamination sources.* In: Zaporozec, A. (ed.) Groundwater Contamination Inventory. IHP-6, Series on Groundwater No. 2. UNESCO.

Lerner, D. N. & Barrett, M. H. 1996. *Urban groundwater issues in the United Kingdom.* Hydrogeology Journal, **4**, 80–89.

Malcolm, E. C., Hiller, J., Foster, L. & Ellis, R. 1996. *Effects of rapidly urbanizing environment on groundwater, Brisbane, Queensland, Australia.* Hydrogeology Journal, **4**, 31–47.

Morton, W. H., Rex D. C., Mitchell J. G. & Mohr P. A. 1979. *Riftward younging of volcanic units in the Addis Ababa region, Ethiopian Rift valley.* Nature, **280,** 284–288.

Tamiru, A. 2001. *The impact of uncontrolled waste disposal on surface water quality in Addis Ababa.* SINET: Ethiopian Journal of Science, **24**, 93–104.

Tamiru, A., Dagnachew, L., Tenalem, A., Yirga, T., Solomon, W. N. & Nuri, M. 2005. *Water quality assessment and groundwater vulnerability to pollution mapping of Addis Ababa water supply aquifers.* UNEP/UNESCO report, Nairobi, Kenya.

Tamiru, A. 2005. *Uncontrolled Groundwater Exploitation From Vulnerable Aquifer Of Addis Ababa, Ethiopia.* International Workshop on groundwater management in arid and semi arid countries. 4–7 April 2005. Cairo, Egypt.

Tamiru, A. (2006) *Sources of heavy metals in the urban environment of Addis Ababa*, Ethiopia Journal of soil and Sediment contamination, Vol. 15, No 6.

WHO 1984. *Guidelines for drinking water quality. Volume 2, Health Criteria and other supporting information.* World Health Organisation, Geneva.

Zanettin, B., Nicoletti, M. & Petrucciani, C. 1978. *The Evolution Of The Chencha Escarpment And The Ganjuli Graben (Lake Abaya) In The Southern Ethiopian Rift.* N. Jour. Geol. Paleont. **8**, 473–490.

Zhang, S., Howard, K., Otto, C., Ritchie, V., Oliver, I. N. S. & Appleyard, S. 2004. *Sources, types, characteristics and investigation of urban groundwater pollutants.* In: Lerner, D. N. (ed.) Urban groundwater pollution. IAH International Contributions to Hydrogeology, **24**.

CHAPTER 16

Groundwater mining: A reality for the Lusaka urban aquifers?

N.H. Mpamba
Ministry of Energy and Water Development, Department of Water Affairs, Lusaka, Zambia

D.C.W. Nkhuwa, I.A. Nyambe, & C. Mdala
The University of Zambia, School of Mines, Geology Department, Lusaka, Zambia

S. Wohnlich
Ruhr University Bochum, Department of Applied Geology, Bochum, Germany

ABSTRACT: The city of Lusaka has historically depended on groundwater from the underlying karstic carbonate and schist aquifers. Inadequate hydrogeological data has hitherto hampered determination of the effects of increasing groundwater abstraction on groundwater levels. Although the recharge estimates vary widely from 8% to 35% of the annual rainfall, groundwater resources availability in terms of quantity and quality, as well as annual recharge and recharge mechanisms are still not well understood. On-going research, using a comparative analytical model for groundwater monitoring in the urban and rural areas of Zambia, gives preliminary evidence of groundwater mining and direct contamination of the Lusaka urban aquifers. In the absence of legal instruments and management tools to enhance the acquisition of groundwater data and information, establishing the capacity of the aquifer to cope with the present and future water demands poses the greatest challenge for the Lusaka city aquifers.

1 INTRODUCTION

Large-scale exploitation of groundwater in the Lusaka urban aquifers dates back to the time when Lusaka was first established as the capital city of Northern Rhodesia, now Zambia. The city has historically depended on groundwater from aquifers of the Lusaka Dolomite Formation as a source of fresh water supply since the 1950s, when the water supply scheme was initiated. At this time, private residential plots, smallholdings and a few industrial sites were also dependent on individual supplies. Hence, Lusaka was once described as the only city in the past or present British territories in Africa that derived its water supply entirely from boreholes (Lambert, 1965).

In the context of the current groundwater development and abstraction trends, the scenario has not changed much. Lusaka Water and Sewerage Company (LWSC), the water utility company responsible for water supply to the city, abstracts approximately 50% of its water requirements from aquifers in the Lusaka urban and adjacent areas and the other 50% is imported as treated surface water from Kafue river, which is located 50 km south of the city (Mtine & Chikama, *personal communication*). Private individuals and industries also abstract substantial quantities of water from these aquifers. Irrigated agriculture in the adjacent areas is another activity that has increased exploitation levels in the same aquifers.

The actual volume of groundwater abstracted annually has not yet been established due to inadequate groundwater data and information, as the Department of Water Affairs (DWA) – the institution mandated to keep records and details on all boreholes drilled in Zambia – cannot execute this task effectively and efficiently due to inadequacies in the legal provisions. DWA only collects and keeps records of basic hydrogeological data gathered during construction of their own boreholes and of those submitted on a voluntary basis to DWA by private drilling companies. Private borehole drilling contractors are not obliged by law to collect and submit borehole records (WRAP, 2005). This has created difficulties for policy and decision-makers to access groundwater data and information especially now, when groundwater resources development appears to be on the increase. The setback with the DWA borehole records is that only descriptive information is given regarding their location. They have no grid reference or Geographic Positioning System (GPS) coordinates, which would facilitate their usage as spatial data to permit informed groundwater development and management decisions to be made.

2 GENERAL INFORMATION ABOUT LUSAKA

The study area is located between UTM Northings $^{82}78431$ and $^{83}08156$, and UTM Eastings $^{6}25256$ and $^{6}60846$ (Figure 1). It is part of the mid-Tertiary peneplain of Central Africa, which occurs at an elevation of 1260 m asl. The area features a sub tropical climate with three distinct seasons: a dry season from mid April to mid August (15–23°C), a hot season from mid August to October (27–38°C) and rainy season from November to mid April during which time the area receives its rains. The average rainfall is 865 mm (WRAP, 2005). Annual Potential Evaporation for Lusaka using PENMAN and THORNTHWAITE methods is 1489 mm/a and 938 mm/a respectively (Nkhuwa, 1996). Three main river basins, Chunga-Mwembeshi, Chongwe and Kafue (Figure 1) drain the Lusaka urban and adjacent areas that are underlain by mainly carbonate rocks. Because of the presence of sinkholes, epikarst and permeable laterite, it is generally rare to find surface streams on the Lusaka carbonate rocks (Lambert, 1965).

2.1 *Geological setting*

Extensive geological mapping and field surveys by Simpson *et al.* (1963), Matheson & Newman (1966) and Cairney (1967) have shown that the Lusaka plateau comprises the Lusaka Dolomite, Cheta and Chunga Formations. The major rocks comprise gneisses and quartzites of the Chunga Formation, schists and quartzites of the Cheta Formation, and thick and extensive sequences of limestones and dolomitic limestones of the Lusaka Dolomite Formation. Because limestones and dolomitic limestones are metamorphosed, it has been suggested that they be called marbles and dolomitic marbles, respectively (Nkhuwa, 2003).

The Lusaka Dolomite and Cheta Formations belong to the Katanga System, while the Chunga Formation constitutes the Basement Complex. The former occupies an area of

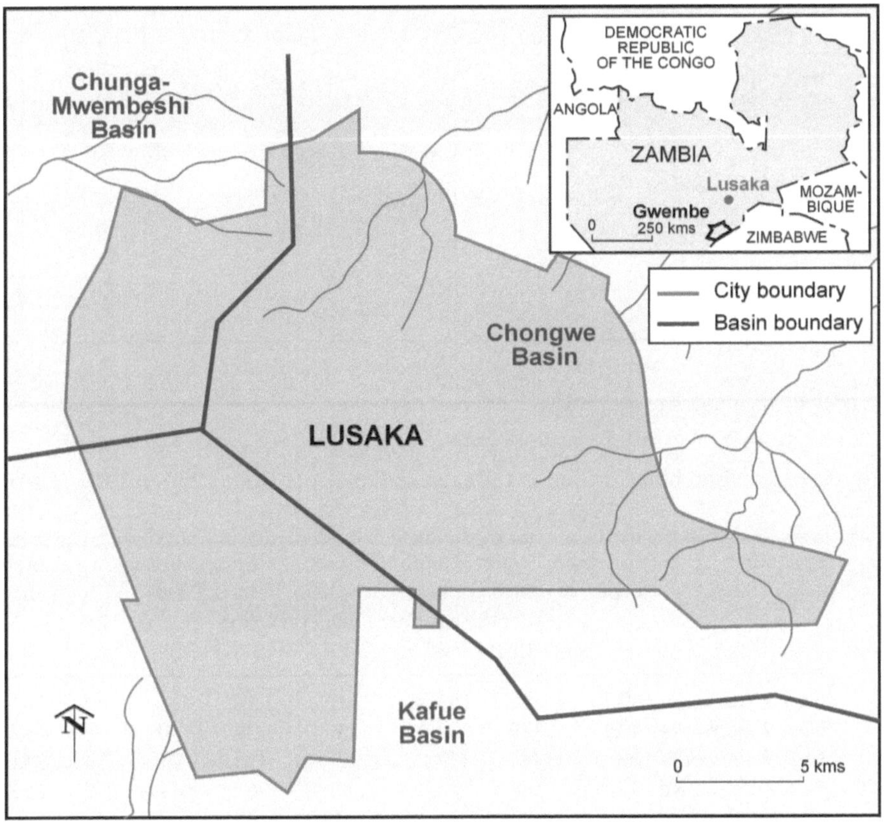

Figure 1. Map of Zambia showing the Lusaka urban area and Gwembe District (insert).

about 470 km^2 in the study area, while Cheta and Chunga Formations occupy an area of about 221 km^2 (Mpamba, 2006).

Tectonic events are suggested to have caused the uplift of the study area to form the current Lusaka Plateau (Figure 2). Recumbent folding accompanied by faulting and thrusting (Simpson *et al.*, 1963) may have caused these tectonic events.

2.2 *Hydrogeological setting*

Major rock outcrops in the study area comprise marbles and dolomitic marbles interbedded with schists and quartzites. These rocks form the city's major aquifers, each with a different yield potential. Marbles have the highest groundwater potential due to the presence of well-developed karstic systems (Nkhuwa, 1996). These rocks conform, in a regional setting, to other Katanga rocks, which outcrop at Chisamba and Kabwe, trending north-westerly, and which possess very high groundwater potential from which these towns derive their water supply (Lambert, 1963). The combined thickness of the Lusaka dolomite and Cheta limestone is not yet established, due to the absence of deep drilling (Nkhuwa, 1996).

These rocks are affected by three sets of discontinuities, which facilitated, during geological time, transmission of water through them. This has resulted in the development of

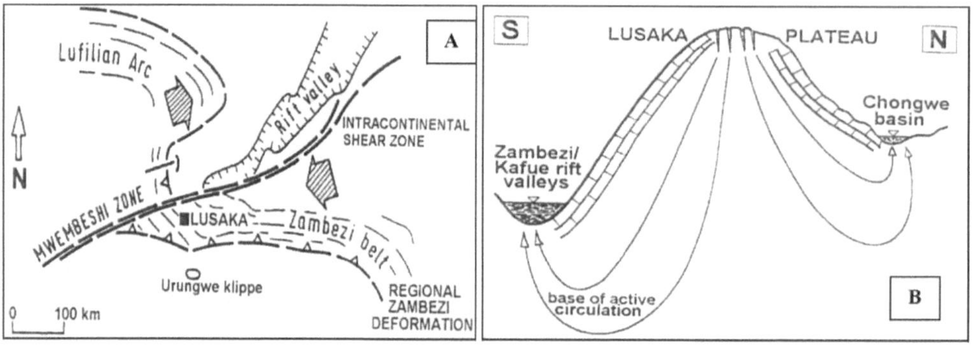

Figure 2. (A) Thrusting and (B) A general uplifting model of the Lusaka due to tectonic events (Drysdall & Smith 1960; Nkhuwa, 1996).

Table 1. Data on boreholes drilled in the Lusaka Dolomite from 1959 to 1961 (Lambert, 1965).

Bh ID	Location	Borehole depth (m)	Water Strike depth (m)	Depth to main aquifer (m)	Static Water-level (m)	Pump Intake (m)	Drawdown (m)	Borehole Yield (l/s)
1	New Waterworks	49.1	30.3	31.2	11.7	42	24	131
2	New Waterworks	49.1	29.1	30.9	15	39	30	43.8
5	New Waterworks	42	39	39	12	23.4	21	87.6
1-17-59	Old Mumbwa Rd	39	4.5	26.4	0.78	6	1.8	26.3
2-17-59	Old Mumbwa Rd	44.7	3	24	0.63	6	3.9	18.9
3-17-59	Old Mumbwa Rd	45	1.8	15	1.2	6	2.4	15.8
4-17-59	Old Mumbwa Rd	60	3.3	36	1.8	6	5.1	34.7
5-17-59	Old Mumbwa Rd	74.1	2.1	12.6	2.1	40.5	40.5	3.7
14-17-59	Beacon C 360	59.1	6.6	36	6.6	32.4	6.3	28.4
2-2-61	Beacon C 361	68.4	8.1	36.3	8.1	39.3	11.7	36.6
Average		53	12.8	28.8	6.0	24.1	14.7	42.8

an integrated system of conduits and solution channels. Other than converting these rocks into an important groundwater store and an important source of water supply to the city, the presence of solution features has transformed this terrain into one with specific and highly complex hydrogeological and environmental conditions that impose enormous restrictions on land use possibilities of the terrain.

3 HISTORICAL AND CURRENT GROUNDWATER DEVELOPMENT

Substantial groundwater abstraction of 43,000 m^3/d was achieved from boreholes constructed in the Lusaka Dolomite from 1959 to 1961 (Lambert, 1963). Examination of the historical borehole details (Table 1) shows that boreholes are characterised by high yields, small drawdowns, water strikes between 15 m and 39 m and an average borehole depth of 53 m. This period marks the start of aquifer exploitation, and the data from these boreholes acts as an excellent baseline from which to examine whether the aquifer is being overexploited.

Between September and November 1961, two major factors affected water-levels in unused boreholes at the Old Water Works (Lambert 1965) (i) groundwater abstraction, and (ii) the construction of a deep drainage network to drain the commercial and industrial areas of Lusaka on the dolomite, which was fully operational during the 1960/61 season.

The effects of groundwater abstraction and groundwater drainage through the deep drainage network, led to a reduction of dewatering of a nearby quarry from 7570 m^3/d to 2838 m^3/d because of a declining groundwater table. This scenario provides an understanding of how quickly the dolomite aquifer responds to the effects of abstraction and induced drainage. Over-pumping and the danger of aquifer depletion attributed to the intense development of farms and residential areas, each with their own borehole, were a source of great concern from the late 1950s to early 1960s (Lambert, 1965).

When compared to the current borehole drilling demand in Lusaka urban area, the same pressures are still at work: increasing demand for water on individual plots that are not connected to the LWSC supply line, or those areas that experience inadequate water pressure. Non-provision of water to some peri–urban and newly developed low-density areas and water shortages are among other reasons for the increasing demand for borehole drilling in the Lusaka aquifers. This also includes the demand for irrigated agriculture and industry.

Table 2 gives some of the historical data for the Lusaka Water and Sewerage Company (LWSC) boreholes and Table 3 information from recently drilled boreholes for the Department of Water Affairs (DWA). The data from tables 1–3 indicate a slight increase in depth to the first and main water strike and also an increase in the depth to the static water-level.

Most of the current groundwater development activities taking place in the Lusaka urban aquifers are demand driven. Rapid growth of the city in the last five decades without town and country planning control, has resulted in industrial and housing units developing on top of highly vulnerable aquifers which should otherwise have been protected. This situation is now difficult to reverse. A number of low- and high-density areas within the city are using on-site sanitation such as septic tanks and pit-latrines to dispose of their excreta. Together with the current intensive generation of solid waste, the situation poses great risks of groundwater contamination in the karstic carbonate aquifers due to anthropogenic activities (Nkhuwa, 1996; Zulu & Nyambe, 2004).

The prevailing situation of not collecting and submitting borehole drilling data and information is attributed to the fact that the current Water Act (Act of 1948) does not include control of groundwater development and abstraction (WRAP, 2003). Hence, exploitation of groundwater resources is not regulated because groundwater has always been regarded as privately owned. This inherent weakness in the law has resulted in failure to carry out sustainable groundwater development, delineate potential recharge areas and set aside areas to support water requirements for domestic, agricultural, industrial and other activities (WRAP, 2003).

4 ESTIMATES OF GROUNDWATER RECHARGE

Estimates of groundwater recharge for the Lusaka plateau vary widely, from 8% to 35% of the annual rainfall (Burdon & & Papakis, 1963; von Hoyer *et al.*, 1978; JICA Report,

Table 2. The LWSC boreholes located in the limestone in Lusaka urban and adjacent areas.

No.	X	Y	Borehole Depth (m)	Borehole yield (l/s)	First Water Strike (m)	Main Water Strike (m)	Pump Discharge (l/s)	Pumping Depth (m)	Drawdown (m)
1	651238	8299585	50.0	27.8	7.5	23.0	26.4	34	5.34
2	651433	8299602	70.0	5.8			16.7	32	15.4
3	650740	8298255	60.0	7.2		28.3	7.2	30	7.93
4	648103	8299406	61.0	6.3	7.2	20.0	12.5	49	
5	641378	8289696	43.0	28.0		30.0	27.8	33	0.89
6	651252	8300339	60.0	13.9			13.9	30	7.36
7	642328	8296266	65.0	50.0	11	11.0	20.0	33	
8	642285	8296608	39.2	13.9			12.5	33	
9	641718	8300369	73.0	1.8		20.0	2.8	27	4.5
10	640120	8297508	56.4	50.0		30.0	20.8	27	
11	627941	8303967	45.0	13.0		30.0	8.3	35	
12	648130	8303606	66.0	22.1	4.4	17.8	22.2	38	20.7
13	647746	8303845	50.0	34.8	6.4	25.5	8.1	38	
14	647850	830381	55.0	5.6	6.7	38.5	8.3	35	
15	640827	8292071	65.0	20.0	10.3	30.0	20.8	40	5.98
16	643412	8294004		5.6	7	43.0	2.2	39	
17	636804	8292631	47.0	38.0	22	30.2	45.0	35	0.62
18	633684	8295028	50.0	40.0			27.8	45	
19	633697	8295028	50.0	41.7			41.9	42	
20	633747	8295030	38.0	27.8			27.8	48	
21	633966	8294882	81.0	13.9	5.4	12.2	20.8	30	
22	633691	8295079	65.0	50.0			50.0	36	
23	641020	8287387	66.0	300			166.7	45	
24	641022	8287386	66.0	300			166.7	44	
25	641497	8291252	65.0	166.7			70.8	35	
26	641498	8291245	70.0	40.0			27.8	46	
Average			58.3	50.9	8.8	26.0	33.7	36.9	7.6

1995; Nkhuwa, 1996). The difference in the estimates could be attributed to the methods used to determine the annual recharge.

1. The annual groundwater recharge value of 8% is based on the renewable groundwater due to rainfall and was estimated based on the results of hydrogeological surveys, groundwater level observations, numerical simulation, surface water analysis and meteorological analysis (JICA Report, 1995).
2. The higher groundwater recharge value of 22% (von Hoyer, 1978) was determined on the basis of hydrological and meteorological analysis that also involved the use of the relative plant-available moisture (RAM) values and was said to be highly dependent on the amount of rainfall.
3. The 23% value by Nkhuwa (1996) was determined on the basis of hydrological and meteorological analysis using PENMAN and THORNTHWAITE methods and the RAM values from von Hoyer *et al.* (1978).

Table 3. List of some boreholes drilled from 2000 to 2005 in marbles in Lusaka urban and adjacent areas (after Mpamba, 2006).

BH ID	X	Y	Borehole Depth (m)	Static water-level (m)	First Water Strike (m)	Main Water Strike (m)	Borehole yield (l/s)	Pumping Depth (m)	Drawdown (m)
1	649398	8299005	40	29	30	36	5	36	0.1
2	640626	8301568	57.5	10.6	25	42	1.5	45	21.8
3	653290	8300603	55.5	16.2	30	36	3.6	42	10.5
4	647805	8302255	28.8	10.2	17.6	24	3	25	14.4
5	650562	8297406	40	13.4	28	36	7.5	34	7.3
6	649552	8296580	49	25.6	3.2	25	4	42	2.2
7	636148	8297720	52	27.6	42	50	1	45	4.2
8	643967	8292134	48	27	36	39	2	42	2.1
9	649183	8287537	37.8	17.62		27	0.7	34	10.4
10	642184	8289247	37	12		36	7	32	16.4
11	650528	8296324	50	20	38	40	5	36	0.7
12	648671	8294923	85.5	11.4	61	78	0.8	37	13.7
13	643522	8291522	60	14.5	18	36	1.8	45	33.5
14	649025	8287046	31.5	16.3	9	24	2.5	24	5.9
15	647126	8291656	49.5	13.1	19	27	4	42	10.5
16	647529	8288381	37.2	13.3	27	32	4	34	10.9
17	636787	8289793	31.5	7.57	5	19	0.8	21	17.9
18	635603	8282098	49.5	17	22	37	2	42	10.3
19	633145	8286790	34	10.4	18	28	3	26	11.3
20	635471	8283473	60	9.33	18	30	1.5	42	15.5
21	633561	8287083	50	10.3	12	24	2.5	42	3
22	628943	8293148	32	4.8	5	18	5.5	24	0.7
23	622452	8295079	42	8.3	12	36	3.5	36	5.17
24	633145	8287116	39.5	10	18	24	5	34	14.4
25	631091	8298213	47	1.8	3	34	2.1	42	22
26	639251	8288242	38.1	13.3	12	12	0.25	36	6
27	631379	8294244	49.5	2.7	18	44	0.5	45	37.4
28	649013	8290750	41.2	7.2	12	30	3	34	23
29	641625	8300213	67.5	15	48	54	1.5	48	23.7
30	641781	8300820	49	8.4	23	39	1.5	45	4.7
31	642897	8286916	57.88	16.2	40	52	0.5	52	32.4
32	634198	8291588	38	3	6.16	30	6	34	13.8
33	627398	8299011	39.8	1.8	3	6	3	25	17.3
34	636979	8291116	40	5.0	4	21	6	35	25.4
35	637857	8293819	48	24	12	32	0.9	30	24.6
36	633270	8287207	37.5	12.14	12	24	5.5	30	3.0
37	636709	8284191	36.5	15.9	18	30	3	27	2.06
38	633582	8279437	50	2.9	28	30	4	48	0.5
39	647850	8288649	42	15.2	30	30	6	27	12.9
Average			45.6	12.8	19.6	32.6	3.1	36.4	12.6

4. The 35% groundwater recharge value (Burdon & Papakis, 1963) was determined based on hydrological and meteorological analysis that also involved comparison of areas in the Mediterranean countries with karstic features like those occurring in the Lusaka marbles.

Other studies include determination of infiltration capacity estimated as 76 mm /hour for soils on the Lusaka Dolomite (Taque, 1969), while effective porosity is in the range of 2.5%–7.5% (Jones & Topfer, 1972). Borehole logs in the marbles show fracturing and karst development extending to an average depth of 80 m (von Hoyer, 1978; Nkhuwa, 1996). Each of the hydrogeological studies carried out in the study area has had its own strengths and weaknesses, but they have been able to compliment each other in improving our understanding of the Lusaka urban aquifers. Even with all the available results from various detailed studies, uncertainties still exist about the groundwater resources in these aquifers in terms of storage, replenishment and side effects arising from their over-exploitation (Nkhuwa, 1996).

The recommendation is that the amount of groundwater withdrawn from the aquifer should not exceed the estimated groundwater recharge of 8% of the annual rainfall (JICA Report, 1995). In the case of the Lusaka plateau, this is calculated as 45 Mm^3/a based on the long-term average annual rainfall of 822 mm/a calculated from meteorological records from 1938/39 to 1993/94 (Nkhuwa, 1996).

5 HYDROGEOLOGICAL STUDIES

Knowledge about the nature of the aquifer system and groundwater system is fundamental when undertaking any groundwater assessment study. Some of the information required may be obtained from hydrogeological, hydrochemical and groundwater level monitoring data. Two areas – Lusaka urban and adjacent areas, and Gwembe District as shown in Figure 1 – were selected for regular collection of groundwater quality and groundwater level data. This paper discusses hydrogeological and groundwater level observation data for the Lusaka urban and adjacent areas only. Groundwater quality data from the two project areas will constitute another paper for a later publication.

The major focus of the study was to examine and identify existing boreholes constructed by the DWA in the study area and abstraction boreholes owned by the LWSC. In addition, two groundwater observation boreholes – one located in the schist and another located in the marble – were constructed with the 50 mm diameter PVC casing pipes and according to the recommended construction standards (Freeze & Cherry, 1979) from which water-level and quality data were generated.

Groundwater quality data was also generated from DWA boreholes with the objective of evaluating the groundwater resource in the study area in terms of its suitability for drinking, irrigation and industrial use. The parameters analysed for in the water samples were the major ions consisting of cations and anions as well selected suites of minor and trace elements. Water samples were collected during well development and pumping tests. In the case of rural areas, water samples were collected during the installation handpumps. Groundwater quality analyses were done according to the objective of the investigation (Kovalevsky *et al.*, 2004). The ultimate aim of the study was to ensure that groundwater resources were being developed sustainably for the socio-economic benefit of Zambia.

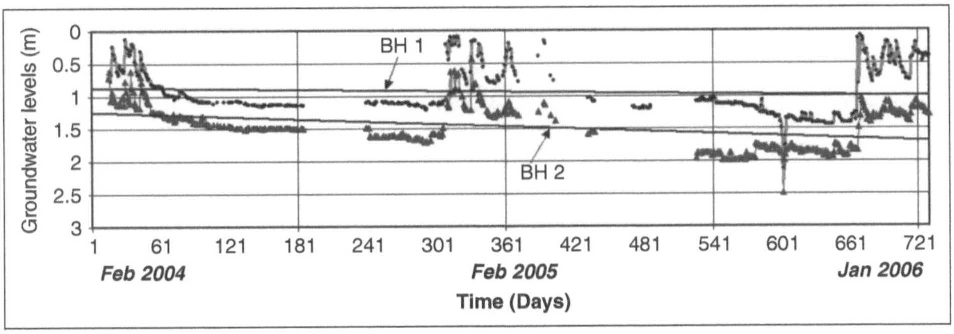

Figure 3. Groundwater levels observation borehole BH1 (schist) and BH2 (marble).

5.1 *Results*

From this research, it was established that there are 1800 private borehole records with the DWA with a total yield of 5760 l/s representing an average of 3.2 l/s per borehole. This translates to a total abstraction of 182 Mm^3 per year if boreholes are pumped continually. However, if the sources are only pumped for 8 hours a day, then the annual abstraction would be 60.5 Mm^3. According to borehole completion reports, boreholes are used for domestic, irrigation and industrial purposes. Actual abstraction is likely to be more than the values estimated here because of pumping from boreholes for which there are no records.

The water utility has 79 production boreholes in the study area with a total yield of 2269 l/s (This translates into a total annual groundwater abstraction of 71 Mm^3 assuming the boreholes are pumped 24 hours per day. The water utility also experiences water losses in the form of unaccounted-for-water from the city water supply network in excess of 50% of its daily total production of about 210,000 m^3 (Mtine &Chikama, *personal communication*).

These estimated abstraction above is already well in excess of the estimated annual recharge of 45 Mm^3/a (assuming 8% of the annual rainfall). Therefore, abstraction from the aquifer may appear to be sustained by groundwater from storage (especially in the marble), and seepage from septic tanks, irrigation water and unaccounted-for-water from the water supply network of the LWSC.

Based on borehole data, an attempt was made to quantify the volume of groundwater in storage. Taking the effective porosity of 2.5% (Jones & Topfer, 1972) for an area of 470 km^2 in the study area, minimum aquifer storage volume in the marble at an average effective aquifer depth of 50 m is estimated as 588 Mm^3. However, not all this water would be available for abstraction (Kovalevsky *et al.*, 2004)

Groundwater monitoring in the aquifer also corroborates the suggestion of groundwater mining (Figure 3). Water-levels show a continuous drop in both the schist and the marble, with the latter being affected the most as depicted by the trend lines for the respective aquifer lithologies. However, Figure 3 also indicates that a marked recovery of groundwater levels within a few days from the onset of every rainy season does occur, but with variations according to different rainfall amounts received annually (Table 4). Rapid aquifer drainage is also evident during and at the end of the rainy season, when levels drop rapidly.

Kriging of static water-levels for the DWA boreholes for the period February 2005 to April 2005 generated the plot for groundwater flow direction shown in Figure 4. From this

Table 4. Annual rainfall amounts at Sheki Sheki DWA Met. Station (after Siwale, 2006).

Hydrological Year	Rainfall amount (mm)	Rain days
2002/2003	845.7	74
2003/2004	749	59
2004/2005	891	51
2005/2006	977.2	61

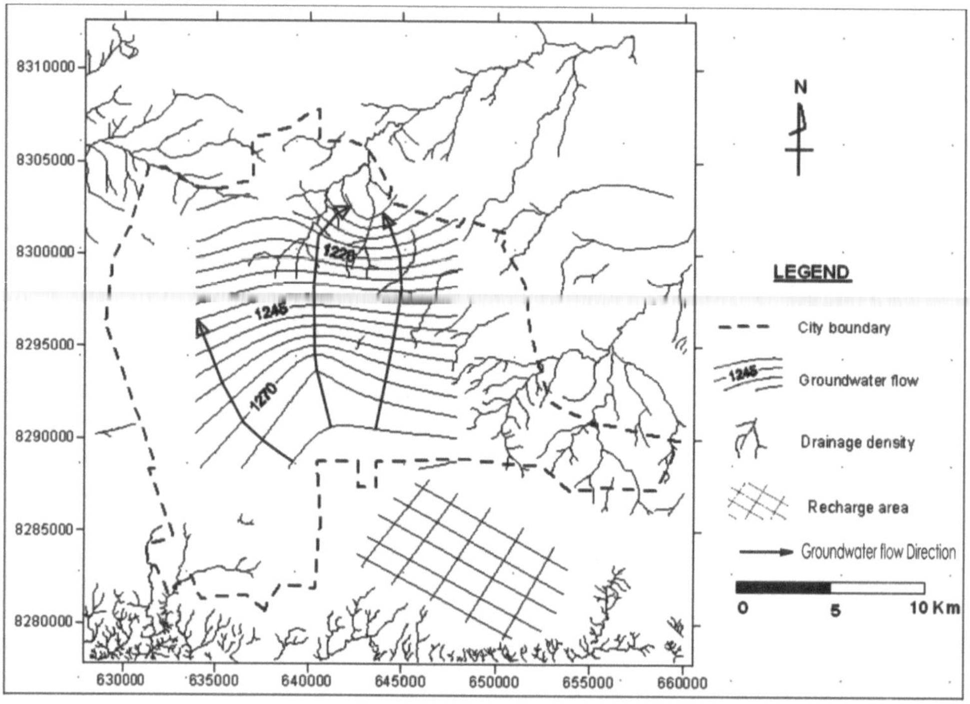

Figure 4. Map showing groundwater flow direction and the main recharge area during the rainy season: February to April 2005.

Figure, flow appears to originate from an area southeast of the city boundary, and strongly directed towards the northwest. Where the water-table intersects with the ground surface, springs have emanated into surface streams, which drain the Lusaka Plateau during periods, when infiltration of rainwater is exceeded by surface runoff.

A plot of rainfall data and groundwater level observation readings (Figure 5) illustrates direct recharge in both the schist and the marble and that these two aquifer lithologies respond similarly. These two observation boreholes are 200 m apart.

Direct recharge occurs, when water infiltrates into the ground and percolates to the groundwater store. Rapid groundwater level response to rainfall is attributed to recharge

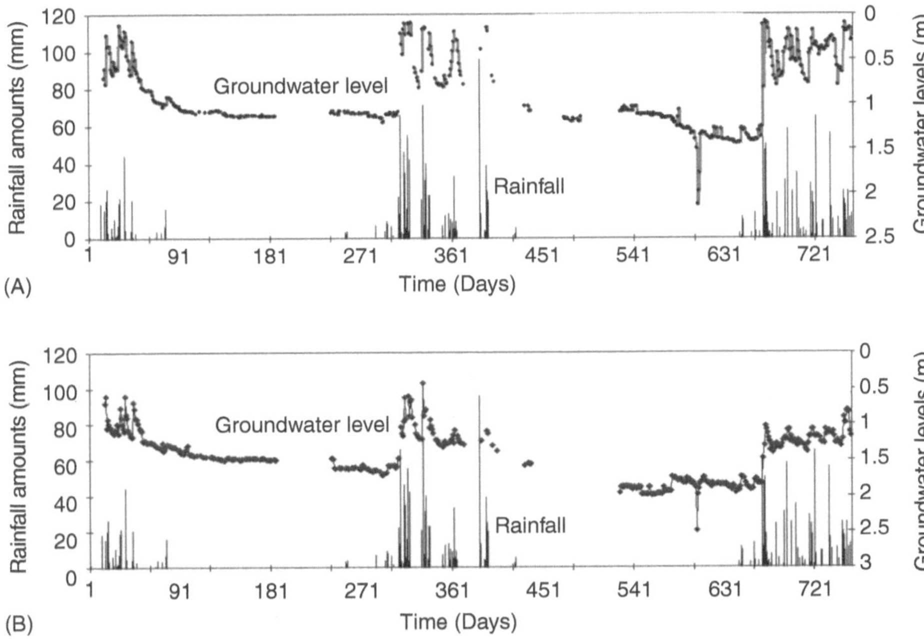

Figure 5. (A) Groundwater levels response to rainfall at BH1 in the schist: 2003/04 to 2005/06 rain season, and (B) Groundwater levels response to rainfall at BH2 in marble: 2003/04 to 2005/06 rain season.

that normally occurs via preferential paths such as extensional fault zones and sinkholes (Xu & Beekman, 2003). In the case of Lusaka Plateau, direct recharge is likely to occur through the laterite cover, fractures and sinkholes. This situation seems to support findings by Taque (1969), that soils on the Lusaka Dolomite have 76 mm / hour infiltration capacity. Rapid entry of water into the subsurface is also a clear indication that the aquifer has little protection from contamination.

The periods of groundwater level rise (December – March) and fall (March – December) are distinct, while small episodic level rises and falls in the water-table are also noticeable between March and December.

6 DISCUSSION

Estimating the water balance for the aquifer is difficult. A major obstacle is calculating actual groundwater abstraction given the current levels of available data and the legal instruments (WRAP, 2005). The current estimated abstraction from both the DWA recorded boreholes and the 79 LWSC boreholes may greatly underestimate the groundwater-mining problem in Lusaka urban aquifers, since they do not include boreholes drilled by the private contractors whose records are not with the DWA. At the same time, there is a great need to verify the correct recharge value for the carbonate karstic aquifers and schist aquifers on the Lusaka Plateau, given the evidence of direct recharge.

Considering that Lusaka urban area has many households using on-site sanitation such as septic tanks and pit latrines, and has a number of irrigation farming activities, it is suggested that the recharge in the study area includes inflows from these sources. In addition, unaccounted-for-water from the LWSC water supply network does constitute substantial recharge, which must be taken into consideration. Outflows from the aquifers consist of abstraction, natural aquifer drainage, evapotranspiration, drainage to streams and springs.

Therefore, in order to protect the identified main recharge area situated south-east of the city from anthropogenic activities, such as residential housing, farming etc, it is imperative to ensure that such activities are avoided in this area.

7 CONCLUSIONS

The study has provided evidence that it is imperative to ensure that hydrogeological data and groundwater level observation data, as well as groundwater quality data (though not presented in this paper) is collected as an integral part of all groundwater resources development projects and programmes. Estimates of groundwater abstraction and the effect of abstraction on Lusaka urban aquifers in relationship to annual recharge has been possible to establish due to groundwater data and information gathered during the study. When collection of such data and information is carried out regularly and is backed by some legal instruments, groundwater management is expected to improve.

During the study, a borehole completion report form has proved to be a practical and useful tool for collection of borehole drilling data especially with the inclusion of GPS coordinates. Since groundwater data and information provides the opportunity to make informed groundwater management decisions, data collection during geophysical surveys and borehole construction as well as submission of reports to relevant government institutions is inevitable and should be encouraged. While the cost involved may limit the extent of data collection, there is no excuse to neglect collection of borehole drilling data during borehole construction when extra costs would only arise towards the purchase of a GPS unit, whose cost is negligible. Groundwater level observation data serves to give information on groundwater quantity as and when abstraction is done. Hence, it qualifies to be collected on a regular basis as well in order to improve our understanding on the effects of progressive groundwater abstraction.

ACKNOWLEDGEMENTS

Data and information used were acquired through groundwater development programmes and projects undertaken in the Ministry of Energy and Water Development by the DWA. The authors would like to thank Mr. Adam Hussen, the Director of the DWA for supporting the research work. Thanks to the Lusaka Water and Sewerage Company for the data, and Deutscher Akademischer Austausch Dienst (DAAD) for financially supporting the short-term research visit to Ruhr University Bochum – Germany, where most of the analyses were carried out. Recognition is also given to the DWA staff and cooperating partners that participated in acquisition of data used in this paper.

REFERENCES

Burdon, D. J. & Papakis, N. 1963. *Handbook of Karst Hydrogeology*, F.A.O., Rome, Italy.

Cairney, T. 1967. The *Geology of the Leopards Hill area: Explanation of Degree Sheet 1528, SE Quarter*. Geological Survey Department Report, **21**, Lusaka, Zambia.

Drysdall, A. R. & Smith, A. G. 1960. *Recumbent folding in the Lusaka Dolomite*. Geological Survey Department Northern Rhodesia, Occasional Paper, **22**, 43–45.

Freeze, R. A. & Cherry, J. A. 1979. *Groundwater*. Prentice-Hall, Inc. Englewood Cliffs, New Jersey USA.

JICA (Japan International Cooperation Agency) 1995. *The Study on the National Water Resources Master Plan in The Republic of Zambia*. Ministry of Energy and Water Development, YEC.

Jones, M. J. & Topfer, K. D. 1972. *The Groundwater Resources of Kabwe Area with Geophysical Notes*. Department of Water Affairs, Lusaka, Zambia.

Kovalevsky, V. S., Kruseman, G. P. & Ruston, K. R. 2004. *Groundwater studies, An International Guide for Hydrogeological Investigations*. UNESCO IHP-VI, Series on Groundwater, **3**, UNESCO, Paris.

Lambert, H. H. J. 1965. *The Groundwater Resources of Zamb*ia. Department of Water Affairs, Lusaka, Zambia.

Matheson, G. D. & Newman, D. 1966. *Geology and structure of the Lusaka area*. Record of the Geological Survey. Zambia, **10**, 10–19.

Mpamba, N. H. 2006. *Comparative Analytical Model for Groundwater Monitoring in the Urban and Rural areas of Zambia. – Groundwater Resources Data and Information* PhD, School of Mines, Department of Geology, The University of Zambia, Lusaka, Zambia.

Nkhuwa, D. C. W. 1996. *Is groundwater management still an achievable task in the Lusaka aquifer*? Proceedings of the XXIX. Conference of the IAH, Bratislava, Slovak Republic, 209–213.

Nkhuwa, D. C. W. 2003. *Hydrogeology and Engineering Geological problems of urban development over karstified marble in Lusaka Zambia*. RMZ Materials and Geoenvironment, **50**, 273–276.

Simpson, J. G., Drysdall, A. R. & Lambert, H. H. J. 1963. *The Geology and Groundwater Resources of the Lusaka Area*. Explanation of degree sheet 1528, NW quarter. Department of Geological Survey Report, **16,** Lusaka, Northern Rhodesia.

Siwale, C. 2006. *Annual rainfall records for Sheki – Sheki Met. Station from 2002/03 to 2005/06 rain seasons*, Department of Water Affairs, Lusaka, Zambia.

Taque, M. 1969. *Artificial recharge at No. 1 borehole area, Lusak*a. WAD, Lusaka, Northern Rhodesia.

Von Hoyer, Kohler, M. & Schmidt, G. 1978. *Groundwater and Management Studies for Lusaka Water Supply*. Part 1 Groundwater study, Hanover, Germany.

WRAP (Water Resources Action Programme) 2003. *Report on the National Water Resources Action Programme Consultative Forum: The Proposed Institutional and Legal Framework for the Use, Development and Management of Water Resources in Zambia*. Ministry of Energy and Water Development, Lusaka, Zambia.

WRAP (Water Resources Action Programme) 2005. *Zambia Water resources Management Sector Report for 2004*. Ministry of Energy and Water Development, Lusaka, Zambia.

Xu, Y. & Beekman, H. E. (eds.) 2003. *Groundwater recharge estimation in Southern Africa*. UNESCO IHP Series, **64**, UNESCO, Paris.

Zulu, J. D. S. & Nyambe, I. A. 2004. *Karstified Lusaka Marble aquifer, Zambia: Implications on Groundwater Contamination and community Management in Kanyama Settlement*. In: abstract Volume, International Geological Congress, Florence, Italy.

CHAPTER 17

Impact of the Casablanca municipal landfill on groundwater resources (Morocco)

A. Fekri, A. Benbouziane & C. Marrakchi
Université Hassan II of Mohammedia Faculté des Sciences Ben M'sik, Casablana, Morocco

M. Wahbi
Université Cadi Ayyad, Faculté des Sciences etTechniques de Gueliz, Marrakech, Morocco

ABSTRACT: Since 1986 the municipal solid wastes produced by the city of Casablanca (Morocco) have been stockpiled in a landfill located in old quarries 10 km from the city's periphery. The base of the landfill comprises fractured quartzite, which were not sealed before dumping the waste. The aquifer is therefore at high risk of contamination from leachate. In 1990, a study detected the beginning of groundwater pollution downstream from the landfill. During the present study, two hydrochemical sampling campaigns were carried out, the first in high groundwater levels, the second when groundwater levels were low. Groundwater information was collected from wells downstream from the landfill. The results showed an important disparity concerning the measured parameters. In addition, water from a certain number of wells no longer meets drinking water supply standards or irrigation standards. A principal component analysis was helpful in distinguishing the contaminated wells and characterizing the pollution plume. The comparison of the latest results with the 1990 study show an advance of the pollution plume towards the city, along a series of faults. The progression of the plume is partly controlled by pumping wells.

1 INTRODUCTION

A study in 1990 detected the beginning of groundwater contamination downstream from the Mediouna landfill, Morocco, only four years after of its start-up in 1986. In the absence of laws governing this branch of industry, and the increasing quantities of waste being produced in the area, no measures were taken to solve this problem.

The aim of this study was: first, to characterize the groundwater quality downstream from the landfill and; second, to determine the limits of the contaminated area, in order to analyze the progression of the leachate plume. To achieve these goals, an analysis of the geological and hydrogeological contexts was carried out, as well as statistical analyses of the hydrochemical data, obtained during two sampling campaigns carried out during 2002.

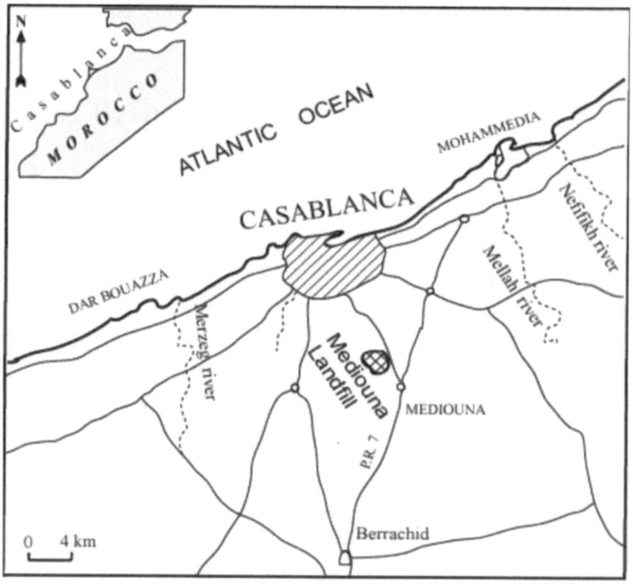

Figure 1. Location of Mediouna landfill.

2 PHYSICAL SETTING

2.1 *Location*

The choice of the landfill site was due to short-sighted economic reasons. This site is easily accessible and has a large volume. Located 10 km southeast of Casablanca (Figure 1), it is composed of 13 quarries, which add up a volume of 3 Mm^3 over 78 hectares, of which 60 are assigned to the landfill. The main road (PR7) passes nearby to connect Casablanca to Marrakech.

Since 1986 the Urban Community of Casablanca began to use the quarries for landfill, without sealing off the substratum or installing any leachate and gas collection system. The landfill consisted of a simple deposit in easily accessible zones, without any cover of the urban waste by soil. This management method lasted until 1995, the date when the Urban Community entrusted a private company to manage the landfill. The main observed change was the waste beginning to be covered with soil and the making of a soil barrier along the main peripheral way, to contain the site. In 1999, a partial wall parallel to the main road was built to hide the landfill. The quantity of waste produced by Casablanca was in the order of 1850 ton/day in 1989; 2680 ton/day in 1996 (Urban community of Casablanca, unpublished data, 1996), and 3800 ton/day in 2003.

2.2 *Climate*

The climate is semi-arid, with a wet season characterized by two of precipitation maximums: the main during December, the secondary during March. The annual average rainfall is approximately 450 mm. The minimum average temperature varies between 7 and 9°C from

December to February, while the maximum average temperature oscillates between 26 and 28°C for September and August. The direction of the dominant winds from November to February is south and southwest and, during the dry season, north or northwest. Wind speed lies between 3.6 and 18 km/h at mid-day and the mode is 6 km/h. An estimation of effective rain by the Turc monthly method for 25 years, gave a value of 60 mm/a. (General Direction Hydraulics, unpublished data, 1991).

3 HYDROGEOLOGICAL SETTING

The primary formations underlying the landfill are Cambrian and Ordovician marine sediments affected by the Hercynian Orogeny (Destombes & Jeannette, 1956). They were compressed and their transformation gave rise to Acadian green schist and quartzite (Ruhard, 1975). They were subsequently eroded to form an extensive peneplain. The Pliocene higher tertiary sector and the quaternary are also marine formations with lumachellic or conglomerate facies covered by calcarenites. They were primarily formed by gastropod shells or mollusk remains cemented by a slightly argillaceous limestone. This formation was exploited in the quarries to provide road construction materials.

Hydrogeological studies show that water circulation is primarily through the quartzite, due to the presence of fractures. The shallow calcarenites, though permeable, are generally dry. Depth to the groundwater across the area varies between 4 and 21 m. Fracturing was identified from locating lineaments on aerial photographs (1987) at 1:20,000 scale. The results show the predominance of two families of discontinuities, one ranging between 20 and 40°N and the other ranging between 120°N and 140°N (Figure 2). This result is consistent with other studies: in the coastal area to which the study sector belongs, the deformation affecting the primary formations is expressed by a system of fault dextral setback with NNE-SSW and E-W directions which are locally associated to a system of fault senestre setback with WNW-ESE direction (Laamrani, 1993).

Piezometric data, established from data taken in 2002, are presented in Figure 3. Two distinct trends can be identified: one is located in the west of the area and presents a strong hydraulic gradient of about 0.4%; the other is located in the east, with a weak hydraulic gradient of about 0.01%, probably inferring higher hydraulic conductivity in this sector.

4 METHODS

The concern about the water quality in this aquifer is due to the presence of rural wells downstream of the landfill, exploited not only for irrigation purposes, but also for local population drinking water supply. The majority of these points are located in topographically lower meadows associated with major fractures. To determine the variation of the groundwater levels during a hydrological cycle; two sampling campaigns were carried out, one in April and the other in September 2002, corresponding respectively to high and low groundwater levels. A chart of the increase in electric conductivity (EC) between 1990 and 2002 is shown in Figure 4. The zone of interest corresponds to the plume limited by a null variation of the EC (see Christensen *et al.*, 1994). Further investigations were carried out in this zone: temperature, pH, EC, and dissolved oxygen were measured in the field and

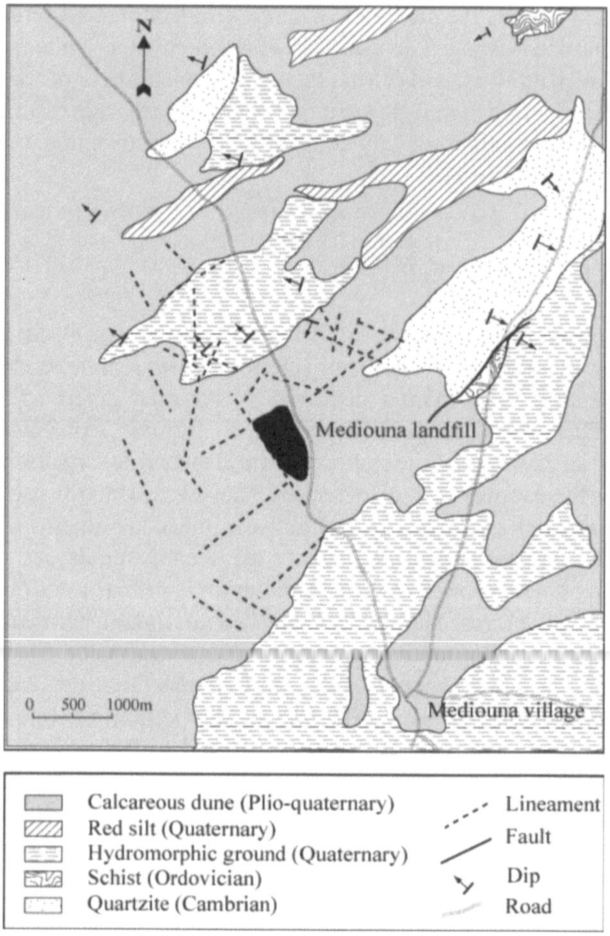

Figure 2. Geological map of the landfill's setting.

samples were taken and sent to the laboratory, according to the methods and standards of the French association of standardization (AFNOR, 1989).

5 RESULTS AND DISCUSSION

The results of the analyses are grouped in Tables 1 and 2. The water pH is neutral or slightly acid for a certain number of points with very high EC values. There is a high variation in EC and major ions as well as an increase in nitrate concentrations, which largely exceed the standards of accepted limits for drinking water on the wells. Al and Cd also exceed drinking water standards (see Tables 1 and 2 and also Chofqi *et al.*, 2003). There is a marked increase in concentration of most parameters in September (when groundwater levels are low, compared to those measured in April. The measured parameters are comparable to those observed in wells monitoring other landfills; e.g. the Marrakech landfill (Hakkou, 2001) and Rabat landfill (Khamlichi *et al.*, 1997).

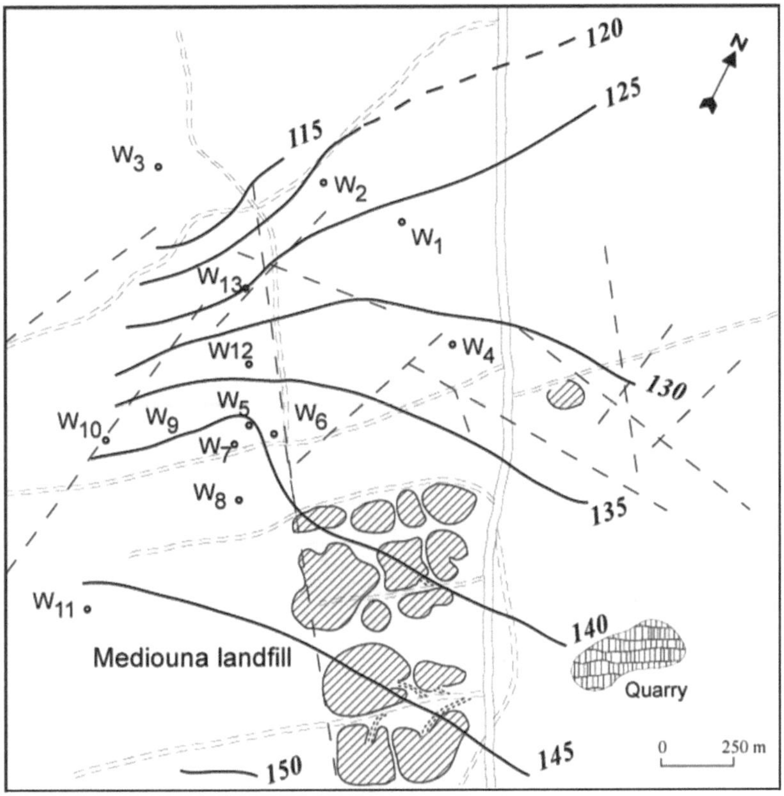

Figure 3. Groundwater levels and location of piezometers in the study area.

5.1 *Principal component analysis*

To further investigate the impact of the leachate landfill on groundwaters, principal component analysis (PCA) was carried out. This analysis makes it possible to visualize the existing relations between the measured parameters and also to group similar wells. The contribution to total variability by the first three axes resulting from the PCA, realized on the values of the twelve parameters obtained for the thirteen wells, is 84%, including 67% for axis 1; 11% for axis 2; 8% for axis 3. Axis 1 is primarily characterized in the positive direction by the major elements, chemical oxygen demand (COD) and EC, which are strongly correlated (Figure 5). Axis 2 is mainly given in the positive direction by the Cr content and Axis 3 is especially marked in the positive direction by the presence of Al. Only two axes were selected since they present 78% of the original variance value.

The projection of the wells on the factorial of axes 1 and 2 shows three groups (see Figure 5):

Group 1: located in the positive side of axis 1, constituted by the wells W4, W5, W6 and W7. It is characterized, at the same time, by a strong content of organic matter expressed in high COD values, and the lowest DO values, and by a strong mineralization expressed by high EC values, which depends on the concentration of the groundwater major elements characteristics. All these points are located near by the landfill zone.

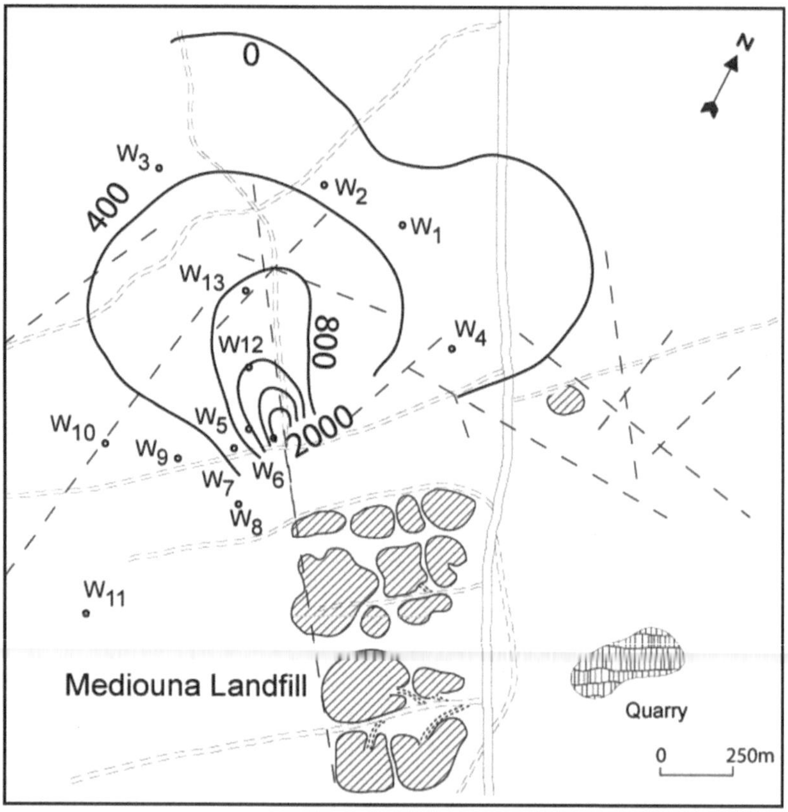

Figure 4. Map electric conductivity (μs/cm) isovariation between 1990 and 2002 showing the increase in electrical conductivity over the period.

Group 2: located in the negative side of axis 1. This group is constituted by the wells W3, W9, W10 and W11, which are characterized by the weakest mineralization and a total absence of organic matter. The levels are similar to those observed in the wells of the non polluted zone during 1990's campaign (Elghachtoul *et al.*, 1992). The location of these wells is shown in Figure 4.

Group 3: located in median position on axis 1, comprising wells W1, W2, W8, W12 and W13. It is characterized by higher EC values than those of group 2 and lower than those of group 1, which starts to be affected by an organic pollution, expressed by low COD values.

Principal Component Analysis was also undertaken for the samples taken in September 2002 representing low groundwater levels (Figure 6). The form is similar to that observed for the high waters data except for the W4 well, which migrated from the first to the third group. The main changes observed in the water of the W4 well are a reduction in EC and organic matter's content. This was not due to the dry season effects, but the consequence of deepening process that this well underwent between the two campaigns. This is normal insofar as pollution decreases according to the depth downstream from landfill (Barker *et al.*, 1986).

Table 1. Physical and chemical characteristics of the groundwater downstream of the Mediouna landfill sampled in April 2002 (mg/l unless otherwise stated).

Wells	T(°C)	pH	E.C (μS cm^{-1})	DO	CD	NO_3	Cl	SO_4	HCO_3	Na	K	Ca	Mg	Al	Cr	Fe	Zn	Pb	Cu	Cd	Mn
W1	20.5	7.58	840	1.5	0	74.35	139.35	29.71	194.96	84.42	4.47	105.4	10.21	0.74	0.09	0.06	0.02	0.05	0.05	0.05	<0.02
W2	20.5	7.42	1060	6.2	48	112.38	183	44.77	196.1	118.71	7.61	132.6	7.78	0.47	0.07	0.05	0.04	0.03	0.09	0.01	0.02
W3	21	7.43	890	6	0	118.3	110.61	25.56	170.8	47.92	2.17	110.8	12.15	0.58	0.04	0.13	0.05	0.06	0.07	0.07	<0.02
W4	18.3	7.01	2170	2.9	190	122.95	695.79	47.31	152.39	174.57	28.1	186.78	31.10	0.57	0.13	0.11	0.04	0.02	0.08	0.00	1.83
W5	19.3	7.09	4000	2.6	202	101.4	992	47.47	576.51	399.51	26.3	432	63.91	0.94	0.02	0.24	0.05	0.05	0.09	0.05	0.38
W6	20.1	7.18	5800	2.1	259	134.05	1237.7	75.08	810.94	585.27	28.4	631.61	94.71	0.73	0.11	0.16	0.09	0.10	0.10	0.06	2.60
W7	19.4	7.26	3630	1.75	125	209.1	781.87	27.22	487.57	320.17	17.1	379.47	73.71	0.25	0.07	0.11	0.04	0.06	0.06	0.04	<0.02
W8	21	7.23	1080	4.5	38	131.2	221.19	17.91	102.03	104.64	4.26	91.8	10.57	0.52	0.05	0.09	0.06	0.05	0.08	0.05	<0.02
W9	20	7.72	710	5.75	0	96.8	148.26	5.09	57.06	45.49	2.28	73.6	4.62	0.61	0.15	0.12	0.03	0.01	0.07	0.05	<0.02
W10	20.2	7.88	635	6.9	0	122.2	132.59	5.75	65.51	74.26	1.61	53.2	5.83	0.97	0.03	0.12	0.04	0.04	0.08	0.03	<0.02
W11	19.4	7.61	1246	5.9	0	194.01	229.78	4.65	90.81	30.04	2.16	218.96	8.69	0.94	0.09	0.23	0.13	0.01	0.08	0.03	<0.02
W12	18.4	7.19	2050	2.6	91	119.48	392.42	22.38	117.07	109.2	4.66	180.8	24.42	0.76	0.09	0.19	0.06	0.01	0.07	0.04	<0.02
W13	18.7	7.18	2180	2.1	32	312.01	314.8	39.88	72.12	95.73	11	224.4	16.89	0.87	0.09	0.20	0.04	0.05	0.08	0.05	<0.02

Table 2. Physical and chemical characteristics of the groundwater downstream of the Mediouna landfill sampled in September 2002 (mg/l unless otherwise stated).

Wells	T(°C)	pH	EC ($\mu S\ cm^{-1}$)	DO	COD	NO_3	Cl	SO_4	HCO_3	Na	K	Ca	Mg	Al	Cr	Fe	Zn	Pb	Cu	Cd	Mn
W1	21.9	7.5	1943	1.07	0	77.97	381.5	74.28	498.22	133.17	3.32	269.60	32.49	0.99	0.10	0.23	0.03	0.04	0.06	0.05	<0.02
W2	21.3	7.51	2180	0.77	32	274.54	348.5	179.3	384.69	183.88	5.82	279.93	31.05	0.81	0.16	0.17	0.05	0.03	0.07	0.01	0.37
W3	21.3	7.39	1900	4.02	0	223.03	353.3	97.61	287.87	110.88	3.16	241.83	26.86	0.87	0.07	0.26	0.07	0.06	0.09	0.07	<0.02
W4	22.1	6.81	1900	2.85	84	64.65	468.4	16.41	368.39	227.33	[illegible]4.34	163.54	27.23	0.45	0.03	0.21	0.07	0.07	0.11	0.06	<0.02
W5	21.7	6.76	5200	0.81	192	199.36	1422	109.3	762.63	523.46	[illegible]0.62	566.27	84.93	0.98	0.19	0.23	0.07	0.01	0.10	0.04	2.46
W6	21.9	6.73	7450	0.95	144	269.55	1937	110.46	907.42	820.66	[illegible]0.45	726.26	114.2	0.93	0.15	0.23	0.05	0.05	0.06	0.07	3.50
W7	22.1	6.91	4750	1.25	154	169.36	1208	92.5	651.48	445.06	[illegible]9.4	487.06	72.15	0.78	0.12	0.20	0.02	0.12	0.06	0.05	<0.02
W8	22.1	6.93	3200	1.95	0	150.49	776.7	75.51	488.27	320.84	5.37	336.04	39.12	0.72	0.05	0.22	0.03	0.15	0.10	0.04	<0.02
W9	21.7	7.06	1900	3.3	0	253.98	337.6	37.62	364.3	100.41	2.92	278.33	18.81	0.81	0.12	0.22	0.12	0.04	0.08	0.04	<0.02
W10	21.5	7.19	1800	4.4	0	240.13	235.7	72.47	482.81	168.09	3.94	203.78	22.80	1.02	0.04	0.15	0.05	0.05	0.07	0.03	<0.02
W11	21.3	7.08	2050	3.88	0	443.17	301.3	108.2	252.08	72.51	2.41	360.25	14.29	1.35	0.07	0.19	0.04	0.03	0.10	0.07	<0.02
W12	21.7	7.04	3450	2.23	72	60.77	942.5	74.37	407.04	274.1	4.11	410.32	55.74	1.09	0.07	0.28	0.03	0.22	0.06	0.08	<0.02
W13	20.6	7.04	3500	2.77	38.4	602.34	720.5	88.95	402.5	171.07	5.55	560.33	42.50	0.86	0.14	0.30	0.05	0.03	0.11	0.04	0.05

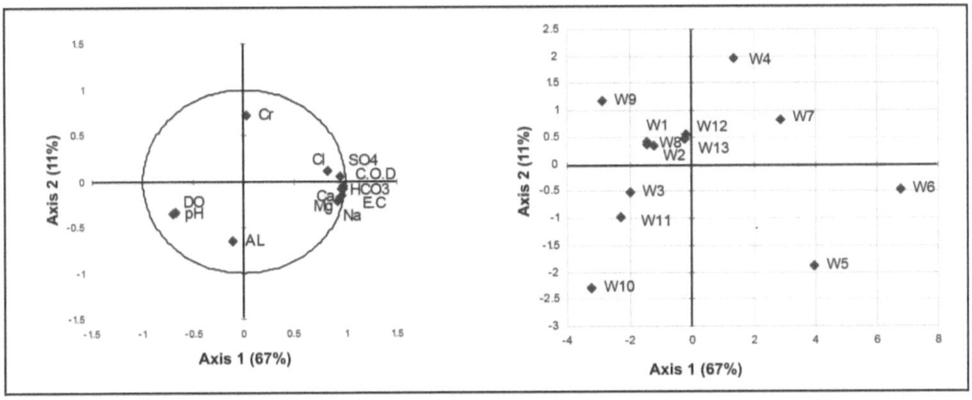

Figure 5. The behaviour of the variables and the samples in PCA (April 2002).

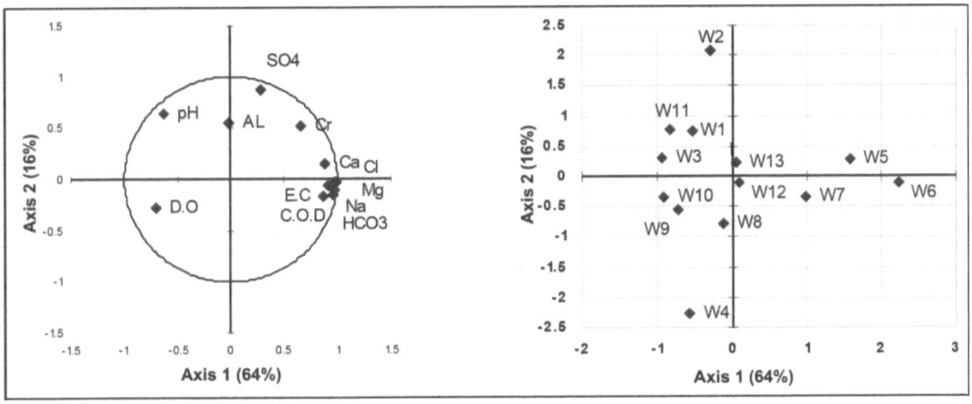

Figure 6. The behaviour of the variables and the samples in PCA (September 2002).

The results of the statistical analyses are interpreted and displayed on Figure 7. The pollution plume, which was highly confined in 1990, has strongly progressed. This progression is controlled by fracturing. In fact, the polluted zone takes the shape of an ellipse around an axis materialized by the F1 fault. The fault acts like a drain that leads the leachate from the landfill downstream. This fault is at the origin of the accentuated pollution of the near wells W5, W6 and W7, where contamination is easily detectable by a simple chemical analysis and the brown water colour. The F2 fault crosses the F1 fault and drains a part of the leachate, which pollutes W4 well. The impact of the leachate drained by this fault weakens as it moves downstream, as indicated by the decreasing EC, and reducing organic matter. Within the downstream of the W13 well, a NNE-SSW fault F3 crosses the F1 fault and deviates a part of the polluting load to degrade the water quality on W2 well, whereas W1 well, closely located to landfill, seems untouched.

Another factor is important in the movement of the pollution. Pumping wells along the F1 fault form a hydraulic barrier against the fast contamination spread. However, with the increased quantity of stored waste, and consequently the growing volumes of leachate,

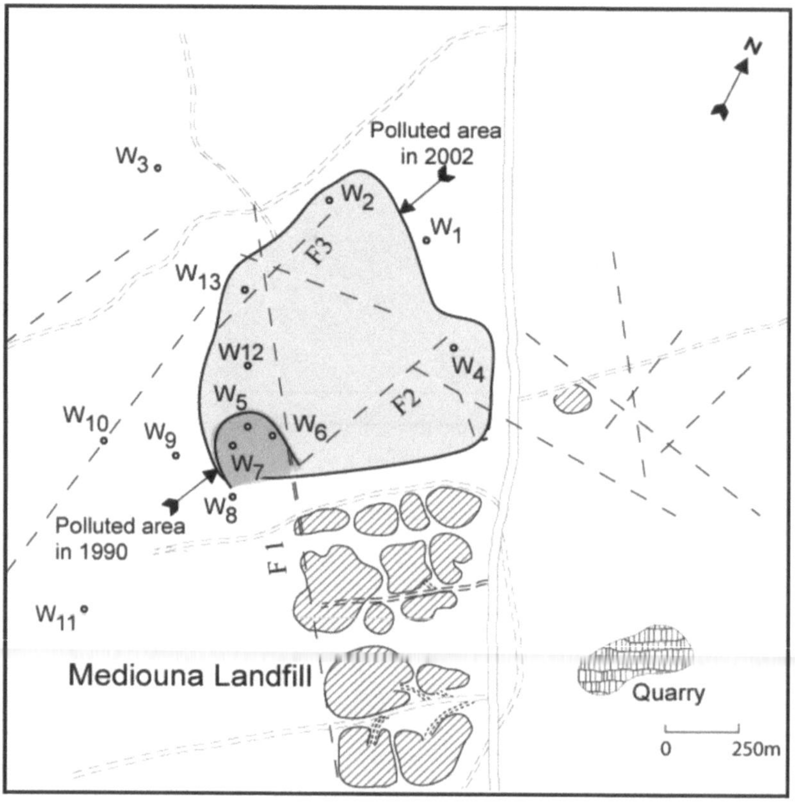

Figure 7. Geographic distribution of the two well groups identified by PCA.

the pumped volumes could not match the leachate production, facilitating the pollution progression towards downstream.

6 CONCLUSIONS AND RECOMMENDATIONS

The Casablanca landfill was established in 1986 in old quarries whose substratum was not sealed, with the result that a plume of leachate has polluted the local aquifer since 1990. The current study highlights the progressive movement of the pollution plume towards the Casablanca town, and the permanent loss of groundwater quality, since the majority of water samples taken from wells are of poor quality.

The analysis of the pollution from two hydrochemical campaigns (at high and low groundwater levels) indicates two main areas of contamination. The first is close to the landfill, where contamination is highly accentuated and accompanied by a strong odour and brown colour of the water extracted from wells. The second, located downstream, is characterized by a primarily inorganic pollution with low organic matter content. The observed anisotropy on the groundwater pollution is due to two factors: direct structural control, where the faults favour pollution progression; and also to the presence of the pumping wells, which constitutes a curtain against the propagation of pollution downstream the landfill. Knowledge of the groundwater quality beyond the landfill could be improved. Recommendations are made to the basin's agency to install a monitoring system, boreholes along the identified faults, in order to act as pollution collector wells. This step will bring an additional accuracy degree, making it possible to choose the best strategy to solve this problem.

REFERENCES

AFNOR, 1989. *Recueil de normes françaises: eau, méthodes d'essai.* [Collection of French standards: water, testing methods]. Deuxième édition, Paris, France.

Barker, J. F., Tessmann, J. S., Plotz, P. E. & Reinhard, M. 1986. *The organic geochemistry of sanitary landfill leachate plume.* Contaminant hydrology, **1**, 171–189.

Chofqi, A., Younsi, A., Lhadi, E., Mania, J., Mudry, J. & Veron, A. 2003. *Pollution d'une nappe phréatique par les métaux lourds du lixiviat de décharge d'El Jadida, Maroc* [Pollution of a groundwater by heavy metals of landfill leachate, Jadida Morocco]. Proc Deuxièmes Journées des Géosc. de l'Environnement. Univers. Ibn Tofail, Kénitra, Maroc.

Christensen, T. H., Kjeldsen, P., Albechten, H. J., Heron, G., Neilson, P. H., Bjerg, P. L. & Hom, P. E. 1994. *Attenuation of landfill leachate polluants in aquifers.* Crit Revue Environ Sci Techno, **24**, 119–202.

Destombes, J. & Jeannette, A. (1956) *Etude géotechnique de la région de Casablanca* [Geotechnical study of the area of Casablanca]. Notes et Mémoires du Service Géologique du Maroc, **130**.

Elghachtoul, Y., Berrada, L. & Lakranbi, S. 1992. *Impact des déchets ménagers sur la qualité de l'eau, cas du grand Casablanca* [Impact of domestic waste on the quality of water, casaof large Casablanca]. Eau et développement, **41**, 54–60.

Hakkou, R. 2001. *la décharge publique de Marrakech: caractérisation des leachate, étude de leur impact sur les ressources en eau et essais de traitement* [Marrakech landfill: caracterization of the leachate, study of their impact on the water resources and tests of traitement]. Thèse de doctorat d'état, Faculté des Sciences et Techniques, Marrakech, Morocco.

Khamlichi, M. A., Lakranbi, S., Kabbaj, M., Jabry, E. & Kouhen, M. 1997. *Etude d'impact de la décharge publique d'Akrach, Rabat, Maroc* [Study of the impact of Akrach lanfill, Rabat, Morocco]. Revue marocaine de génie civil, **68**, 17–31.

Laamrani, A. 1993. *Relations déformations – déplacements le long de failles hercynienne: système de Bouznika et système du Cherrat-Benslimane et du cherrat-Yquem* [Relation deformation – displacement along hercyniennes faults : system of Bouznika and system of Cherrat – Benslimane and the Cherrat – Yquem]. Thèse de 3ème Cycle, Univ. MohammedV, Rabat, Maroc.

Ruhard, J. P. 1975. *Chaouia et plaine de Berrechid. Ressources en eau du Maroc* [Chaouia and plain of Berrechid]. Notes et Mémoires du Service Géologique du Maroc, **231**.

Groundwater Chemistry and Recharge

CHAPTER 18

Groundwater in Africa – palaeowater, climate change and modern recharge

W.M. Edmunds

Oxford Centre for Water Research, Oxford University Centre for the Environment, Oxford University, UK

ABSTRACT: Groundwaters of known age contained in major aquifer systems in the African sedimentary basins are of specific value in determining low resolution (±1000 a) characteristics of past climates, specifically palaeotemperature, air mass origins, humid/arid transitions and rainfall intensity. Results from both northern and southern Africa indicate the predominance of a westerly Atlantic air flow during the Late Pleistocene. Greater aridity during the Last Glacial Maximun (LGM) over most of northern Africa is recorded by the absence of dated groundwaters. An intensification of the African monsoon during the Early Holocene is apparent from isotopically light groundwaters found especially in Sudan. Maximum cooling around the LGM of 5–7°C is recorded in the noble gas recharge temperatures from Africa. Modern recharge can be readily identified from the chemical and isotopic signatures (Cl, $\delta^{18}O$ and ^{3}H) in the unsaturated zone and in shallow groundwaters. The results indicate the non-renewability of many groundwater sources now being exploited across the arid and semi-arid regions of Africa.

1 INTRODUCTION

The semi-arid and arid regions of Africa contain a number of paradoxes regarding water resources and their availability. They contain some of the largest fresh water reserves in the world, yet these are located in aquifers in remote areas or at depths that do not allow the water to be exploited economically. The arid and semi-arid regions currently overlying these basins, such as the Sahara, Sahel and the Kalahari in recent times still sustain significant populations dependent on rain-fed agriculture and the accessible shallow groundwater. Prior to the onset of the present day arid conditions around 4500 years BP the present desert regions were much wetter and sustained populations with access to surface waters occurring as large lakes and perennial rivers.

The same regions are now under renewed threat. Signs of modern climate change have probably been seen already in the prolonged droughts in the Sahel region since the early 1970s which led to population migration, desertification of marginal areas and famine. Yet it is in areas of Sub-Saharan Africa including the Sahel, that some of the most rapid population growth is predicted to occur in the foreseeable future. Compounding the problem are predictions that climate change in coming decades will have the severest impacts on drier regions of Africa. The extreme event frequency and magnitude, a feature of the monsoon affected semi-arid regions, will increase even with a small increase in temperature and will become greater at higher temperatures (IPCC, 2004). The combined impact of rapid population growth and climate change are certain to have a disproportionate effect on Africa (Arnell, 2004), compounded by the large uncertainties due to the inadequate data for the semi-arid tropics. All these factors have severe consequences for water availability.

In present times groundwater forms the primary drinking water source in the arid and semi-arid regions of Africa, since river flows are unreliable and the few large freshwater lakes are either ephemeral (e.g. Lake Chad) or no longer exist. Water exploitation has changed dramatically over the past few decades by the introduction of advanced drilling technology for groundwater (often available alongside oil exploitation), as well as the introduction of mechanised pumps. This has made attractive the development of readily available shallow groundwater resources as well as introducing new artesian resources to areas located in traditional discharge areas such as the oases, with significant drawdowns of the water tables and consequent loss and damage of ecosystems.

In many parts of Africa, therefore, an illusion and raised expectations have been created in a generation or so, of the availability of plentiful groundwater; yet in practice, falling water levels testify to an over-development. This in turn implies an inadequate scientific understanding of the resource, the failure of scientists to convey their message, or lack of societal and political awareness to act on the scientific evidence pointing to the lack of sustainability of the resources and their management. In addition to the issues related to quantity, quality issues, especially in shallow environments, exacerbate the situation. This is because the natural groundwater regime, established over long time scales, has developed chemical (and age) stratification in response to recharge over a range of climatic regimes and geological controls. Drilling cuts through the natural quality layering and abstraction from boreholes may lead to deterioration in water quality with time as water is drawn either from low transmissivity strata, or is drawn down from the near-surface where saline waters may have accumulated.

The main objective of this paper is to draw attention to the non-renewability of the groundwater resources in much of Africa, focusing especially on the Sahara – Sahel region. In this area water being used for development is mainly palaeowater, recharged over a geological timescale with relatively small quantities of modern water available which in turn allow only small scale activities. In this paper the isotopic and chemical signatures of the water contained in African groundwaters are reviewed since they provide clear evidence of the boundaries between the modern (renewable) and palaeowater (non-renewable) resources.

2 GROUNDWATER RECORDS IN AFRICAN SEDIMENTARY BASINS

Africa is underlain by crystalline basement rocks of the African Shield upon which sedimentary basins of ages ranging from the Cambrian to the Quaternary were superimposed. The tectonic setting of Africa has favoured the creation of large sedimentary basins overlying the Shield, as well as the northern Sahara Platform, containing continental sediments, whilst marine sediments are mainly restricted to the present coastal areas. The separation of the African from the South American plate led to the creation of sedimentary basins in coastal areas of western and southern Africa. In northern Africa, sedimentary basin formation has been controlled by block faulting, with a general NW-SE orientation, from the Paleozoic to the Mesozoic, as a result of alternating regional compression and tension of the basement, combined with a clockwise rotation of the African continent (Klitsch, 1970). Separation of the Sahara Platform from the Tethys geosynclinal area along the Mediterranean axis took place during the late Carboniferous or Permian lasting until the Jurassic culminating in the Tertiary, the last major basin formation coinciding with the formation of the Atlas Mts. The separation of Africa from Asia took place during the Tertiary lasting until the present day and evidence of volcanism from this period associated with older structural elements is found in several parts of northern and eastern Africa. In southern Africa the formation of sedimentary basins such as the Karoo took place in post-Jurassic times in association with the separation of the Gondwana and Antarctic plates.

The geology of northern Africa is shown as schematic cross sections in Figures 1a and 1b which distinguish the large sedimentary basins in the Sahara and North Africa from the areas to the south where sedimentary thicknesses are much lower. In the Sahara region several overstepping basins separated by unconformities are found mainly containing continental sediments, although the marine transgressions of the Tertiary era produced interbedded marine and continental sediments. South of the Sahara, marine sediments are restricted to the immediate coastal region.

The largest basins of northern Africa extend to depths in excess of 3 km and these may contain fresh groundwater related to the very pure siliceous lithology and the continuity of groundwater flow over a geological timescale (Pallas, 1980). Groundwater flow is likely

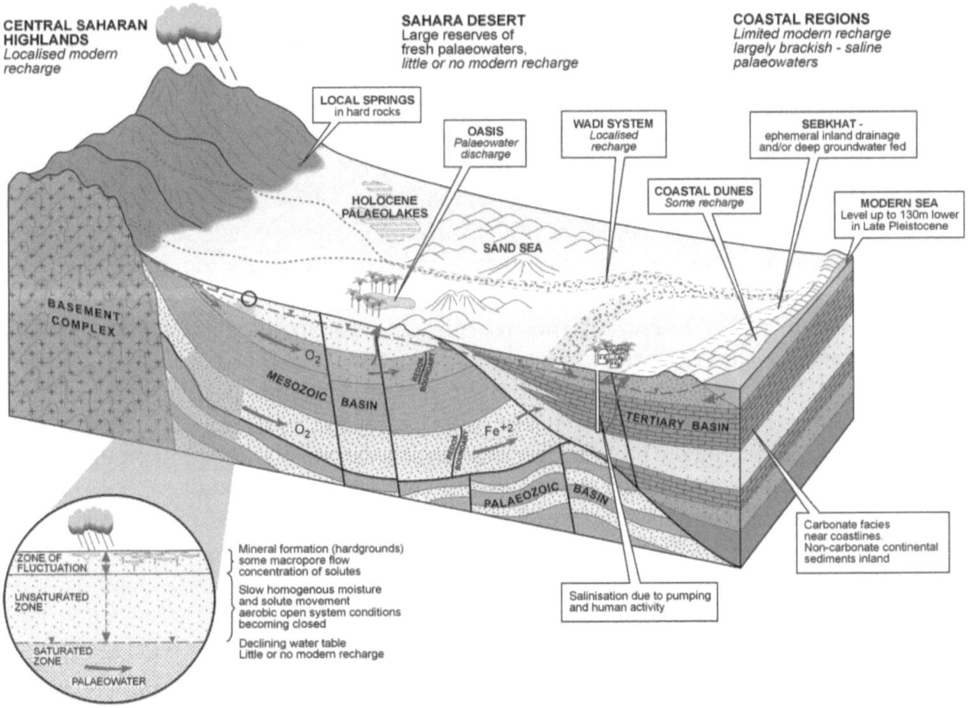

Figure 1a. Schematic representation of the sedimentary basins in Saharan Platform of North Africa together with the present day landscape features.

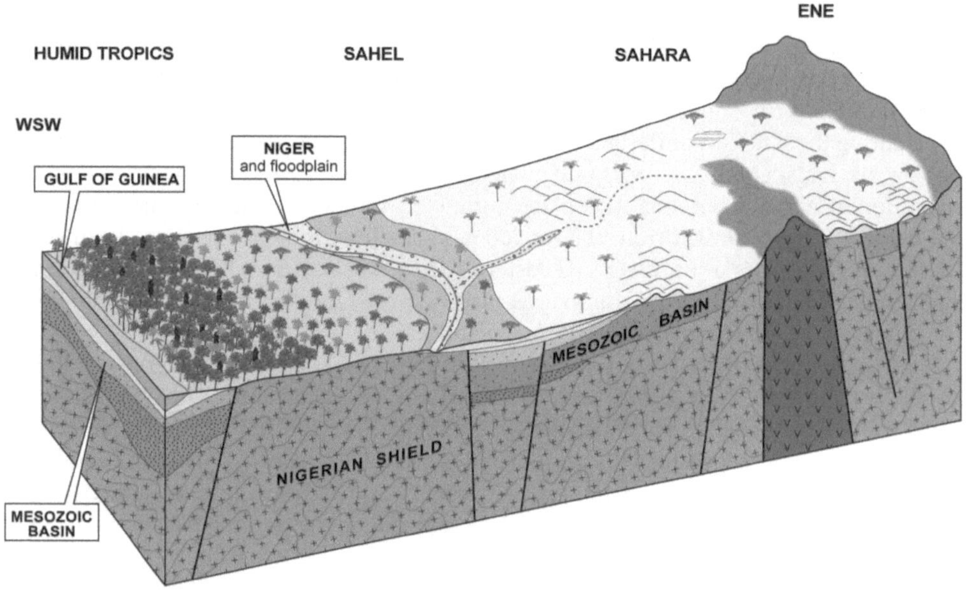

Figure 1b. Schematic representation of the sedimentary basins in the Sahel and coastal West Africa together with the present day landscape features.

to be transgressive across the basins (Pallas & Salem, 2000). Much of northern Africa is overlain by Quaternary or Recent sands and related unconsolidated sediments. This gives rise to the modern landscape where regional groundwater discharge areas from the basins are emphasized by sebkhats (Figure 1a). Large rivers were important features of the Holocene landscape in north Africa (Pachur, 1980; Pachur & Kröpelin, 1987) and these flowed until some 4 ka BP but now are represented by ephemeral wadis. The only major river apart from the Nile which crosses the semi-arid area of northern Africa is the Niger (Figure 1b). The unconsolidated nature of the surficial Quaternary and Recent sediments in the Sahel is favourable for groundwater recharge but only where the rainfall exceeds 200 mm/a (IAEA, 2001).

Large sedimentary basins are widespread beneath the Sahara and Sahel (Figure 2). The basins are among the largest in the world after the Great Artesian Basin of Australia (Radke *et al.*, 2000). The African basins have been the subject of exploration for hydrocarbons during the past half century, during which time the hydrogeology has also been moderately well characterized despite the remoteness of the region. However there is now an excellent data base from isotopic and chemical investigations and the principal studies that have been carried out with special emphasis on the hydrochemistry and palaeoclimatic reconstruction are indicated in Figure 2.

3 GEOCHEMICAL AND ISOTOPIC TOOLS FOR GROUNDWATER ARCHIVE STUDIES

Inert tracers (stable isotopes ($\delta^{18}O$, $\delta^{2}H$), together with noble gases, NO_3, Cl, Br/Cl, are the main palaeoenvironmental indicators used in groundwater studies and retain different

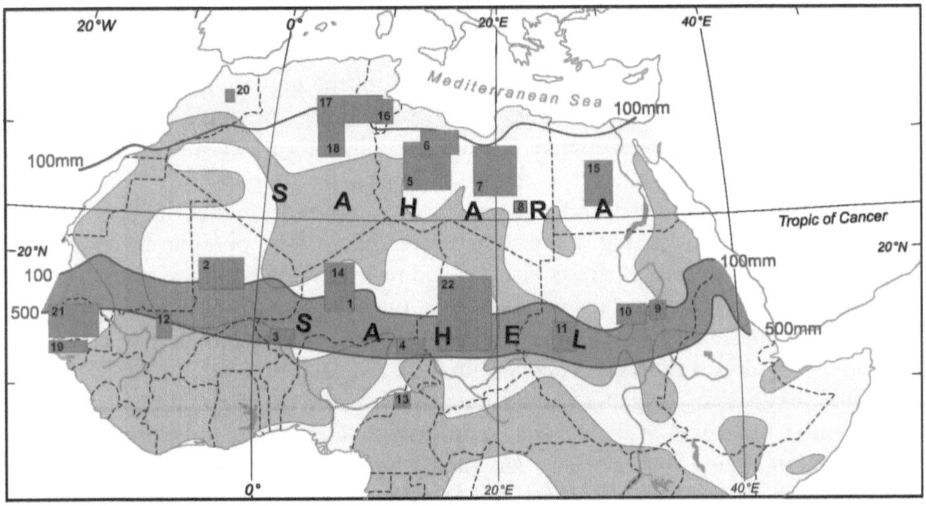

Figure 2. The distribution of sedimentary basins in North Africa (light shading) and the location of studies used as data sources for the present paper: 1) and 2) Azaouad (Fontes *et al.*, 1991); 3) Illumeden (le Gal La Salle, 1992); 4) Chad Basin; 5) Western Libya (Salem *et al.*, 1980); 6) Northern Libya (Srdoc *et al.*, 1980); 7) Sirt Basin (Edmunds & Wright, 1979); 8) Kufra Basin (Edmunds & Wright, 1979); 9) Butana region (Darling *et al.*, 1987); 10) Kordofan (Groening *et al.*, 1993); 11) Darfur (Groening *et al.*, 1993); 12) Kolokoni-Nara (Dincer *et al.*, 1983); 13) Garoua (Njitchoua *et al.*, 1993); 14) Irhazer (Andrews *et al.*, 1993); 15) Western Desert (Thorweihe, 1982); 16) and 17) Continental Intercalaire (Guendouz *et al.*, 1998); 18) Complexe Terminal; 19) Casamance (Faye *et al.*, 1993); 20) Saïs (Kabbaj *et al.*, 1978); 21) N Senegal (Faye, 1994); 22) Chad basin (UNESCO 1972).

information on the input conditions (Herczeg & Edmunds, 1999). The importance of these tracers for reconstruction is greatly enhanced if absolute chronometers such as radiocarbon are available. Indirect evidence of the palaeohydrology in the Late Pleistocene and Holocene may be also be deduced from a variety of indicators contained in dated lake sediments (Fontes & Gasse, 1991; Gasse, 2000; Hoelzmann *et al.*, 2000). However, unlike most palaeoenvironmental indicators, such as ice cores or tree rings, which contain high-resolution information, data available in large groundwater bodies are of low resolution (typically ± 1000 a). This is due to the advection or dispersion of any climatic input signal in the water body. Additionally, many groundwater data are obtained from pumped samples, mixing stratified water of different ages. Even the absence of dated waters over some time interval may indicate periods of drought (Sonntag *et al.,* 1978). The correlation between groundwater records and aeolian deposition in semi-arid/arid regions can also provide complementary evidence of wet and dry intervals (Edmunds *et al.*, 1999; Swezy, 2001).

Radiocarbon is the primary tool for reconstruction of groundwater records which depend upon a reliable chronology. Over the timescales of interest ($10^3 - >10^6$ a) other specialised options for absolute dating, such as ^{39}Ar and possibly ^4He, also exist (Loosli *et al.*, 1999). Chlorine-36 and Krypton-81 have also recently been applied as potential age indicators in the North African aquifers (Sturchio *et al.*, 2004; Guendouz & Michelot, 2006) although for the focus on Holocene records and the Pleistocene/Holocene transition these other tools are inappropriate.

Many caveats apply to the use of ^{14}C in groundwaters due to difficulties of knowing input conditions and the water-rock interactions involved, especially where carbonate minerals are present along flow paths (Clark & Fritz, 1997). In the predominantly non-carbonate aquifers of many large basins, however, age correction may be applied with caution to provide calibration of the flow sequence. Changes in radiocarbon activities along flow lines, expressed as pmc (percent modern carbon), allow relative timescales to be established, especially using $\delta^{13}C$ and supporting geochemical evidence to model the sources of the carbon. However due to mixing and/or reaction it may not be possible to resolve ages within a few thousand years. Care is also needed to verify the closed system conditions in the aquifers since other crustal sources of CO_2 may be present (Andrews *et al.*, 1994). The $\delta^{13}C$ values may also be used to infer near surface phenomena including reactions in the soil and with surface crusts and changes in C_3-C_4 vegetation types.

The wide variations in oxygen-18 and deuterium observed in precipitation at the present day are sensitive indicators of change and complexity in temperature, precipitation patterns and air mass circulation, especially in the Sahara/Sahel region (Dray *et al.*, 1983). Past rainfall stored as palaeo-groundwater, together with the other hydrological archives such as ice, provide evidence of former climatic conditions (Rozanski *et al.*, 1997). Climatic changes are expressed primarily as: (i) isotopic depletion relative to modern groundwaters with reference to the meteoric water line; (ii) change in the deuterium excess, signifying changes in humidity in the air mass as it detaches from its primary oceanic source moving over arid regions; and (iii) local condensation and evaporation effects within clouds or in falling rain. Oxygen isotopic enrichment due to near-surface evaporation is also found in many groundwaters signifying that rates of aquifer recharge are likely to be low.

Chloride, in contrast to water itself which may be lost by evaporation, is conservative and so changes in chloride concentrations with time, where these can be measured, are usually good indicators of changes in aridity/wetness. The combined use of chloride and the stable isotopes of water ($\delta^{18}O$, $\delta^{2}H$), moreover, provides a powerful technique for studying past environments in groundwaters. Over continental areas groundwater solutes are dominantly of atmospheric origin and are concentrated in proportion to evaporation during recharge. The large freshwater reserves in some basins of modern arid zones are therefore a priori indicators of wetter climates. With lowered sea levels during the Pleistocene and lasting until 8.5 ka BP, there was also the opportunity for freshwater to advance offshore relative to the present day coastlines and to displace saline formation waters (Edmunds & Milne, 2001). The present-day distribution of groundwater salinity therefore provides clues to climate variations. In present day shallow aquifers higher salinity is mainly a legacy of the onset of more arid conditions during the past 4000 years.

Noble gas contents (Ne, Ar, Kr, Xe) of groundwater (corrected for excess air) under closed system conditions reflect the annual mean air temperature. They form the most reliable indicator of palaeo-temperature in groundwaters and may be used to help interpret the significance of changes in the stable isotope ratios, which may not always be related to temperature (Andrews, 1993; Andrews *et al.*, 1994; Stute & Schlosser, 1993; Stute & Talma, 1998).

Nitrate remains inert in the presence of dissolved oxygen and may retain the signature of the environmental conditions at the time of recharge (Edmunds & Gaye, 1997). In Africa high nitrate concentrations, quite frequently exceeding World Health Organisation (WHO) limits for drinking waters, are considered to result naturally from fixing by leguminous vegetation. In reducing waters, the former existence of higher nitrate concentrations may be determined by enhanced N_2/Ar ratios (Andrews *et al.*, 1994). The ratio of Br/Cl may also be used to fingerprint the atmospheric (as well as geological) origins of the Cl (Edmunds, 1996; Davis *et al.*, 1998) and Br enrichment usually indicates biomass decay, including especially contributions from forest fires (Goni *et al.*, 2001).

The unsaturated zone may also, under favourable circumstances, contain records of past environment and climate at decadal to millennial scale resolution, mainly as variations in salinity and in stable isotope enrichments in percolating waters (Edmunds & Tyler, 2002). Such records are particularly found in porous media in areas of low moisture flux, notably beneath modern arid or semi-arid areas. The resolution of unsaturated (vadose) zone records will depend on the dispersion of the signal (Cook *et al.*, 1992) but decadal scale records may be retained, as in west Africa, over one or two hundred years (Gaye & Edmunds, 1996), or up to the millennial scale over the Late Pleistocene (Tyler *et al.*, 1996).

4 ORIGINS OF PALAEOWATERS IN THE NORTH AFRICAN SEDIMENTARY BASINS

Groundwaters containing radiocarbon as well as oxygen and hydrogen stable isotope compositions have been compiled and reviewed for Africa by Edmunds *et al.* (2003) and a resumé is given here. Age corrections were not used for the main analysis in view of the uncertainties that may be introduced. Plots of radiocarbon activities with data taken from the cited studies, expressed as per cent modern carbon (pmc), were used to provide a relative timescale to evaluate the significance of the stable isotope ($\delta^{18}O$) compositions. In these plots three timescales, Modern, Holocene and Late Pleistocene are recognized. A comparison of three areas (NW Africa and NE Africa and Sahel region of Nigeria, Niger and Cameroun, taken from the wider study and illustrating the northern Africa groundwater evolution is shown in Figures 3, 4 and 5. These results are then compared with the published accounts from southern Africa including the Stampriet aquifer in Namibia.

The Saïs Plain, northern Morocco, contains records of dated groundwaters from a Quaternary lacustrine carbonate sequence (Kabbaj *et al.*, 1978) with $\delta^{13}C$ values (-13 to $-17‰$) indicating little or no reaction with the carbonate matrix (Figure 3). In Senegal, data are from the Cretaceous Maastrichtian aquifer with a few results also from the overlying Oligo-Miocene (Faye, 1994; Faye *et al.*, 1993). For Mali, results from Fontes *et al.* (1991) contain a transect from the Niger River near Toumbouctou northwards. Only those data from north of the Azaouad Ridge are included here since those to the south are considered to have recharged from northwards (Holocene) flooding of the Niger. Also included are data from western Mali on the line from Koulikoro north to Nara (Dinçer *et al.*, 1984) where the aquifers are mainly found in sandstones and schists of Cambrian age; the $\delta^{13}C$ values are all more negative than $-10‰$.

The northeast Africa region is dominated by the extensive Nubian Cretaceous Sandstone aquifer in the Kufra Basin, northern Sudan and the Western Desert of Egypt. The Nubian system is then overstepped in Libya by Tertiary sedimentary basins, which, like the Nubian system, are mainly continental in origin but with marine facies (containing brackish water) nearer to the coast (Figure 2). Data used here (Figure 4) are from freshwaters from the continental aquifers, which preserve mainly initial carbon isotope inputs with little or no modification by water–rock interaction, although some Holocene waters (Edmunds & Wright, 1979) may (as described above) have reacted with active calcretes, affecting age

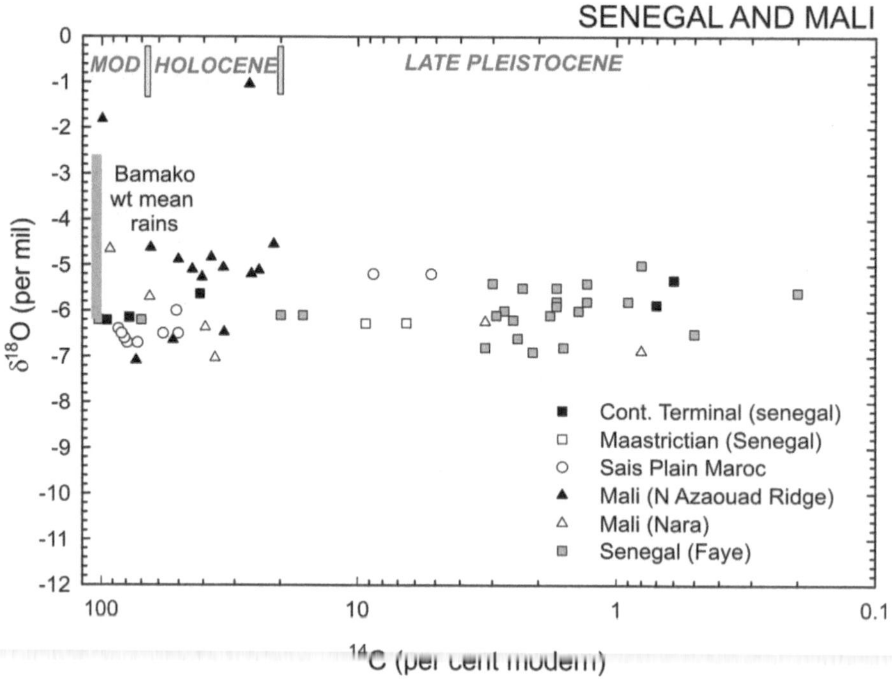

Figure 3. The change in $\delta^{18}O$ during the late Pleistocene and Holocene for Senegal Morocco and Mali as shown by dated groundwaters (^{14}C as percent modern carbon).

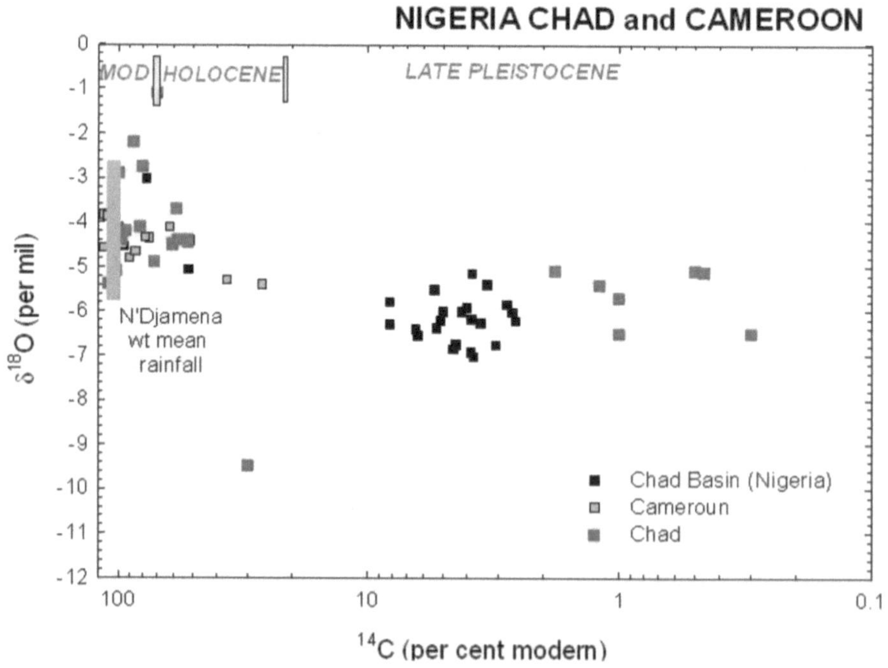

Figure 4. The change in $\delta^{18}O$ during the late Pleistocene and Holocene for Nigeria, Chad and Cameroon as shown by dated groundwaters (^{14}C as percent modern carbon).

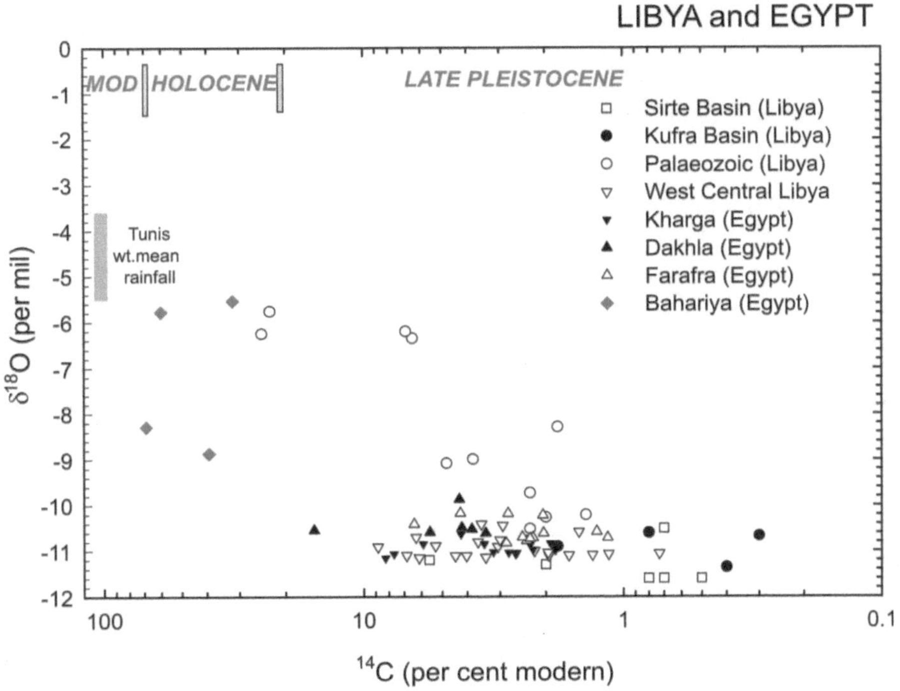

Figure 5. The change in $\delta^{18}O$ during the late Pleistocene and Holocene for north east Africa (Libya and Egypt) as shown by dated groundwaters (^{14}C as percent modern carbon).

correction. Data from northern Libya are from the Sirt Basin (Edmunds & Wright, 1979), and from the Murzuq and other basins in western and central Libya (Srdoé *et al.*, 1980; Salem *et al.*, 1980). Data from Egypt (Thorweihe, 1982) are from the phreatic or semi-confined aquifers feeding the major oases of the Western Desert in the Nubian sandstone, which may be compared with data from southern Libya (Kufra) in Edmunds & Wright (1979). Data from the same aquifer in northern Sudan (Figure 4) come from Darling *et al.* (1987) (Butana), as well as from Darfur and Kordofan (Groening *et al.*, 1993).

Pleistocene groundwaters (Figure 5) are found in the Quaternary Middle and Lower Zone aquifers of Nigeria, near Lake Chad which were emplaced during lowered levels of Lake Chad, but groundwaters of early Holocene age are not found (or preserved) here, recharge having been prevented by the Holocene rise in Lake Chad lake levels (Edmunds *et al.*, 1997; 1999) or subsequently displaced by modern rain. Dated groundwaters in Chad are available from the original studies on the basin by UNESCO (1972). In Cameroon only groundwaters of Holocene age have been recognised (Nijtchoua *et al.*, 1993) in sandstones of the Garoua region.

These plots illustrate the significant differences in climate and recharge characteristics of the late Pleistocene and Holocene compared with the present day. An overall trend is found in Late Pleistocene palaeowater $\delta^{18}O$ composition from west to east from around −6‰ to −10.5‰, first observed by Sonntag *et al.* (1978). This describes an overall continental effect with an Atlantic moisture source moving from west to east, the air mass having evolved according to Rayleigh (closed system) fractionation with residual isotopically lighter rains further to the east. The former widespread extent of these rains recorded in groundwaters beneath the present Sahara points to the shift of the Atlantic jet stream well to the south during most of the Late Pleistocene, as well as a weakening of the southwest monsoon. The more enriched isotopic signatures of groundwaters in Nigeria/Cameroon, compared with waters to the north suggests that, during part of the Late Pleistocene, monsoon rains derived from the Gulf of Guinea still prevailed south of 16°N. In Senegal, the similarities of the isotopic compositions reflect the continuity of the maritime influence but with recharge derived from the southwest monsoon activity of variable strength.

Late Pleistocene palaeowaters of northeast Africa show trends on a north to south basis, seen most clearly in the groundwaters in Libya and Egypt. All these waters are also isotopically much lighter than in western Africa. Groundwater in the Kufra Basin (Nubian Sandstone) is isotopically the lightest in north Africa (−11.5‰) and this compares with the Sirte Basin to the north where the palaeowaters are some 3‰ more enriched in $\delta^{18}O$ (Edmunds & Wright, 1979). Groundwaters from the Egyptian oases also show distinctive compositions (Thorweihe, 1982) lying within the range −10 to −11‰. Thus each sedimentary basin seems to have a distinctive composition, which supports the likelihood of local evolution of groundwater within each basin, forming at the extreme of the evolution of the Atlantic air mass source. The effects of runoff from the central Saharan mountains may also have contributed to isotopically light runoff.

A gap in the record exists in most of the data-sets for groundwaters with ^{14}C of approximately 5–15 pmc. This was first noticed for North African ^{14}C data sets by Geyh & Jäkel (1974) and demonstrated for groundwaters by Sonntag *et al.* (1978, 1980). This is interpreted as an arid interlude, coinciding with the Last Glacial Maximum (LGM) in Europe.

Holocene groundwaters show markedly different properties across the continent both in relation to those from the Late Pleistocene, as well as trends during the Holocene itself. In Morocco, the Holocene groundwaters are isotopically lighter (around 1–2‰ in $\delta^{18}O$) relative to those from humid periods of the Late Pleistocene. In Senegal there is a similar if subdued tendency less marked than to the north. This may be explained by the maritime situation of both countries, where the influence of the Atlantic moisture (and sea-surface temperature effects) is felt directly. The relative isotopic enrichment during the Late Pleistocene may be the result of the change in the ocean composition due to the lighter isotopes being enriched in the ice caps. This is the only noticeable effect over the past 30 ka in the coastal areas and the otherwise similar isotope compositions demonstrate the constancy of the westerly Atlantic air masses, and the south westerly monsoons (which varied in intensity), over the whole period.

Elsewhere in northern Africa, modern and most Holocene groundwaters have isotopic compositions which are significantly enriched relative to the Late Pleistocene pluvial periods. However the compositions of the Early Holocene groundwaters are depleted when compared to the modern rains. The Holocene, on the basis of extensive evidence from other proxy data, was a period with short (millennial scale) duration, but intense wet phases, which were not synchronous across the continent, from 11.4 to 5.4 ka BP (Gasse, 2000), prior to the desiccation that led to the present day conditions commencing around 4.5 ka BP. The groundwater isotopic composition therefore suggests the greater intensity of the rainfall in the Holocene as compared with the present-day rainfall, due to a greater strength and northward movement of the African monsoon.

5 CONTINENTAL INTERCOMPARISONS

The results from North Africa can be compared with those from the semi-arid regions of South Africa and in particular the Karoo sequence typified by the Stampriet (Aoub) aquifer of Namibia and the Uitenhage artesian aquifer near to the southern coast of South Africa, most recently reported by Stute & Talma (1998) but drawing also on work by Heaton *et al.* (1986).

The results in Figure 6 summarise the oxygen stable isotope and noble gas trends in dated waters from the Stampriet and Uitenhage aquifers where age increases and geochemical processes evolve systematically downgradient in the sandstone aquifer (Tredoux & Kirchner 1981). Radiocarbon ages were corrected on the basis of $\delta^{13}C$ values of groundwater bicarbonate, and assumptions concerning the values of the initial solid phase and soil CO_2.

The different patterns exhibited by the $\delta^{18}O$ in the two aquifers are rather striking. The Uitenhage aquifer conforms to the typical lower (more negative) $\delta^{18}O$ values for groundwater recharged during glacial times, although the depletion is only around 1‰ compared with some of the much larger separations seen in the more continental aquifers in northern Africa. In the Stampriet aquifer however there is an enrichment of around 1.3‰ during the

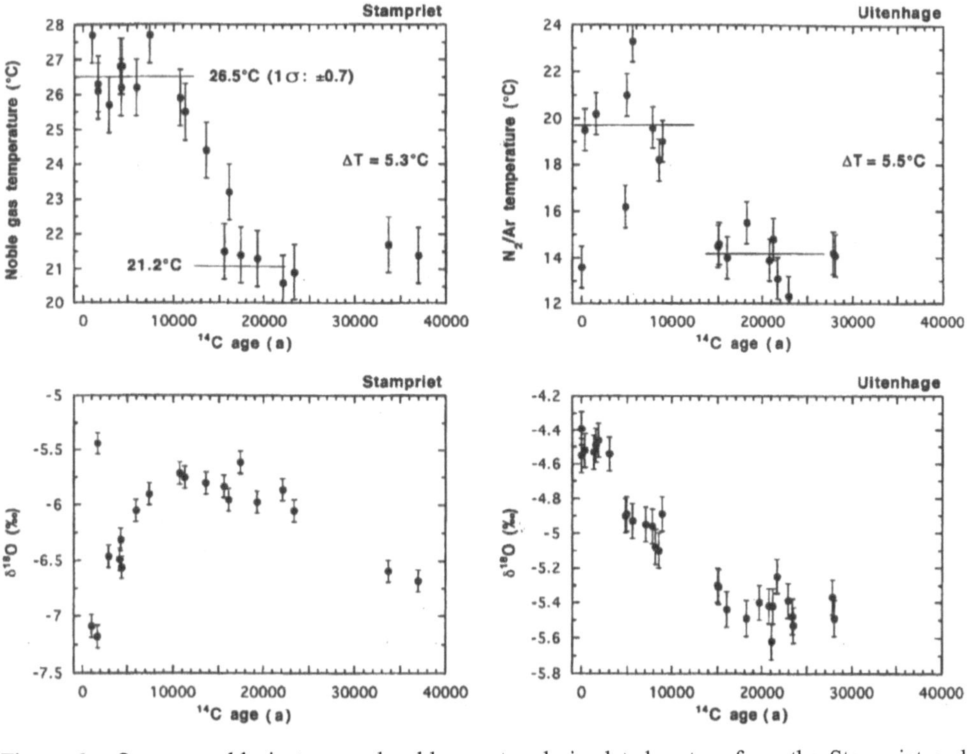

Figure 6. Oxygen stable isotope and noble gas trends in dated waters from the Stampriet and Uitenhage aquifers (Stute & Talma 1998).

LGM. This pattern is explained (Stute & Talma 1998) by a dominance of Atlantic moisture during the LGM.

The South African results compare well with those from West Africa (Figure 3) and also with trends observed from Europe (Loosli *et al.*, 1999; Edmunds, 2005). The Late Pleistocene groundwaters in northern Africa record evidence of cooler climates and significant recharge prior to the LGM. Air mass circulation over Africa during the Late Pleistocene was significantly different from the present day with evidence, shown clearly in the groundwater archive, of a reinforcement and southward shift of the Atlantic westerly flow across the present Sahara during the period. A corresponding decline of monsoon rains occurred at this time over West Africa and the Sahara with the onset of a period of aridity. Evidence then is found for a northward extension of the African monsoon, with increased intensity and a significant increase in rainfall, notably during the Early to Mid-Holocene coinciding with a retreat of the Atlantic system to the north. In coastal Portugal, as in Morocco, however the lack of any stable isotope depletion indicates the constancy of the southwest Atlantic air circulation at medium latitudes over the whole of the Late Pleistocene and Holocene, as well as proximity to the oceans near the Atlantic seaboard.

The continental palaeoclimatic changes and cooling are confirmed by the noble gas recharge temperature (NGRT) evidence. Palaeotemperature measurements using noble gases have now been carried out particularly in Africa, Europe and America, where large sedimentary basins occur and where samples of groundwater from pumped boreholes may be obtained (preserving pressures sufficient to avoid loss of gases). Earlier results were reviewed by Stute & Schlosser (1993) but results are available from Niger (Le Gal La Salle, 1992), Senegal (Faye, 1992), Algeria (Guendouz *et al.*, 1998), Nigeria (Edmunds *et al.*, 1999), Namibia and South Africa (Stute & Talma, 1998), Oman (Weyhenmeyer *et al.*, 2000), Portugal (Carreira *et al.*, 1996) and NE Brazil (Stute *et al.*, 1995). These have been calculated relative to the mean annual ground temperature as represented by present day near-surface groundwater temperatures. It is found that the average global cooling at and immediately prior to the last glacial maximum (samples measured over the range approximately 30 to 20 ka BP) was between 5–6°C. However there is evidence from continental areas such as northern Nigeria as well as Niger that the cooling was 1–2°C degree cooler (6–7°C).

The Late Pleistocene and Holocene climate and hydrology in Africa is consistent with palaeo-groundwater archives in Europe (Bath *et al.*, 1979; Stute & Deák, 1979; Edmunds & Milne, 2001) where significant replenishment of aquifers took place during the pluvial periods and where general cooling of around 5–6°C is recorded in the NGRT. In continental Europe an absence of dated groundwaters during the period approximately 10–18 ka correlates with a recharge gap corresponding to permafrost cover ahead of the continental ice sheet.

In Portugal, as in Morocco, there is some evidence which may indicate a degree of continuity of recharge in coastal regions. Evidence of cooling is preserved in the NGRTs, indicating the constancy of the Atlantic air circulation at medium latitudes over the whole of the Late Pleistocene and Holocene for the maritime coastal areas. Evidence from southern Africa (Stute and Talma 1998) also shows similar characteristics in the Stampriet aquifer, as for example in Portugal and Morocco, consistent with an Atlantic source of moisture and with consistent amount of global cooling. The Uitenhage aquifer however shows more evidence of continentality and the absence of radiocarbon data for a period of some 5000 years may reflect a degree of aridity as seen in areas of northern Africa, further from the Atlantic source of moisture.

6 GROUNDWATER – PALAEOWATER OR MODERN RECHARGE?

There is clear indication from the geochemical and isotopic records that emplacement of groundwater across most of Africa in areas and at depths accessible for exploitation, has taken place on a geological timescale with turnover times of tens of thousands of years, corresponding mainly to the much wetter climates over the late Pleistocene. Following periods of aridity in more continental areas significant recharge took place also during the pluvial periods of the Early and mid-Holocene and water of this generation can be identified at shallow depths overlying older waters. This can be seen well for example in Libya, where recharge along lines of formerly flowing wadis still is preserved as linear lenses (Edmunds & Wright, 1979), in Niger related to flooding of the Niger River northwards into the Azaouad depression (Fontes *et al.,* 1991) and in Sudan, where there is evidence of recharge during the Holocene in the phreatic aquifer (Edmunds *et al.*, 1992; Edmunds, 2003).

Evidence for modern recharge is derived from both the chemical and isotopic signatures. Studies of unsaturated zone profiles have been carried out in the Sahel region (Edmunds *et al.*, 1992; Gaye & Edmunds, 1996; Edmunds *et al.*, 1999), as well as extensive studies in Botswana (Beekman *et al.*, 1997; Selalo, 1998; de Vries *et al.*, 2000). These studies use the chloride mass balance to quantify recharge rates knowing the composition of chloride in the modern rainfall, rainfall amounts and also the mean (steady state) concentrations of recharge in the unsaturated zone. Under favourable conditions, where fairly homogeneous unsaturated sediments occur, the chemical and isotopic data may also be used to follow recharge history at the decadal scale (Cook *et al.*, 1992), often confirmed by instrumental records of rainfall patterns over the past 100–130 years. Whilst it has been possible to achieve this in northern Africa, it has proved difficult in much of southern Africa since the presence of hard grounds and calcretes has caused dispersion of the recharge signal and also led to some degree of bypass flow rather than the homogeneous piston flow demonstrated in the Sahel region. A good example of the evidence for modern recharge is shown (Figure 7) by the record from Louga, Senegal (Gaye & Edmunds, 1996). Here a record of several decades is preserved in the unsaturated Quaternary sands where the mean annual recharge is found to be 34 mm a^{-1}. The record of the prolonged Sahel drought of the 1970s and 1980s is also recorded in the higher Cl in the interstitial moisture as well as in the record of enriched (more positive) stable isotope signals. The timescale is then confirmed by tritium which provides a well-defined maximum corresponding to the 1963 peak in thermonuclear testing, also confirming the homogeneous (piston flow) movement in the profile.

From the various studies across Africa modern recharge is likely to be insignificant where the mean annual rainfall is below 200 mm/a. This is confirmed widely from studies world-wide (IAEA, 2001). However focused recharge may occur at the present day along wadi lines and other favourable locations during ephemeral intense rainfall events in more arid regions with less than 200 mm/a.

7 SUMMARY

From studies conducted during the past 30 years or more there is now an excellent data base for groundwater occurrence in semi-arid and arid regions of Africa, aided particularly by

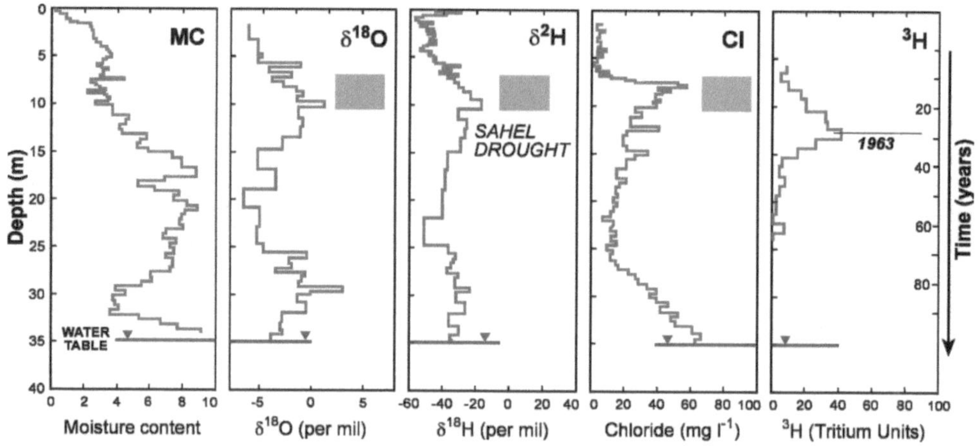

Figure 7. Groundwater recharge over the past 70 years (Louga Senegal) as demonstrated from chemical and isotopic indicators in unsaturated zone moisture profiles. Using a rainfall of 290 mm/a, mean annual Cl of 2.8 mg/l and mean unsaturated zone Cl of 23.6 mg l^{-1}, a decade scale recharge of 34 mm a is indicated. Tritium and stable isotope data support this timescale and qualitatively support the conclusions.

the search for hydrocarbons in the sedimentary basins. For many areas the geological and hydrogeological records are sparse and isotopic and geochemical information provides good evidence for assessing the recharge conditions, groundwater movement and the renewability or non-renewability of the water resources. Groundwaters of known age contained in major aquifer systems in the African sedimentary basins are of specific value in determining low resolution (±1000 a) characteristics of past climates, specifically palaeotemperature, air mass origins, humid/arid transitions and rainfall intensity. Results from northern Africa indicate the predominance of a westerly Atlantic air flow during the Late Pleistocene; in north Africa as the moisture moved east the recharge was characterized by a continental effect of about 4.5‰ $\delta^{18}O$. Greater aridity is recorded by the absence of dated groundwaters over most of northern Africa during the LGM. An intensification of the African monsoon during the Early Holocene is apparent from isotopically light groundwaters found especially over Sudan. Maximum cooling around the LGM of 5–7°C is recorded in the noble gas recharge temperatures from Africa. Evidence from North African groundwaters is matched by complementary data from northern Europe, where a recharge gap is found corresponding to permafrost conditions. Similar patterns of recharge in the southern hemisphere are also found to those in the north of the continent, where the influence of Atlantic westerly air masses provided continuous recharge during the late Pleistocene and Holocene, yet the south coast of Africa shows to some extent a continental effect, away from the west coast influence.

Groundwater development across arid and semi-arid regions must be based on the knowledge that much of the resource is non-renewable. This is clear in many sedimentary basins from the falling water tables and declining artesian pressures. In the Sahel region and in areas with greater than 200 mm/a rainfall it can be shown from chloride mass balance and isotopic indicators that the resource can be renewable.

REFERENCES

Andrews, J. N., Lee, D. J. 1979. *Inert gases in groundwater from the Bunter Sandstone of England as indicators of age and palaeoclimatic trends*. Journal of Hydrology, **41**, 233–252.

Andrews, J. N. 1993. *Isotopic composition of groundwaters and palaeoclimate at aquifer recharge*, In: Isotope techniques in the study of past and current environmental changes in the hydrosphere and atmosphere. IAEA, Vienna, 271–292.

Andrews, J. N., Fontes, J., Aranyossy, J. F., Dodo, A., Edmunds, W. M., Joseph, A. & Travi, Y. 1994. *The evolution of alkaline groundwaters in the Continental Intercalaire aquifer of the Irhazer Plain, Niger*. Water Resources Research, **30**, 45–61.

Arnell, N. W. 2004. *Climate change and global water resources: SRES emissions and socio-economic scenarios*. Global Environmental Change, **14**, 31–52.

Bath, A. H., Edmunds, W. M. & Andrews, J. N. 1979. *Palaeoclimatic trends deduced from the hydrochemistry of a Triassic sandstone aquifer*. International Symposium on Isotope Hydrology, Vienna, Vol 2, 545–568. IAEA-SM-228/27.

Beekman, H. E., Selaolo, E. T. & De Vries, J. J. 1997. *Groundwater Recharge and Resources Assessment in the Botswana Kalahari*. GRES-II project final report, Lobatse.

Carreira, P. M., Soares, A. M. M., Marques da Silva, M. A., Araguas, L. A., Rosanski, K. 1996. *Application of environmental isotope methods in assessing groundwater dynamics of an intensively exploited coastal aquifer in Portugal*. In: Isotopes in Water Resources Management, Vol 2. IAEA, Vienna, 45–58.

Clark, I. D. & Fritz, P. 1997. *Environmental Isotopes in Hydrogeology*. Lewis, Baton-Rouge, 328pp.

Cook, P. G., Edmunds, W. M. & Gaye, C. B. 1992. *Estimating palaeorecharge and palaeoclimate from unsaturated zone profiles*. Water Resources Research, **28**, 2721–2731.

Darling, W. G., Edmunds, W. M., Kinniburgh, D. G. & Kotoub, S. 1987. *Sources of recharge to the Basal Nubian Sandstone Aquifer, Butana Region, Sudan*. In: Isotope Techniques in Water Resources Development. IAEA, Vienna, 205–224.

Davis, S. N., Whittemore, D. O. & Fabryka-Martin, J. 1998. *Uses of chloride/bromide ration in studies of potable water*. Ground Water, **36**, 338–350.

De Vries, J. J., Selaolo, E. T. & Beekman, H. E. 2000. *Groundwater recharge in the Kalahari, with reference to paleo-hydrologic conditions*. Journal of Hydrology, **238**, 110–123.

Dincer, T., Dray, M., Zuppi, G. M., Guerre, A., Tazioli, G. S. & Traore, S. 1984. *L'alimentation des eaux souterraines de la zone Kolokani-Nara au Mali*. In: Isotope Hydrology, 1983: Proceedings of the International Symposium on Isotope Hydrology in Water Resources Development, IAEA, Vienna, 341–365.

Dray, M., Gonfiantini, R. & Zuppi, G. M. 1983. *Isotopic composition of groundwater in the southern Sahara*. In: Palaeoclimates and palaeowaters: a collection of environmental isotope studies. IAEA, Vienna, 187–199.

Edmunds, W. M. 1999. *Groundwater nitrate as a palaeo-environmental indicator*. In: Armannsson, H. (ed.), Geochemistry of the Earth's Surface. Proc 5th International Symposium on the Geochemistry of the Earth's Surface. Reykjavik, Iceland, Balkema. Rotterdam, 35–38.

Edmunds, W. M. 2001. *Investigations of the unsaturated zone in semi-arid regions using isotopic and chemical methods and applications to water resources problems*. In: Isotope-based assessment of groundwater renewal in water-scarce regions, TECDOC-1246. IAEA Vienna, 7–22.

Edmunds, W. M. 2003. *Hydrogeochemical processes in arid and semi-arid regions – focus on North Africa*. In: Simmers, I. (ed.) Understanding Water in a Dry Environment: Hydrological Processes in Arid and Semi-Arid Zones, UNESCO International Contributions to Hydrology 23. Balkema, pp 251–287.

Edmunds, W. M. 2005. *Groundwater as an archive of climatic and environmental change*. In: Aggarwal, P. K. & Gat J Froehlich, K. (eds.) Isotopes in the Water Cycle: past, present and future of a developing science. Springer, 341–352.

Edmunds, W. M. & Wright, E. P. 1979. *Groundwater recharge and palaeoclimate in the Sirte and Kufra basins, Libya.* Journal of Hydrology, **40**, 215–241.

Edmunds, W. M., Gaye, C. B. & Fontes, J-Ch. 1992. *A record of climatic and environmental change contained in interstitial waters from the unsaturated zone of northern Senegal.* In: Isotope Techniques in Water Resources Development, 1991. IAEA. Vienna, 533–549.

Edmunds, W. M. & Gaye, C. B. 1997. *High nitrate baseline concentrations in groundwaters from the Sahel.* Journal Environmental Quality, **26**, 1231–1239.

Edmunds, W. M. & Milne, C. J. (eds.) 2001. *Palaeowaters in Coastal Europe: evolution of groundwater since the Late Pleistocene.* Geological Society London Special Publication, **189**.

Edmunds, W. M., Fellman, E. & Goni, I. B. 1999. *Lakes, groundwater and palaeohydrology in the Sahel of NE Nigeria: evidence from hydrogeochemistry.* Journal of the Geological Society London, **156**, 345–355.

Edmunds, W. M. & Tyler, S. W. 2002. *Unsaturated zones as archives of past climates: towards a new proxy for continental regions.* Hydrogeology Journal, **10**, 216–228.

Edmunds, W. M., Dodo, A., Djoret, D., Gasse, F., Gaye, C. B., Goni, I. B., Travi, Y., Zouari, K. & Zuppi, G. M. 2004. *Groundwater as an archive of climatic and environmental change. The PEP-III traverse.* In: Battarbee, R. W., Gasse, F. & Stickley, C. E. (eds.). Past climate variability through Europe and Africa, Developments in Palaeoenvironmental Research series: Kluwer Dordrecht, 279–306.

Faye, A. 1994. *Recharge et palaeorecharge des aquifères profonds du bassin du Senegal. Apport des isotopes stables et radioactifs de l'environnement et implications palaeohydrologiqes et palaeoclimatiques.* Thesis Doc ès Sciences. Université Cheikh Anta Diop. Dakar.

Faye, A., Tandia, A. A., Travi, Y., Le Priol, J. & Fontes, J-Ch. 1993. *Apport des isotopes de l'environnement à la connaissance des aquifères de Casamance (extreme sud du Sénégal).* In: Les Ressources en Eau au Sahel. Tech. Doc. 721, IAEA, Vienna, 123–132.

Fontes, J-Ch., Gasse, F. 1991. *PALHYDAF (Palaeohydrology in Africa) program: objectives, methods, major results.* Palaeogeography Palaeoclimatology Palaeoecology, **84**, 191–215.

Fontes, J-Ch., Andrews, J. N., Edmunds, W. M., Guerre, A. & Travi, Y. 1991. *Palaeorecharge by the Niger River (Mali) deduced from groundwater chemistry.* Water Resources Research, **27**, 199–214.

Fontes, J-Ch., Stute, M., Schlosser, P. & Broecker, W. S. 1993. *Aquifers as archives of palaeoclimate.* Eos, **74**, 21–22.

Gasse, F. 2000. *Hydrological changes in the African tropics since the Last Glacial Maximum.* Quaternary Science Review, **19**, 189–211.

Gaye, C. B. & Edmunds, W. M. 1996. *Intercomparison between physical, geochemical and isotopic methods for estimating groundwater recharge in northwestern Senegal.* Environmental Geology, **27**, 246–251.

Geyh, M. A. & Jäkel, D. 1974. *Spätpleistozäne und Holozäne Klimageschichte der Sahara aufgrund zugänglicher ^{14}C Daten.* Z. Geomorphol, **18**, 82–98.

Goni, I. B., Fellmann, E. & Edmunds, W. M. 2001. *Rainfall geochemistry in the Sahel region of northern Nigeria.* Atmos. Environ., **35**, 4331–4339.

Groening, M. C., Sonntag, C. & Suckow, A. 1993. *Isotopic evidence for extremely low groundwater recharge in the Sahel zone of Africa.* In: Thorweihe, U. (ed.) Geoscientific Research in North-east Africa. Balkema, Rotterdam, 671–676.

Guendouz, A., Moulla, A. S., Edmunds, W. M., Shand, P., Poole, J., Zouari, K. & Mamou, A. 1998. *Palaeoclimatic information contained in groundwaters of the Grand Erg Oriental, N. Africa.* In: Isotope Techniques in the Study of Past and Current Environmental Changes in the Hydrosphere and Atmosphere. IAEA, Vienna 555–571.

Guendouz, A. & Michelot, J-L. 2006. *Chlorine-36 dating of deep groundwater from northern Sahara.* Journal of Hydrology, **328**, 572–580.

Heaton, T. H. E., Talma, A. S. & Vogel, J. C. 1986. *Dissolved gas palaeotemperatures and ^{18}O variations derived from groundwater near Uitenhage, South Africa*. Quaternary Research, **37**, 203–213.

Herczeg, A. L. & Edmunds, W. M. 1999. *Inorganic ions as tracers*. In: Cook, P. G. & Herczeg, A. L. (eds.) Environmental Tracers in Subsurface Hydrology Kluwer. Boston, 31–77.

Hoelzmann, P., Kruse, H-J. & Rottinger, F. 2000. *Precipitation estimates for the eastern Saharan palaeomonsoon based on a water balance model of the West Nubian palaeolake basin*. Global and Planetary Change, **26**, 105–120.

IAEA, 2001. *Isotope-based assessment of groundwater renewal in water-scarce regions*. TECDOC-1246, IAEA, Vienna.

IPCC 2004. *Third Assessment Report*. Intergovernmental Panel on Climate Change, www.ipcc.ch/wg2sr.pdf

Klitsch, E. 1970. *Die Struktur Geschichte der Zentralsahara – neue Erkentnisse zum Bau und zur Palaeographie eines Tafellandes*. Geologische Rundschau, **59**, 459–527.

Kabbaj, A., Zeryouhi, I., Carlier, Ch. & Marce, A. 1978. *Contribution des isotopes du milieu a l'étude de grands aquifères du Maroc*. In: International Symposium on Isotope Hydrology, Vol.2, IAEA, Vienna, 491–524.

Le Gal La Salle, C. 1992. *Circulation des eaux souterraines dans l'aquifère captif du Continental Terminal – Bassin des Illumeden, Niger*. Thesis Doc. ès Sci., Univ. Paris Sud (Orsay), Paris, France.

Loosli, H. H., Lehmann, B. & Smethie, W. M. 1999. *Noble gas radioisotopes (^{37}Ar, ^{85}Kr, ^{39}Ar, ^{81}Kr)*. In: Cook, P. G. & Herczeg, A. L. (eds.) Environmental Tracers in Subsurface Hydrology, Kluwer, Boston, 379–396.

Loosli, H. H., Aeschbach-Hertig, W., Barbecot, F., Blaser, P., Darling, W.G., Dever, L. Edmunds, W. M., Kipfer, R., Purtschert, R. & Walraevens, K. 2001. *Isotopic methods and their hydrogeochemical context in the investigation of palaeowaters*. In: Edmunds, W. M. & Milne, C. J. (eds.) Palaeowaters of Coastal Europe: evolution of groundwater since the Late Pleistocene, Geological Society London Special Publications, **189**, 193–212.

Mazor, E. 1972. *Palaeotemperatures and other hydrological parameters deduced from noble gases dissolved in groundwaters, Jordan Rift Valley, Israel*. Geochim. Cosmochim. Acta, **36**, 1321–1336.

Njitchoua, R., Fontes, J-Ch., Dever, L., Naah, E. & Aranyossy, J. 1993. *Recharge naturelle des eaux souterraines du bassin des gres de Garoua. (Nord Cameroun)*. In: Les Ressources en Eau au Sahel. TECHDOC-721, IAEA, Vienna, 133–146.

Pachur, H-J. 1980. *Climatic history in the Late Quaternary in southern Libya and the western Libyan desert*. In: Salem, M. J. & Busrewil, M. T. (eds.), The Geology of Libya, Academic Press, London, 781–788.

Pachur, H. J. & Kröpelin, S. 1987. *Wadi Howar: Palaeoclimatic evidence from an extinct river system in the southeastern Sahara*. Science, **237**, 298–300.

Pallas, P. 1980. *Water resources of the Socialist Peoples Libyan Arab Jamahiriya*. In: Salem, M. J. & Busrewil, M. T. (eds.), The Geology of Libya. Academic Press, London, 539–593.

Pallas, P. & Salem, O. 2000. *Water resources utilisation and management of the Socialist People Arab Jamahiriya*, In: Regional aquifer systems in arid zones: managing non-renewable resources: Proceedings of the International Conference, Tripoli, Libya, 1999. UNESCO, Paris, 147–172.

Radke BM, Ferguson J, Cresswell RG, Ransley TR, Habermehl MA (2001) Hydrochemistry and implied hydrodynamics of the Cadna-Owie-Hooray Aquifer Great Artesian Basin. Bureau of Rural Sciences. Australia.

Rozanski, K., Johnsen, S. J., Schotterer, U. & Thompson, L. G. 1997. *Reconstruction of past climates from stable isotope records of palaeo-precipitation preserved in continental archives*. Hydrological Sciences Journal, **42**, 725–745.

Salem, O., Visser, J. H., Dray, M. & Gonfiantini, R. 1980. *Groundwater flow patterns in the western Libyan Arab Jamahirya evaluated from isotopic data*. In: Arid-zone Hydrology: Investigations with Isotopic Techniques, IAEA, Vienna, 165–179.

Selaolo, E. T. 1998. *Tracer Studies and Groundwater Recharge Assessment in the Eastern Fringe of the Botswana Kalahari.* PhD thesis, Vrije Universiteit Amsterdam, Printing and Publishing Co., Gaborone, Botswana.

Sonntag, C., Klitsch, E., Lohnert, E. P., Munnich, K. O., Junghans, C., Thorweihe, U., Weistroffer, K. & Swailem, F. M. 1978. *Palaeoclimatic information from D and ^{18}O in 14C-dated North Saharian groundwaters; groundwater formation from the past.* In: Isotope Hydrology, IAEA, Vienna, 569–580.

Sonntag, C., Thorweihe, U., Rudolph, J., Lohnert, E. P., Junghans, C., Munnich, K. O., Klitsch, E., El Shazly, E. M., Swailem, F. M. 1980. *Isotopic identification of Saharan groundwaters; groundwater formation in the past.* Palaeoecol. Africa, **12**, 159–171.

Srdoc, D., Sliepcevic, A., Obelic, B., Horvatincic, N., Moser, H. & Stichler, W. 1980. *Isotope investigations as a tool for regional hydrogeological studies in the Libyan Arab Jamahiriya.* In: Arid-zone Hydrology: Investigations with Isotopic Techniques. IAEA, Vienna, 569–580.

Sturchio, N. C., Du, X., Purtschert, R., Lehmann, B. E., Sultan, M., Patterson, L. J., Lu, Z-T., Muller, P., Bigler, T., Bailey, K., O'Connor, T. P., Young, L., Lorenzo, R., Beker, R., El Alfy, Z., El Kaliouby, B., Dawood, Y. & Abdellah, A. M. A. 2004. *One million year old groundwater in the Sahara revealed by krypton-81 and chlorine-36.* Geophysical Research Letters, **31**, DOI: 10, 1029/2003GL019234.

Stute, M. & Deák, 1989. *Environmental isotope study (14C, 13C, 18O, D, noble gases) on deep groundwater circulation systems in Hungary with reference to palaeoclimate.* Radiocarbon, **31**, 902–918.

Stute, M., Schlosser, P. 1993. *Principles and applications of the noble gas paleothermometer,* In: Swart, P. K., Lohmann, K. C., McKenzie, J. & Savin, S. (eds.), *Climate Change in Continental Isotopic Records*, Am. Geophys. Union. Geophys. Monograph, **78**, 89–100.

Stute, M., Forster, M., Frischkorn, H. J., Serejo, A., Clark, J. F., Schlosser, P., Broecker, W. S., Bonani, G. 1995. *Cooling of tropical Brazil (5° C) during the last glacial maximum.* Science, **269**, 379–383.

Stute, M. & Schlosser, P. 1999. *Atmospheric noble gases.* In: Cook, P. G. & Herczeg, A. L. (eds). Environmental Tracers in Subsurface Hydrology. Kluwer, Boston, 349–377.

Stute, M. & Talma, S. 1998. *Glacial temperatures and moisture transport regimes reconstructed from noble gases and $d^{18}O$, Stampriet aquifer, Namibia.* In: Isotope Techniques in the Study of past and current Environmental Changes in the Hydrosphere and Atmosphere. IAEA, Vienna, 307–318.

Swezey, C. 2001. *Eolian sediment responses to late Quaternary climatic changes: temporal and spatial patterns in the Sahara.* Palaeogeog. Palaeoclimatol. Palaeoecol, **167**, 119–155.

Thorweihe, U. 1982. *Hydrogeologie des Dakhla Beckens (Ägypten).* Berliner Geowiss. Abh. (A), **38**, 1–58.

Tredoux, G. & Kirchner, J. 1981. *The evolution of the chemical composition of artesian water in the Aoub Sandstone (Namibia/South West Africa).* Trans. Geol. Soc. S. Africa, **84**, 169–175.

Tyler, S. W., Chapman, J. B., Conrad, S. H., Hammermeister, D. P., Blout, D. O., Miller, J. J., Sully, M. J. & Ginani, J. N. 1996. *Soil-water flux in the southern Great Basin, United States: Temporal and spatial variations over the last 120 000 years.* Water Resources Research, **32**, 1481–1499.

UNESCO 1972. *Investigations of groundwater in the Lake Chad basin.* UNESCO, Paris.

Vaikmäe, R., Vallner, L., Loosli, H. H., Blaser, P. C. & Juillard-Tardent, M. 2001. *Palaeogroundwater of glacial origin in the Cambrian-Vendian aquifer of northern Estonia.* In: Edmunds, W. M. & Milne, C. J. (eds.) Palaeowaters in Coastal Europe: evolution of groundwater since the Late Pleistocene. Geological Society London Special Publications, **189**, 17–27.

Weyhenmeyer, C. E., Burns, S. J., Waber, H. N., Aeschenbach-Haertig, W., Kipfer, R., Loosli, H. H. & Matter, A. 2000. *Cool glacial temperatures and changes in moisture source recorded in Oman groundwaters.* Science, **287**, 842–845.

CHAPTER 19

Estimating groundwater recharge in the southwestern sector of the Chad basin using chloride data

I.B. Goni

Department of Geology, University of Maiduguri, Maiduguri, Borno State, Nigeria.

ABSTRACT: Groundwater is the perennial source of supply in the southwestern sector of the Chad basin. Thus, quantifying the rate of recharge is fundamental to the sustainable management of this resource. Chloride concentrations in rainfall, the unsaturated zone and groundwater were used to estimate the diffuse recharge flux using a steady-state chloride mass balance approach. Average Cl concentrations in rainfall measured from three stations over eight years is 1.63 mg/l (although modelling gives a lower long term value of 0.65 mg/l). Eight unsaturated zone Cl profiles were obtained and Cl concentrations in the regional groundwater were measured for over 400 samples from wells and boreholes. An average recharge rate of 40 mm/a was estimated from the unsaturated zone profiles. The regional rate obtained using Cl concentrations in groundwater is slightly higher, estimated at 50 mm/a, which also takes into account other mechanisms of recharge such as from river channels, pools, depressions and regional flow that bypass the unsaturated zone. The estimated rate of recharge reduces to 20 mm/a when the modelled Cl in rainfall is used as input data, this is considered a long term minimum recharge rate to the region. Therefore, estimated recharge rate to the groundwater in this region from the Cl data range from 20–50 mm/a. Analysis of abstraction in the area indicates that this rate can sustain present day abstractions for domestic use in villages via dug wells. This implies that the observed decline in the water table in recent years is largely due to a concurrent reduction in rainfall due to drought and consequent reduction in recharge.

1 INTRODUCTION

A rational approach to the management and development of groundwater resources requires knowledge of the origin of these resources and rates of their replenishment. This information represents basic input data for quantitative modelling, which increasingly is being used to assist groundwater management. The use of geochemical data (especially conservative solutes like Cl) provides a powerful tool for the investigation of modern hydrological processes and to reconstruct the historical recharge regime of large systems with extended groundwater residence times. The chemical methods are based on tracing geochemical signal through the components of hydrologic cycle, and in the process estimate rate of groundwater recharge (qualitatively and/or quantitatively). These methods are becoming increasingly more important in recharge studies especially in semi-arid and arid regions because determination of recharge using conventional water balance approach is often difficult or impossible (Allison *et al.*, 1994).

The chemical tracers most commonly used in recharge studies are, ^{3}H, ^{14}C, ^{36}Cl, ^{15}N, ^{18}O, ^{2}H, ^{13}C and Cl. Of these, the first three are radioactive, with their concentrations in the hydrological cycle greatly modified by nuclear testing. Both ^{3}H and ^{36}Cl from atmospheric testing have been used for soil water tracing and recharge studies (Zimmerman, *et al.*, 1967; Gvirtzman & Margaritz, 1986; Philips, *et al.*, 1988). Incidentally, ^{3}H, ^{2}H and ^{18}O, are the best tracers for the movement of water because they form part of the water molecule itself. Chloride is a conservative ion and has mobility similar to that of water molecule with one important exception, where water molecule is removed by evapotranspiration, Cl is concentrated in the residual solution. Thus, unsaturated zone Cl profiles provide an excellent technique for estimation of recharge in semi-arid and arid regions. Validation that the Cl is from atmospheric sources is however necessary.

The study area lies in the south western part of the Chad basin, which is in the Sahel region of Africa, to the west of Lake Chad (Figure 1). The Chad basin has apparently been a structural depression since the early Tertiary period and, therefore, has been subjected to land subsidence and sedimentation rather than erosion. Sedimentation in the basin led to the deposition of stratigraphical formations of arenaceous and/or argillaceous lithology. All the arenaceous layers or beds of the formations in the basin are potential aquifers, although the youngest of the sequences (the Chad Formation) contains the principal identified aquifers in the basin. This formation was deposited in or near a large ancestral Lake Chad during late Tertiary and Quaternary times on an uneven surface. The Chad Formation dips gently east and northeast toward Lake Chad in conformity with the slope of the land surface (Figure 2). Except for a belt of alluvial deposits around the edge of the basin, the Formation is of lacustrine origin and consists of thick beds of clay intercalated with irregular beds of sand, silt, and sandy clay. It occupies approximately 152,000 km^2 of northeastern Nigeria (Oteze & Fayose, 1988) and is underlain by the Tertiary Kerri-Kerri Formation, Cretaceous sedimentary rocks, and the crystalline basement complex. The lithological logs from the area are highly variable and one of the outstanding problems is the understanding of the hydrologic links between the aquifers and thus whether modern direct recharge is reaching the main groundwater system at depth.

Recent studies have shown that the water-table of the phreatic aquifer and piezometric head in deep artesian aquifers are rapidly declining and thus raising questions on the overall sustainability of the groundwater resources. The main control on sustainable resource is the balance between the rates of recharge and discharge. These fundamentals are not clearly understood in this region, and indeed in most semi-arid and arid regions. Little work has been carried out to assess the total water balance, and recharge and abstraction rates are poorly defined. There is the absolute need to better manage the existing groundwater resources in a sustained manner, as perennial surface water is almost non-existent. Since sustainable management especially from the context of water supply planning requires knowledge of the renewal rate, the estimation of groundwater recharge thus becomes fundamental.

Despite the indication of modern recharge in the region, the amount and distribution are not well known, neither are the relationships between the modern hydrological cycle and groundwater storage well understood. This paper estimates the amount of groundwater recharge to the south western part, which represents the Nigerian sector, of the Chad basin. Chloride data from rainfall, the unsaturated zone and groundwater were used to estimate

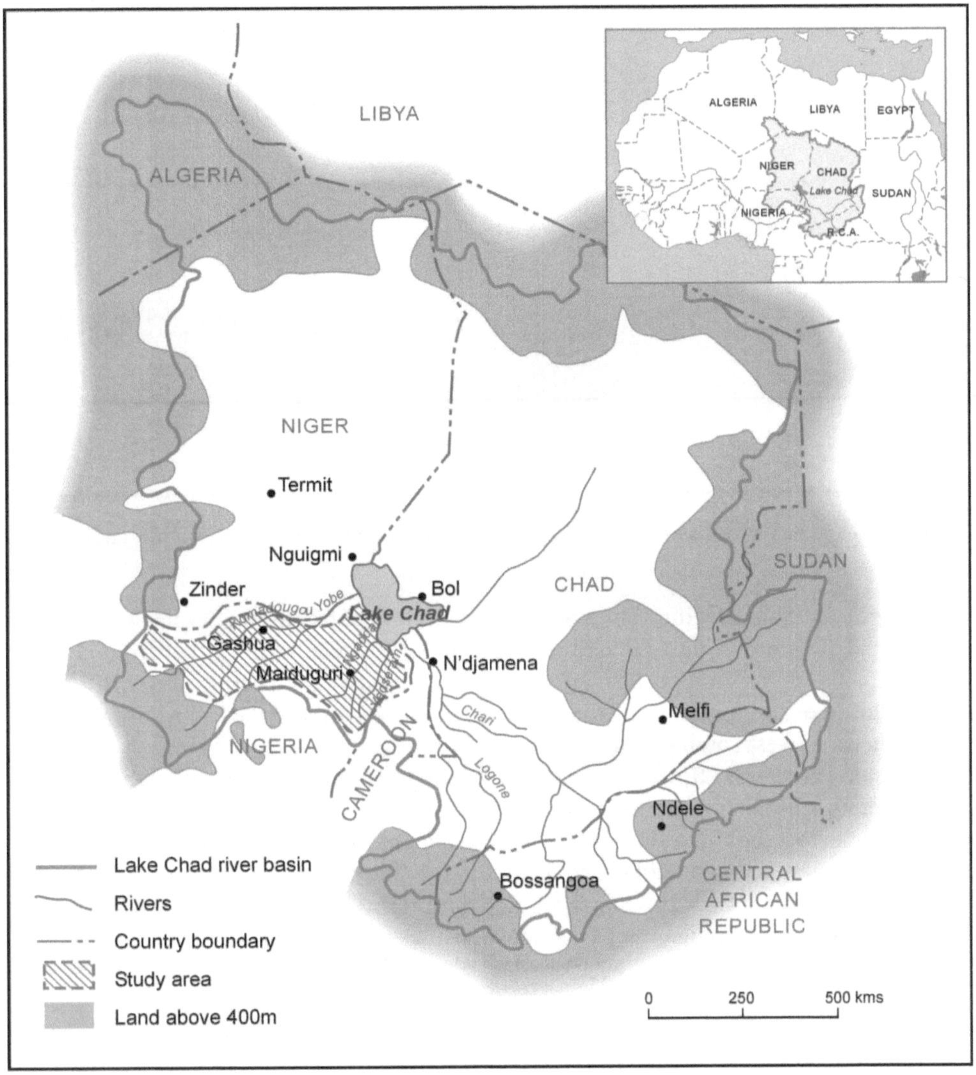

Figure 1. Map of the Chad basin showing the study area in the southwestern part (adapted from Schneider, 1989).

the diffuse recharge flux using a steady-state chloride mass balance approach (Allison & Hughes, 1978; Edmunds & Walton, 1980; Gaye & Edmunds, 1994; Goni, 2002).

2 CLIMATOLOGICAL SETTING

The rainfall distribution over the southwestern Chad basin and indeed over western Africa is determined by the position of the meteorological equator and its two associated structures, the ITF (Inter Tropical Front) and the ITCZ (Inter Tropical Convergence Zone). The ITCZ rarely exerts its influence over continental areas north of latitude of 12°N, the southern

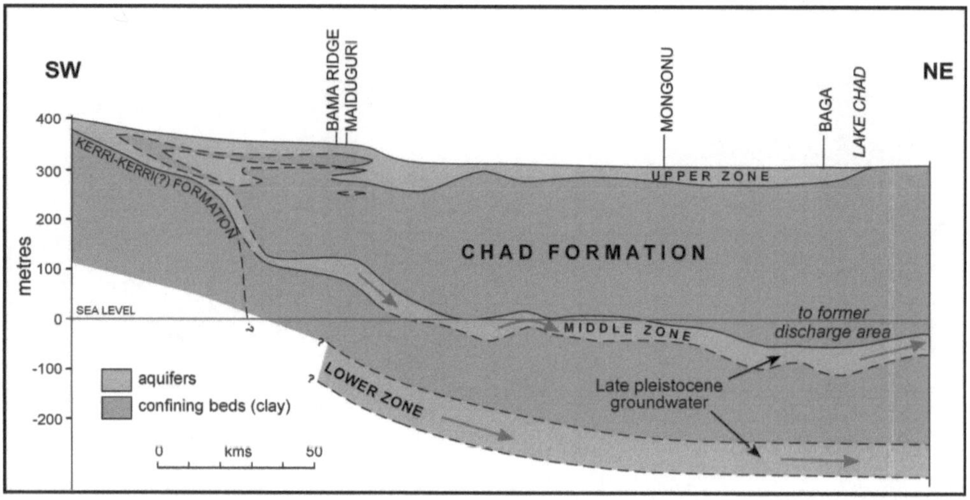

Figure 2. Geological cross section of the Chad Formation with the three aquifer zones indicated.

boundary of the Sahel region. Thus Sahelian rainfall depends almost exclusively on the position and structure of the ITF, and is mostly of convective origin, either from isolated cumulonimbus or from cloud formations that often evolve in the form of squall lines. Such squall lines are a feature of the Sahelian rainy season and they move in a general east/southeast to west/southwest direction (Lebel *et al.*, 1992).

In the study area, this climatic regime results in a long dry season (October to May) and a shorter rainy season (June to September), which are related to seasonal winds. During the winter months the cool, dry, dust-laden "harmattan" blows from the Sahara in the north, bringing low humidity, cool nights and warm days. In the summer months, moisture-laden winds blows from the Gulf of Guinea in the south, bringing higher humidity, rains, and more uniform diurnal temperature. The monsoon advances from the south so that the rains start earlier, are heavier and last longer southwards, although in general there is high spatial and temporal variability over the entire area. The present day rainfall at the Maiduguri station for the 2005 season is 862 mm, about 27% higher than the long term (1915–2005) average of the station of 626 mm (Figure 3).

3 HYDROGEOLOGICAL SETTING

The Plio-Pleistocene Chad Formation and the younger overlying Quaternary sediments are the main source of groundwater in the study area. The Chad Formation is essentially an argillaceous sequence in which minor arenaceous horizons occur (Barber, 1965), and the formation shows considerable lateral and vertical variability in lithology. Barber and Jones (1960) have named three clearly defined arenaceous horizons of the Chad Formation as the Upper, Middle and Lower Zone aquifers (Figure 2). The Lower and Middle Zones are confined, whereas the Upper Zone ranges from confined to semi-confined and unconfined in places. The Upper Zone sands are considered to be lake margin, alluvial fans or deltaic sediments related to sedimentation in and around Lake Chad, which has varied considerably

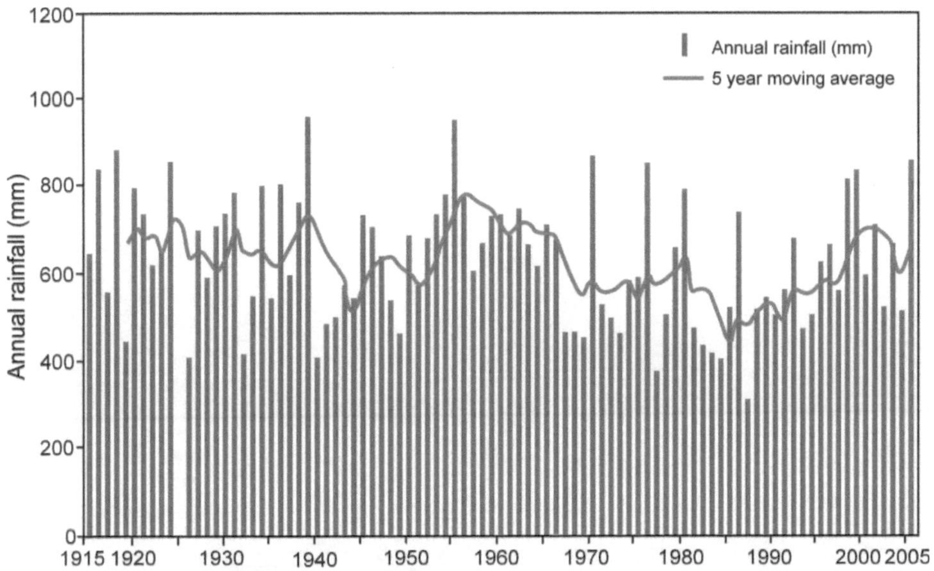

Figure 3. Annual rainfall distribution and its five year moving average for the Maiduguri station.

in size throughout the Quaternary (Durand, 1995). The clays are mainly lake deposits laid down under non-turbulent conditions and are most extensive near to the present day lakeshore. The lithological logs from the area are highly variable and it is currently not possible to present a typical stratigraphic profile of the near surface Formations. Around Maiduguri (type locality of the Chad Formation) the Upper Zone aquifer includes not only a surface zone of recent sands with an unconfined water table but deeper layers of sands from the Chad Formation, complexly intercalated between clays, and partially confined by the clays. Beacon Services Limited (1979) further subdivided the Upper aquifer system into three zones, an Upper A Zone under water table conditions and underlying B and C Zones which are semi-confined and/or confined in places.

4 METHODOLOGY

Samples of rainfall were collected from rain gauges by local meteorological observers at 3 sites in the south western Chad basin (Figure 4) – at Kaska (13°57.18'N, 10°80.02'), Garin Alkali (12°48.97'N, 11°03.07'E) and Maiduguri (11°51.88'N, 13°13.25'E) – on a storm-event basis for the rainy seasons of 1992 to 2004 (although with some gaps due to difficult logistical problems). The rainfall amount was measured at the end of each event and sample immediately poured into a Nalgene® bottle to minimise evaporation.

Hand augering was used to drill eight unsaturated zone profiles in the region (Figure 4). Samples were obtained and homogenised in a thick nylon bag over 0.25 m intervals for the first 10 m, and over 0.5 m interval thereafter to the total depth of profile. The homogenised samples were then subsampled into glass Kilner jars, which were sent to the British Geological Survey (BGS) laboratory, Wallingford, United Kingdom for analysis. Moisture contents are measured gravimetrically and chemical analyses were carried out on samples obtained either by centrifugation (Kinniburgh & Miles 1983) or by elutriation with distilled water.

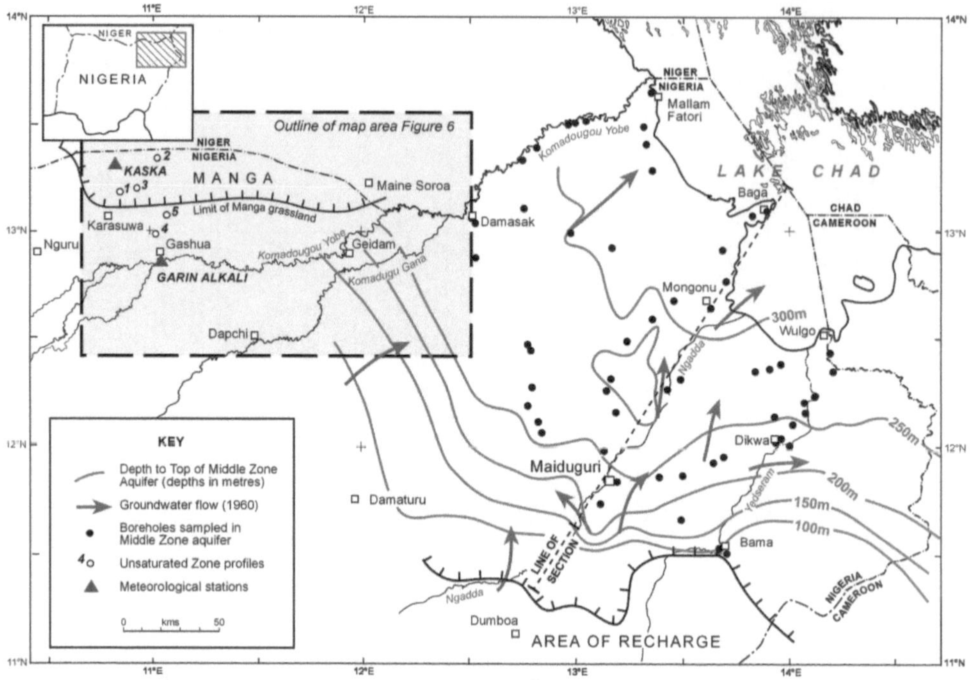

Figure 4. The study area showing the meteorological stations, unsaturated zone profiles and other hydrologic parameters (Modified from Edmunds *et al.*, 1999).

Surface water from Komadugu Yobe River, its tributaries and the Lake Chad were sampled at different points during the dry season. Groundwater samples from approximately 400 wells tapping the Upper A Zone aquifer and shallow boreholes were also collected with the assistance of North East Arid Zone Development Programme (NEAZDP) staff from villages across NE Nigeria. Most of these wells were located on the outskirts of the settled areas and away from areas of possible contamination from domestic/industrial sources. All were in daily use and all produced fresh water. Sampling of the surface water and groundwater were carried out in such a way that two samples were collected in polythene bottles – both filtered through Whatman 0.45 um membrane filters and one of the samples acidified to 1% with HNO_3. Samples were properly sealed and sent to BGS (United Kingdom) and GSF – National Research Centre for Environment and Health (Germany) laboratories for the chemical analysis. Chloride, bromide and nitrate concentrations were determined by automated colorimetry, while ICP-OES was used for other inorganic constituents.

5 RESULTS AND DISCUSSION

5.1 *Rainfall*

Rainfall chemistry has been measured in this study as an input to estimate present day rates of recharge. Results for weighted mean chloride deposition relative to total rainfall measured at Kaska, Garin Alkali and Maiduguri stations for the period 1992 to 1997 and 2000 & 2004 respectively are presented in Table 1.

Table 1. Weighted mean chloride values for rainfall stations in this study.

Station and year	Rainfall (mm)	Weighted mean Cl (mg/l)
Kaska 1992	320.5	2.8
Kaska 1993	327.3	1.3
Garin Alkali 1992	549.4	1.6
Garin Alkali 1995	614.5	3.4
Garin Alkali 1996	297.5	0.6
Garin Alkali 1997	226.7	0.7
Maiduguri 2000	502.1	2.0
Maiduguri 2004	349.8	0.6

Weighted mean chloride for each station = Σ [(individual event total/season total) × Cl (individual event)]

The accumulation of chloride occurs evenly throughout the season, after an initial high deposition at the beginning. This accumulation pattern is temporally and spatially uniform; however, localized convective events may change the general deposition pattern (Goni *et al.*, 2001). This supports the assumption that the long term average chloride in precipitation is generally constant; an assumption used for chloride recharge estimation methods. Although, it is note-worthy that concentrations in rainfall are the greatest source of error in this method of recharge estimation.

The analysed rainfall chloride includes a component of dry deposition as an artefact, producing bias in the recharge estimates. The rain gauge probably acts as a preferential collector of dry deposition, which does not always settle on the land surface (Edmunds *et al.*, 2002). For the long term, however, there will be some smoothing in the signal so that use of the average Cl value of 1.63 mg/l (here over eight years) may be appropriate for recharge estimation. However, values obtained from the modelling of Cl in rainfall gave an average of 0.65 mg/l (Goni, 2002), which implies that in this region the effective Cl deposition on the land surface is lower than that measured in rain gauges. Possible explanation could be that the rain gauge collects dust storms commonly associated with rainfall in this region, which do not settle on the land surface (Goni, 2002). Cl measured using an automatic precipitation collector that avoids the aerosols deposits before rainfall gave a value of 0.3 mg/l for the 1996 rains at Banizoumbou, in the Sahelian Savannah of Niger (Galy-Lacaux & Modi, 1998) similar to the modelled value. Measurement of the dry deposition at the same site (Banizoumbou) gives 0.7 mg/l Cl, constituting about 70% of total deposition. The 1.63 mg/l Cl measured in this study is therefore likely to be the total deposition, in which case the 0.65 mg/l model value is the approximate precipitation component of the total.

5.2 *Unsaturated zone*

Rainwater falling during the summer monsoon event has a widely varying intensity and chemical composition, and on entering the soil undergoes evaporation and drives geochemical reaction. In the sandy soils of semi-arid and arid zone, strong weathering takes place in the upper 1.5 m (Gaye & Edmunds, 1996) and this may give rise to a zone with higher concentration of Cl and other solutes. In this zone mineralisation occurs due to sequences of

Table 2. Characteristics of the eight unsaturated zone chloride profile sites and the calculated rate of recharge (Goni, 2002).

Profile	Geology	Depth (m)	No. Samples (n)	Rainfall P (mm)	Mean Cl C_s (mg/l)	Mean annual recharge R_d (mm/a)
(1) GM 1	Manga dune sands	*15.5	51	400	14	50
(2) MD 1	Manga dune sands	22.5	65	400	21	34
(3) MN 1	Manga dune sands	*16.5	53	400	41.5	17
(4) N-TM	Alluvial sands	*18.75	58	400	11.7	60
(5) W-WGR	Alluvial sands	*19.25	60	400	17.7	40
(6) MG	Fixed dune sands	16.25	53	600	29.5	36
(7) KA 1	Manga dune sands	*15.5	50	400	18.3	39
(8a) MF (upper)	Fixed dune silt	0.0–3.0	10	400	47	15
(8b) MF (Lower)	Lacustrine silt/Clay	3.0–16.0	42	400	2892	0.2

* Profile has reached the water table.

seasonal wetting and drying, which modifies the input chemistry for some elements resulting from water-rock interaction, although Cl and some other elements remain inert. Macropore (by-pass) flow is restricted to this upper horizon. Below this depth (perhaps up to 2.5 m in vegetated areas) a steady-state chloride profile generally develops, which represents the output from the soil zone and within which any incoming rainfall variations are well mixed.

Interstitial water samples from the unsaturated zone were obtained by both centrifugation and elutriation to investigate the infiltration of rainfall, rates of movement (residence times) and hence the recharge rates to the groundwater. Chloride is used as the principal variable since it remains inert during recharge process and (unlike water) is conserved in the unsaturated zone. Provided all Cl is atmospherically derived its concentration in the unsaturated zone may be used to estimate recharge (Gaye & Edmunds, 1996; Allison *et al.*, 1994). In Table 2 results of recharge and residence time calculations are shown for eight unsaturated zone profiles from the study area. Depth distributions of Cl in these profiles are shown in Figure 5.

Note that the precipitation of 600 mm for the MG profile in Table 2 is higher than the rest. MG is located at latitude 12°N (where higher annual rainfall is expected) while the other profiles are situated north of latitude 13°N.

The direct recharge R_D is calculated using the formula $R_D = PC_P/C_S$ where P is the regional rainfall, C_P is the spatially-averaged Cl concentration in rainfall and C_S is the mean unsaturated zone Cl concentration below the weathering/mixing zone (Edmunds *et al.*, 1988).

The mean chloride concentrations in the eight profiles from the unsaturated zone of the study area range from 12–42 mg/l (Goni, 2002) when the MF (Lower) profile is not considered. Corresponding recharge rates derived from the profiles range from 15 to 60 mm/a, representing 4 to 15% of the rainfall over the past three decades in an area previously thought to have little or no recharge (IWACO, 1985). The MF (Lower) profile is composed of low permeability clayey deposit, where direct recharge is insignificant. In the Manga

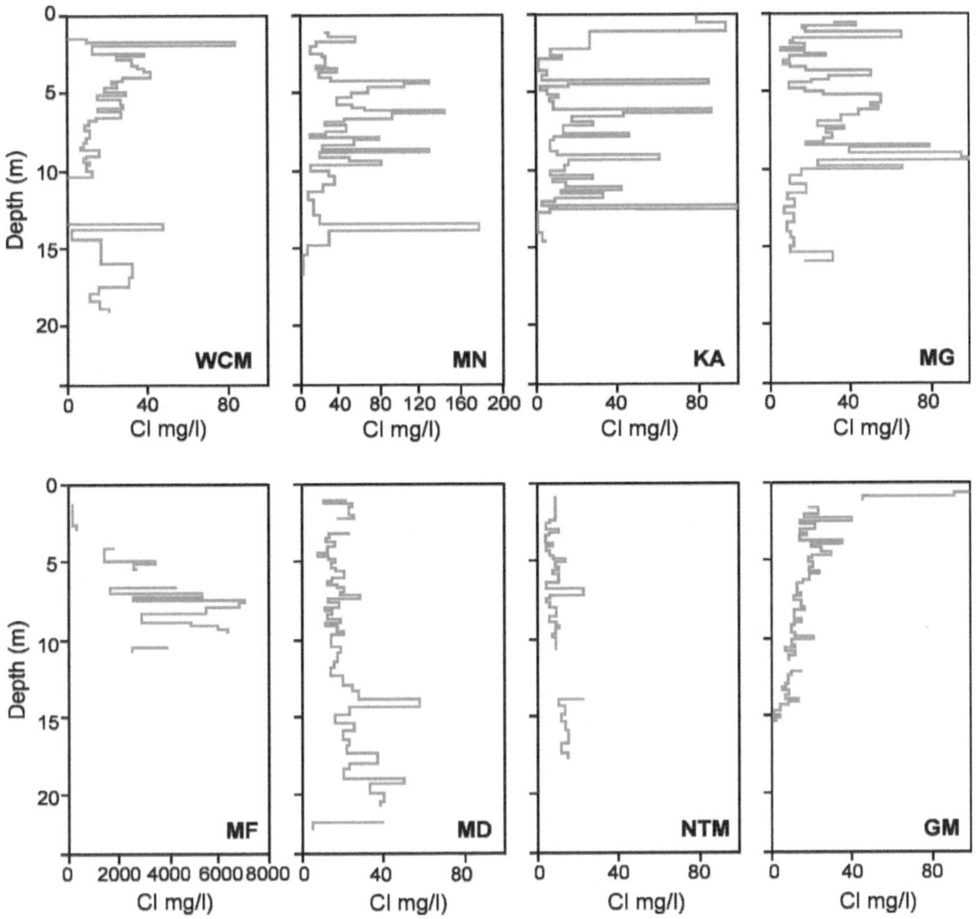

Figure 5. Depth distribution of Cl concentration in the unsaturated zone from depth profiles.

Grassland dunes the average recharge is 40 mm/a (Goni, 2002). These values are slightly lower than recharge estimates (60 mm/a) derived for the Manga Grassland from the water balance of the Kajemarum oasis (Carter, 1994). However the chloride profile data are long term estimates spanning up to 21 years and therefore give an estimate of the recharge rates over the time scale of the recent Sahel drought; values of recharge at least 30% higher might be expected under the wetter climatic periods which characterised the long term.

5.3 *Shallow groundwater*

The hydrogeochemical distribution of chloride measured in wells and boreholes is shown in Figure 6. The concentrations range widely from 0.5–96 mg/l, with an average of 16 mg/l. Most of this represents water sampled directly from the water-table. This mean value may be considered as a minimum long term (e.g. 30 years) average recharge value since it may include sites influenced by contamination or by palaeosalinity. However, the concentration range reflects the heterogeneity in recharge rates, resulting from differences in soil type and

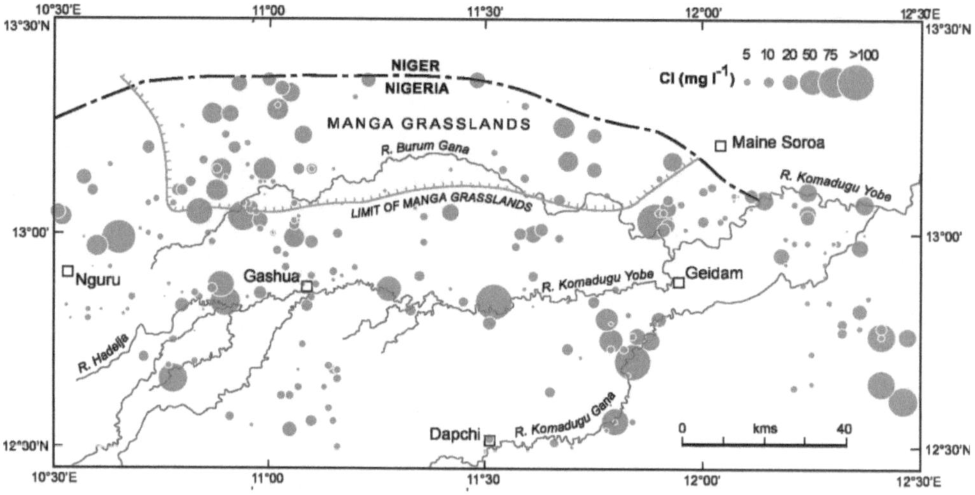

Figure 6. Regional distribution of Cl in the shallow groundwater of the study area (after Edmunds *et al.*, 1999).

depth, vegetation type and coverage, as well as topographic slope variations. Some lower values reflect preferred infiltration routes; no geographical bias to Cl is observed. Some of the high values probably relate to salinity acquired near playas or modern lakes, although, notably, no very high Cl samples occur, suggesting that saline lakes (or their vestiges) are rare and very local in extent (Edmunds *et al.*, 2002).

From the evidence obtained from the unsaturated zone, it is most likely that the water-table Cl concentrations represent atmospheric inputs and, therefore may be used to determine recharge rates at the regional scale. Applying the Cl mass balance equation ($R_d = C_pP/C_g$), taking the regional average precipitation at 500 mm, regional average Cl in precipitation of 1.63 mg/l and the average Cl in the saturated zone of 16 mg/l (from the 410 analysis of groundwater), an estimated long term recharge rate of 50 mm/a is obtained. This higher estimate at a regional scale, as compared to the profile results, could be attributable to additional sources of recharge apart from the direct infiltration via the unsaturated zone; the average of the direct infiltration has been estimated at 40 mm/a using the same equation. Edmunds *et al.* (1999) have identified the other possible sources of recharge as ponding and surface runoff. The wells in Lambawa and Abadam show clearly that they are hydrologically connected with the nearby Yobe River. The similarity in concentrations of conservative Cl between these waters supports this assertion, and also adds weight to the view that other sources of recharge apart from through the unsaturated zone exist. Importantly, this implies a very large groundwater resource of approximately 10.2 Ml km^2/a for the sand-covered part of the Sahel region (Goni, 2002).

These recharge estimates are derived using the measured Cl in rainfall, which has a mean value of 1.63 mg/l. However, Goni *et al.*, (2005) obtained a lower Cl concentration in rainfall of 0.65 mg/l from modelling. Therefore, if the average modelled Cl value is used, the mean regional recharge rate reduces to 20 mm/a. Therefore, present day recharge rates for the study area are estimated to range between 20–50 mm/a. This is in agreement with the range (20–50 mm/a) reported by Favreau *et al.* (2002), for the southwest of Niger, an area with a similar semi-arid landform.

5.4 *Conceptual model of recharge*

A conceptual model of recharge for this area is presented in Figure 7. The model is based on 152,000 km^2 as the area of study, an average rainfall of 500 mm (although a higher amount of 1200 mm is used for the south and southwest) and Cl concentration in rainfall of 0.65 mg/l is used as the regional long-term value (Goni *et al.*, 2005). The Cl in the groundwater of the south and southwest is 2 mg/l, in river water and depressions it is estimated at 3 mg/l and in the unsaturated zone it is 22 mg/l (an average from the profiles in this study).

The conceptual model (Figure 7) recognises three mechanisms of recharge and attempts to evaluate the contributions of each mechanism to the groundwater of the region (Goni 2005). From the model, considerable lateral recharge (~390 mm/a) via regional groundwater flow is taking place from the south and southwest. This appears reasonable considering the high rainfall (over 1200 mm) and fractured nature of the bedrock in the area, which favours rapid infiltration. The model estimates that only 1% of the area (about 1520 km^2) has this recharge amount and therefore its annual contribution to the overall regional groundwater recharge is about 4 mm (Goni, 2005).

Surface water recharge, for example via river channels, depressions, pools etc., is estimated at 108 mm/a over about 8% of the study area (12,160 km^2); which also includes fossil river channels. In the Hadejia and Hadejia-Nguru wetlands recharge to the shallow aquifer via floodwater infiltration has been estimated to range between 73–197 mm, with an average of 132 mm (Goes, 1999). Adeniji (1991) estimates substantial annual quantity of water (14.8 Mm^3) that flows from Lake Chad to recharge the Upper aquifer of the Nigerian sector of the basin. The annual contribution from these surface sources of recharge to the regional groundwater is about 8.5 mm (Goni, 2005). Hydrological depressions have been mention in the Chad basin with one in the Nigerian sector around Gubio town near the Lake Chad (Schneider, 1989). This explains the flow from the Lake Chad to recharge the Upper aquifer.

From the model, approximately 70% of the study area (about 106,400 km^2) receives direct vertical recharge via the unsaturated zone. Using the Cl concentrations in the unsaturated zone (average of 22 mg/l), recharge rate is estimated at 15 mm/a. This contributes about 10.5 mm to the regional groundwater recharge annually. For 21% of the area (e.g. the clayey areas in the eastern part of the study area) there is no recharge.

The model estimates total regional recharge rate of 23 mm/a, which is in reasonable agreement with the long term spatially average regional recharge rate of 20 mm/a obtained by Cl mass balance calculations (Goni, 2002). Carter (1994) and Carter *et al.* (1994) use piezometric fluctuations of boreholes normal to the interdune lakes and playas to indicate that flow gradients are towards the lake, which implies present day recharge through the dunes. In their studies, recharge rate is estimated at 60 mm/a, the high rate is as a result of the loose unconsolidated dune sands overlying the area. Also in the southwest of Niger with similar semi-arid landform, present day recharge rates of between 20–50 mm/a have been reported (Favreau *et al.*, 2002). These results reinforce the concept of active modern day recharge in the region; dispelling the earlier assumptions that effective recharge by precipitation is zero.

The recharge estimates obtained from the modelling appear to be lower than those obtained from the Cl mass balance equation and piezometric fluctuations. The model results of recharge are considered as the minimum long-term average over the region. The conceptual model will require more accurate data demarcating areas covered by the various recharge mechanism (rivers, depressions and high lands in the south and southwest) for it

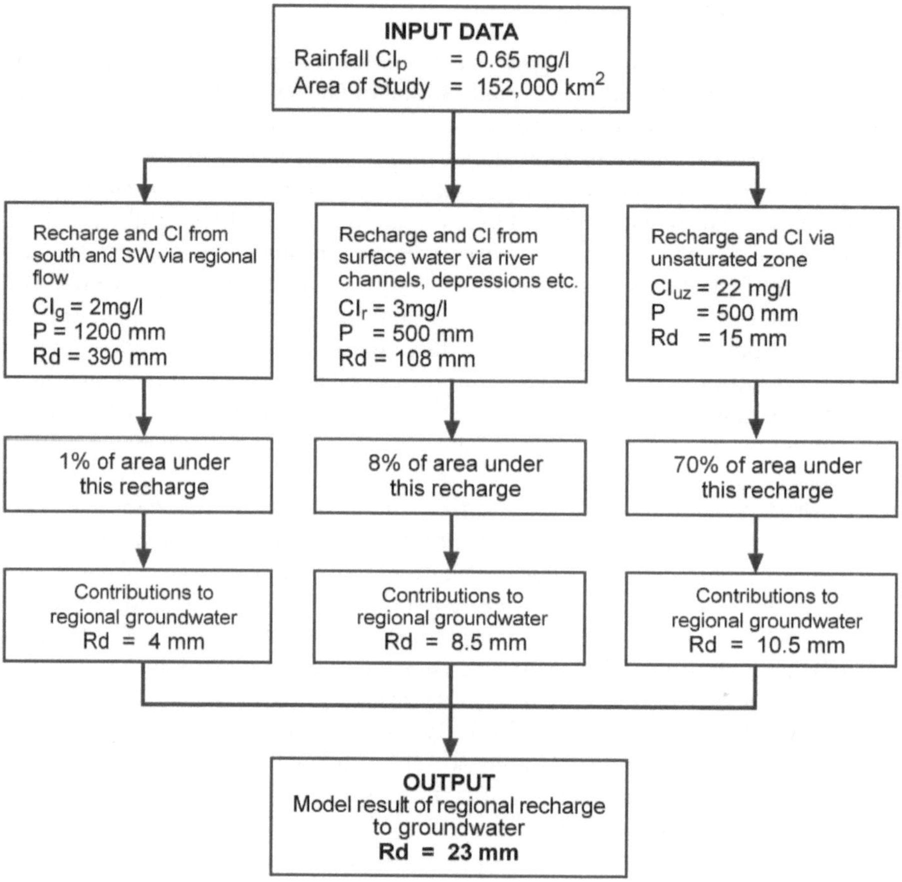

Figure 7. Conceptual model showing the amount of recharge and Cl deposition to the regional groundwater of the study area from the different recharge mechanisms.

to be fine-tuned. However, the good agreement in regional recharge values between the model and estimates from other methods gives an indication that it is a reasonable concept.

Overall, the geochemical data have shown that the shallow aquifer in the Nigerian sector of the Chad basin receives present day recharge estimated at 20–50 mm/a. This is considerably in excess of the local demands (approximately 2.4 mm/a) in the rural communities (NEAZDP, 1990). This demand represents only a small abstraction via hand-dug wells for domestic purposes, with the technology itself imposing limitation on the quantity that can be withdrawn.

It thus appears that the shallow aquifer receives considerable recharge that can sustain present day abstraction levels of the hand-dug wells commonly used in the villages. However, caution must be exercised in introducing mechanised pumping for urban or rural supplies. Such changes should be preceded by detailed studies. The observed decline in the water table in recent years should most probably be due to the reduced rainfall during the Sahel drought and consequent reduction in recharge, although human impacts may also be significant.

6 CONCLUSION

Geochemical data were used to estimate rates of recharge to the shallow groundwater in the Nigerian sector of the Chad basin. Chloride concentrations in rainfall for stations in the study area give a mean value of 1.63 mg/l, although a much lower value of 0.65 mg/l was obtained from modelling.

Estimates of groundwater recharge obtained from the Cl mass balance method using the eight profiles from the unsaturated zone produce an average value of 40 mm/a. The regional recharge estimates are also obtained using the same approach with Cl concentration in groundwater, and these give a mean recharge rate of 50 mm/a, indicating additional sources of recharge bypassing the unsaturated zone. The conceptual model recognises three recharge mechanisms and estimates total regional recharge rate of 23 mm/a to the groundwater in the south western Chad basin, using lower Cl input of 0.65 mg/l from rainfall. Stable isotope data were also used to demonstrate that present day rainfall is recharging the shallow groundwater.

The renewal rate of the shallow groundwater in this region is estimated at between 20–50 mm/a, which is significant and can sustain current rate of abstraction by dug wells (2.4 mm/a). But it probably cannot sustain mechanised pumping. The future of this region lies in the sound management of the shallow groundwater, since the deep confined aquifers contain palaeowater that has not probably been coupled to the present hydrological cycle. In this water scarce region demand management strategy must be adopted aimed at reducing usage and eliminating wastage.

REFERENCES

Adeniji, F. A. 1991. *Groundwater management and recharge potentials in the Chad basin of NE Nigeria.* Nigerian Journal of Water Resources, **1**, 29–48.

Allison, G. B., Gee, G. W. & Tyler, S. W. 1994. *Vadose-zone techniques for estimating groundwater recharge in arid and semi-arid zones.* Soil Society of America Proceedings, **58**, 6–14.

Allison, G. B. & Hughes, M. W. 1978. *The use of environmental chloride and tritium to estimate total local recharge to an unconfined aquifer.* Australian Journal of Soil Research, **16,** 181–195.

Barber, W. 1965. *Pressure water in the Chad Formation of Bornu and Dikwa emirates, north-eastern Nigeria.* Bulletin of the Geological Survey of Nigeria, **35**.

Barber, W. & Jones, D. G. 1960. *The geology and hydrology of Maiduguri, Bornu province.* (unpublished records) Geological Survey of Nigeria.

Beacon Services Ltd. 1979. *Maiduguri water supply, Hydrogeological report Vol. 1, Hydrogeology and future exploitation of groundwater.* Consulint Int. S.r.l., 158pp.

Carter, R. C. 1994. *The groundwater hydrology of the Manga Grassland north east Nigeria; importance to agricultural development strategy for the area.* Quarterly Journal of Engineering Geology, **27**, 73–83.

Carter, R. C., Morgulis, E. D., Dottridge, J. & Agbo, J. U. 1994. *Groundwater modelling with limited data: a case study in a semi-arid dunefield of northeast Nigeria.* Quarterly Journal of Engineering Geology, **27**, 85–94.

Durand, A. 1995. *Quaternary sediments and climates in the central Sahel.* African Geoscience Rreview, **2**, 323–614.

Edmunds, W. M., Darling, W. G. & Kinniburgh, D. G. 1988. *Solute profile techniques for recharge estimation in semi-arid and arid terrain.* In: Simmers, I. (ed.) Estimation of natural groundwater recharge, D. Reidel Publishing, 139–157.

Edmunds, W. M., Fellman, E. & Goni, I. B. 1999. *Lakes, groundwater and palaeohydrology in the Sahel of NE Nigeria: evidence from hydrogeochemistry*. Journal of the Geological Society, London, **156**, 345–355.

Edmunds, W. M., Fellman, E. Goni, I. B. & Prudhomme, C. 2002. *Spatial and temporal distribution of groundwater recharge in northern Nigeria.* Hydrogeology Journal, **10**, 205–215.

Edmunds, W. M. & Walton, N. R. G. 1980. *A geochemical and isotopic approach to recharge evaluation in semi-arid zones- past and present.* In: Application of Isotopic techniques in Arid Zone Hydrogeology. Proc. Advisory Group Meeting, IAEA Vienna, 1978, 47–68.

Favreau, G., Leduc, C., Marlin, C., Dray, M., Taupin, J., Massault, M., Le Gal La Salle, C. & Babic, M. 2002. *Estimate of Recharge of a Rising Water Table in Semiarid Niger from ^3H and ^{14}C Modeling.* Ground Water, **40**, 144–151.

Galy-Lacaux, C. & Modi, A.I. 1998. *Precipitation chemistry in the Sahelian Savanna of Niger*, African Journal of Atmospheric Chemistry, **30**, 319–343.

Gaye, C.B. & Edmunds, W.M., 1996. *Inter comparison between physical, geochemical and isotopic methods for estimating groundwater recharge in north western Senegal.* Environmental Geology, **27**, 246–251.

Goes, B. J. M., 1999. *Estimate of shallow groundwater recharge in the Hadejia-Nguru Wetlands, semi-arid Northeast Nigeria.* Hydrogeology Journal, **7**, 294–304.

Goni, I. B. 2002. *Realimentation des eaux souterraines dan le secteur Nigerian du bassin du lac Tchad: approche hydrogeochimique*. These de Doctorat d'etat, Universite d'Avignon, France.

Goni, I. B. 2005. *Conceptual model of recharge in the Nigerian sector of the Chad basin using Cl mass balance.* African Geoscience Review. **12**, 61–68.

Goni, I. B., Fellman, E. & Edmunds, W. M. 2001. *A geochemical study of rainfall in the Sahel region of northern Nigeria.* Atmospheric Environments, **35**, 4331–4339.

Goni, I. B., Travi, Y. & Edmunds, W. M. 2005. *Estimating groundwater recharge from modelling unsaturated zone chloride profiles in the Nigerian sector of the Chad basin.* Journal of Mining and Geology, **41**, 123–130.

Gvirtzman, H. & Margaritz, M. 1986. *Investigation of water movement in the unsaturated zone under an irrigated area using environmental tritium.* Water Resources Research, **22**, 635–642.

IWACO 1985. *Study of the water resources in the Komadougou Yobe basin.* Report for the Nigeria-Niger joint commission for cooperation, Niamey, Niger. IWACO BV Rotterdam.

Kinniburgh, D. G. & Miles, D. L, 1983. *Extraction and chemical analysis of interstitial water from soils and rock.* Environmental Sciences Technology, **17**, 362–368.

Lebel, T., Sauvageot, H., Hoepffner, M., Desbois, M., Guillot, B. & Hubert, P. 1992. *Rainfall estimation in the Sahel: the EPSAT-NIGER experiment.* Journal of Hydrological Sciences, **37**, 201–215.

NEAZDP, 1990. *Groundwater Resources Report.* Report of the North East Arid Zone Development Programme, Garin Alkali, Gashua, Nigeria.

Oteze, G.E. & Fayose, S.A. 1988. *Regional development in the Hydrogeology of Chad basin.* Water Resources, **1**, 9–29.

Phillips, F. M., Mattick, J L., Dural, T. A., Elmore, D. & Kubik, P. W. 1988. *Chlorine-36 and tritium from nuclear weapons fallout as tracers for long term liquid and vapour movement in desert soils.* Water Resources Research, **24**, 1877–1891.

Schneider, J. L. 1989. *Geologie et Hydrogeologie de la republique du Tchad.* These de Doctorat d'etat, Universite d' Avignon, Vol. I.

Zimmermann, U., Munnich, K.O. & Reother, W. 1967. *Downward movement of soil moisture traced by means of hydrogen isotopes.* In: Stout, G. E. (ed.) Isotope Techniques in the Hydrologic Cycle. From Symposium on Isotope Techniques, Urbana, Ill., Nov. 10–12, 1965. American Geophysical Union, Washington, D. C., 28–36.

CHAPTER 20

Groundwater in North and Central Sudan

P. Vrbka
Dieburger Str. 108, Groß-Zimmern, Germany

R. Bussert
Technische Universität Berlin, Institut für Angewandte Geowissenschaften, Berlin, Germany

O.A.E. Abdalla
Sultan Qaboos University, College of Science, Department of Earth Sciences, AlKhod, Sultanate of Oman

ABSTRACT: North and Central Sudan represents the hydrogeological interface between the sedimentary Blue and White Nile Basins and the large Nubian Aquifer System of North Sudan. The average annual precipitation increases from north (<50 mm) to south (380 mm). Stable isotopes of >320 water samples reveal three possible mechanisms of groundwater recharge: a) natural recharge along wadis and in lowlands during floods, b) man-induced infiltration of irrigation water, and c) natural river bank infiltration along rivers. Groundwater more than 25 km away from rivers carries the isotopic signature of 'palaeo-groundwater' in many places. Since the Early Holocene 'climatic optimum', direct recharge by precipitation is considered as negligible in the northern region. In the southern region of the investigated area there are indications of a modern recharge. However, the spatial significance and the quantity are unknown and further research is needed. The biggest challenges for the future are quantifying the available renewable resource and managing groundwater abstraction sustainably.

1 INTRODUCTION

The purpose of the present investigation is to shed light on the groundwater resources of North and Central Sudan, on the origin and recharge, principally by means of the study of the environmental isotopes ^{18}O and ^{2}H. Owing to the complexity of hydrogeological research, most of the studies carried out in the region so far have been rather localized, small-scale and scattered. For example, Bonifica (1986, 1987) provided unpublished reports of hydrogeological studies in the area north of Dongola (see Figure 1). Kheir (1986) concentrated his investigation on the region around Dongola. The work of Jacob (1990), carried out under the auspices of the IAEA (International Atomic Energy Agency), was carried out around Atbara, Shendi and in the Gezira south of Khartoum. Vrbka (1996) conducted research in the area between Khartoum, Wadi Muqaddam and Wadi el Milk, whereas Abdalla (1999) investigated the hydrogeology of sedimentary basins south-west of Khartoum.

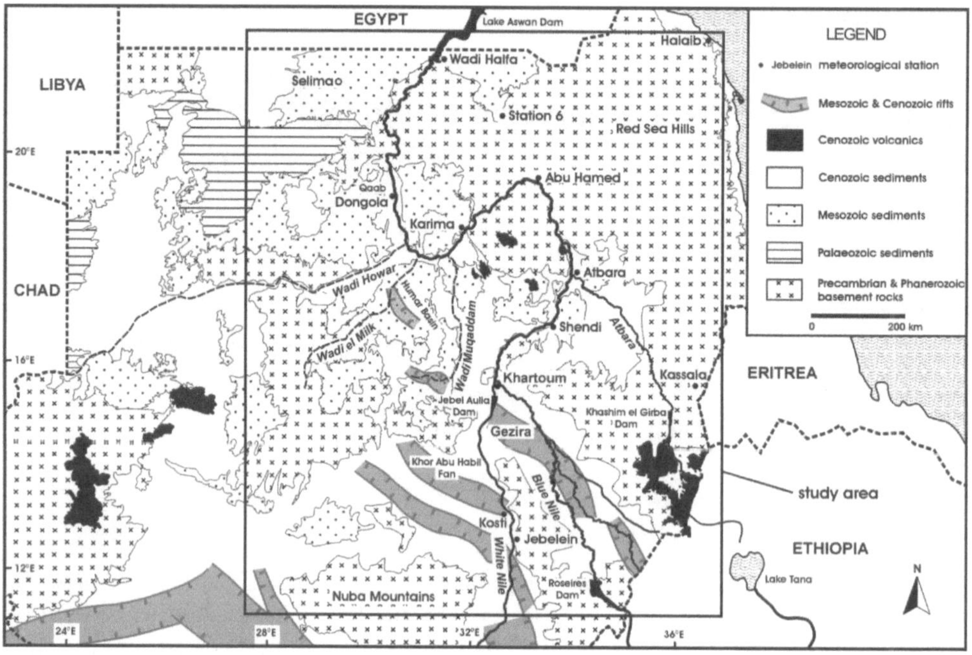

Figure 1. Simplified geological map of the study area, with meteorological stations, 'sampling areas' and discussed localities. Geological map based on Schandelmeier & Reynolds (1997), GRAS (2004)

In this paper, the authors consider a region covering 1000 km in east-west and 1200 km in north-south direction, in total an area of 1.2×10^6 km^2 (Figure 1), whereas the entire Sudan covers approximately 2.5×10^6 km^2. Hence, this study is an approach to integrate results of local and scattered studies into a comprehensive, regional description.

With the objective to present climatological data, twelve (out of approximately 130) representative meteorological stations have been selected (Figure 2): Halaib, on the shore of the Red Sea; Wadi Halfa at the border to Egypt; Station 6, a railway station in the eastern Nubian desert; Abu Hamed, Dongola, Karima, Atbara and Shendi at the River Nile; Kassala at the border to Eritrea; Khartoum at the confluence of the Blue and the White Nile; Kosti and Jebelein at the White Nile. Since it is not possible to show all water sampling locations (>320) in Figure 1, 'sampling areas' with a high density of sampling points are defined and described in the following: East of the Blue Nile; between the White and Blue Nile (Gezira); the area around Khor Abu Habil; around Khartoum, Shendi and Atbara; south of Wadi Muqaddam, an area with basement outcrops; the Wadi Muqaddam and the northern part of Wadi el Milk; the northern Nile bend between Karima, Dongola and the Oasis Selima. Thus, large areas in northern Sudan with full desert climate are included; for position of sampling points see: Bonifica (1986); Kheir (1986); Vrbka *et al.* (1993); Vrbka (1996); Abdalla (1999).

As emphasis is on climate and hydrology, a brief summary of hydrogeological characteristics is given next. The depth-to-groundwater only amounts to a few meters in the Nile valley, but may reach >250 m in remote and elevated areas, e.g. in the zone of the Humar

Basin. In the northern part the general flow direction of groundwater is towards the north, with a gradient in the range of 10^{-3} to 10^{-4}. In the 'southern region' the flow situation is more differentiated (Vrbka 1996; Abdalla 1999). Porosities of the Cretaceous sediments can reach up to 0.25 (25%). The hydraulic conductivities, from pumping test analyses, range from 10^{-6} to 10^{-4} m/s.

2 RESULTS AND DISCUSSION

2.1 *Climate*

The rainfall of twelve selected stations situated between 12°N and 22°N is shown in Figure 2. The record may be rather short and interrupted as for Station 6 (record from 1950–1988, N = 32), but may be also long and continuous as for Khartoum (1899–1996, N = 98). The stations are grouped into a 'Northern region' and a 'Southern region'. In the north, the elevation of the ground starts with 2 m asl in Halaib and rises to 183 m asl in Wadi Halfa, then ascends from 380 m asl in Khartoum to 500 m asl in Kassala. An elevation difference of only 5 m is recorded along the White Nile valley between Khartoum (380 m) and Jebelein (385 m). However, in some areas the elevations rise above 700 m a.s.l., e.g. Humar Basin, Red Sea Hills, and Nuba Mountains.

In the north, precipitation is as little as 3 mm/a in Wadi Halfa at the border to Egypt and amounts to 28 mm/a in Karima at the northern bend of the Nile. At the Red Sea shore in Halaib, the long-term record is 33 mm/a, whereas Dongola in the Sahara desert receives as little as 18 mm/a, a similarly low value as in Karima. In the south there is a significant increase: Atbara (71 mm/a), Shendi (101 mm/a), Kassala (298 mm/a), Kosti (380 mm/a), Jebelein (377 mm/a).

Precipitation generally occurs as summer monsoon rain, with a maximum in August. The prevailing winds blow from the north during winter, whereas southern directions dominate during summer. Only in the north, where different wind patterns are encountered, the maximum precipitation occurs during other months, e.g. in Halaib – on the Red Sea shore – in November. The mean monthly temperatures range between 23°C in January and 34°C in June, but reach extreme values close to 0°C or up to 50°C on single days. The average relative humidity varies from about 20% to 30%, and the evapotranspiration (PICHE) may reach values >6000 mm/a.

The long-term rain pattern is demonstrated by using the longest available record, that of station Khartoum, which starts in 1899 (Figure 3). There is a significant variance with shifts between less than 50 mm and more than 350 mm for single years. The mean for the shown period is 153 mm/a, whereas the median is 138 mm/a. It is striking that both the maximum value of 415 mm/a and the minimum values of only 4 mm/a have been observed during the past two decades. The minimum value of as little as 4 mm/a occurred twice, in 1984 and 1990; the maximum value of 415 mm/a in 1988 caused strong flooding. The 10-year mean also has strong shifts, with the biggest 'jump' between the second (107 mm/a) and the third decade shown (223 mm/a). Based on the present data record, there is a trend towards lower values during the 20th century. Although low precipitation is a continuous problem especially for the northern region, the irregularity of rainfall – with the possibility of a drought in certain years – makes the situation even worse for the rural areas, where no irrigation water may be available.

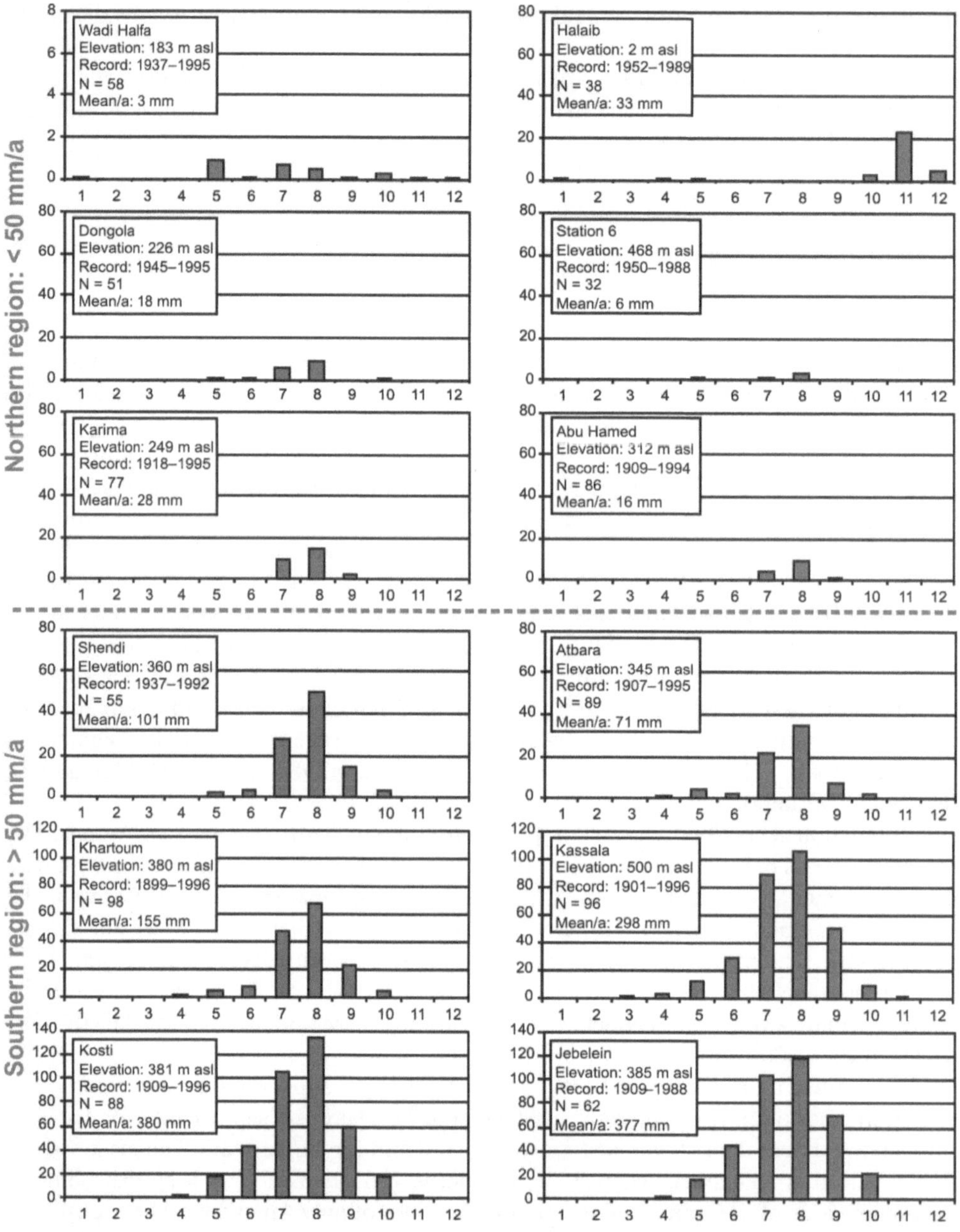

Figure 2. Monthly precipitation in mm of selected stations named in Figure 1; 1 = January, 12 = December; scales differ (Source: Agrometeorology Group-FAO-SDRN; NOAA)

2.2 *Geological setting*

North and Central Sudan consist of large areas of Precambrian basement rocks, broad sedimentary basins of Mesozoic to Cenozoic age, solitary Palaeozoic and Mesozoic anorogenic ring complexes, isolated domains of Late Cretaceous to Cenozoic volcanic rocks,

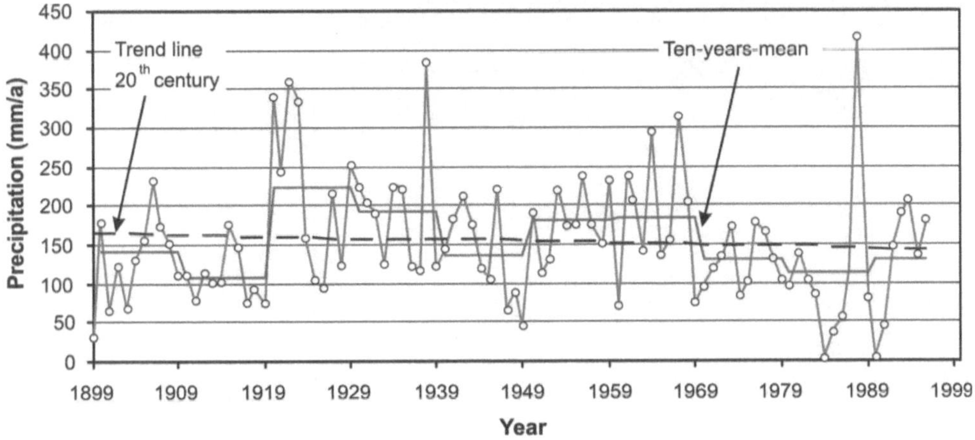

Figure 3. Data of the climatic station Khartoum, with trend line for the 20th century and ten-years-means (Source: Agrometeorology Group-FAO-SDRN; NOAA)

and an extensive cover of Quaternary sediments (Figure 1). The sedimentary basins and their Mesozoic to Cenozoic sediments constitute the major aquifers, whereas the basement rocks, the anorogenic ring complexes and the volcanic rocks mainly form minor aquifers or aquitards.

The Precambrian basement rocks consist of gneisses, granites, schists, quarzites and marbles, with minor intercalations of metagabbros and serpentinites. Most of the basement areas east of the Nile belong to the Arabian-Nubian Shield and are Neoproterozoic (Pan African) in age. In contrast, the majority of the basement areas west of the Nile form part of the East Saharan craton and represent pre-Neoproterozoic crustal domains that were partly rejuvenated by deformational and magmatic processes during the Neoproterozoic (Schandelmeier *et al.*, 1987, 1990). The Proterozoic basement was intruded by anorogenic ring complexes of Palaeozoic or Mesozoic age which are built up of granites, syenites and volcanites (Vail, 1985). During most of the Palaeozoic and the early Mesozoic, the study area was dominated by erosion and/or non-sedimentation (Klitzsch, 1984, 1986). Sedimentary rocks of Palaeozoic age are therefore largely missing in the study area. In the aftermath of the Hercynian orogeny, during the late Palaeozoic and the early Mesozoic, sedimentation in Northern Sudan was restricted to a continental basin framed by uplifts.

The evolution of intracratonic rift basins started during the Late Jurassic (Bosworth, 1992; McHargue, *et al.*, 1992; Genik 1993), most probably in response to plate tectonic processes such as the opening of the Central Atlantic Ocean. Rifting lasted from the Late Jurassic until the Early Tertiary, with three stages of active rifting (Bosworth, 1992), but most basins represent single-cycle rifts. The larger basin structures are composed of half-grabens linked by transfer zones or interbasinal highs. Those in Northern Sudan are in excess of 3.5 km (e.g. Humar Basin), the basins in South Sudan possibly more than 10 km deep (Schull, 1988; McHargue, *et al.*, 1992).

The sediments exposed in Northern Sudan are mostly Cretaceous (Albian-Santonian) in age and either belong to the Wadi el Milk Formation or to the Wadi Howar Formation (Klitzsch & Wycisk, 1987; Wycisk *et al.*, 1990). The Wadi el Milk Formation encompasses deposits in the region between Khartoum and Dongola, and the Wadi Howar Formation

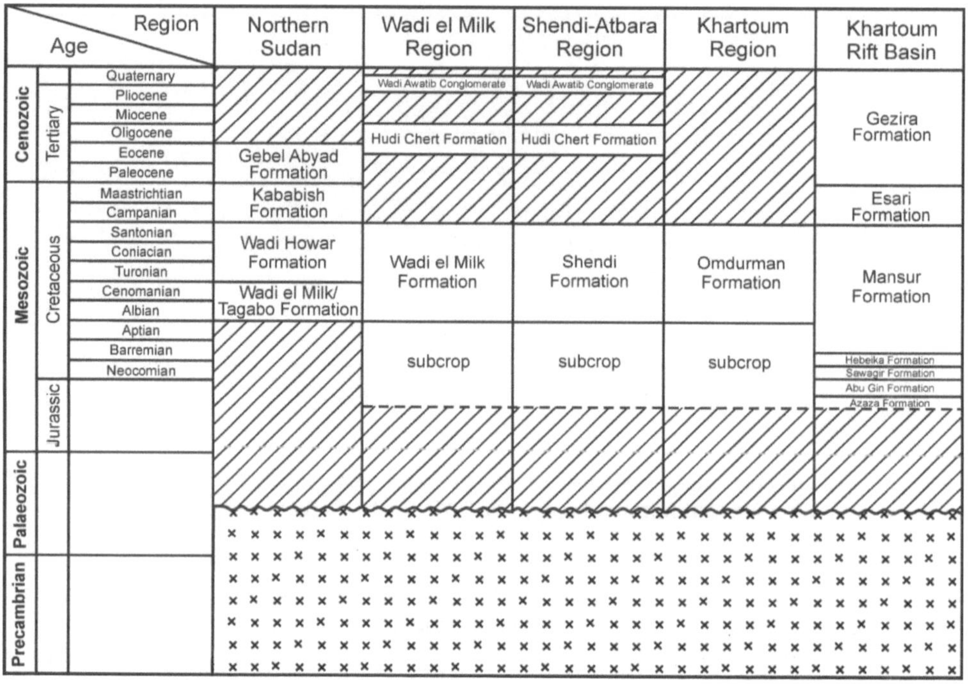

Figure 4. Stratigraphy and sedimentary sequences in the study area based on Wycisk *et al.* (1990), Awad (1994)

sediments in the surrounding of the Gebel Abyad (Figure 4). These formations are laterally equivalent and lithologically similar. They are dominated by tabular and trough cross-bedded fluvial sandstones forming extensive sheet- and wedge-shaped sandstone bodies, deposited in sand-rich braided rivers. Stacked successions of tabular cross-bedded sandstones associated with point bar sequences might represent deposits of unconventional rivers that alternated between braided and high-sinuosity channel patterns. More rarely, mixed-load point bar sequences are intercalated into fine-grained flood plain deposits and indicate an alluvial plain environment dominated by meandering rivers that were accompanied by large, long-lived flood plains. Fine-grained deposits of the Wadi el Milk Formation exposed in the Humar Basin region represent meandering river, flood plain, lacustrine and pedogenic sediments and were named Wadi Abu Hashim Member by Bussert (1998). Local equivalents to both the Wadi el Milk Formation and the Wadi Howar Formation in the Shendi-Atbara and the Khartoum region were named Shendi Formation and Omdurman Formation by Whiteman (1970). The Shendi Formation is dominated by rather fine-grained cross-bedded fluviatile sandstones and contains iron-oolite horizons as well as kaolinitic palaeosoils (Germann *et al.*, 1990). In the subsurface of the Khartoum Basin, equivalent sediments are represented by alluvial fan and floodplain deposits of the Mansur Formation (Wycisk *et al.*, 1990).

After rifting ceased in the Tertiary, some of the sedimentary basins were inverted and/or uplifted, whereas other basins continued to subside, especially those of Central and South Sudan. In these basins, sedimentation proceeded on alluvial fans and in lakes. In the Khartoum Basin, the Blue Nile terminated in form of an inland delta, the Gezira alluvial

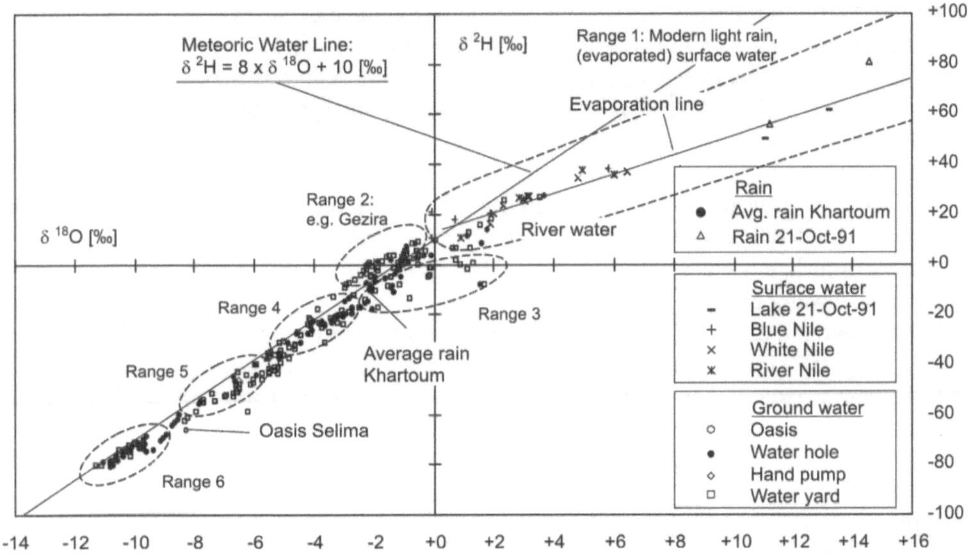

Figure 5. Isotopic signature of >320 water samples from Sudan collected between 12°N and 22°N

fan. The very early fan deposits of the Gezira fan are dominated by sand- and gravel-rich channel deposits (Williams & Adamson, 1980; Adamson *et al.*, 1982), but the Lower Gezira Formation, Paleocene/Oligocene to Miocene in age, was deposited mainly by meandering streams, whereas the Upper Gezira Formation, Miocene to Quaternary in age, was sedimented partly in braided channels (Awad & Bireir, 1993; Awad, 1994). The Holocene Gezira Clays represent floodplain deposits (Adamson & Williams, 1980).

Dating the connection of the Egyptian and the Sudanese Nile segments is rather controversial. Hassan (1976), Butzer (1980) and Said (1993) assigned the Pleistocene as date of the formation of the 'modern Nile', whereas Berry & Whiteman (1968), Burke & Wells (1989) and Adamson *et al.* (1993) suggested a Tertiary or even Upper Cretaceous age. Uplift of the Ethiopian Plateau started during the Oligocene (McDougall *et al.*, 1975), and should have resulted in an immense inflow of water into Sudan and therefore might have helped to establish a 'connected Nile'. Sedimentological and palaeontological evidences in Egypt nevertheless fail to prove any 'equatorial' influence before the Pleistocene (Hassan, 1976; Said, 1981; Issawi & McCauley, 1992). During the Tertiary period, all rivers sourced in the Ethiopian highlands appear to have terminated within the Sudan.

During the Quaternary, the climate fluctuated between arid and humid periods. A terminal Pleistocene humid phase is documented by lake carbonates and increased groundwater recharge (Thorweihe, 1986; Pachur *et al.*, 1990). During the Last Glacial Maximum, arid climate condition caused the formation of dune fields (Gläser, 1987). The late Pleistocene/early Holocene climate was again humid as documented by widespread lacustrine and fluviatile deposits (Pachur *et al.*, 1990; Kröpelin, 1999).

2.3 *Isotopes in water samples*

In Figure 5 more than 320 water samples of localities between 12°N and 22°N, collected during the last three decades, are plotted. These represent precipitation, (ephemeral) lake water, Blue Nile, White Nile and River Nile water, groundwater from hand-dug water holes, from handpumps and boreholes (water yards). A profound knowledge of relevant information such as date of sampling, characteristics of sampling points as depth-to-groundwater, lithology and position of screens is necessary for the interpretation of such large number of samples. As the 'isotopic range' represents level of transition from precipitation to groundwater, identifying depth of water samples is important for result interpretation. The groundwater from water yards with deep screens represents the deepest levels, whereas the screens of handpumps are usually tapping the shallow groundwater. Groundwater extracted from hand-dug water holes instead reveals the characteristics of the 'surface' of the groundwater body.

For comparison and a better understanding, the isotopic means of modern rain in Khartoum (1962–1978) is plotted in Figure 5: $\delta\,^{18}O = 2.1‰$; $\delta\,^{2}H = -10.0‰$. Precipitation in Khartoum is isotopically enriched compared to SMOW ($\delta\,^{18}O$ and $\delta\,^{2}H = 0‰$). This is attributed to the fast northward movement of the wet equatorial monsoon air masses coming from the south, and to the high evapotranspiration during the rainy season causing almost a quantitative 'recycling' of surface moisture back into the atmosphere with relatively little isotopic fractionation. In Figure 5 the spread for $\delta\,^{2}H$ ranging between $+80‰$ and $-80‰$ separates waters of different origin. The depicted 'ranges' plotting more or less along the global Meteoric Water Line (Figure 5) are discussed below.

In the upper right corner, samples of modern light rain and lake water are plotting within **range 1**. These samples are isotopically enriched when compared to SMOW, the average values of rain in Khartoum and the groundwater samples. Samples from the Blue Nile, White Nile and River Nile also plot within range 1. The river water samples, collected at different times of the year (e.g. at the beginning and at the end of a rainy season), plot along an evaporation line with a slope of approx. 4. The isotopically enriched samples of light rain, which were collected at the end of the rainy season in 1991 west of Khartoum, and those of ephemeral lakes formed during the rainy season, show values between $\delta\,^{2}H + 40‰$ and $\delta\,^{2}H + 80‰$. However, in the area west of Khartoum – or other areas – no groundwater with comparable isotopic signature was found. This shows that this surface water does not reach the groundwater table, or if it does, the amount is too small to induce a significant change to the isotopic signature of the groundwater. It is apparent that most of the groundwater samples plotted show no relation to isotopically enriched modern waters.

As an interesting finding, water samples from shallow hand-dug water holes in the upper reaches of wadis, e.g. those draining from the elevated Humar Basin Plateau (>700 m asl) to the Wadi Muqaddam (300–350 m asl), are enriched in $\delta\,^{2}H$. Nonetheless, the deeper groundwater in these areas is strongly depleted (see range 5). Hence, the main regional aquifer seems to be locally overlain by perched groundwater bodies. However, the highest deuterium values $\delta\,^{2}H > +20‰$ were observed in several boreholes east of the Gebel Aulia Lake (Vrbka *et al.*, 1993). These findings are a direct indication of a continuous and intense infiltration from the White Nile, which is impounded at the Gebel Aulia dam south of Khartoum forming a lake more than 100 km long and several km wide. Water from hand-dug wells in the Khor Abu Habil alluvial fan may also plot in range 1, as found by Abdalla (1999).

The water samples of **range 2**, mainly collected east of the Blue Nile, between the two Niles south of Khartoum in the Gezira irrigation scheme, and along the River Nile (Shendi and Atbara sampling area), form a dense cluster just *above* the average modern rain in Khartoum. But they also plot close to the samples of the Blue Nile. There is a shift of the isotopic signature within the Nile waters due to seasonality. It is striking that the Blue Nile water and the groundwater samples in this cluster have in common an δ excess $\delta\,^2H > +10‰$, as they plot *above* the Meteoric Water Line, too. Thus the samples from the Gezira indicate a strong admixture of Blue Nile and River Nile water to the local groundwater. The more depleted samples of ranges 4 to 5, instead, tend to plot *below* the Meteoric Water Line with a δ excess $\delta\,^2H < +10‰$.

Many water samples of **range 3** were collected mainly in the area around Dongola east and west of the River Nile. Here also, especially in the east, irrigation schemes similar to those in the Gezira exist, which run along the Nile. Groundwater west of Dongola flows towards the Qaab Depression indicating recharge from the Nile. The samples of range 3 are mainly taken from boreholes and are densely clustered around the average modern rain in Khartoum. However, the isotopic influence is caused by the water from the River Nile, which is used for irrigation and for flooding the fields. The precipitation in the area of Dongola is about 18 mm/a, which is very low. Moreover, it does not rain every year: during the record years from 1945–1995 there were 10 years without precipitation (Agrometeorology Group-FAO-SDRN; NOAA). It is evident that these samples plot just *below* range 2 having a δ-excess $\delta\,^2H < +10‰$ and the dotted ellipse of the range is somewhat inclined. This is due to some samples shifting away from the Meteoric Water Line. There is a strong evaporation effect on the groundwater, partly also due to the shallow depth-to-groundwater, locally <10 m below ground level.

A few water samples within **range 4** stem from boreholes along the course of the River Nile (Shendi, Atbara area, eastern bank), but some samples are also taken from boreholes in the Gezira, more than 30 km away from the Blue Nile. Unfortunately, the position of the screens and therefore sampling depth below the water-table are often unknown. In the Nile valley, wells are mostly situated not more then 5 km away from the river banks and samples from these wells are isotopically enriched compared to pure 'palaeo groundwater' (range 6). Therefore, a strong mixture with Nile water is obvious. The share of the Nile water may range between 40% and 85%, as calculated by Vrbka *et al.* (1993).

The isotopic signature of range 4 is also found relatively frequently in hand-dug wells along wadi courses away from the Niles (e.g. tributaries to Wadi Muqaddam). Recharge by direct infiltration via wadi systems into the unconsolidated Quaternary sediments occurs locally during flash floods. If there is a relatively high frequency, these floods may contribute significantly to groundwater recharge; however, at a very local scale. Thus, again, perched and mostly shallow aquifers can be identified locally, overlying the deeper regional aquifer. For most areas – especially in the north – a rather low flood frequency must be assumed.

Along the major wadi, Wadi Muqaddam (Figure 1), groundwater with $\delta\,^2H = -40‰$ to $-60‰$ (**range 5**) can be found sporadically in patches, adjacent to the 'palaeo groundwater' (range 6). Most of the samples are taken from water yards tapping the main aquifer body at greater depth. Some samples from relatively deep (some 10 m to >100 m) hand-dug wells have a comparable signature. These samples in range 5 may represent palaeo groundwater partly mixed with relatively younger flood events, average values have been formed during the last thousand years. However, there is also the probability that some of these waters were formed during a transition time some thousands years ago when the climate was shifting from more humid, cooler conditions to the modern dry and hot climate.

Samples taken in the south, in the vicinity of 14°N, indicate relatively younger recharge events, based on stable isotopes and ^{14}C-values. In the area south of Wadi Muqaddam and south-west of Khartoum, a combination of relatively high precipitation (>200 mm/a), high surface runoff due to less permeable basement rocks and relatively small depth-to-groundwater (<40 m) seems to favour local groundwater formation. This is also indicated by modern rain collected by Abdalla (1999) plotting within range 4.

The samples in **range 6** show a very strong depletion as compared to SMOW, the Khartoum rain, river and other surface water. According IAEA (2001), precipitation of this isotopic signature has not yet been found today on the African continent when considering the weighted annual $\delta\,^{2}H$– or $\delta\,^{18}O$-values. Instead, values around $\delta\,^{2}H$–70‰ can be found in rain in the vicinity of latitudes 45°N or 45°S. Thus, the groundwater values below $\delta\,^{2}H$–60‰ mainly represent deep groundwater formed during cooler and wetter climatic conditions than those prevailing in Sudan today. It is significant that similar isotopic signatures were found in boreholes with deep screens – thus characterizing the deeper parts of the main aquifer – and in adjacent water holes characterizing the uppermost part or just the very surface of the main aquifer. In these areas the 'rather old and isotopically depleted groundwater' starts directly at the top of the main water body. These findings were corroborated by Tritium and ^{14}C data (Vrbka, 1996). This type of groundwater is prevailing in the northern part of the study area, north of 15°N, along Wadi Muqaddam and Wadi el Milk, hence, in general a certain distance away from the Nile valley. The water sample taken at the Selima Oasis also shows a strong depletion but plots due to a small isotopic enrichment away from the 'palaeo groundwater'. This is due to evaporation, shifting the sample along a regression line towards less depleted values.

3 DISCUSSION

In this study, the environmental isotopes ^{18}O and ^{2}H have acted as very useful indicators of the origin of groundwater and of recharge mechanisms. In the northern region of the study area, the isotopic depletion is up to $\delta\,^{2}H = -81$‰, whereas in the southern region groundwater with an isotopic signature around $\delta\,^{2}H = -80$‰ has not been reported yet. The southward trend towards less depleted values might be due to moisture coming mostly from the south rather than from the west, as assumed for the northern region. To test this hypothesis, isotopic data of groundwater from areas south of 12°N is needed (both groundwater data from boreholes and from hand-dug wells) to compare its isotopic composition at different depths. At present, however, the authors do not have any information regarding isotopic groundwater composition south of the study area. In case that rather old groundwater with isotopic signature around $\delta\,^{2}H = -80$‰ could be found in places farther south, at locations with a lower depletion, a significant recharge by younger precipitation could be assumed.

Several studies have proven that significant groundwater recharge occurs along perennial river courses, principally through river bank filtration. Other areas of groundwater recharge are large irrigation schemes, e.g. in Gezira and Dongola, where Nile water is used for irrigation purposes. In cases where irrigation schemes also use groundwater, recycling of the groundwater recharge from rivers occurs. Due to evaporation effects such recharge may lead in the long term to salinity increase and isotopic enrichment. The strong influence of river water on the groundwater is traceable as far as 20–25 km away from the rivers. Flow distance of the groundwater and its age, based on ^{14}C, correspond to a groundwater linear velocity of 5–6 m/a. Moving 5–6 km/ka would take the groundwater 3500 to 5000 years to cover a distance of 20–25 km. This time span corresponds well with the end of the last humid period in North-East Africa. Thus, in the vicinity of the Niles the groundwater may have a high proportion of river water which is continuously infiltrating with changing isotopic signature since the end of the last humid period. Farther away from the Nile direct recharge during the past humid period is the main contributor. In some areas irrigation channels transport river water to further remote farm land. In these cases, channel losses during transport contribute to groundwater recharge at distances even greater than 25 km from the rivers.

Although recharge from direct infiltration of modern precipitation is evident from environmental isotopes, this recharge is only local, sporadic, temporal and difficult to quantify. Minor amounts and irregular spatial distribution demonstrate this recharge to be insufficient to secure any long term water supply of proposed new settlements in dry areas. Therefore, alternative water supplies to cover demands in settlements away from the Nile and its tributaries should be considered.

In some regions of the world there is a tendency to resolve energy and water problems through the construction of river dams. At first, it seems rewarding to build embankment dams along the rivers at suitable places, partly to impound the river water well above the groundwater table und thus to induce artificial groundwater recharge by river bank filtration. Several dams already exist in the study area, e.g. Jebel Aulia Dam, Roseires Dam, Sennar Dam, Khashm el Girba Dam, and Aswan Dam. Since 2004, the Merowe Dam is under construction and is expected to be completed by 2008/09. North of Dongola another dam, the Kajbar Dam, is proposed. Although the benefits of 'damming the Nile' might be immediate, they are not everlasting, as the solution of one problem may cause others (e.g. reservoir sedimentation, water quality problems, downstream erosion, ecosystem damages) and a sustainable solution, if any is possible, is shifted into the future. Due to the trend of building river dams, local people and irrigation schemes will continue to depend mainly on river water brought by conduits or pipelines from barrages along the Nile. The settlements in the dry regions of Sudan will nevertheless continue to depend partly on the use of fossil groundwater. Wherever possible, the groundwater, which is usually of good quality, should be used for human consumption, and the river water primarily for irrigation purposes.

A deviation of the Nile, a sort of channelling or 'cascade of lakes' between Khartoum and Korti through the Wadi Muqaddam, could eventually allow relief to the densely populated Khartoum area. This could permit use of the area between the Wadi Muqaddam and the Nile, to release the population pressure from the fast growing areas at the confluence of the two Niles. The technological ventures required are challenging and their acceptance by the population and their sustainability regarding the ecosystem are unknown and need to be investigated. However, the increasing social and environmental pressures induced by the population address new questions and drive the search for sustainable solutions.

ACKNOWLEDGEMENTS

The authors thank the different institutions and authorities, namely IAEA (Vienna), GRAS and RWC (Khartoum), University of Technology Berlin, DAAD (Bonn) which made field work, data collection and publication possible.

REFERENCES

Abdalla, O. A. 1999. *Groundwater hydrology of the west-central Sudan: hydrochemical and isotopic investigations, flow simulation and resources management*. PhD, University of Technology, Berlin.

Adamson, D., McEvedy, R. & Williams, M. A. J. 1993. *Tectonic inheritance in the Nile basin and adjacent areas*. Israeli Journal of Earth Sciences, **41**, 75–85.

Adamson, D. A. & Williams, M. A. J. 1980. *Structural geology, tectonics and the control of drainage in the Nile basin.* In: Williams, M. A. J., & Faure, H. (eds.) The Sahara and the Nile, 225–252.

Adamson, D. A., Williams, M. A. J. & Gillespie, R. 1982. *Palaeogeography of the Gezira and of the lower Blue and White Nile valleys.* In: Williams, M. A. J. & Adamson, D.A. (eds.) A Land Between Two Niles, 165–219.

Agrometeorology Group-FAO-SDRN; NOAA Climatic data. The presented climatic data were extracted and prepared from a *.dat file by the authors.

Awad, M. Z. 1994. *Stratigraphic, Palynological and Paleoecological Studies in the East-Central Sudan (Khartoum and Kosti Basins), Late Jurassic to Mid-Tertiary.* Berliner Geowissenschaftliche Abhandlungen (A), **161**, 1–163.

Awad, M. Z. & Bireir, F. E. R. A. 1993. *Oligo-Miocene to Quaternary palaeoenvironment in Gezira area, central Sudan.* In: Thorweihe, U. & Schandelmeier, H. (eds.) Proceedings of the International Conference on the Geoscience Resources in NE Africa, 465–470.

Berry, L. & Whiteman, A. J. 1968. *The Nile and the Sudan.* Geographical Journal, **134**, 1–33.

Bonifica 1986. *Hydrogeological studies and investigations in Northern Sudan.* Supporting Report No 1, 2, 8, 9, 16, 17. Bonifica, Società per azioni, Italy.

Bonifica 1987. *Hydrogeological studies and investigations in Northern Sudan.* Bonifica, Società per azioni, Italy.

Bosworth, W. 1992. *Mesozoic and early Tertiary rift tectonics in East Africa.* Tectonophysics, **209**, 115–137.

Burke, K. & Wells, G. L. 1989. *Trans-African drainage system of the Sahara: was it the Nile?* Geology, **17**, 743–747.

Bussert, R. 1998. *Die Entwicklung intrakratonaler Becken im Nordsudan.* Berliner Geowissenschaftliche Abhandlungen (A), **196**, 1–329.

Butzer, K. W. 1980. *Pleistocene history of the Nile Valley in Egypt and Lower Nubia.* In: Williams, M. A. J. & Faure, H. (eds.) The Sahara and the Nile, 253–280.

Genik, G. J. 1993. *Petroleum geology of Cretaceous-Tertiary rift basins in Niger, Chad, and Central African Republic.* Bulletin of the American Association of Petroleum Geologists, **77**, 1405–1434.

Germann, K., Fischer, K. & Schwarz, T. 1990. *Accumulation of lateritic weathering products (kaolins, bauxitic laterites, ironstones) in sedimentary basins of northern Sudan.* Berliner Geowissenschaftliche Abhandlungen (A), **120**, 109–148.

Gläser, B. 1987. *Altdünen und Limnite in der nördlichen Republik Sudan als morphogenetisch-paläoklimatischer Anzeiger.* Akademische Wissenschaften Göttingen., Hamburg.

GRAS 2004. *Geological Map of the Sudan. 1:2,000,000.* Geological Research Authority of the Sudan, Khartoum.

Hassan, F. A. 1976. *Heavy minerals and the evolution of the modern Nile.* Quaternary Research, **6**, 425–444.

Issawi, B. & McCauley, J. 1992. *The Cenozoic rivers of Egypt: The Nile problem.* In: Adams, B. & Friedman, R. (eds.) The Followers of Horus: Studies in Memory of Michael Allen Hoffman, 105–122.

IAEA 2001. *GNIP Maps and Animations.* International Atomic Energy Agency, Vienna.

Jacob, H. 1990. *Groundwater research in selected areas of the Northern Sudan using environmental isotopes* (^2H, ^3H, ^{18}O). Technical Report to the IAEA, Vienna.

Kheir, O. M. 1986. *Hydrogeology of the Dongola Area, Northern Sudan.* Berliner Geowissenschaftliche Abhandlungen (A), **74**, 1–81.

Klitzsch, E. 1984. *Northwestern Sudan and bordering areas: geological development since Cambrian time.* Berliner Geowissenschaftliche Abhandlungen (A), **50**, 23–45.

Klitzsch, E. 1986. *Plate tectonics and cratonal geology in Northeast Africa (Egypt, Sudan).* Geologische Rundschau **75**, 755–768.

Klitzsch, E. & Wycisk, P. 1987. *Geology of sedimentary basins of northern Sudan and bordering areas*. Berliner Geowissenschaftliche Abhandlungen (A), **75**, 97–136.

Kröpelin, S. 1999. *Terrestrische Paläoklimatologie heute arider Gebiete: Resultate aus dem Unteren Wadi Howar (Südöstliche Sahara/Nordwest-Sudan)*. In: Klitzsch, E. & Thorweihe, U. (eds.) Nordost-Afrika: Strukturen und Resourcen, 446–506.

McDougall, I., Morton, W. H. & Williams, M. A. J. 1975. *Age and rates of denudation of the Trap Series basalts at Blue Nile gorge, Ethiopia*. Nature, **254**, 207–209.

McHargue, T. R., Heidrick, T. L. & Livingston, J. E. 1992. *Tectonostratigraphic development of the Interior Sudan rifts, Central Africa. Geodynamics of Rifting II. Case History Studies on Rifts: North and South America and Africa*, Tectonophysics, **213**, 187–202.

Pachur, H. J., Kröpelin, S., Hoelzmann, P., Goschin, M., Altmann, N. 1990. *Late Quaternary fluvio-lacustrine environments of western Nubia*. Berliner Geowissenschaftliche Abhandlungen (A), **120**, 203–260.

Said, R. 1981. *The Geological Evolution of the River Nile*. Springer Verlag, Berlin.

Said, R. 1993. *The River Nile*. Pergamon Press, Oxford.

Schandelmeier, H. & Reynolds, R. 1997. *Palaeogeographic-Palaeotectonic Atlas of North-East Africa, Arabia, and Adjacent Areas*. Balkema, Rotterdam.

Schandelmeier, H., Klitzsch, E., Hendriks, F. & Wycisk, P. 1987. *Structural development of North-East Africa since Precambrian times*. Berliner Geowissenschaftliche Abhandlungen (A), **75**, 5–24.

Schandelmeier, H., Utke, A., Harms, U. & Küster, D. 1990. *A review of the Pan-African evolution in the NE Africa: towards a new dynamic concept for continental NE Africa*. Berliner Geowissenschaftliche Abhandlungen (A), **120**, 1–14.

Schull, T. J. 1988. *Rift basins of interior Sudan: Petroleum exploration and discovery*. Bulletin of the American Association of Petroleum Geologists, **72**, 1128–1142.

Thorweihe, U. 1986. *Isotopic identification and mass balance of the Nubian Aquifer System in Egypt*. Berliner Geowissenschaftliche Abhandlungen (A), **72**, 87–97.

Vail, J. R. 1985. *Alkaline ring complexes in Sudan*. Journal of African Earth Sciences, **3**, 51–59.

Vrbka, P. 1996. *Hydrogeologische und isotopenhydrologische Untersuchungen zu regionalen Problemen der GW-Neubildung, der GW-Zirkulation und des Wasserhaushaltes im Nordsudan*. Berliner Geowissenschaftliche Abhandlungen (A), **186**, 1–158.

Vrbka, P., Jacob, H., Fröhlich, K. & Salih, M. A. 1993. *Identification of Groundwater Recharge Sources in Northern Sudan Using Environmental Isotopes*. Journal of Environmental Hydrology, **1**, 8–16.

Whiteman, A. J. 1970. *Nubian Group: origin and status*. Bulletin of the American Association of Petroleum Geologists, **54**, 522–526.

Williams, M. A. J. & Adamson, D. A. A. 1980. *Late Quaternary depositional history of the Blue and White Nile rivers in central Sudan*. In: Williams, M. A. J. & Adamson, D. A. A. (eds.) The Sahara and the Nile, 281–304.

Wycisk, P., Klitzsch, E., Jas, C. & Reynolds, O. 1990. *Intracratonal sequence development and structural control of Phanerozoic strata in Sudan*. Berliner Geowissenschaftliche Abhandlungen (A), **120**, 45–86.

CHAPTER 21

Hydrochemical and stable isotopes compositions of saline groundwaters in the Benue Trough, Nigeria

M.N. Tijani

Department of Geology, University of Ibadan, Ibadan, Nigeria

ABSTRACT: Hydrochemical and stable isotope profiles of 64 saline groundwaters samples from 20 different locations in the Benue Trough, Nigeria are presented and discussed in relation to the source of primary salinity with respect to the inland extension of the Gulf of Guinea during the Cretaceous period. Average electrical conductivity (EC) values of the saline groundwaters in the lower and middle region of the trough are 56,300 μS/cm and 22,200 μS/cm respectively. The saline groundwaters are Na—Cl type enriched in Ca and Sr and depleted in Mg and SO_4 relative to a seawater evaporation trend and have $\delta^{18}O$ values of -5.99‰ and δ^2H values of up to -40‰. The interpretations of the hydrochemical data strongly support salt/halite dissolution as the primary source of salinity and rule out evapo-concentrated and hydrothermal sources. However, the lower $\delta^{18}O$ and δ^2H values for most of the analyzed samples also suggest that the dissolution of halite/disseminated salts are related to influx of seawater during the inland transgressive extension of the Gulf of Guinea and subsequent remobilization of fossil/connate formation water rather than mineralization of the infiltrated meteoric water. The observed hydrochemical characters indicate that the original marine chemistry has been modified by water-rock interactions, involving dolomitization and enrichment of Ca, through cation exchange process.

1 INTRODUCTION

Occurrences of saline groundwaters in the Cretaceous Benue Trough of Nigeria have been known for a long time. The saline groundwaters occur in ponds, as springs and in dug-holes (Figure 1) and are characterized by a varied salinity; in some locations, salinity of 2 or 3 times greater than that of seawater is not uncommon. Like the evolution of the Benue Trough itself, the occurrence of these brines resulted in a number of hydrogeological studies over many decades. The central focus of most of the studies is related to the genetic source(s) of the salinity of the groundwaters, which has generated some controversy apparently due to the lack of proof of evaporites/salt deposits within the trough. Three models of saline water evolution have emerged as outlined in Uma & Loehnert (1992) and summarized as follows:

1. Connate or fossil/seawater source as marine interstitial formation waters (Tattam, 1943; Offodile, 1976; Uzuakpunwa, 1981).

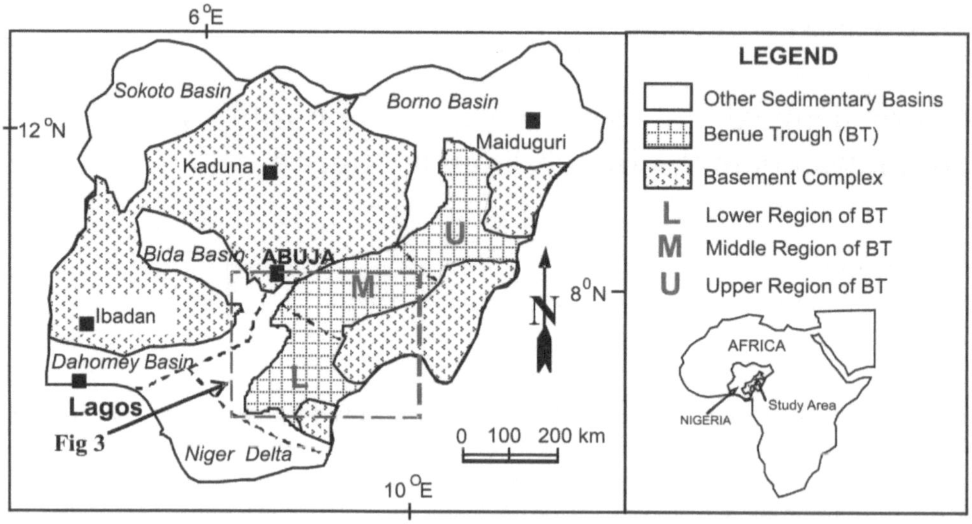

Figure 1. Location and geological setting of the Benue Trough, Nigeria.

2. Evaporite/salt deposit source, at depths either as lenses or diapiric structures (Orajaka, 1972; Ford, 1980; Egboka & Uma, 1986).
3. Hydrothermal source associated with Pb-Zn-Baryte mineralization within the Benue Trough (McConnel, 1949; Farrington, 1952; Olade, 1976; Akande *et al.,* 1988).

However, recent studies, based not only on hydrochemical data but also on isotope data seem to generally support the genetic evolution related to fossil seawater/connate formation water (Uma *et al.*, 1995; Loehnert, *et al.*, 1996; Tijani *et al.*, 1996). Nonetheless there is a need for a composite hydrochemical evolution model that will not only account for the sources of the primary salinity, but also account for the possible hydrochemical evolution of the saline waters and relationships with other tectonic elements within the trough. Hence, this study outlines the hydrochemical and stable isotope characteristics of the saline groundwaters in the Cretaceous Benue Trough (Nigeria). The primary goal is to assess the plausibility of existing evolution models in relation to possible transgression imprints of the Gulf of Guinea.

There are several other mechanisms (such as shale membrane filtration, seawater evaporation, salt/halite dissolution and water rock interactions) that have been advocated worldwide in respect to genetic evolution of brines in sedimentary basins (Chave, 1960; Clayton *et al.*, 1966; Billings *et al.*, 1969; Hanshaw & Coplen, 1973; Carpenter, 1978; Frape & Fritz, 1982 among many others). This study presents a preliminary attempt of evaluation and interpretation of chemical and isotope data in order to develop a conceptual hydrochemical evolution model of saline water that will account for the primary sources of salinity and at the same time fit into the overall geotectonic and stratigraphic settings of the Benue Trough.

1.1 *Location and Geological Framework*

The Benue Trough is an 800 km long and 100–150 km wide, NE-trending, intra-continental Cretaceous rift-basin geographically divided into three regions, i.e. the lower (southern), middle and upper (northern) region of the Benue Trough (Figure 1). The tectonic evolution

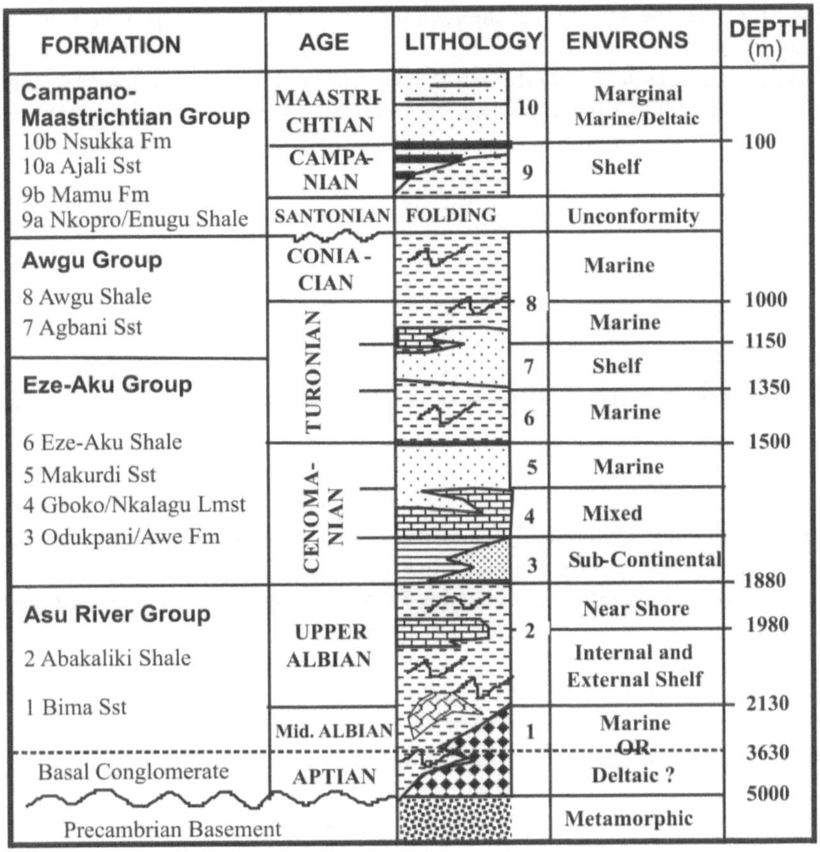

Figure 2. Generalized stratigraphy of the Benue Trough, Nigeria.

of the Benue Trough is associated with the break-up of Gondwanaland; the separation of Africa from South America and the opening of the Gulf of Guinea and South Atlantic Ocean during Early Cretaceous time.

Sedimentation and stratigraphical settings are characterized by transgressive and regressive cycles, starting in pre-Albian or mid-Albian with the Asu-River Group that lies directly over the Precambrian basement followed by subsequent Cretaceous sediments up to the Maastrichtian (Burke *et al.*, 1970). The sedimentary filling in this rift-like basin ranges from about 2000 m thick in the upper/northern region to over 6000 m in the southern (lower) region. The sedimentary filling ranges from predominantly shale to various degrees of interlayering of moderately to well indurated shale, mudstone, sandstone and even limestone and has been affected by two main tectonic episodes in the pre-Turonian and Santonian periods. The generalized stratigraphical framework of the Benue Trough with inferred depositional palaeoenvironments are shown in Figure 2, while further details on general geology, stratigraphy and tectonics of the Benue Trough are presented elsewhere in Offodile (1976), Olade (1975), Wright (1976), Ofoegbu (1984) and Benkhelil (1986).

In terms of the hydrogeological setting, the saline groundwater in the Benue Trough is primarily confined to the predominantly shaly marine sequence of the Asu-River Group (Albian-Cenomanian), the Eze-Aku Group and the Awgu Shale Formation (Turonian-Coniacian). Groundwater usually occurs at the surface in the form of springs, ponds, and dug-holes. However, the variation in the local lithology at the outcrop sites from predominantly shale in the lower region to sandstones (with intercalation of shale/limestone units) in the middle region suggests that the saline groundwaters are not lithologically controlled (Uma & Loehnert, 1992). Rather the occurrences and up-fluxing movements are controlled by a possible convective force generated by the infiltrating recharge water and/or hydrostatic force responsible for the displacement and upward movement of the saline water through the fracture systems within the trough.

2 MATERIAL AND METHODS

A number of sampling operations were undertaken, involving the determination or measurement of field parameters, followed by laboratory hydrochemical analyses of the samples (for the major and minor elements) using standard procedures (Stednick, 1991; DVWK, 1992) at the hydrogeochemical laboratory of the Geologic Institute, University of Muenster, FRG. Temperature and electrical conductivity were measured in the field with a digital WTW-Conductivity meter model LF/92 while pH was also measured with a WTW-meter, model pH/91. Samples for the analyses of major cations and minor elements (Na^+, K^+, Ca^{2+}, Mg^{2+}, Ba^{2+}, Sr^{2+}) were collected in sterilized plastic bottles, acidified and preserved in a refrigerator prior to analyses using ICP-OES spectrophotometer. For the anions, un-acidified samples were also collected in sterilized plastic bottles and preserved in refrigerator prior to analyses. Cl^- and SO_4^{2-} were measured by the ion chromatographic method while HCO_3 and Br were determined through titration methods. For this study about sixty-four data sets from twenty locations with varied replicate samples (Table 1) covering at least two field-sampling operations were used in the evaluation. Further details regarding sampling and data evaluation were presented elsewhere in Tijani (2004).

For the measurement of oxygen and hydrogen isotope ratios, separate samples were also collected in sterilized plastic bottles, while the measurements were undertaken at the specialized laboratory of *Hydroisotop GmbH,* Schweitenkirchen, Germany. Both $^{18}O/^{16}O$ and $^2H/H$ were measured using conventional mass spectrometric methods and the results are reported in ‰ with respect to VSMOW standard. Measurement precision was ±0.15‰ for $\delta^{18}O$ and ±1.5‰ for δ^2H. Locations of sample sites within the Benue Trough are shown in Figure 3.

3 RESULTS

Summary data of the hydrochemical analyses for the saline groundwater samples are presented in Table 1 along with data on seawater chemistry from the Gulf of Guinea. The pH values range from 5.8–7.6 in both regions with no definite trend among the different saline groundwater samples. Temperature of the saline groundwater ranged generally between 24.7 and 36.8°C in both regions with the exception of 2 warm saline springs (Akiri and Awe) in the middle region with an average temperature of about 40°C. Total dissolved ions (TDI) ranged between 5263 and 88,800 mg/l (median 33,910 mg/l) and between 6577 and 27,257 mg/l (median 8841 mg/l) respectively for the lower and middle region of the Benue Trough (Table 2). This implies that the dissolved ions in the analyzed saline waters represent about $^1/_7$ to $2^1/_2$ times that of seawater salinity in the lower region and to $1^1/_2$ times of seawater salinity in the middle region.

Table 1. Summary of hydrochemical analyses results and seawater data.

L/No.	Locality	N	Temp. °C	pH	EC	TDI	Ca	Mg	Na	K	Ba	Sr	HCO_3	Cl	SO_4	Br
Lower Region of the Benue-Trough																
1	Enyigba shaft	4	29.2	6.8	11,205	6,808	90.3	21.4	2,124	71.7	19.3	9.3	456.1	2,990	–	1.0
7	Ishiagu	3	30.9	7.0	10,225	6,141	278.1	34.4	1,809	104.2	7.4	12.5	267.9	3,477	90.2	2.1
12	Uburu	2	30.2	6.3	48,850	35,273	550.0	96.0	13,500	310.0	4.8	28.8	225.0	20,550	–	5.5
14	Okposi SL	4	30.4	7.1	52,890	38,597	479.4	59.9	15,152	319.3	26.5	46.5	202.3	22,310	32.0	7.0
17	Lokpanta	1	32.5	5.8	79,500	33,256	514.0	56.5	11,984	295.0	38.5	51.8	41.5	20,240	–	33.1
22	Olachor	2	27.5	6.7	70,600	50,423	2,433.0	225.8	16,830	458.4	184.9	265.2	217.8	29,338	315.0	30.3
23	Okpenyi	2	24.7	6.7	61,900	44,926	2,078.5	190.7	14,143	414.7	146.0	212.6	280.9	27,210	175.1	20.8
24	Abachor	4	26.6	5.7	81,825	58,860	3,145.8	317.7	18,917	441.8	243.4	329.1	215.1	34,776	330.3	16.5
28	Ijegu	6	26.6	6.8	80,350	58,128	2,191.3	204.9	19,068	379.1	215.8	254.1	244.0	35,174	441.3	21.1
30	Gabu	2	32.1	7.2	52,900	37,403	1,371.0	156.4	11,503	295.1	244.6	162.2	136.0	23,490	21.0	11.8
	Average	30	28.3	6.7	56,319	38,452	1,407.4	144.1	12,951	305.7	125.7	153.2	250.7	22,773	245.0	13.2
Middle Region of the Benue-Trough																
34	Awe I	6	32.0	6.8	13,226	6,900	103.3	29.1	2,635	65.2	2.0	7.9	476.6	3,570	16.5	1.8
35	Awe II (WS)	6	39.9	6.6	15,774	8,675	130.7	33.9	3,331	78.1	3.9	10.2	546.6	4,523	12.1	3.2
36	Awe III	3	36.8	7.6	15,407	9,443	238.1	51.6	3,617	107.8	9.8	23.6	635.8	4,693	22.9	–
38	Ribi (SP)	3	28.9	6.9	28,170	20,058	426.1	76.3	6,635	226.9	74.7	42.2	414.8	12,148	–	3.8
40	Azara SP	6	30.5	6.4	33,868	20,302	471.9	91.9	7,049	257.5	65.6	114.7	370.8	11,867	–	5.7
44	Akiri (WS)	2	40.7	7.4	10,300	5,997	101.4	18.9	2,191	74.7	18.0	8.1	471.1	3,097	–	2.4
45	Keana (SL)	1	36.2	7.5	34,000	21,421	465.1	68.4	7,755	369.8	59.5	41.0	653.3	12,000	–	5.5
46	Kanje	4	31.8	6.5	14,850	8,784	199.2	46.6	3,108	118.4	10.6	14.9	185.3	5,100	–	3.1
52	Arufu (SL)	1	29.1	6.9	22,500	14,582	434.4	54.6	5,063	226.4	10.1	21.1	490.4	8,190	–	4.5
55	Akwana (SL)	2	35.1	6.8	55,900	37,201	1,113.5	121.5	13,280	590.2	19.9	58.8	446.5	21,350	335.0	10.1
	Average	34	34.0	6.8	22,242	13,568	304.5	56.0	4,854	170.8	28.1	36.5	448.1	7,640	45.6	4.7
Seawater (Gulf of Guinea)							400.0	1,350	10,500	380.0		8.0	142.0	19,000	2,700	67.0

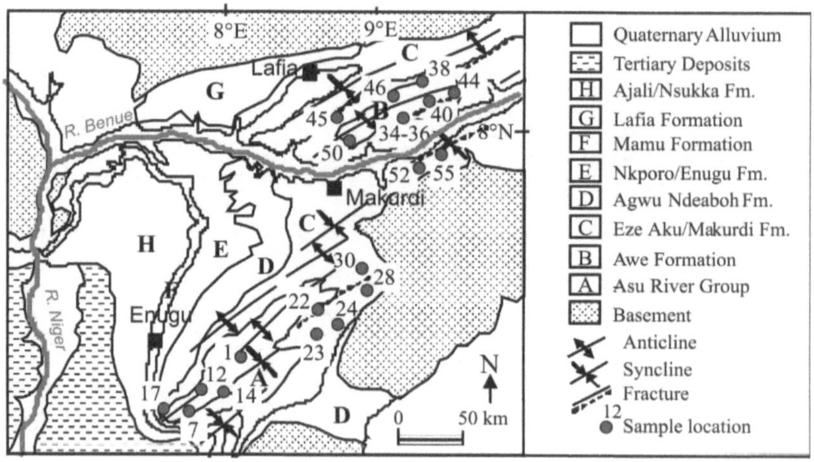

Figure 3. Location of saline groundwater sampling sites in the Benue Trough. Nigeria.

For the major cations, Ca concentrations range from 42–4432 mg/l in samples from the lower region while those from the middle region range between 54–1465 mg/l. Mg exhibits lower concentration with values of 12–502 mg/l and 18–134 mg/l for samples from lower and middle regions respectively. Expectedly, saline water samples exhibit much higher Na concentrations with values of 1429–29,072 mg/l (median; 11,790) for the lower region and 1999–17,311 mg/l (median; 3517 mg/l) for the middle region. However, K concentrations are generally lower with values of 20–778 mg/l for the lower region and 55–701 mg/l for the middle region. For the anions, chloride ion is the most dominant with concentration of 2152–57,688 mg/l and 2340–26,400 mg/l for sample from lower and middle regions respectively. HCO_3 and SO_4 ions exhibit concentrations of 38–697 mg/l and 5–910 mg/l respectively for saline water samples from both the lower and middle regions.

In general, the profiles of major ions as graphically presented in the *Schoeller* plot (Figure 4), indicate a predominantly Na-Cl water type, with Na and Cl representing about 75% and 85% respectively of the total dissolved ions in all of the analyzed samples from both regions of the Benue Trough. Also, the chemical profiles revealed a general enrichment of calcium, relative to seawater composition, whereas magnesium and sulphate are depleted compared to modern seawater (see Table 1 and Figure 4).

The results of stable isotope analyses as presented in Table 2 show δ^2H values in the range of −25.6 to +2.5‰ and −40.7 to +5.7‰ for the lower and middle region respectively, while $\delta^{18}O$ values range from −5.15 to +0.82‰ and −5.99 to +1.27‰ for the lower and middle region respectively. A closer look at the data revealed that most of samples from the middle region have generally more negative values of $\delta^{18}O$ and δ^2H compared to those from the lower region. However, the surface saline ponds/lakes, from both regions, exhibit positive values of $\delta^{18}O$ and δ^2H, which may be interpreted as indications of dilution or mixing of the primary water source with recent meteoric waters. Furthermore there is a significant shift of about 10–15‰ between the δ^2H values of most samples from the lower and middle region along the global meteoric waterline. In other words, the more negative values δ^2H for most samples from the middle region compared to those from the lower region may be attributed to possible depletion of δ^2H through continental and latitude

Table 2. Results of stable isotope (δ^2H and $\delta^{18}O$) analyses.

S/No.*	Description	TDI (mg/l)*	$\delta^{18}O$	δ^2H
Lower Region of the Benue-Trough				
1/90	Enyigba S-I	5263	−3.91	−23.8
1/94	Enyigba S-I	8588	−3.86	−21.3
7/94	Ishiagu II	6504	−4.25	−24.6
12/90	Uburu	60,079	−0.56	−7.4
12/91	Uburu	10,467	0.82	1.5
14/91	Okposi SL	54,871	−1.25	−7.5
14/95	Okposi SL	52,158	−1.04	−7.4
17/95	Lokpanta	33,256	−1.66	2.5
22/94	Olachor	17,149	−2.68	−10.4
23/95	Okpenyi	61,690	−4.63	−15.4
24/94	Abachor	88,799	−4.47	−21.3
28/94	Ijegu I	85,592	−5.15	−25.6
29/94	Ijegu II	23,619	−0.15	−6.0
30/94	Gabu	40,242	−3.30	−16.6
Middle Region of the Benue-Trough				
34/90	Awe I	8095	−5.99	−40.7
34/94	Awe I	7714	−5.48	−36.7
35/93	Awe II (WS)	7093	−5.84	−34.5
35/94	Awe II (WS)	10,861	−5.76	−38.2
38/91	Ribi (SP)	10,673	−1.48	−8.7
38/94	Ribi (SP)	25,458	−5.13	−31.6
40/93	Azara I (SP)	18,406	−4.32	−22.9
40/94	Azara I (SP)	21,649	−4.06	−24.1
44/94	Akiri (WS)	6577	−5.13	−33.2
45/95	Keana (SL)	21,421	−3.27	−6.0
46/91	Kanje I	7071	1.27	5.7
47/90	Kanje II	8469	−0.82	−11.1
52/95	Arufu (SL)	14,582	−4.38	−20.5
56/95	Akwana II (SL)	27,257	−4.08	−15.6

*S/No. = Sample number / year of sampling; *TDI = Total dissolved ions

effects. Such depletion in isotopic ratio is often linked to palaeo synsedimentary formation water or recharge water incorporated into the basin when the climate was cooler than the present day (Hitchon & Friedman, 1969; Land & Prezbindowski, 1981). In summary, saline groundwaters in the Benue Trough are characterized by negative (depletion) values of $\delta^{18}O$ and δ^2H especially for the more concentrated samples, while the less concentrated surface saline ponds are characterized by positive values of $\delta^{18}O$ and δ^2H which is an indication of possible modifications through dilution and/or evaporation processes.

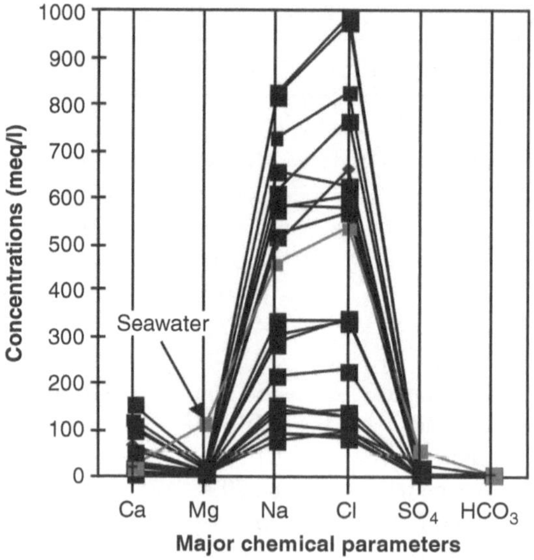

Figure 4. Schoeller diagram plots of the major chemical profiles.

4 DISCUSSION AND INTERPRETATIONS

4.1 *Characteristics of the saline water*

From the composition profiles of the major chemical ionic parameters (see Figure 4), it can be seen that Na and Cl ions accounts for not less than 85% of the TDI in most of the analyzed samples which implies that the saline groundwater in the Benue Trough can be generally characterized as Na—Cl type. In addition, with the exception of bicarbonate, there are generally high positive correlations between the major chemical parameters and chloride as shown in Figure 5. This trend is an indication that the saline groundwaters in the Benue Trough are characterized by similar ionic distribution irrespective of the degree of salinity of the different samples. However, a closer evaluation using the Sulin (1946) classification scheme revealed two sub-divisions; namely Na—Cl water type and Na—Ca—Cl water type. While the primary salinity of both water types can be attributed to a marine source, it is obvious that the latter group must have evolved as a consequence of possible water-rock interactions. It is also interesting to note that most of the samples with Na—Ca—Cl character are from locations close to the margin of the trough and/or basement rock boundaries (see Figure 3). The implication is that the basin margins are usually characterized by carbonate deposits, as exemplified by the Gboko and Nkalagu limestone units in the Benue Trough and hence favour dolomitization reactions that can lead to the evolution of the observed Na—Ca—Cl water type.

A graphical plot of the values in the conventional $\delta^{18}O$ versus $\delta^{2}H$ diagram (Figure 6) show that most of samples from the middle region are generally more negative compared to those from the lower region. The general upward shift of the plotted points along the Global Meteoric Water Line (GMWL) may be an indication of dilution or mixing of the primary water source with recent meteoric waters (Craig, 1969). In addition, some samples from both regions follow a regression line defined by $\delta^{2}H = 5.5*\delta^{18}O–1.5$ which indicates enrichment

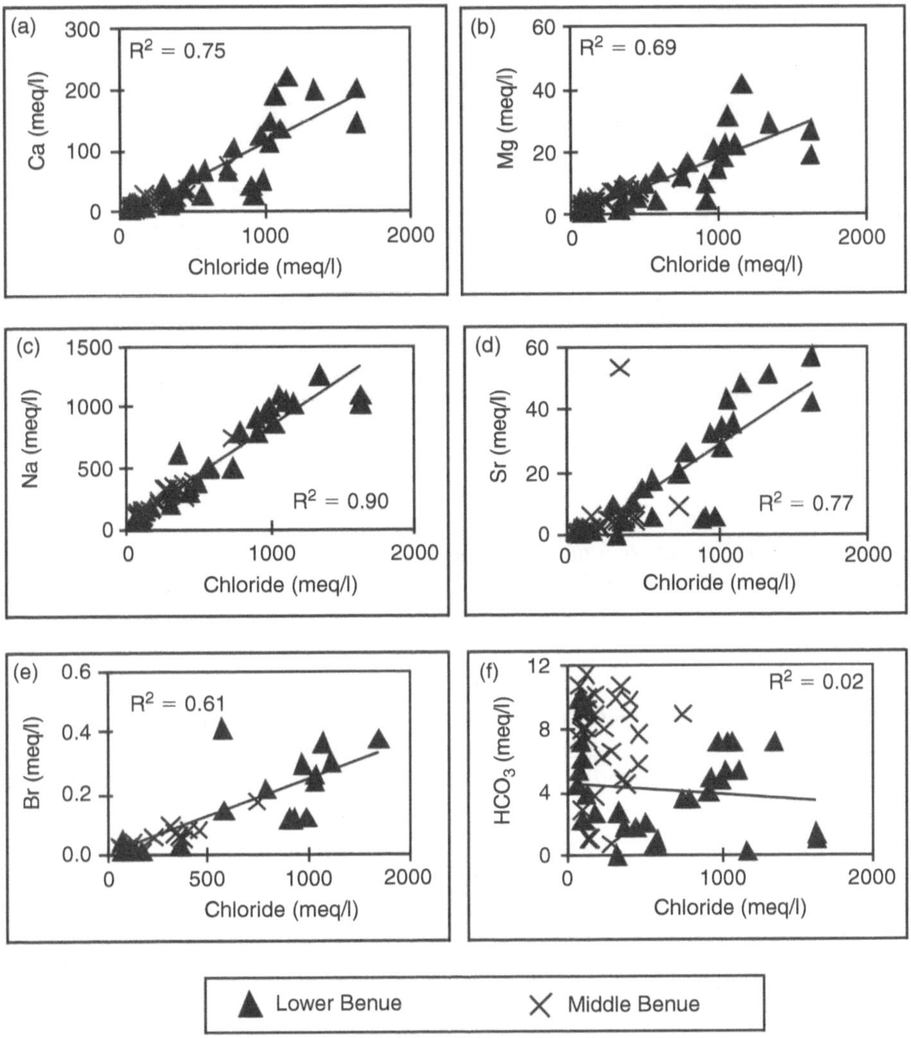

Figure 5. Plots of major ionic species against chloride concentration.

through the evaporation process. Interestingly, this latter group represents samples from surface saline ponds/lakes, which are most likely to be influenced by direct inputs of recent precipitation and atmospheric evaporation process. Nonetheless, two broad groups can be identified; first and most striking are concentrated samples from both regions with negative $\delta^{18}O$ and $\delta^{2}D$ values, implying a possible indication of a palaeo/fossil water source and the second group which constitutes those surface saline ponds/lakes with positive stable isotope ratios, thereby suggesting dilution / evaporation process (see Figure 6).

In summary, the primary salinity of the saline groundwaters in the Benue Trough can be characterized as Na—Cl type most of which had evolved to Na—Ca—Cl type and depleted in Mg and SO_4 relative to seawater. This is a strong indication that the original marine chemical characters of the primary solution have evolved to the present-day brine/saline groundwater through water-rock interactions. Understanding such modifying/controlling

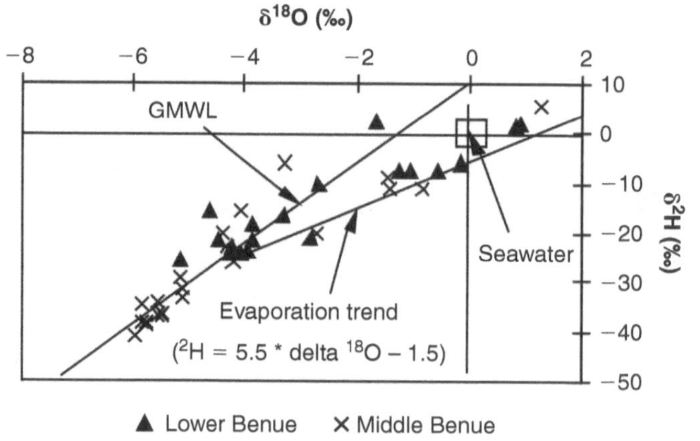

Figure 6. Plot of δ^2H versus $\delta^{18}O$.

chemical processes is a key to unravelly the genetic source of primary salinity in the saline groundwaters in the Benue Trough.

4.2 *Relationship to fossil seawater / connate formation water*

The basic fact that the sedimentation processes in the Benue Trough were related to a series of marine transgression-regression cycles lends credence to the assumption of fossil (Cretaceous) seawater as a possible source of the primary salinity in the saline waters. The assumption is that fossil (Cretaceous) seawater from the Gulf of Guinea, as formation water has evolved to the present groundwater during post-depositional compaction. Therefore, it is expected that the saline groundwaters in the Benue Trough should exhibit a chemical character similar to that of seawater. A direct hydrochemical comparison of the saline groundwater samples with standard seawater (using concentration factors and ionic ratios), as used in similar studies elsewhere (Carpenter & Miller, 1969; Sanders, 1991), is employed in this study. As presented in Table 3, chloride-normalized concentration factors (nCF) with respect to Na have values of approximately 1.0 irrespective of the degree of salinity, with an average value of 1.08 and 1.25 for the lower and middle regions respectively. This is an indication of the close relation between Na and Cl ions.

However, for Ca and Sr, the nCF values are considerably >1 (av. of 2.5 and 10 respectively, for both regions) while the nCF values for Mg are generally <1 (see Table 3). This trend points to the fact that the original marine chemistry of the seawater has been modified to produce saline groundwaters that are enriched in Ca and Sr on the one hand and depleted in Mg on the other hand relative to the primary seawater chemistry. It should be noted that the use of standard seawater as reference, even though the focus here is Cretaceous seawater, is justified by the fact that Chave (1960) had noted that there is hardly any variation in the concentrations of major chemical parameters of seawater since the Cambrian period.

Further evaluation, using ionic ratios between the different chemical parameters, as presented in Table 4, generally show rNa/K, rNa/Cl values comparable to that of seawater. However, rNa/Cl with values >1 may be attributed to excess Na ion input through possible ion exchange process, while rMg/Cl molar ratios unlike rCa/Cl and rSr/Cl, are considerably

Table 3. Summary of chloride-normalized concentration factors with respect to major ionic parameters.

nC-Factors*	Lower Region		Middle Region	
	Range	Mean	Range	Mean
nCF-Ca	1.06–5.12	2.70	1.42–3.96	1.78
nCF-Mg	0.02–0.37	0.10	0.07–0.20	0.11
nCF-Na	0.74–1.98	1.08	0.79–2.02	1.25
nCF-K	0.38–2.08	0.81	0.56–1.77	1.10
nCF-Sr	2.75–24.41	12.90	3.64–87.03	9.41
nCF-Br	0.050–0.481	0.110	0.050–0.302	0.067

*nCF = Chlorine normalised Concentration Factor

Table 4. Summary of major ionic ratios compared with that of seawater.

Ionic ratios	Lower Region		Middle Region		Standard Seawater*
	Range	Mean	Range	Mean	
Na/Cl	0.63–1.69	0.92	0.67–1.72	1.16	0.853
Na/K	19.4–118.6	69.6	35.3–95.8	55.5	46.97
Ca/Cl	0.28–0.19	0.101	0.025–0.148	0.067	0.037
Ca/Na	0.024–0.634	0.132	0.017–0.153	0.065	0.044
Mg/Ca	0.130–1.070	0.232	0.156–1.544	0.399	5.533
Sr/Cl	0.0009–0.008	0.005	0.001–0.030	0.003	0.0003
Sr/Ca	0.016–0.168	0.045	0.022–0.420	0.047	0.009
Cl/Br	1,387–13,729	5,444	2,194–13,729	5,182	649

*Data from Collins, 1975.

lower than that of seawater. The overall trend of the ionic ratios is a further confirmation of the earlier inference, that the saline groundwaters in the Benue Trough are enriched in Ca and Sr and depleted in Mg compared to seawater. Moreover it has been observed that neither Cretaceous seawater nor any Cretaceous sabkha are likely to have had compositions characterized by the low rMg/Ca ratio (Land & Prezbindowski, 1981). Hence, it can be conclude that the ionic character of precursor Cretaceous seawater (apparently from the Gulf of Guinea) must have been modified.

In addition, similar ionic and chemical profiles in brines and mineralized waters elsewhere have been attributed to dolomitization as well as recrystallization of calcite/aragonite (Collins, 1975; Stoessel & Moore, 1983; Sanders, 1991; Wilson & Long, 1993). Therefore, using the same line of argument, the observed depletion-enrichment trend observed in the saline groundwaters from the Benue Trough could be related to modifications of the primary Cretaceous seawater chemistry through water/rock interactions involving dolomitization or recrystallization of carbonates. In this case, the removal of Mg from and release of Ca into the solution/fluid phase are attributed to the dolomitization reaction

($2CaCO_3 + Mg^{2+} \rightarrow CaMg(CO_3)_2 + Ca^{2+}$), while recrystallization of aragonite may be responsible for the release and enrichment of Sr due to the fact that more Sr ions are said to be incorporated into the crystal structures of aragonite than that of calcite (Sass & Starinsky, 1979). Also the presence of limestone in parts of the trough supports the plausibility of the dolomitization process. Furthermore, the low SO_4 concentration alongside the enrichment of Ca may be attributed to possible SO_4 reduction process. Due to the decrease in solubility of oxygen in water as salinity increases, anoxic condition which favors SO_4 reduction do develop rather easily in brines. However, Drever (1988) noted that the net effect of SO_4 reduction is the conversion of SO_4 into equivalent amount of alkalinity, hence leading to enrichment of the solution with calcium.

Another support for the fossil seawater source is related to the low $\delta^{18}O$ and δ^2H values (see Table 2 and Figure 6). These low $\delta^{18}O$ and δ^2H values of the analyzed saline groundwaters are possible indication of a fossil/connate source and/or old fluid system, in this case, remobilized formation waters moving upwards through fracture systems rather than downward mineralization (salinization) of infiltrating meteoric water. Dolomitization and other related diagenetic changes are thought to have taken place during tectonic related movement or remobilization. In summary, there are strong indications in support of fossil (Cretaceous) seawater source for the saline groundwaters in the Benue Trough. Nonetheless, there are serious reservations as to whether such fossil seawater/connate formation water may be adequate to sustain the current flow of saline springs and ponds.

4.3 *Relationship to evaporite / salt dissolution*

Although the presence of a mappable evaporite deposit in the Benue Trough is still a controversial issue, the presence of isolated lenses and/or disseminated salts within the thick marine sediments of the Benue Trough cannot be completely ruled out. Moreover, Petters (1978 & 1982) noted the formation of isolated epi-continental seas within the sub-basins during the early transgression of the Gulf of Guinea during the mid-Albian time; hence one can speculate about the possible evaporation of Cretaceous seawater and formation of evaporites. Then, the question is whether the evolution of the saline groundwater in the Benue Trough is related to (i) evapo-concentrated seawater (liquid brine) or (ii) dissolution of solid salt/halite units initially formed through evaporation of Cretaceous seawater from the Gulf of Guinea. Either of the two mechanisms has been invoked in several studies of basinal saline groundwaters (Nativ, 1996; Martel *et al.*, 2001), though Chi & Savard (1997) noted that these two processes are not mutually exclusive. In evaluating the plausibility of these processes, the Na-Cl-Br relation was employed, especially due to the preferential partitioning of Br between the residual fluid phase and salt/halite crystals during the evaporation of seawater.

Studies have shown that evaporation of seawater and subsequent halite precipitation results in a residual saline groundwater characterized by low Na/Cl and low Cl/Br ratios, whereas saline groundwater originating from halite dissolution is characterized by relatively high Na/Cl ratio ($\sim$1) and high Cl/Br ratio (McCaffrey *et al.*, 1987). Therefore, for the saline groundwaters from the Benue Trough, the relatively high Na/Cl ratio of 0.76–1.05 and the high Cl/Br ratio of 2194–13,729 (see Table 4) favour halite dissolution processes. The graphical plots of molar ratio of Br/Cl and Na/Cl against chloride as presented in Figure 7a and b respectively, also oppose possible influence of evaporation process as the analyzed data fall outside the evaporation trend. Furthermore, saline waters from evaporated seawater

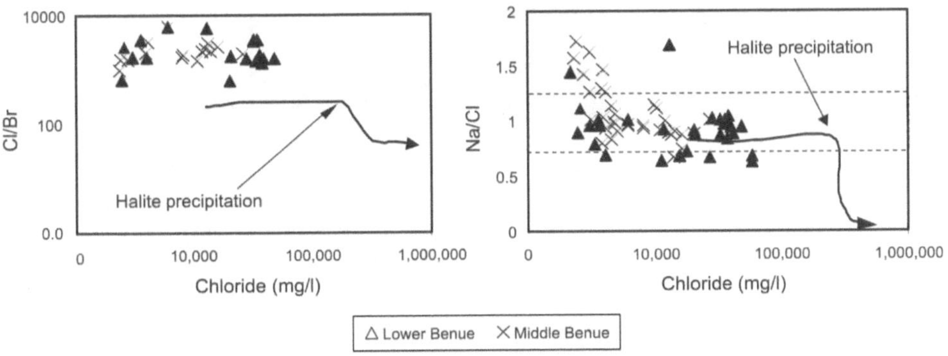

Figure 7. Graphical plots of rCl/Br ratio (left) and rNa/Br ratio (right) against chloride concentrations alongside with seawater evaporation trend.

are generally characterized by positive $\delta^{18}O$ and δ^2H whereas the isotopic characters of those from the dissolution of salt/halite are dependent on the isotope composition of the dissolving water (Knauth & Beeunas, 1986). Figure 6 may initially appear to support evaporation by the plot of some data along an evaporation line defined by $\delta^2H = 5.5 * \delta^{18}O–1$. Though an indication of evaporative fractionation, all these samples involved are surface saline ponds which leads to the conclusion that positive values (enrichment) is related to mixing with meteoric water, or modern evaporation.

In addition, Figure 7a reveals a high Cl/Br ratio with respect to evaporated seawater, thus suggesting halite dissolution just like the Na/Cl value of approximately 1 (Figure 7b). However, the excess of Na (above 1) could be possibly due to exchange process. The plausibility of an exchange reaction is strongly supported by the occurrences of a number of shale units in the geological sequence of the Benue Trough (see Figure 2). Nonetheless, further evaluation as presented in Figure 8, indicates addition of alkalis (mostly Na) through cation exchange process, especially for most of the samples from the middle region, while most of those from the lower region exhibit excess of Ca. Therefore, the excess of Na can be attributed to the cation exchange process, whereas the enrichment of Ca can be attributed to interplay of sulphate reduction and dolomitization reactions.

An indication of salt/halite dissolution is also highlighted by a plot of Na/Cl against Br/Cl (Walter *et al.*, 1990; Horita *et al.*, 1991) as shown in Figure 9, where all of the saline groundwater samples fall within the zone of halite dissolution/recrystallization, though the linear spread along the dissolution line is an indication of salinity differences among the analyzed samples. It should be noted that a simple salt dissolution process alone could not account for the enrichment of Ca and Sr as well as the associated low rMg/Ca values in the analyzed saline groundwaters. Therefore, if salt / halite dissolution is actually a generating mechanism for the saline groundwater in the Benue Trough as suggested by the observed chemical and isotopic characters then, modification of the saline groundwaters by water/rock interactions (dolomitization and cation exchange reactions) must have been responsible for the observed enrichment-depletion trend among the divalent cations.

Another issue lies with lack of known appreciable salt beds in the trough, though disseminated/intergranular salts within the geologic units have been mentioned (Uma & Loehnert, 1992) and also highly favoured by the geological and statigraphical settings of the Benue Trough itself. However, lack of mappable salt beds/deposits cannot be an argument against

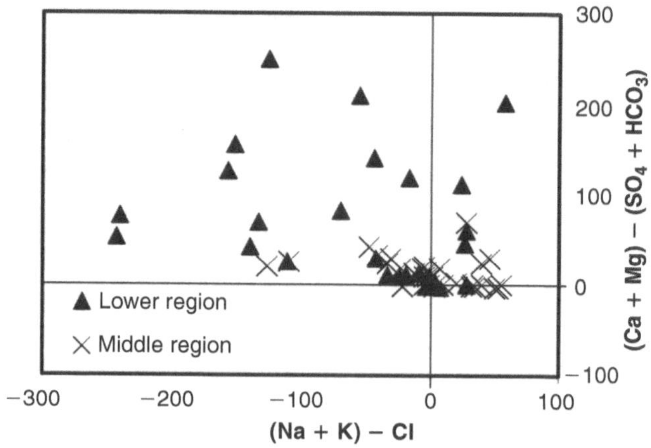

Figure 8. Graphical plot of exchange parameters depicting exchange process.

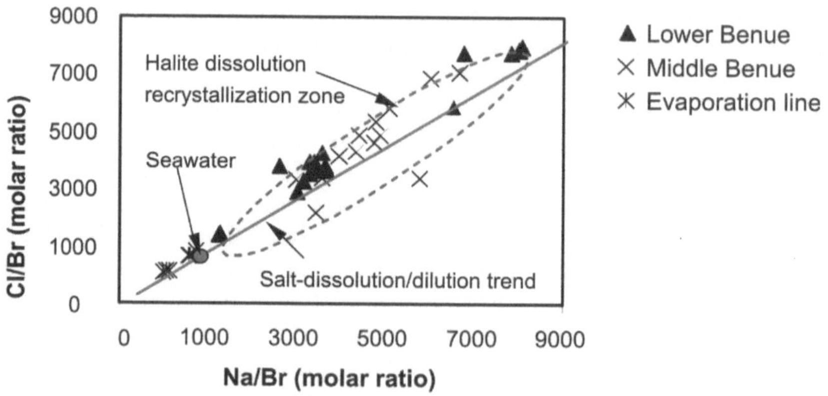

Figure 9. Graphical plot of rNa/Cl against rBr/Cl with the evaporation-dissolution trend.

a halite dissolution model since Ranganathan (1991) and Nativ (1996) noted that absence of massive halite deposits has not necessarily eliminate halite dissolution as a generating mechanism in the past. The fact that high Cl/Br molar ratio (av. 5300) greater than that of seawater (649) for all the analyzed samples and plotting of the data within the halite dissolution zone (see Figure 9) are classical indications of halite dissolution. Moreover, since it is difficult to imagine forming intergranular / disseminated halite salts (under transgression) without evaporation it can be assumed that halite formation must have taken place at one time or the other. In addition, to increase the Cl/Br ratio six to eight times (av. 5300 compared to 649 for seawater) will require at least a 6-fold addition of Cl ion to seawater; a situation, which can only be possible through massive halite dissolution. Hence, the lack of confirmed salt beds may thus be consequences of re-dissolution by the recurrent transgression/seawater flooding during the depositional cycles within the Benue Trough. This is consistent with the low $\delta^{18}O$ and $\delta^{2}H$ ratios for most of the analyzed samples, which suggest ancient dissolving solvent, in this case Cretaceous seawater from the Gulf of Guinea and/or remobilized fossil/connate formation waters rather than infiltration of meteoric water.

4.4 *Relationship to hydrothermal mineralized fluids*

The close spatial association between the saline groundwaters and the mineralized veins in the Benue Trough in many of the outcropping areas has led the advocates of a hydrothermal model to favour a genetic relation of the saline groundwater with the hydrothermal fluids despite limited chemical data. However, recent studies (Uma & Loehnert, 1992; Tijani *et al.*, 1996; Uma, 1998) are of the opinion that the relationship is only hydraulic involving the movement of fluids along the mineralized fracture systems. Even if the observed low SO_4 in the saline groundwater can be linked to the sulphate reduction in relation to Pb-Zn-Baryte mineralization in parts of the Benue Trough, the low $\delta^{18}O$ isotope ratio seems to oppose a hydrothermal fluid source since hydrothermal fluids would have been enriched in ^{18}O by exchange with rocks. Also extremely low concentrations of Pb and Zn (Tijani *et al.*, 1996) even in more concentrated samples point to the unlikelihood of this model.

Furthermore, the possible influence of convective (driving) force of the magmatic/intrusive activity, associated with the mineralization, on the movement/ circulation of the saline groundwaters is questionable due to the fact that nowhere is known to be thermally/magmatically active within the trough. Hence, the occurrences of two warm saline springs in the middle region can be attributed to the normal geothermal gradient due to considerable thickness of the sediments in the trough (>5000 m) rather than any active magmatic activity. Therefore, it can be concluded that there are lines of evidence against the hydrothermal fluid source model. However, it is quite clear that the spatial association between the saline groundwaters and the mineralized veins are hydraulic rather than genetic. The genetic implication of the association between the saline groundwater and mineralized veins is still open to discussion pending further detail studies, e.g. fluid inclusions and age dating etc. of both saline waters and the mineralized vein materials.

5 CONCLUSIONS

The saline groundwaters in the Benue Trough, like other basinal saline groundwaters around the world are characterized by Na-Cl ions with enrichment of Ca, Sr and depletion of Mg, SO_4 and low $\delta^{18}O$ and $\delta^{2}H$ values, which strongly indicate modified fossil seawater as found elsewhere as the source of saline groundwater. Undoubtedly this fossil seawater can be associated with the inland transgressive extension of the Gulf of Guinea during the Cretaceous. However, there is doubt as to whether such modified connate/fossil seawater would be sufficient to sustain the flow of the saline springs and ponds over the past years/decades without any other contributing source. Also, based on the evaluations and interpretations of the hydrochemical and stable isotopes data, there is evidence against both evaporation process (residual brine) (see Figures 5 and 6) and hydrothermal fluid (residual mineralized brine) models. However, the Cl/Br molar ratio of 2194–13,729 (av. 5300), which greatly exceeded the marine value of 649 strongly support dissolution of disseminated and/or halite beds as a regionally acting factor on the primary source of salinity. The lack of observed evaporites / halite salt beds within the Benue Trough notwithstanding, the halite dissolution model is favoured by the prevalent chemical and stable isotope profiles.

The overall implication is that the genesis of the saline groundwaters in the Benue Trough is related to halite dissolution as a regional source of primary salinity. Such halite beds and/or disseminated salts obviously formed during the regressive phases of the depositional cycles must have been re-dissolved either by influx of seawater during the inland transgressive extension of the Gulf of Guinea or by remobilized fossil/connate formation water. The prevalent present-day chemical and stable isotope profiles are reflections of hydrochemical evolution of the initial marine characters through water/rock interactions involving dolomitization and cation exchange reactions. Consequently, the overall evaluation is summarized in a tentative hydrochemical evolution model as presented in Figure 10.

This graphical summary outlines the hydrochemical evolution model of the saline groundwater in relation to the geological, stratigraphical, and tectonic settings of the Benue Trough. However, it should be noted that the hydrochemical evolution scheme as presented in Figure 10 is by no means conclusive, rather it is liable to modification, subsequent to further studies involving radioactive age dating (using ^{3}H, ^{14}C and ^{36}Cl), fluid inclusions and palaeo-hydrogeologic studies. These coupled with hydrochemical modeling will form the basis of further study which is expected to come up with a composite model of evolution that will fit into the overall geotectonic and stratigraphic settings of the Benue Trough.

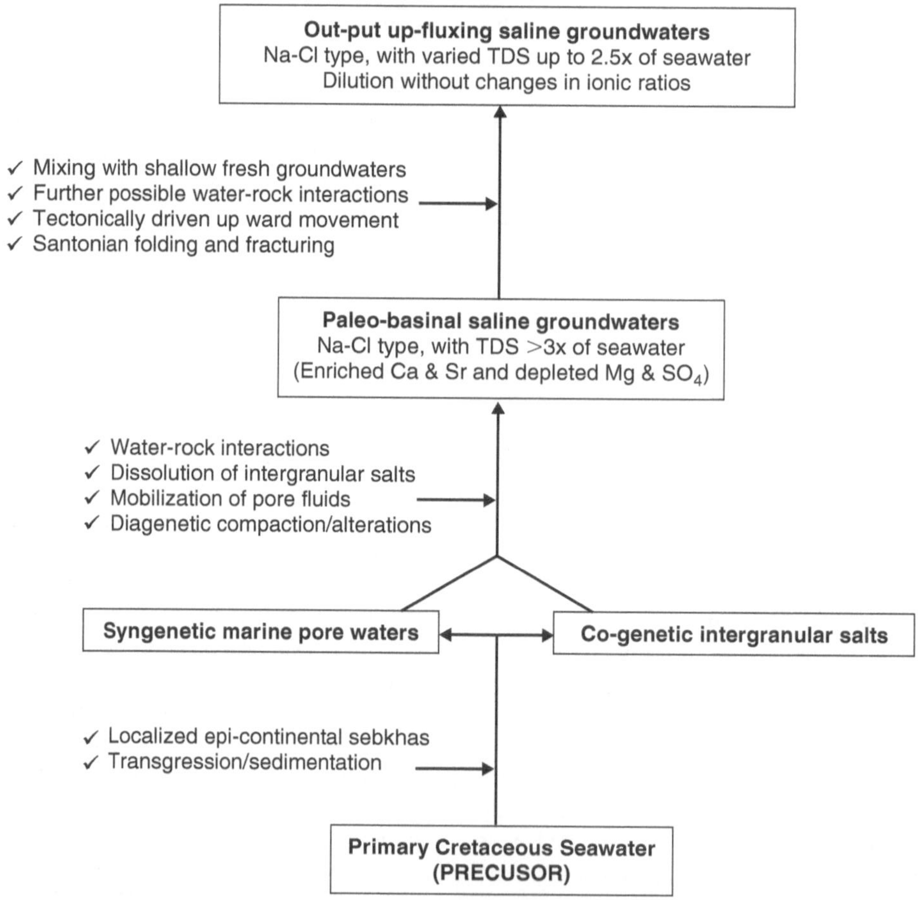

Figure 10. Flow chart depicting the proposed hydrochemical evolution trend/model for saline groundwater in the Benue Trough, Nigeria.

ACKNOWLEDGEMENTS

The author acknowledges the financial supports of VW-Foundation, Hannover, Germany for this study and that of German Academic Exchange Service (DAAD) for the authors' stay in Germany. I also thank Prof. E. Rosenthal for the useful comments and suggestions on the initial version of this manuscript. While recognizing with many thanks, the usual encouragement of Prof. E.P. Loehnert (Muenster, Germany), the author woiuld like to dedicate this work to the blessed memory of Dr. K.O. Uma (UNN, Nsukka, Nigeria), who died in November, 2003, for his professional guidance and advice during the course of field operations for this study.

REFERENCES

Akande, S. O., Horn, E. E. & Reutel, C. 1988. *Mineralogy, fluid inclusion and genesis of the Arufu and Akwana Pb-Zn-F mineralization, Middle Benue Trough, Nigeria.* Journal of African Earth Sciences, **7**, 167–180.

Benkhelil, J. 1986. *Structure and Geodynamics Evolution of the intracontinental Benue-Trough (Nigeria).* Thesis, University of Nice/Publication of Elf (Nigeria) Ltd. 202pp.

Billings, G. K., Hitchon, B. & Shaw, D. R. 1969. *Geochemistry and Origin of formation waters in the western Canada sedimentary basin II: Alkali metals.* Chemical Geology, **4**, 211–223.

Burke, K. C., Dessauvagie, T. F. J. & Whiteman, A. J. 1970. *Geological history of the Benue Valley and the adjacent areas.* In: Dessauvagie, T. F. J. & Whiteman, A. J. (eds), African Geology. Ibadan University Press, Nigeria, 187–205.

Carpenter, A. B. 1978. *Origin and chemical evolution of brines in sedimentary basins.* Oklahoma Geological Survey Cir, **79**, 60–77.

Carpenter, A. B. & Miller, J. C. 1969. *Geochemistry of saline subsurface water, Saline County (Missouri).* Chemical Geology, **4**,135–167.

Chave, K. E. 1960. *Evidence on history of seawater from chemistry of deeper subsurface waters of ancient basins.* AAPG Bulletin, **44**, 357–370.

Chi, G. & Savard, M. M. 1997. *Sources of basinal and Mississippi Valley-type mineralizing brines: mixing of evaporated seawater and halite-dissolution brine.* Chemical Geology, **143**, 121–125.

Clayton, R., Friedman, I., Graf, D., Mayeda, P., Meets, W. & Shimp, N. F. 1966. *The origin of Saline formation waters: Isotopic composition.* Journal of Geophysical Research, **71**, 3869–3882.

Collins, A. G. 1975. *Geochemistry of Oilfield Brines.* Elsevier, Amsterdam, The Netherlands, 496pp.

Craig, H. 1969. *Standard for reporting concentrations of deuterium and oxygen-18 in natural waters.* Science, **133**, 1833–1834.

Drever, J. I. 1988. *The Geochemistry of Natural Waters. 2nd Edition.* Prentice Hall, Englewood Cliff, New Jersey, USA, 437pp.

DVKW 1992. *Entnahme und Untersuchungsumfang von Grundwasserproben.* DVWK-Schriften, 128, Verlag Paul Parey, Hamburg, Germany.

Egboka, B. C. E. & Uma, K. O. 1986. *Hydrochemistry, contaminant transport and their tectonic effects in the Okposi-Uburu salt lake area, Imo State, Nigeria.* Hydrological Science Journal, **31**, 205–221.

Farrington, I. L. 1952. *A preliminary description of the Nigerian Lead-Zinc field.* Economic Geology, **47**, 583–608.

Ford, S. O. 1980. *The economic mineral resources of the Benue Trough.* Earth Evolution Science, **1**, 154–163.

Frape, S. K. & Fritz, P. 1982. *The chemistry and isotopic composition of saline groundwaters from Sudbury Basin, Ontario.* Canadian Journal of Earth Sciences, **19** (1), 645–661.

Hanshaw, B. B. & Coplen, T. B. 1973. *Ultrafiltration by compacted clay membrane II: Sodium ion exclusion at various ionic strengths*. Geochim Cosmochim Acta, **37**, 2311–2327.

Hitchon, B. & Friedman, I. 1969. *Geochemistry and origin of formation waters in the western Canada sedimentary basin I: Stable isotopes of hydrogen and oxygen*. Geochim Cosmochim Acta, **33**, 1321–1349.

Horita, J., Friedman, T. J., Lazar, B. & Holland, H. D. 1991. *The composition of Permian seawater*. Geochim Cosmochim Acta, **55**, 417–432.

Knauth, L. P. & Beeunas, M. A. 1986. *Isotope geochemistry of fluid inclusions in Permian halite with implications for the isotope history of ocean water and origin of saline formation waters*. Geochim Cosmochim Acta, **50**, 419–433.

Land, L. S. & Prezbindowski, D. R. 1981. *The origin and evolution of saline formation water, Lower Cretaceous carbonates, South-central Texas, USA*. Journal of Hydrology, **54**, 51–74.

Loehnert, E. P., Tijani, M. N. & Uma, K. O. 1996. *Evolution and origin of saline groundwaters in the Benue Trough, Nigeria*. Zbl Geol Palaont Teil I Heft, **7/8**: 739–756.

Martel, A. T., Gibling, M. R. & Nguyen, M. 2001. *Brines in the Carboniferous Sydney Coalfield, Atlantic Canada*. Applied Geochemistry, **16**, 35–55.

McCaffrey, M. A., Lazar, B. & Holland, H. D. 1987. *The evaporation path of seawater and the co-precipitation of Br and K with halite*. Journal of Sedimentary Petrology, **57**, 928–937.

McConnel, R. B. 1949. *Notes on Lead-Zinc deposits of Nigeria and Cretaceous stratigraphy of Benue Trough and Cross River Valleys*. Geological Survey of Nigeria, Report **752**.

Nativ, R.1996. *The Brine underlying the Oak-Ridge Reservation, Tennessee, USA: Characterization, Genesis, and Environmental implications*. Geochim Cosmochim Acta, **60**, 787–801.

Offodile, M. E. 1976. *The Geology of the Middle Benue, Nigeria*. Publication of the Palaeontological Institute of the University of Uppsala, Special Volume 4, Uppsala, 166pp.

Ofoegbu, C. O. 1984. *A model for the tectonic evolution of the Benue Trough of Nigeria*. Geol Rundschau, **73**, 1007–1018.

Olade, M. A. 1975. *Evolution of Nigeria's Benue Trough (Aulacogen): A tectonic model*. Geological Magazine, **112**, 575–581.

Olade, M. A. 1976. *On the genesis of lead-zinc deposits in Nigeria's Benue rift (Aulacogen): a re-interpretation*. Journal of Mining and Geology, **13**, 20–27.

Orajaka, S. O. 1972. *Salt-water resources of East Central State of Nigeria*. Journal of Mining and Geology, **7**, 35–41.

Petter, S. W. 1978. *Stratigraphic evolution of the Benue Trough and its implications for the Upper Cretaceous paleogeography of West Africa*. Journal of Geology, **86**, 311–322.

Petter, S. W. 1982. *Central West African Cretaceous – Tertiary benthic foraminifera and stratigraphy*. Palaeontographica Abt A, **179**, 1–104.

Ranganathan, V. 1991. *Salt diffusion in interstitial waters and halite removal from sediments: Examples from Red Sea and Illinois basins*. Geochim Cosmochim Acta, **55**, 1615–1625.

Sanders, L. L. 1991. *Geochemistry of formation waters from the Lower Silurian Clinton Formation (Albion Sandstone), Eastern Ohio*. AAPG Bulletin, **75**, 1593–1608.

Sass, E. & Starinsky, A. 1979. *Behaviour of strontium in subsurface calcium chloride brines, southern Israel and Dead Sea rift valley*. Geochim Cosmochim Acta, **43**, 885–895.

Stednick, J. D. 1991. *Wildland water quality sampling and analysis*. Academic Press Inc. San Diego, USA, 216pp.

Stoessel, R. K. & Moore, C. H. 1983. *Chemical constraints and origins of four groups of Gulf Coast reservoir fluids*. AAPG Bulletin, **67**, 896–906.

Sulin, V. A. 1946. *Water of petroleum formation in systems of natural waters*. Costoptekhizdar, Moscow.

Tattam, C. H. 1943. *Preliminary report on the salt industry in Nigeria*. Geological Survey of Nigeria, Report **778**.

Tijani, M. N., Loehnert, E. P. & Uma, K. O. 1996. *Origin of saline groundwaters in the Ogoja area, Lower Benue Trough, Nigeria.* Journal of African Earth Sciences, **23** (2), 237–252.

Tijani, M. N. 2004. *Evolution of saline waters and brines in the Benue-Trough, Nigeria.* Applied Geochemistry, **19** (9), 1355–1365.

Uma, K. O. 1998. *The brine fields of the Benue Trough, Nigeria: a comparative study of geomorphic, tectonic and hydrochemical properties*. Journal of African Earth Sciences, **26** (2), 261–275.

Uma, K. O. & Loehnert, E. P. 1992. *Research on the saline groundwaters in the Benue-Trough, Nigeria: preliminary results and projections.* Zbl Geol Palaeont Teil I; Heft, **11**, 2751–2756.

Uma, K. O., Tijani, M. N. & Loehnert, E. P. 1995. *Hydrochemical Research on the origin of saline groundwaters in the Benue Trough, Nigeria.* Final Technical Report 1/65602, VW-foundation, Hannover, Germany.

Uzuakpunwa, A. B. 1981. *The geochemistry and origin of the evaporite deposits in the southern half of the Benue Trough.* Earth Evolution Science, **2**, 136–139.

Walter, L. M., Stueber, A. M. & Huston, T. J. 1990. *Br-Cl-Na systematics in Illnois Basin fluids: Constraints on fluid origin and evolution*. Geology, **18**, 315–318.

Wilson, T. P. & Long, D. T. 1993. *Geochemistry and isotope chemistry of Michigan Basin brines: Devonian formations.* Applied Geochemistry, **8**, 81–100.

Wright, J. B. 1976. *Origin of the Benue Trough: a critical review.* In: Kogbe, C. O. (ed.), Geology of Nigeria, Elizabethan Publication Company, Lagos, 309–317.

CHAPTER 22

Hydrochemical and isotopic characteristics of coastal groundwater near Abidjan (southern Ivory Coast)

M.S. Oga

Université de Cocody, UFR des Sciences de la Terre et des Ressources Minières, Abidjan Côte d'Ivoire

Hydrologie et Géochimie isotopique, Laboratoire Interactions et Dynamiques des Environnements de Surface, Université de Paris-Sud, Orsay, France

C. Marlin, L. Dever & A. Filly

Hydrologie et Géochimie isotopique, Laboratoire Interactions et Dynamiques des Environnements de Surface, Université de Paris-Sud, Orsay, France

R. Njitchoua

Hydrotrace, Cergy, France

ABSTRACT: This study of recharge to the Continental Terminal and Quaternary aquifers in the southern Ivory Coast was carried out using both hydrochemical and isotopic methods. Solutes in the groundwater in these aquifers originate mostly from atmospheric contributions and the hydrolysis of silicate minerals. In places, the water quality has deteriorated by nitrate pollution and/or salt-water intrusion from the lagoons. The comparison of the isotopic composition in rain waters with the groundwaters indicates that infiltration by rainwater constitutes the main source of groundwater recharge in the greater Abidjan area. Recharge to the aquifers can occur all year round, not just in the rainy season during which runoff exceeds infiltration. Lack of temporal and spatial variability of heavy isotopes of groundwater is indicative of a good mixing of waters from different recharge episodes and ambient groundwater which is consistent with the porous nature of the aquifer. The isotopic balance of ^{13}C composition of Total Dissolved Inorganic Carbon (ranging from -26.7 to -11.4‰ PDB) and soil CO_2 (from -25.3 to -11.6‰ PDB) indicates that biogenic gas produced under forest cover in the area of Abidjan (mean $\delta^{13}C = -25.0$‰ PDB) and under Savanna around Dabou ($\delta^{13}C = -11.6$‰ PDB) plays an important role in the carbon mineralization of the groundwaters of the Continental Terminal and the Quaternary aquifers. High ^{14}C values (between 92.7 and 114.7 pmC) and the ^{3}H content ($\sim$8.4 TU) show that recharge to the Continental Terminal and the Quaternary aquifers is a recent phenomenon. The residence times computed from ^{14}C activities and tritium content vary from modern to about five hundred years.

1 INTRODUCTION

Abidjan is the largest city in Ivory Coast (4 millions people, more than a fourth of the total country's population which was 15 millions in 1998 according to the National Statistic

Institute of Ivory Coast). It is also an industrial area along the Atlantic Ocean coastline. Like other coastal cities of Western Africa, the recent increase of population has resulted in environmental problems such as deterioration of drinking water in quality and quantity (Boukari *et al.,* 1996; Asubiojo *et al.*, 1997; Ibe & Njemanze, 1998; Kouadio *et al.*, 1998; Ajahi & Umoho, 1998; Banoeng-Yakubo, 2003; Jourda, 2003, 2004; Soro, 2003). Although there is a large drainage network in this area, the water supply of the greater Abidjan area is mainly from two shallow sedimentary aquifers: the Continental Terminal aquifer (CT), the most important groundwater reservoir and the Quaternary aquifer along the coast. Together, the aquifers constitute a regional unconfined system.

The potentiometric data of the CT aquifer has recently shown a general lowering of the water table. This may be due to intensive pumping of groundwater and rainfall variability in the Ivory Coast. The water supply, estimated at 67 Mm^3 in 1991 increased to 99 Mm^3 in 2001, with an increase rate of 2.9 Mm^3/a (data from annual reports of SODECI, "Société de Distribution d'Eaux de Côte d'Ivoire" 1991–2002). The deeper water-levels are now affecting pumping rates, which have has progressively decreased, especially in some districts of Abidjan (i.e. Angré, Abobo, etc). To solve this problem, the government of the Ivory Coast financed a study in order to improve the existing hydrodynamical model of the Abidjan CT (SOGREAH, 1996). Historical excessive exploitation of the CT aquifer has induced contaminated water in the depression of Banco Bay (static level of −13.7 m asl, electrical conductivity of 4.2 mS/cm, NH_4^+ content of 1.6 mg/l; Loroux 1978). Locally, an intrusion of marine/brackish water and/or lagoonal water into the coastal aquifer is evident (i.e. RAN-Plateau borehole with a flow rate of 250 m^3/h and with an EC of 12 mS/cm; Loroux 1978), showing that overexploitation of the aquifer progressively induces a displacement towards the continent of the sea-water/fresh water interface. This situation observed in boreholes close to the Ebrié Lagoon and those of the southern area of the lagoon has constrained the SODECI Company to abandon the exploitation of the boreholes in these areas. Moreover, the water quality is affected by human and industrial activities and the development of agriculture in the suburbs of Abidjan (Aghui & Biémi, 1984; Bado, 1992; Edoukou, 1992).

Most hydrogeological studies in western Africa have focused on semi-arid areas where rural water supply is the main concern. Using environmental isotopes, numerous researches have been undertaken on the recharge of aquifers under semi-arid conditions and its variability due to climatic changes and increasing human impact (e.g. Le Gal La Salle *et al.*, 2001; Favreau *et al.*, 2002). Concerning the coastal aquifers of western Africa little is known about the Ivory Coast compared to neighbouring countries: e.g. Nigeria (Edet & Okereke 2002), Ghana (Archampoing & Hess 2000) and Benin (Boukari *et al.*, 1996). The development and management of groundwater resources in the Abidjan area require an ability to understand and identify the source of recharge to the groundwater system. The objective of this paper is to discuss the recharge and contamination of the CT and the Q aquifers using ^{18}O, ^{2}H, chemical tracers and ^{3}H and ^{14}C.

2 GEOLOGICAL AND HYDROGEOLOGICAL SETTING

The greater region of Abidjan constitutes the central part of a coastal sedimentary basin which covers a surface of 16,000 km^2 between the latitudes of 5°00 and 5°30 N and the longitudes of 3°00 and 6°00 W (Figure 1). The climate of the study area is sub-equatorial characterized by two rainy seasons (March–July and September–November) separated by two relatively dry periods. The annual mean rainfall is between 1500 and 2000 mm for an average temperature of 27°C (monthly temperature between 24 and 30°C). The vegetation is clear forest near the coastline and becomes dense further inland.

Three geomorphological units can be distinguished: the High plateaus (40–50 to 100–120 m asl) represented by outcrops of CT in the North of Ebrié Lagoon, the Low plateaus (8–12 m asl) constituted by outcrops of the Quaternary deposits in the South of Lagoon and the coastal plains. The Ebrié Lagoon is connected to the Atlantic Ocean by the Vridi Canal and is mainly fed by fresh water discharging from the Comoé River and some small coastal rivers, which are secondary sources i.e. Agnéby and Mé). In addition to the rivers, rainfall represents 10% of the recharge to the fresh water (Varlet, 1978).

The geology of the Ivory Coast is dominated by Precambrian bedrock. Along the coastline, the southern area is characterized by a narrow sedimentary basin representing only 3% of the total surface of the country. This coastal basin is fault-controlled Cretaceous/Quaternary strata dipping towards the Guinea Gulf. Inland, the thickness of the basin is probably in the order of several kilometers (Spengler & Delteil, 1966).

The Continental Terminal formations are the Tertiary deposits dated from Miocene to Pliocene; they are mainly composed of sands, clays, sandstones and indurate lateritic layers, and localized in the north of the Ebrie Lagoon (max 160 m). The Quaternary formations composed of sands and silts are located in the southern part of Ebrie Lagoon (max 140 m). In the centre of the basin, there are transgressive and unmatched series of the Maestrichtien (upper Createaceous) rocks which are located under the CT formations. They are made up of sandy limestone and sands. Apart from the centre, the CT formations lay directly over the Precambrian rocks of the study area.

In the sedimentary basin of the Ivory Coast, the geological entities, which are of hydrogeological interest, are as follow:

- Two types of aquifer exist in the Quaternary deposits: marine sands (Nouakchottien) and fine sands (Oogolien). Groundwater in the Nouakchottien aquifer is very vulnerable to pollution because its piezometric surface is close to the ground surface.
- In the Tertiary formations, both fluviatile (CT3) and clay (CT4) sands are hydrogeologically important. The aquifer (CT3) may be confined when there is a clay level in the roof of CT3; when there is no clay level between CT3 and CT4 deposits, the CT3 and the CT4 are interconnected.

The hydraulic conductivity of the CT aquifer is variable due to lateral changes in the grain size of water-bearing sediments (10–100 m/d in the sands and the sandstones and 0.01 to 0.1 m/d in the clayey sands). The transmissivity values are between 100 and 10,000 m^2/d, with porosity values ranging from 0.05 to 0.20. The regional groundwater flow occurs from north to south, i.e. towards the lagoon. The hydraulic conductivity of the Quaternary aquifer ranges between 3 and 100 m^2/d. The hydraulic gradient increases up to 3‰ close to the lagoon. The flow rates are low compare to that of the CT aquifer: 2–22 m^3/h for Q aquifer and 7–338 m^3/h^- for CT aquifer.

Deeper in the basin, around 200 m below the ground surface, the Maestrichtian carbonates and sandstones constitute a confined aquifer. Only the borehole of SADEM Company (total depth of 191 m) draws its waters from this aquifer. This aquifer is artesian with a potentiometric surface at +27 m above the sea level.

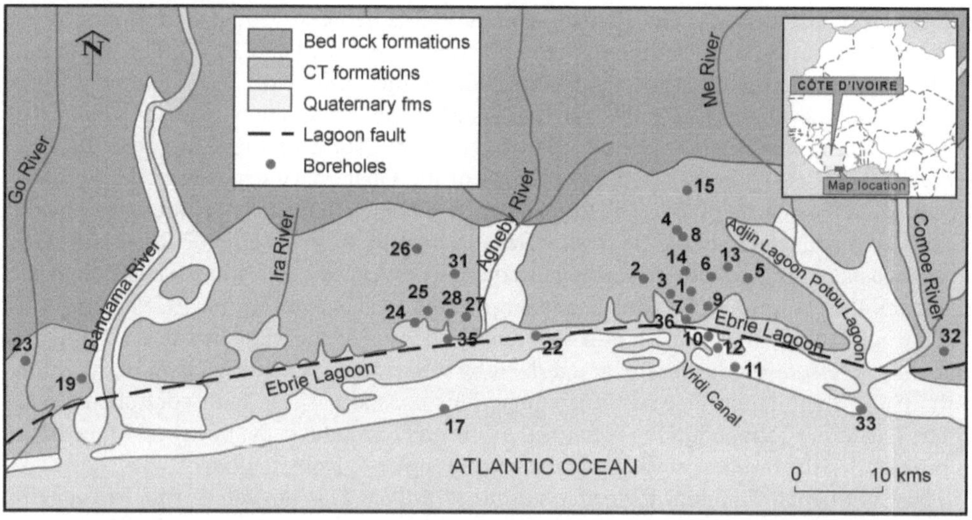

1) Adjame **2)** Niangon **3)** Zone ouest **4)** Anonkoua **5)** Riviera N. **6)** Zone est **7)** Plateau **8)** Coquivoire **9)** Hotel Ivoire **10)** Solibra **11)** Bima **12)** Capral **13)** Riviera C. **14)** Zone nord **15)** Anyama **17)** Jacqueville **19)** Lahou **23)** Yocoboue **24)** Bodou **25)** Bouboury **26)** Lopou **27)** Dabou **28)** Pass **31)** Orbaf **32)** Bonoua **33)** Bassam **22)** Jacqueville Ebrie Lagoon **35)** Dabou Ebrie Lagoon **36)** Abidjan Ebrie Lagoon

Figure 1. Location of the study area and the sampled wells.

3 MATERIALS AND METHODS

3.1 *Hydrochemistry*

Twenty six water wells were sampled in August 1996 for isotopic and chemical analysis of water: 21 samples from the Continental aquifer and 5 samples from the shallow aquifer of the Quaternary. Samples were collected along two North-South transects, A and B (Figure 1). These wells are used for drinking water and were being pumped during the sampling period. In addition, three water samples were collected in the Ebrié Lagoon during two periods, firstly in the dry season (March 1995) under the wharf of Dabou and under the Houphouet Boigny bridge in Abidjan, and secondly just after the long rainy season (August 1996) in Jacqueville. Soil gas (CO_2) was also collected under different vegetation in the great Abidjan region.

The parameters analysed or measured were: pH, temperature (°C), electrical conductivity (μS/cm), alkalinity, sodium (Na^+), potassium (K^+), calcium (Ca^{2+}), magnesium (Mg^{2+}), chloride (Cl^-), nitrate (NO_3^-), sulphate (SO_4^{2-}), and silica (SiO_2). Groundwaters filtered (0.45μm) in the field were collected in 250 ml polythene bottles for geochemical analyses. As the pH is low ($3.50 \leq pH \leq 6.58$) in the great Abidjan groundwater, it was deemed not necessary to acidify water samples. For the evaluation of groundwater chemistry, thermodynamic calculations were used (Rollins, 1987) and silicate stability diagrams. X-ray diffraction analyze was carried out on soil samples from the Continental Terminal.

To understand the origin of the mineralisation of groundwaters, water-rock interaction was studied using the diagrams of partial balances Na_2O-Al_2O_3-SiO_2-H_2O, K_2O-Al_2O_3-SiO_2-H_2O and CaO-Al_2O_3-SiO_2-H_2O. The description of the secondary processes of mineralisation was carried out with tests of correlation between the major ions Cl^- vs K^+ and Cl^- vs Ca^{2+}. Another diagram showing the correlation between Cl^- and Na^+ contents with the distance from the Ebrié Lagoon was also plotted. The mixture rate between groundwater and lagoon waters was also assessed. The Cl^- balance equation can be defined by:

$$[Cl^-]_{sample} = a[Cl^-]_{CT} + (1-a)\,[Cl^-]_{Ebriélagoon}$$

Where:

$[Cl^-]_{sample}$ is the chloride content of the sample,
$[Cl^-]_{CT}$is the mean of chloride content of CT groundwater (except Plateau, Adjamé and Pass samples),
$[Cl^-]_{lagoon}$ is the chloride content of the Ebrié Lagoon ($[Cl^-]$lagoon = 790.5 mg/ in 1996, *i.e.* around 4% of the chloride content of sea water).
a is the proportion of 'pure' CT groundwater in the mixture.

3.2 *Isotopic methods*

Twenty-nine unfiltered water samples were collected in 20 ml glass bottles with poly-sealed lids for stable isotopes analysis (^{18}O and ^2H). The method of Epstein and Mayeda (1953) was used for the preparation of samples for $\delta^{18}O$ analysis. Hydrogen gas extraction was done by the Coleman *et al.* (1982) method using zinc reduction at around 500°C. The stable isotope analyses of water samples were obtained on a VG Sira 10 spectrometer for the oxygen and on a collector mass spectrometer for the deuterium at Laboratoire d'Hydrologie et de Géochimie Isotopique (LHGI), Orsay (France). Reproducibility is better than 0.2‰ for $\delta^{18}O$ and about 2‰ for δ^2H. The results for both isotopes are expressed in per mill (‰) deviation from the V-SMOW standard using the δ-scale.

Samples for tritium analysis were collected in 0.5 litre polyethylene bottles and preserved for analysis. Twenty-nine were collected for ^{14}C and twenty-two of them were analysed for ^3H. Tritium contents were measured by the electrolytic enrichment method (Kaufman & Libby, 1954) at Thonon (Centre de Recherche en Géodynamique, Université de Paris 6, France).

The carbon isotope analyses were done on the Total Dissolved Inorganic Carbon (TDIC) of samples. The ^{13}C content of groundwater was measured using a mass spectrometer at LHGI. The ^{14}C of TDIC measurements were undertaken at LHGI using the conventional method of Fontes (1971); for samples with low carbon content, 1 litre of groundwater was sampled. The ^{14}C activities of this water were measured at Gif-sur-Yvette, using an accelerator mass spectrometer.

Tritium and ^{14}C activities are respectively expressed in Tritium Units (1 TU = 1 atom of ^3H in 10^{18} atoms of ^1H) and in percent modern carbon (pmC), whereas carbon-13 content values are given in δ‰ notation relative to Pee Dee Belemnite (PDB), standard. Analytical errors are within ±0.2‰ for $\delta^{13}C$ and vary between 0.4 and 0.9 TU for ^3H and between 0.5 and 1.6 pmC for ^{14}C. The determination of groundwater residence times using both tritium and ^{14}C measurement is based on the following equation:

$$t = \frac{T_{0.5}}{Ln2} Ln\left(\frac{A_0}{A_t}\right)$$

In which:
t is the residence time,
$T_{0.5}$ the half-live of the considered radioisotope $T_{0.5} = 5730 \pm 30$ years (Godwin 1962) for ^{14}C and $T_{0.5} = 12.43$ years for ^{3}H.
A_t and A_0 are observed activity at time t and initial activity, respectively.

4 RESULTS

4.1 *Hydrochemistry*

Geochemical results are shown in Table 1 for chemical composition. Groundwater from CT displays low values of pH (3.5–5.4) and low mineralization. Electrical conductivity values range from 19 to 55 μS/cm and TDS values range from 9 to 47 mg/l. Three data points show higher mineralization (146 mg/l for Adjamé 8, 130 mg/l for Plateau C4 and 81 mg/l for Pass). The representative median of the CT groundwater are 4.3 for pH, 58 μS/cm for EC and 39 mg/l for TDS, respectively. The groundwater temperatures range between 25.5 and 28.6°C with a median value at 26.7°C. Groundwater temperature are locally lower than that of the annual mean air temperature (27°C). This may be due to preferential recharge at the end of the main rainy season when the air temperature is low. The average temperature of July and August is as cold as 25°C. The dominant ions are Cl^- and NO_3^- for anions and Na^+ for cations (Figure 2). Locally, some samples are enriched in other ions: HCO_3^- and K^+ for Orbaf, HCO_3^- and Ca^{2+} for Zone Ouest 1. These data points are not different in terms of pH. Slight, but significant chemical differences exist between the eastern and western part of the aquifer: groundwater in the urban area of Abidjan is characterized by relatively high concentrations in NO_3^- (13.4 mg/l), Cl^- (7.3 mg/l) and Na^+ (6.6 mg/l) compared with groundwater in the rural area of Dabou (NO_3^- 4.5 mg/l; Cl^- 4.4 mg/l and Na^+ 3.1 mg/l). This implies a difference in TDS (45.5 mg/l in the east and 29.6 mg/l in the west).

Water from different locations within the Quaternary aquifer is of similar physical and chemical features. However, groundwater from Q aquifer is slightly more mineralized than CT (85 to 138 mg/l with a median value of 115 mg/l). The groundwater also presents higher temperature (ranging from 26.6 to 30.4°C), especially in wells having a deeper water-level (20–50 m below soil surface). The pH values range from 3.8 to 6.6. Nitrate or bicarbonate are the dominant anions and the Na^+ or Ca^{2+} are the dominant cations (Figure 2).

4.2 *Oxygen-18 and deuterium*

The oxygen and hydrogen isotopic compositions of groundwater are shown in Table 2. The isotopic values of CT samples are quite homogeneous within a small range from −3.4 to −2.4‰ for ^{18}O and from −13.7 to −6.2‰ for ^{2}H. The medians are −3.0‰ for ^{18}O and −10.3‰ for ^{2}H. More enriched groundwater samples were collected from Coquivoire and Bonoua in the Abidjan area. Most samples collected in Abidjan (East CT) show $\delta^{18}O$ and $\delta^{2}H$ values slightly more depleted than those from Dabou area (West CT). The stable isotope data of Q groundwater range from −2.9 to −2.2‰ for ^{18}O and from −12.6 to −3.6‰ for ^{2}H with a median of −2.7‰ for ^{18}O and −10.3‰ for ^{2}H. Grand Bassam contains the most enriched groundwater ($\delta^{18}O = -2.2$‰ and $\delta^{2}H = -3.6$‰). They are similar to those of the CT aquifer.

Table 1. Physical and chemical composition of groundwater and Ebrié Lagoon water.

Location	Code	Aquifer	Total deep (m)	T °C	EC µ S.cm^{-1}	pH	Eh (mV)	Cl	NO_3^-	SO_4^{-2}	HCO_3^-	Na^+ $^-$(mg.l^{-1})	K^+	Mg^{+2}	Ca^{+2}	Al^{+3}	SiO_2	TDS	IB (%)
Lagoon (1)	22	nd	nd	27.1	2900	7.41	nd	790.5	0.0	91.3	372.7	456.6	36.9	60.9	24.5	0.00	17.9	1851	5.6
Lagoon (2)	35	nd	nd	nd	nd	nd	nd	2704.9	0.0	500.4	nd	nd	nd	nd	–	–	–	–	–
Lagoon (3)	36	nd	nd	nd	nd	nd	nd	15599.4	0.0	1857.3	nd	nd	nd	nd	–	–	–	–	–
Adjamé Nord	1	CTE	98.6	27.0	233	4.11	497	20.4	69.0	1.9	0.6	24.5	6.7	1.1	4.8	0.95	16.3	146	0.3
Zone Ouest	3	CTE	59.8	26.1	55	4.96	512	5.0	8.2	3.6	7.0	3.4	1.3	0.5	4.3	0.06	13.3	47	1.1
Anonkoua	4	CTE	120.5	25.9	42	4.58	527	4.9	4.9	0.6	0.6	4.8	1.3	0.5	0.9	0.04	15.0	33	4.3
Nord Riviera	5	CTE	76.0	25.9	36	4.34	527	4.9	2.7	1.1	0.1	2.8	0.4	0.6	0.2	0.00	13.7	27	1.2
Zone Est	6	CTE	105.6	26.0	40	3.94	515	4.8	5.0	1.0	0.7	4.5	0.8	0.8	0.8	0.04	10.2	29	4.3
Plateau	7	CTE	87.8	27.5	222	4.09	487	27.2	48.9	3.3	0.2	24.9	5.3	1.8	5.4	0.04	13.1	130	0.1
Coquivoire	8	CTE	–	26.9	28	4.33	527	4.0	3.3	0.0	0.8	3.3	0.2	0.4	0.6	0.00	10.4	23	1.9
Hôtel Ivoire	9	CTE	80.0	26.6	37	4.40	557	4.5	3.0	1.1	0.7	3.9	0.5	0.3	0.5	0.00	13.8	28	1.4
Niangon	2	CTE	69.4	26.0	36	4.33	463	4.7	3.5	0.8	2.5	2.5	0.8	0.3	0.3	0.07	14.9	30	2.6
Riviera Centre	13	CTE	129.3	26.0	39	3.99	607	4.0	4.8	0.6	2.6	3.2	0.2	0.2	0.2	0.05	12.4	28	1.2
Zone Nord	14	CTE	120.0	25.5	39	4.34	627	4.4	3.9	0.5	0.9	2.9	0.2	0.3	0.2	0.04	10.6	24	2.4
Anyama Adjame	15	CTE	63.0	26.7	54	4.17	567	2.7	12.5	1.4	0.9	2.4	1.0	0.7	1.2	0.07	10.2	33	3.3
Bonoua	32	CTE	94.0	26.4	31	4.48	577	2.8	4.7	0.6	0.4	2.9	0.4	0.0	0.2	0.05	1.2	13	0.5
Grand Lahou	19	CTW	54.7	26.7	33	5.36	471	4.5	0.0	2.8	0.8	2.8	0.2	0.3	1.2	0.04	11.6	24	1.3
Yocoboué	23	CTW	21.8	26.8	38	4.50	592	4.8	1.0	1.9	2.5	3.9	0.9	0.3	1.1	0.05	19.2	36	1.5
Bodou	24	CTW	–	27.3	27	4.44	563	2.2	1.0	2.0	0.3	1.6	0.1	0.0	0.0	0.05	11.5	19	1.1
Bouboury	25	CTW	42.6	27.4	21	4.33	537	2.1	0.3	1.3	1.2	1.8	0.5	0.1	0.1	0.04	1.8	9	0.7
Lopou	26	CTW	76.3	27.3	19	3.50	630	1.8	0.3	1.2	1.4	1.3	0.2	0.1	0.1	0.02	7.0	13	2.1
Dabou	27	CTW	82.8	28.6	26	4.26	598	1.9	0.0	2.4	1.7	1.4	0.2	0.2	0.2	0.04	11.9	20	2.3
Pass	28	CTW	10.0	27.7	138	4.25	467	15.6	32.0	3.9	0.2	10.2	6.6	1.6	3.5	0.48	6.8	81	2.4
Orbaf	31	CTW	97.2	27.1	28	4.42	537	2.2	1.1	2.7	5.6	1.7	1.4	0.4	0.9	0.05	18.4	34	1.9
Grand Bassam	33	Q	36.6	26.6	208	6.58	27	16.2	8.8	0.8	73.2	16.9	3.1	3.9	4.6	0.08	4.3	132	9.1
Solibra	10	Q	30.0	28.8	217	5.19	253	35.8	9.7	15.6	22.9	25.6	5.3	2.5	6.0	0.00	12.3	136	2.8
43ème BIMA	11	Q	34.5	29.2	140	6.46	−32	8.4	0.0	4.2	49.4	7.8	1.7	4.0	3.4	0.05	6.2	85	8.2
Capral	12	Q	48.5	29.5	224	5.12	558	21.2	60.2	7.7	0.5	17.7	3.2	1.7	15.5	0.05	10.0	138	0.8
Jacqueville	17	Q	20.3	30.4	162	3.78	545	13.7	42.9	1.5	1.0	13.3	2.5	3.1	4.1	1.42	2.4	86	4.1

Notes: All groundwater samples were collected in August 1996. The data points are located on Figure 1. Water from the Ebrié Lagoon was sampled in August 1996 in Jacqueville (Lagoon 1) and in March 1995 in Dabou (Lagoon 2) and in Abidjan (Lagoon 3). “nd” stands for “not determined”. “CTE” stands for “Continental Terminal East CTW” stands for “Continental Terminal West.

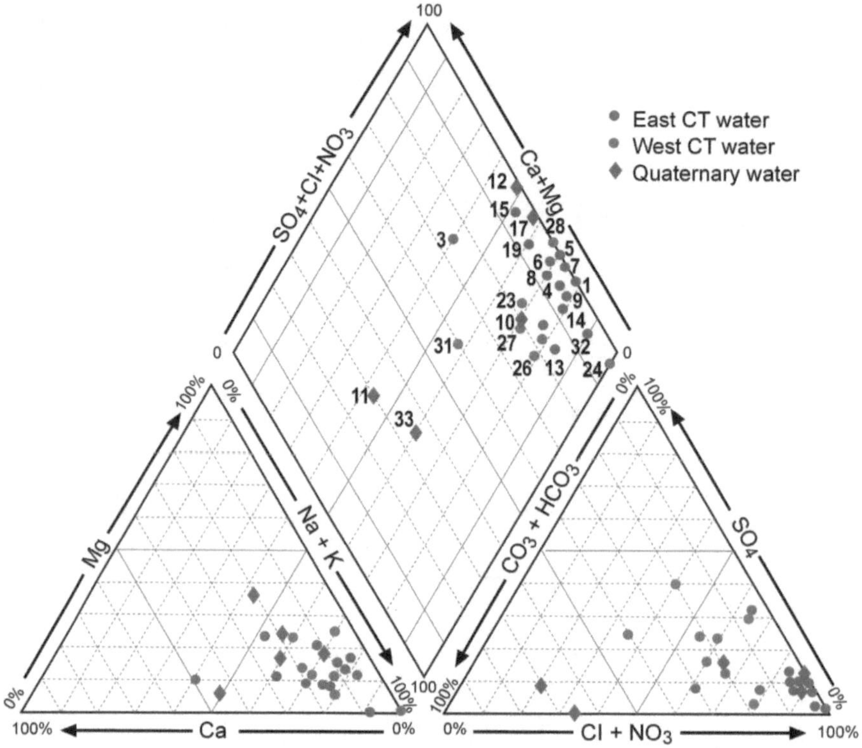

Figure 2. Piper diagram of the greater Abidjan area's groundwater samples.

4.3 *Tritium*

The tritium content ranges up to 8.4 TU. The median value of groundwater from CT and Q aquifers is 3.7 ± 1.6 TU. Only two samples of the 22 measurements show tritium contents below the detection limit (0.8 TU). This indicates that the recharge to the coastal aquifers is mainly modern, occurring during or after the thermonuclear tests. Relatively high ^3H values have also been obtained by Acheampong & Hess (2000) from the coastal aquifers in Ghana. To the east, the CT aquifer becomes slightly younger than the one to the west (median: 2.8 ± 0.4 TU to the east and 2.0 ± 0.4 TU to the west). However, the differences are not significant.

4.4 *Carbon isotopes*

The ^{13}C and $\delta^{14}C$ values for samples are shown in Table 2. Groundwater samples from CT and Q display high activities for ^{14}C (median at 102.6 ± 7.0 pmC). These data are consistent with the tritium data and indicate recent recharge to the coastal aquifer. The western part of the CT aquifer presents a median radiocarbon activity at 99.6 ± 6.6 pmC, while on the east the groundwater's radiocarbon activity is above 100 pmC (104.9 ± 5.3 pmC). These differences may be significant. To the east, the $\delta^{13}C$ (TDIC) of groundwater of CT range from −26.7‰ to −20.2 ‰ with a median of −24.4‰. Those obtained on the western part of CT range from −25.0 to −11.4‰ with a median of −17.3‰ PDB. However, the wide

Table 2. Isotopic composition of groundwater and Ebrié Lagoon water

Location	Sample number	$\delta^{18}O$ ‰	$\delta^{2}H$ VSMOW	$\delta^{13}C$ ‰ PDB	$A^{14}C$ pmC	±	^{3}H TU	±	$\delta^{13}C_{CO2}$ ‰ PDB	log pCO_2	TDIC mM
Lagoon (1)	22	−1.5	−0.4	nd	nd	nd	nd	nd	nd	nd	
Lagoon (2)	35	0.6	3.9	nd	nd	nd	nd	nd	nd	nd	
Lagoon (3)	36	0.9	5.7	nd	nd	nd	nd	nd	nd	nd	
Adjamé	1	−3.0	−10.0	−21.5	113.0	0.8	5.0	0.4	−20.5	−1.2	2.2
Zone Ouest	3	−3.0	−13.2	−24.8	106.5	1.2	3.5	0.7	−24.1	−1.1	2.8
Anonkoua	4	−2.9	−11.8	−24.9	102.1	0.7	3.3	0.8	−24.0	−1.8	0.5
Nord Riviera	5	−3.1	−12.0	−26.5	100.5	1.3	5.0	0.5	−25.5	−2.2	0.2
Zone Est	6	−3.0	−9.5	−25.8	99.9	0.9	5.0	0.8	−24.8	−1.1	2.9
Plateau	7	−2.8	−13.0	−20.9	112.4	1.3	5.9	0.5	−19.9	−1.7	0.7
Coquivoire	8	−2.4	−6.5	−23.3	104.0	1.2	nd	nd	−22.3	−1.4	1.3
Hotel Ivoire	9	−3.1	−10.6	−26.0	97.6	1.0	3.8	0.4	−25.0	−1.5	1.0
Niangon	2	−3.0	−8.0	−26.7	102.5	1.4	2.8	0.6	−25.7	−0.9	4.2
Riviera Centre	13	−3.2	−13.7	−25.6	100.2	1.1	nd	nd	−24.5	−2.1	0.3
Zone Nord	14	−3.1	−11.5	−26.5	104.9	1.2	nd	nd	−25.5	−1.4	1.5
Anyama Adjame	15	−2.7	−10.4	−24.3	113.3	1.3	4.0	0.4	−23.3	−1.1	2.4
Bonoua	32	−2.5	−6.2	−20.2	106.5	0.8	3.4	0.6	−19.2	−1.9	0.5
Grand Lahou	19	−3.4	−10.7	−22.1	97.2	1.2	8.4	0.5	−21.8	−2.4	0.1
Yocoboué	23	−2.6	−7.6	−25.0	114.7	1.1	3.1	0.5	−24.1	−1.1	2.9
Bodou	24	−3.1	−11.4	−13.6	97.7	0.8	3.3	0.4	−12.6	−1.9	0.4
Bouboury	25	−2.6	−8.9	−13.5	92.7	0.8	3.2	0.4	−12.5	−1.2	2.1
Lopou	26	−2.6	−9.4	−11.4	100.0	1.6	2.6	0.4	−10.4	−0.3	16.6
Dabou	27	−2.9	−7.4	−12.8	101.1	1.3	0.0	0.8	−11.8	−1.0	3.3
Pass	28	−3.0	−7.4	−23.1	96.2	0.8	4.7	0.4	−22.2	−1.9	0.4
Orbaf	31	−2.6	−10.3	−17.2	97.0	1.1	0.0	0.8	−16.3	−0.6	7.6
Grand Bassam	33	−2.2	−3.6	−13.1	84.3	0.9	2.0	0.4	−17.7	−1.7	1.9
Solibra	10	−2.6	−11.9	−19.0	102.7	1.2	nd	nd	−18.5	−0.8	5.5
43ème BIMA	11	−2.8	−12.6	−22.3	100.5	1.3	4.6	0.6	−26.2	−1.7	1.4
Capral	12	−2.9	−10.8	−19.4	108.9	1.2	3.2	0.7	−18.9	−1.5	1.1
Jacqueville	17	−2.7	−7.0	−22.2	110.8	0.8	2.5	0.4	−21.2	−0.8	5.2

Sampling for isotopic measurements of groundwater was performed in August 1996. Water from the Ebrié Lagoon was sampled in August 1996 in Jacqueville (Lagoon 1), in March 1995 in Dabou (Lagoon 2) and in Abidjan (Lagoon 3).

range of data may be related to the differences in pH between all waters. In order to compare all $\delta^{13}C$ values, $\delta^{13}C$ of CO_2 at isotopic equilibrium with TDIC were calculated (Table 2). A large difference can be seen between the median values of eastern and western parts of the CT aquifer (−16.5‰ to the west and −23.4‰ to the east). In order to verify if $\delta^{13}C$ TDIC is at equilibrium with soil CO_2, soil CO_2 under different vegetal covers was sampled for $\delta^{13}C$ measurement (range from −25.3 to −11.6‰ PDB, Table 3). This comparison is necessary to discuss the origin of dissolved carbon and thus to calculate apparent residence times of groundwater from radiocarbon (mean residence time of inorganic carbon is assumed to be

Table 3. ^{13}C contents in soil CO_2 under different vegetation covers in the great Abidjan area The results are expressed in ‰ PDB.

$\delta^{13}C$ of soil gas (CO_2)	Sample 1	Sample 2	Mean
Forest	−24.8	−25.1	−25.0
Natural savanna	−11.6	nd	−11.6
Bush	−18.8	−18.6	−18.7
Rubber tree (*Hevea brasiliensis*)	−22.7	−23.1	−22.9
Palm tree (*Elaeis guineensis*)	−25.3	−25.2	25.3
(Banana tree (*Musa sp*))	−23.1	nd	−23.1

'nd' stands for 'not determined'.

representative of that of the water). The $\delta^{13}C$ of soil CO_2 produced under the Banco forest ranges from −25.1 and −24.8‰. The Banco forest is the remnant of south Ivory Coast primary forest located in the northern part of Abidjan city. Hevea trees (*Hevea brasiliensis sp.*) and palm trees (*Elaeis guineensis sp.*) also produce CO_2 depleted in ^{13}C (−22.9‰, and −25.3‰ PDB, respectively). Even if the banana tree (*Musa sp.*) is a C4-type plant, its CO_2 is depleted in ^{13}C (−23.1‰ PDB). In soils of natural bush and savanna, $\delta^{13}C$ values of CO_2 are higher, at −18.7 and −11.6‰ PDB, respectively.

5 DISCUSSION

5.1 *Origin of major ions*

Both the CT and Q groundwaters contain small amounts of solutes but are significantly different in terms of TDS (median TDS of 39 mg/l for CT aquifer and of 115 mg/l for Q aquifer). The ranges of chemical compositions are consistent with those reported by previous authors (Jourda, 1987; Soro, 1987; Tapsoba, 1990) for CT and Q aquifers.

Groundwater is characterized by rather high calculated pCO_2 values (from $10^{-2.4}$ to $10^{-0.3}$ atm). These high values are consistent with adjacent coastal aquifers whose conditions are similar to those of Cotonou aquifer (Boukari *et al.*, 1996). The studied groundwater does not exhibit high concentration of bicarbonate (median of 1 mg/l) as expected from the calculated pCO_2. At the recorded pH (3.5 to 6.5), DIC is mainly represented by H_2CO_3. The median content in DIC is 2.6 mM, which remains low. Under such a sub-equatorial warm and humid climate, the inorganic C geochemistry in groundwater is probably strongly controlled by production of biogenic CO_2 in the unsaturated zone. Oxidation of organic matter in the unsaturated zone may be invoked (Boukari *et al.*, 1996). The trend existing between DIC and the depth of water-table seems to corroborate this hypothesis: the deeper the water-table, the less the water enriched in DIC. At depth, the groundwater tends to be less acidic and less mineralized in terms of inorganic carbon. The determinant parameter controlling the DIC content and the pH of groundwater seems to be the proximity to the soil and the plant root system. As the urbanization of Abidjan has considerably changed during the last decade, it is difficult to establish the correlation between the natural past vegetal cover and the geochemical characteristics of the groundwater.

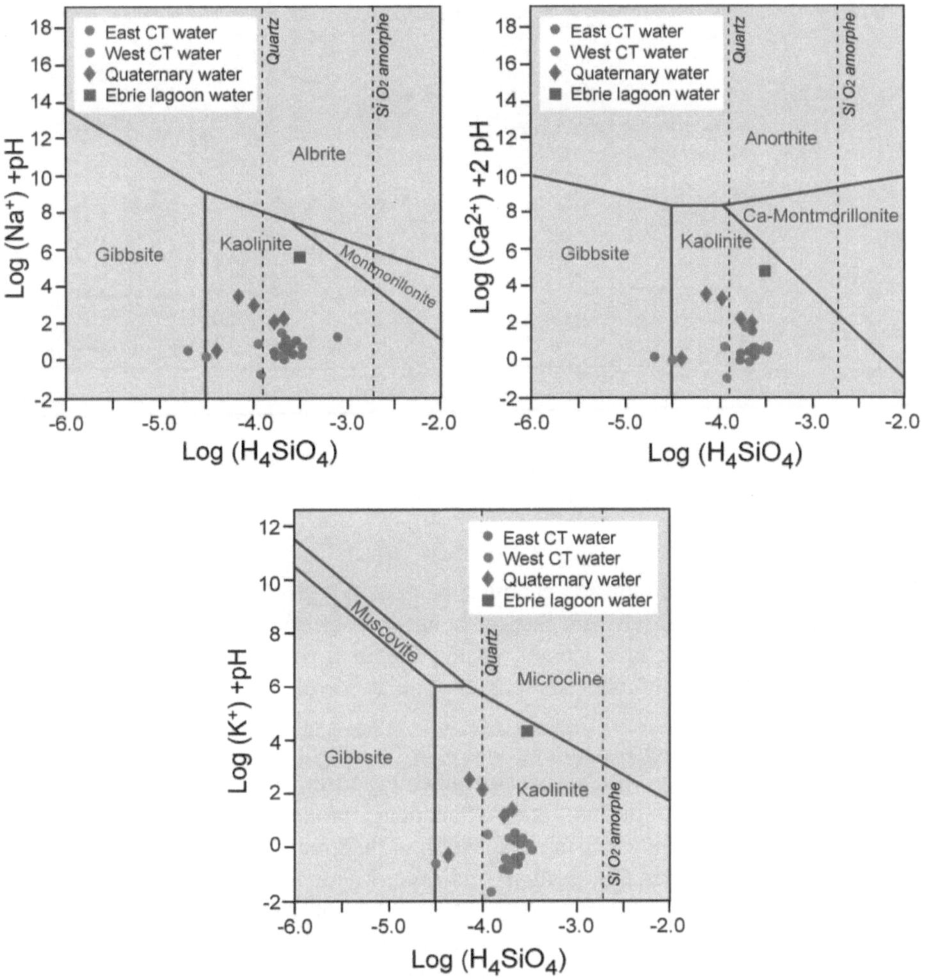

Figure 3. Diagram of the greater Abidjan area water stability in the system of partial balance Na_2O-Al_2O_3-SiO_2-H_2O,$K_2OAl_2O_3$-SiO_2-H_2O and CaO-Al_2O_3-SiO_2-H_2O.

The high pCO_2 values and the warm and humid climatic conditions of southern Ivory Coast are expected to favour the hydrolysis of aluminosilicate minerals. In the stability diagrams of anorthite, microcline and albite, the stability fields of gibbsite, kaolinite, microcline and Ca or Na-montmorillonite are indicated (Figure 3). XRD analyses performed on bulk sample show that the Continental Terminal sands are dominated by quartz, albite and anorthite, kaolinite and illite. In Figure 3, all the investigated groundwaters are located in the kaolinite stability domain far from the geochemical equilibrium with the rocks, including the amorphous silica (Oga, 1998).

The groundwater ions mainly derive from the atmospheric input and the acid weathering of minerals of the skeleton of the aquifer such as aluminosilicates including albite, anorthite and microcline. As dissolution of aluminosilicate minerals is a slow process, the low TDS of groundwater may be interpreted as the consequence of short residence time of solutions

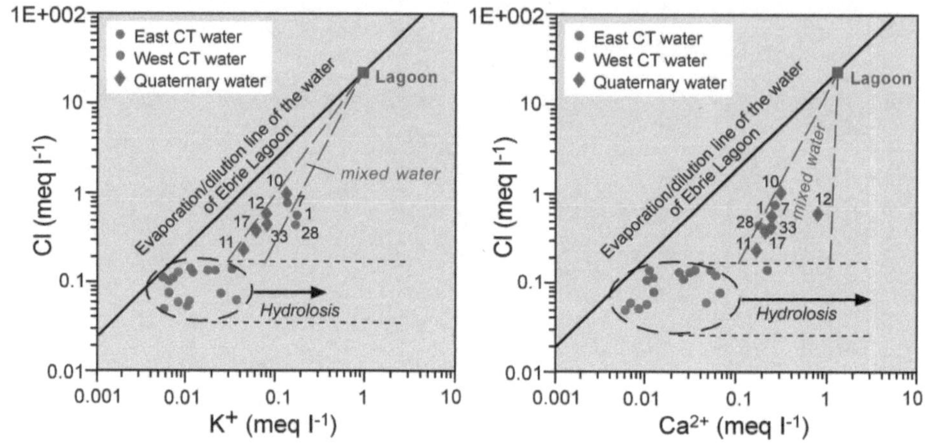

Figure 4. Correlation between Cl^- and K^+ (left) and Cl^- and Ca^{2+} (right) of the great Abidjan area's groundwater samples and the lagoon evaporation line.

within the aquifers. The groundwater chemical composition changed slightly. Moreover, all groundwaters are undersatured with respect to amorphous silica. The low mineralization of the groundwater and the homogeneity of the chemical composition reveal that all the groundwater may come from modern recharge. This is corroborated by ^{3}H and ^{14}C results which are presented below.

All the groundwater from the unconfined aquifers are undersaturated with respect to calcite (−8.5 to −1.9) and with respect to other evaporite minerals. Consequently, no carbonate dissolution exists in this system. Secondary processes such as mixing and the pollution; modify the groundwater mineralisation in the great Abidjan.

The relationships between Cl^- content and other major ions (K^+, Ca^{2+}) is shown in Figure 4, with the indication of the Lagoon Water Dilution Line (LWDL). The first case corresponds to the mineralization of groundwater controlled by the atmospheric input and water-rock interaction. The second case is concerned with the points that contact the water lagoon samples. The proportion of lagoon water in the mixture increases up to 4% at Solibra. Values between 2 and 3% are obtained in groundwater from boreholes of Adjamé, Plateau, Pass, Capral and Grand Bassam.

The diagrams of Figure 5 show that the variations of Cl^- and Na^{+2} contents are related to the distance from the Ebrié Lagoon, i.e. from the ocean. Two groups were distinguished:

1. The first group (I), 4 to 16 km from the lagoon shoreline, is only composed of CT groundwaters having low, homogeneous Cl^-, and Na^+ contents ($\leq$5 mg/l), except for 2 boreholes sampled in Abidjan city (Adjamé and Plateau) and for one domestic well (Pass);
2. The second group (II), 0 to 4 km from the lagoon shoreline is concerned with samples from both the Q and the CT aquifers. The Cl^- and Na^+ contents vary according to the boreholes sampled. The groundwaters, which are mixed with the lagoon water, show nitrate contamination, except for the points 10, 11 and 33.

The nitrate contents in rainwater (annual mean of 2.0 mg/l, Oga, 1998) cannot explain the high contents observed in some groundwater samples (NO_3^-: 32 to 69 mg/l) which also

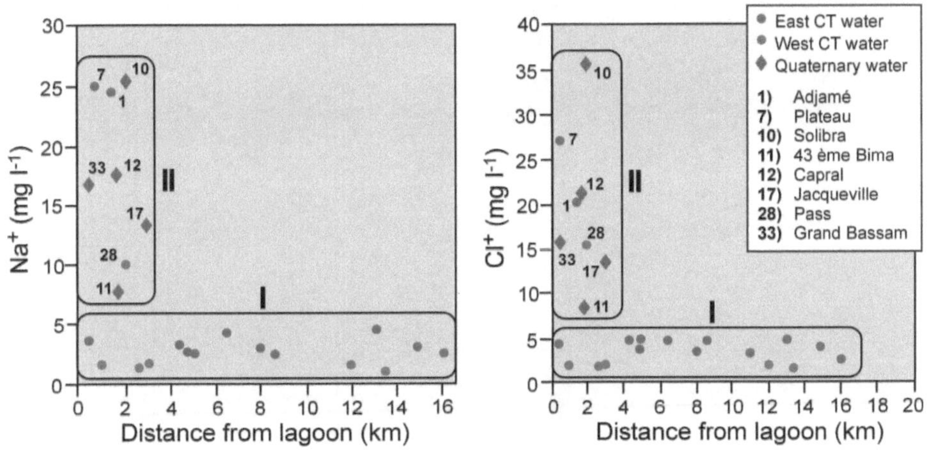

Figure 5. Evolution of the sodium and chlorides contents in Abidjan groundwater samples according to two North-South transects.

indicate high TDS values (from 81 to 146 mg/l). The nitrate contamination of water in the intensively urbanized districts of Abidjan (Plateau, Adjamé and Capral) and Jacqueville may come from the disposal of waste water and organic matter. The local pollution affects the quality of groundwater in the great Abidjan area. Therefore, the contamination at about one hundred meters at Adjamé (98.6 m) and Plateau (87.8 m), is a clue of the vulnerability of phreatic aquifer in this area. Concerning the domestic well of Pass (West CT), high NO_3^- content is rather related to the manures used for the fertilization of the soils in the banana plantations around the well (depth to water at 10 m, Oga 1998).

According to Tapsoba (1995), groundwater from the CT aquifer (NO_3^- 50–60 mg/l) are more exposed to nitrate pollution than water from a borehole (NO_3^- from 6 to 55 mg/l) because they are shallow, around 10 m. More recent studies indicate that the underground water pollution of the CT and of Q by nitrate is not related to the depth (Oga, 1998). Nitrate pollution is an important phenomenon, especially in the large African Capital cities because of urbanization, deforestation and the development of peri-urban agriculture. Studies carried out in the Ivory Coast (Faillat & Rambaud, 1988; Faillat, 1990) indicated that the origin of nitrate contamination is the deforestation releasing nitrogen retained by the root of the trees. Elsewhere in West Africa, Groen *et al.* (1988), Joseph & Girard (1990), Girard (1993) and Boukari *et al.* (1996) stressed the extent of the phenomenon in Burkina-Faso, Niger and Bénin, respectively.

5.2 *Origin and recharge processes of CT and Q aquifers*

The ^{18}O an ^{2}H values in this study are also close to the values obtained in the Ivory Coast by Jourda (1987) from CT groundwater of the Abidjan area (−3.2 to −3.8‰; n = 4), by Soro (1987) and groundwaters from the Mé basin (−3.2‰; n = 2) and by Tapsoba (1995) in the Dabou area (−3.7‰; n = 3). Archampoing and Hess (2000) reported similar mean values for shallow groundwater from a sedimentary basin of Ghana (mean $\delta^{18}O$ of −3.0; mean $\delta^{2}H$ of −14.0‰).

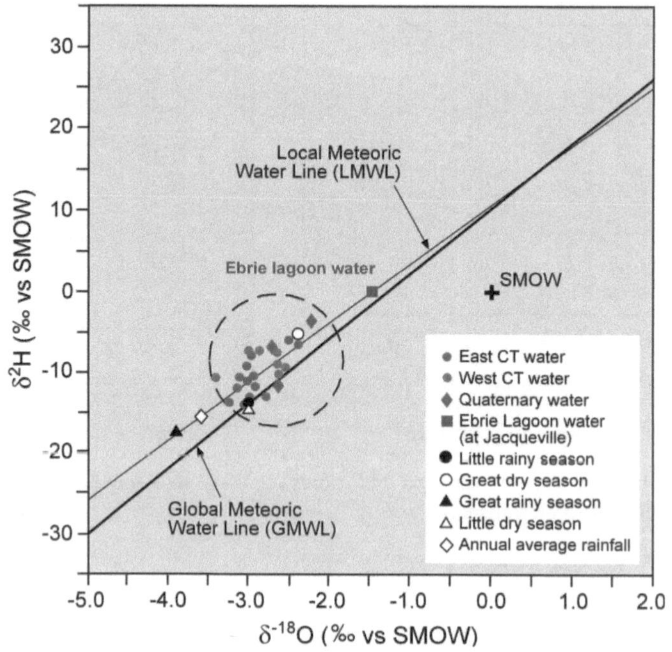

Figure 6. Correlation line between δ^2H and δ^{18}O of the groundwater of the great Abidjan area. The Local Meteoric Water Line (LMWL) is given by the regression line $\delta^2H = (7.3 \pm 0.2) \times \delta^{18}O + (10.6 \pm 0.2)$ *cf* Oga (1998). The Global Meteoric Water Line (GMWL) is given by the regression line $\delta^2H = 8 \times \delta^{18}O + 10$ *cf* Craig (1961).

All isotope data of groundwater are within the range of rainfall ^{18}O and ^{2}H contents sampled during one year, from September 1995 to August 1996. The mean value of rainfall δ^{18}O and δ^2H weighted by rainfall amount is −3.6‰ and −15.6‰, respectively (Oga 1998).

The relationships between δ^{18}O and δ^2H are shown in Figure 6. On this plot, the Global Meteoric Water Line ($\delta^2H = 8\ \delta^{18}O + 10$; Craig 1961) is reported. Moreover, the Local Meteoric Water Line, ($\delta^2H = (7.3 \pm 0.2)\ \delta^{18}O + (10.6 \pm 0.2)$) established by Oga (1998) for the great Abidjan area is shown. All the groundwaters are located along the GWML; this indicates that the recharge of aquifers of the great Abidjan region is mainly derived from rainwater.

The points representing the average water contents of the short dry and humid seasons as well as those of the large dry season localize themselves to proximity of those of groundwater. This shows that the rain from these seasons participates in the groundwater recharge. During the long rainy season, minimum recharge takes place; certainly, this is due to the important streaming phenomenon during the substantial rainfall. Water of the CT and Q aquifers is identical from the isotopic point of view, which seems to indicate that this water would have the same origin.

Spatial homogeneity of ^{18}O and ^{2}H values results from a good mixing between the waters of different recharge episodes within the aquifer. This observation is consistent with the porous matrix of the aquifer and the existence of a relatively important dynamic reserve of the sheet of water.

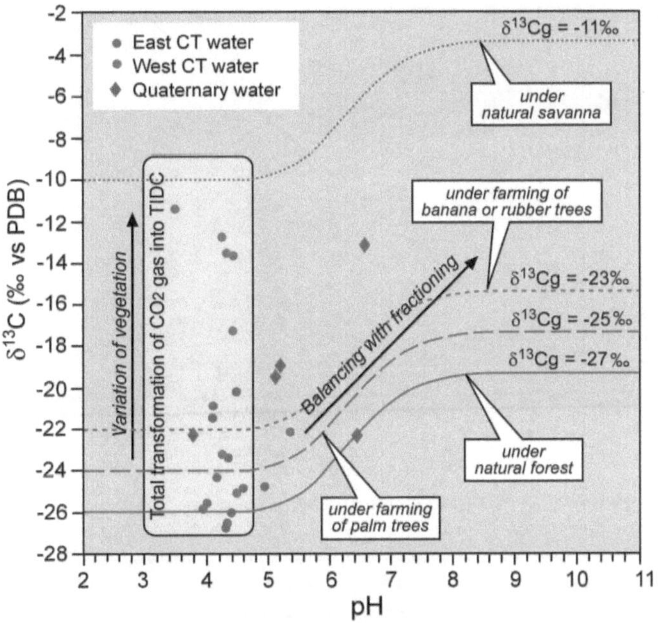

Figure 7. Evolution of ^{13}C content in relation to the pH of groundwater in open system (Wigley, 1975).

This vertical recharge of the CT and the Q aquifers is confirmed by the results of the analysis of ^{13}C (‰ PDB) of Total Dissolved Inorganic Carbon (TDIC). The distribution of the samples in the diagram (^{13}C vs pH, Figure 7) and the calculated $\delta^{13}C$ indicates that carbon-13 of the TDIC has a biogenic origin.

It also shows that the carbon mineralisation of the liquid phase is carried out in a system opened to soil CO_2 (Oga, 1998). The TDIC comes from the biogenic CO_2 dissolution since all the groundwater are in isotopic equilibrium under open conditions with biogenic CO_2 generated either by the forest cover in the region of Abidjan or the savanna in the region of Dabou (Table 3).

Indeed, the ^{13}C contents of the TDIC show a range of variation extended enough from −11.4 to −26.6% PDB and differ slightly from the contents of ^{13}C of the soil gas measured in the area and which varies according to the type of plants of −25.3% to −11.6%. This is in agreement with the pH of this water and the low value of the factor of enrichment between H_2CO_3; the dominant carbonaceous species in solution, and CO_2 gas.

Groundwaters of the CT and Q display high activities, from 92 to 115 pmC for ^{14}C and from 0 to 8.4 UT ^3H contents, which suggest an active recharge system to the two coastal aquifers. There is no relationship between total depth of the boreholes and the tritium and ^{14}C contents. This implies a good homogeneity of water within each aquifer and probably an absence of stratification of water ages in the costal aquifers as indicated by the concentrations of major ions and the stable isotopes.

The high ^{14}C activities (92–115 pmc) and ^3H contents of the CT and the Q correspond to mean residence times between now and 500 years, except Grand Bassam sample. The groundwater polluted by nitrate and mixed with lagoon water shows a strong ^{14}C and/or ^3H activity.

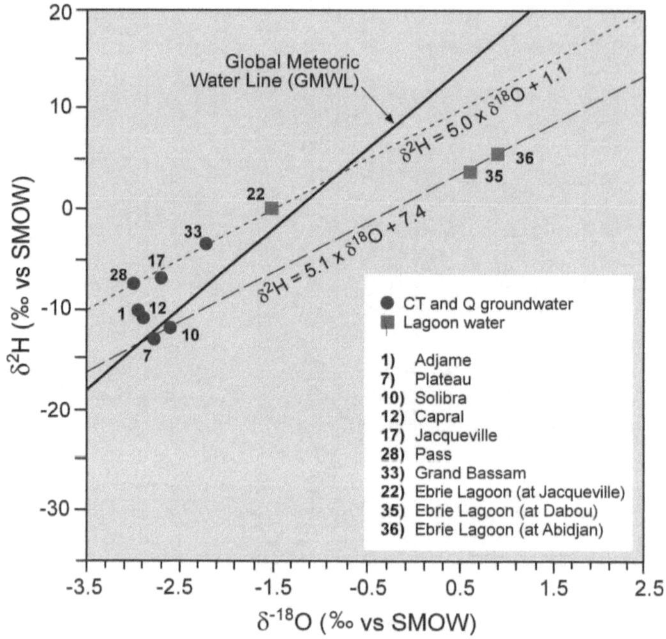

Figure 8. Relationship between δ^2H and $\delta^{18}O$ contents of groundwater (with $[Cl^-] > 10\,mg \cdot L^{-1}$) and lagoon water.

To clarify the relation between the groundwater sampled at Pass, Plateau, Adjamé, Grand Bassam, 43eme Bima, Jacqueville, Solibra and Capral and the one of the lagoon, the ^2H and ^{18}O contents of groundwater and lagoon water were compared in Figure 8 (δ^2H vs $\delta^{18}O$). The lagoon water is sampled in different seasons. The ^2H ^{18}O relationship defined by the equation $\delta^2H = 5.1\delta^{18}O + 7.4$ ($R = 0.99$) indicates a mixing between the groundwater in Pass, Jacqueville, Grand Bassam and those of the Ebrié lagoon sampled in August 1996. Those of Plateau and Solibra boreholes define another line of mixing with the lagoon sampled in March 1995. The second line, which equation line is $\delta^2H = 5.0\ \delta^{18}O + 1.1$ ($R = 0.99$), is parallel with the first one. Water of Capral and Adjamé is located between the two lines defined previously. This water contains the two water components of the lagoon, with a component "dry season" more marked in Adjamé and a component "wet season" at Capral.

The contents of ^2H and ^{18}O confirm the mixing of groundwater collected in the sedimentary aquifer with water from the lagoon in accordance with the of chemistry.

6 CONCLUSIONS

Solutes in groundwater in Abidjan come from atmospheric contributions (input) and natural weathering of silicate minerals such as albite, anorthite, etc. In coastal areas, the increase of groundwater salinity is related to lagoon water intrusion (brackish water coming from the dilution of seawater) and/or nitrate pollution. However, this intrusion is quantitatively insignificant (from 1% to 4%) and modifies the water quality of groundwater in the coastal aquifers to only a minor extent. The high concentrations of NO_3^- (32 to 69 mg/l) in some groundwaters are due (related) to anthropogenic pollution in the urban environment and agricultural pollution in rural areas. This pollution is the consequence of extensive urbanization of the area and the unceasingly increasing pressure of the demand for water in the great Abidjan area. It can reach some 100 m depths and remains a particular concern for the authorities which must take action to abandon the boreholes at risk.

The aquifers of the CT and Q contain young and vulnerable water. Direct infiltration remains the principal mechanism of groundwater recharge in the great Abidjan area. Consequently, a variation of rainfall in this area will affect the recharge to the aquifers.

The homogeneity of groundwater of the Continental Terminal and Quaternary aquifers is suggested by the concentrations in major ions and the contents of stable (^{18}O and ^{2}H) and radioactive (^{3}H, ^{14}C) isotopes. In perspective, the improvement of the present results is necessary. The isotopic study of rainwater (^{18}O, ^{2}H, ^{14}C, ^{3}H) in the great Abidjan region will allow characterization and validation of the aquifer input.

ACKNOWLEDGEMENTS

This study was supported by the Laboratoire d'Hydrologie et de Géochimie isotopique (LHGI) of Paris XI University at Orsay (France). We would like to express our recognition to the personnel in charge of the LHGI. We also thank the technicians of Sodeci Company from the Ivory Coast for their assistance in sampling, Dr Z. Ettien from the University of Bouake in the Ivory Coast, and anonymous reviewers for useful comments that improved the manuscript.

REFERENCES

Acheampong, S. Y. & Hess, J. W. 2000. *Origin of shallow groundwater system in the southern Voltaian sedimentary basin of Ghana : an isotopic approach.* Journal of Hydrology, **233**, 37–53.

Aghui, N. & Biémi, J. 1984. *Géologie et hydrogéologie des nappes de la région d'Abidjan et risques de contamination.* Ann. Univ Nat. de Côte d'Ivoire, série C (Sciences), **20**, 313–347.

Ajahi, O. & Umoho, A. 1998. *Quality of groundwater in the coastal plain sands aquifer of the Akwa Ibom State, Nigeria.* Journal of African Earth Sciences, **27**, 259–275.

Asubiojo, O. I., Nkono, N. A., Ogunsa, A. O., Oluwole, A. F., Ward, N. I., Akanle, O. A. & Spyrou, N. M. 1997. *Trace elements in drinking and groundwater samples in Southern Nigeria.* The Science of Total Environment, **208**, 1–8.

Bado, M. 1992 *Les paramètres de la pollution de l'eau, les effets, mesures et contrôles.* In: Journées techniques sur la pollution et les risques des activités économiques. Service de l'inspection des installations classées, Abidjan, BAD, 30 juin-3 juillet 1992, 64–78.

Banoeng-Yakubo, B. K. 2003. *Assessment of pollution status and vulnerability of water supply aquifers of keta, Ghana.* Project UNEP/UNESCO/UN-HABITAT/ECA Activity report .December 2003.

Boukari, M., Gaye, C. B., Faye, A. & Faye, S. 1996. *The impact of urban development on coastal aquifers near Cotonou, Benin.* Journal of African Earth Sciences, **22**, 403–408.

Coleman, M. L., Shepherd, T. J., Durham, J. J., Rousse, J. E. & Moore, G. R. 1982. *Reduction of water with zinc for hydrogen isotope analysis.* Analytical Chemistry, **54**, 993–995.

Craig, H. 1961. *Standard for reporting concentrations of deuterium and oxygen-18 in natural waters.* Science, **133**, 1833–1834.

Edet, A. E. & Okereke, C. S. 2002. *Delineation of shallow groundwater aquifers in the coastal plain sands of Calabar area (Southern Nigeria) using surface resistivity and hydrogéological data.* Journal of African Earth Sciences, **35**, 433–443.

Edoukou, G. D. 1992. *La pollution de l'eau: cas des établissements industriels d'élevage et des abattoirs*. In: Journées techniques sur la pollution et les risques des activités économiques; service de l'inspection des installations classées, Abidjan, BAD, 30 juin-3 juillet 1992, 144–168.

Epstein, S. & Mayeda, T. K. 1953. *Variations of the* $^{18}O/^{16}O$ *ratio in natural waters*. Geochimica Cosmochimica Acta, **4**, 213–224.

Faillat, J. P. 1990. *Origine des nitrates dans les nappes de fissures de la zone tropicale humide. Exemple de la Côte d'Ivoire*. Journal of Hydrology, **113**, 231–264.

Faillat, J. P. & Rambaub A. 1988. *La teneur en nitrate des nappes de fissures de la zone tropicale humide en relation avec les problèmes de déforestation*. Comptes Rendus de l'Académie des Sciences, Paris, **306**, 1115–1120.

Favreau, G., Leduc, C., Marlin, C., Dray, M., Taupin, J. D., Massault, M., Le Gal La Salle, C. & Babic, M. 2002. *Estimate of recharge of a rising water table in semiarid Niger from* ^{3}H *and* ^{14}C *modelling*. Ground Water, **40**, 144–151.

Fontes, J. C. H. 1971. *Un ensemble destiné à la mesure de l'activité du radiocarbone naturel par scintillation liquide*. Revue de Géographie Physique et de Géologie Dynamique, **13**, 67–86.

Girard, P. 1993. *Techniques isotopiques (15N, 18O) appliquées à l'étude des nappes des altérites et du socle fracturé de l'ouest africain – Etude de cas: L'Ouest du Niger*, Thèse Doct. Université de Québec à Montréal (UQAM), Canada.

Godwin, H. 1962. *Half-life of radiocarbon*. Nature, **195**, 984.

Groen, J., Schumann, J. B. & Geirnaert, W. 1988. *The occurrence of high nitrate concentration in groundwater in villages in Northwestern Burkina Faso*. Journal of African Earth Sciences, **7**, 999–1009.

Ibe, K. M. & Njemanze, G. N. 1998. *The impact of urbanisation and protection of water resources, Oweri, Nigeria*. Journal of Environmental Hydrology, **6**, paper 9, October 1998, http:/www.hydroweb.com.

Joseph, A. & Girard, P. 1990. *Etude de la pollution en nitrate des aquifères du socle : exemple de la nappe de Niamey*. Ministère de l'Hydraulique du Niger.

Jourda, J. R. P. 2004. *Qualité des eaux souterraines de la nappe d'Abidjan*. Bulletin N° 4. Juin 2004.

Jourda, J. R. P. 2003. *Evaluation de la pollution et de la vulnérabilité des aquifères des grandes cités urbaines d'Afrique*. Projet UNEP/UNESCO/UN-HABITAT/ECA. Rapport sur les activités en Côte d'Ivoire. Rapport à mi-parcours de la 2ème phase. Décembre 2003.

Jourda, J. R. P. 1987. *Contribution à l'étude géologique et hydrogéologique de la région du Grand Abidjan (Côte d'Ivoire)*. Thèse Doct. Univ. Scient. Techno. et Med. de Grenoble.

Kaufman, S. & Libby, W. F. 1954. *The natural distribution of tritium*. Physical Review, **93**, 1337–1344.

Le Gal La Salle, C., Marlin, C., Leduc, C., Taupin, J. D., Massault, M. & Favreau, G. 2001. *Renewal rate estimation of groundwater based on radioactive tracers* (^{3}H,^{14}C) *in an unconfined aquifer in a semi-arid area, Iullemeden Basin, Niger*. Journal of Hydrology, **254**, 145–156.

Loroux, B. F. E. 1978. *Contribution à l'étude hydrogéologique du bassin sédimentaire côtier de Côte d'Ivoire*. Thèse Dr 3ème cycle, Université de Bordeaux I, No 1429, Talence 1978.

Kouadio, L. P., Abdoulaye, S., Jourda, P., Loba, M. & Rambaud, A. 1998. *Conséquences de la pollution urbaine sur la distribution d'eau d'alimentation publique à Abidjan*. Cahier de l'Association Scientifique Européenne pour l'Eau et la Santé, **3**, 1–75.

Oga, M. S. 1998. *Ressources en eaux souterraines dans la région du Grand Abidjan (Côte d'Ivoire): Approches hydrochimique et isotopique*. Thèse Doct. Univ Paris XI, Orsay.

Rollins, L. 1987. PC WATEQ DAGEN: an IBM and PC-compatibles adaptation of WATEQF. User's guide, Woodland, California.

SODECI 1991 – 2002. *Rapports techniques SODECI de 1991 à 2002*. Siège social Treichville, Abidjan Côte d'Ivoire.

SOGREAH, 1996. *Etude de la gestion et de la protection de la nappe assurant l'alimentation en eau potable d'Abidjan. Etude sur modèle mathématique. Rapports de phase 1 et 2*. République de Côte d'Ivoire, Ministère des Infrastructures Economiques, Direction et Contrôle des Grands Travaux (DCGTx).

Soro, N. 2003. *Gestion des eaux pour les villes africaines : évaluation rapide des ressources en eau souterraine/occupation des sols*. UN-HABITAT. Avril 2003 (Rapport final).

Soro, N. 1987. *Contribution à l'étude géologique et hydrogéologique du Sud-Est de la Côte d'Ivoire, Bassin versant de la Mé*. Thèse Doct. Univ. Scient. Techno. et Med. de Grenoble.

Spengler, A. & Delteil, J. R. 1966. *Le bassin secondaire-tertiaire de Côte d'Ivoire. Bassins sédimentaires du littoral Africain*. Symposium 1ère partie, Littoral Atlantique, Association Des Services Géologiques Africains, 99–113.

Tapsoba, A. S. 1995. *Contribution à l'étude géologique et hydrogéologique de la région de Dabou (sud de la Côte d'Ivoire): Hydrochimie, isotopie et indice de vieillissement des eaux souterraines*. Thèse Doct. Univ. Nat. de Côte d'Ivoire.

Tapsoba, A. S. 1990. *Etude géologique et hydrogéologique du bassin sédimentaire de la Côte d'Ivoire: Recharge et qualité des eaux dans l'aquifère côtier (région de Jacqueville)*. Mémoire de DEA, Université Cheik Anta Diop Dakar, 65 pp.

Varlet, F. 1978. Le régime de la lagune Ebrié, Côte d'Ivoire : Traits physiques essentiels, *Travaux et Documents ORSTOM,* **83**, 162p.

Wigley, T. M. L. 1975. *Carbon-14 dating of groundwater from closed and open systems*. Water Resources Research, **11**, 324–328.

CHAPTER 23

Hydrogeochemistry of a fractured aquifer in the Ogoja/Obudu area of SE Nigeria

A. Edet

Department of Geology, University of Calabar, Calabar, Nigeria

B. Ekpo

Department of Pure and Applied Chemistry, University of Calabar, Calabar, Nigeria

ABSTRACT: The hydrogeochemistry of parts of two tectonic terrains, the Obudu plateau and Benue trough were studied during a dry season period (November 2002, December 2002, and January 2003). This was to determine the process controlling the water chemistry and to assess the quality of water. The results show that, based on ionic ratios, precipitation and water-rock interaction (silicate weathering) are the main controlling factors contributing to solute concentration in these areas. The main water types in the area include Ca—Na—HCO_3—Cl, Ca—Na—HCO_3, Na—HCO_3—Cl, and Na—Cl. Most of the parameters considered are below international acceptable limits for drinking, domestic and irrigation purposes. However, iron and manganese seems to be a problem with concentrations above the recommended limits. Using Sodium Absorption ratio (SAR), percent sodium (% Na) and residual sodium carbonate (RSC) as criteria, the waters appear suitable for irrigation purposes. During the dry season the water quality deteriorates.

1 INTRODUCTION

The dry season, which last from late October to early March in Nigeria, is generally marked by high temperature, low precipitation, and high evapotranspiration. This phenomenon generally leads to a decrease in aquifer recharge, decline in groundwater levels, and deterioration of water quality as observed in some other parts of the globe (e.g. Kazemi, 2004; Daessie, *et al.*, 2005). The present area of investigation in southeastern Nigeria is gradually experiencing an increase in population and human activity alongside infrastructural development which is primarily due to the promotion of tourism by the state government. Such infrastructures include the construction of an airstrip, roads, and the reactivation of the Obudu Cattle ranch.

Water for local consumption and industrial use is extracted mainly from boreholes and wells. Early studies indicated a wide range in total dissolved solids (TDS) concentrations, varying from 23.6 to 326 mg/l (Edet & Okereke, 2005). Despite this there is no programme by the government to monitor the quality of potable water. The present work reports on the result of work carried out to determine the processes controlling the water chemistry and to assess the water quality for drinking, domestic and agricultural purposes during a dry

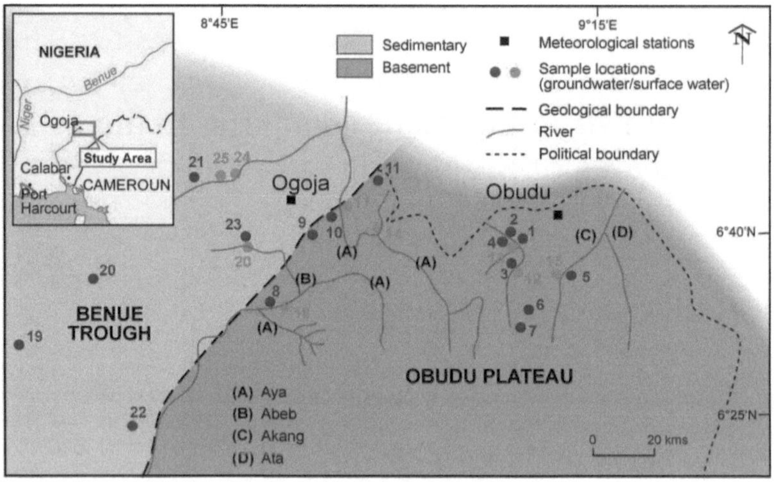

Figure 1. Parts of Obudu Plateau and Benue Trough showing sample locations.

period. This study serves as a guide to promote any initiatives by stakeholders to start a monitoring programme as a prelude to proper management of the resources in view of the likely increases in demand with the development of tourism.

2 LOCATION AND CLIMATE

The study area straddles the Obudu plateau and the Benue Trough (Figure 1). The Obudu plateau ranges in altitude from about 150 to 2000 m above the adjoining Benue Trough, which is less than 200 m above sea level (Edet & Okereke, 2005). The area is dissected by numerous drainage systems, the main ones being the Aya, Abeb, Akang, Ata and Okpanku Rivers (Figure 1).

The area experiences the tropical type climate with wet (April - October) and hot dry (November – March) seasons. Approximately 95% of the annual rainfall is received during the wet season while the remaining 5% is accounted for during the dry season. The average annual rainfall for the Obudu Met Station (1995 to 2000) is 2100 mm and 1800 mm for the Ogoja Met Station (1977 to 1999). Available data for temperature for the Ogoja Met Station (1989–2002) show that the max temperature for the dry months ranged from 33.1 to 35.8°C compared to the wet months with temperature in the range 30.2 to 34.1°C.

3 GEOLOGY AND HYDROGEOLOGY

Geologically, the area is underlain by Precambrian basement complex (Obudu Plateau) and Cretaceous sedimentary rocks (Benue trough), Figure 1. The major rock types within the Obudu Plateau include gneisses, schist, metaperidotites, amphibolites, quartzite, charnokites, granites, dolerite, and pegmatite intrusions. These rocks belong to the three major lithological divisions of the Nigerian basement, namely: the migmatite gneiss complex, the metavolcano-sedimentary series and Pan African Older Granites (Ekwueme & Kroener, 2006). Overlying the basement rocks in the Benue Trough are sedimentary rocks. The major rock types are sandstone, shale, mudstone, and limestone belonging to the Asu

River Group and Eze-Aku Formation. The thickness of sediments here has been estimated to range from 2000 to 6000 m (Tijani, 2004). Two major tectonic events affected the area during the pre-Turonian and post Santonian periods. This gave rise to numerous faults and folds with characteristics NW-SE and N-S fracture systems. Details of geology of the study area are included in Umeji & Fitch (1988), Ekwueme (1991, 1998), Petters & Ekweozor (1982), Ofoegbu (1990), Idowu & Ekweozor (1993), Ekwere & Ukpong (1994), and Tijani *et al.* (1996).

The water-bearing unit in the plateau is the regolith. The thickness of the regolith varies up to 90 m. Measured static water-levels vary from 5 to 15 m (Edet & Okereke, 2005). Existing data for the sedimentary area show that hydraulic properties vary greatly. Yields as high as 30–67 m^3/hour have been recorded within the sandstones. Depths to the groundwater table range from ground-level to about 15 m below the ground surface (Uma *et al.* 1990). The shales are baked and fractured in the area. This allows for the storage, flow and abstraction of water. Groundwater levels within the shale varied from 2 to 13 m. Transmissivity in successful boreholes in the shales have been measured in the range of 10–26 m^2/d (Edet, 1993).

4 MATERIAL AND METHODS

A total of seventy eight (78) water samples were collected from twenty six (26) locations between the months of November 2002 and January 2003. Of this number, 18 locations (consisting of 11 groundwater and 7 surface water samples) were from the basement terrain of Obudu plateau. These samples are designated as 1 to 11 for groundwater and 12 to 18 for surface water respectively. Samples from 8 locations (made up of 5 groundwater and 3 surface water samples) were collected from the sedimentary terrain and are designated as 19 to 23 and 24 to 26 for the groundwater and surface water respectively (Figure 1).

The samples were collected in polyethylene bottles and kept cool until analysis. Various physical parameters were measured in the field using standard equipment. These include Temperature and conductivity (WTW 96), dissolved oxygen (Oxguard Handy MK 11 electronic meter) and pH/Eh (WTW pH 90 meter) measurements were made in the field. Samples for Biological Oxygen Demand (BOD) were collected in 250 ml brown bottles and incubated for 5 days. The BOD was measured as the difference between initial oxygen concentration in sample and the concentration after five days incubation in BOD bottles (APHA, 1989). The COD were determined trimetrically.

The samples were filtered through a thin polycarbonate membrane with 0.45 μm pore size. Samples for heavy metal analysis were acidified to $pH < 2$. The chemical analyses were based on the standard methods presented in APHA (1989). All the samples were assessed for charge balance and were all within the acceptable range of ± 5.

The heavy metals concentrations were determined by means of Atomic Absorption Spectrophotometer (AAS) Perkin Elmer model 2380.

5 RESULTS AND DISCUSSIONS

Details and a summary of the physicochemical, biological parameters and heavy metals for the different geologic terrains for the different sample periods for both groundwater and surface water are given in Table 1.

Table 1. Summary physicochemical and biological parameters (including Fe and Mn) in water for the study area.

	Groundwater Obudu plateau (n = 33)			Surface water Obudu plateau (n = 21)					Groundwater Benue trough (n = 15)					Surface water Benue trough (n = 9)						
	Mean	Med	Min	Max	SD	Mean	Med	Min	Max	SD	Mean	Med	Min	Max	SD	Mean	Med	Min	Max	SD
Temp	28.4	29.0	22.7	30.5	1.7	28.0	28.5	23.0	30.6	2.0	28.7	29.0	26.0	29.6	1.1	28.9	29.5	26.0	30.5	1.5
pH	6.2	6.2	5.0	8.1	0.8	6.4	6.4	5.4	7.4	0.6	6.2	6.0	5.2	8.3	0.9	6.5	6.6	5.8	7.2	0.5
EC	162.0	153.7	50.0	338.8	80.9	60.7	60.4	22.7	96.8	22.6	411.8	403.0	25.0	865.0	258.8	60.3	63.0	40.1	72.1	10.8
DO	6.0	6.2	3.2	8.2	1.4	6.5	6.5	5.0	8.1	0.7	6.4	6.2	4.2	8.9	1.2	6.1	6.0	5.4	7.2	0.5
BOD	2.5	1.2	0.0	12.4	3.4	1.5	1.2	0.5	3.8	0.9	0.8	0.6	0.1	2.5	0.7	0.6	0.7	0.4	0.8	0.1
COD	0.2	0.1	0.0	1.2	0.3	0.5	0.4	0.2	1.2	0.3	0.2	0.2	0.1	0.4	0.1	0.4	0.3	0.2	0.8	0.2
Eh	83.6	85.0	65.0	98.0	8.0	83.1	81.0	50.0	121.0	14.8	92.7	93.0	70.0	125.0	15.7	89.3	80.0	75.0	124.0	15.6
TDS	76.8	68.5	22.0	162.8	39.1	32.5	32.1	13.6	58.5	12.4	203.9	194.2	16.7	425.4	128.9	45.9	40.2	30.4	72.4	14.8
TH	38.4	35.7	13.4	100.0	17.8	23.3	16.6	4.8	55.5	18.3	63.3	62.3	40.3	94.3	17.6	17.1	16.6	11.4	22.3	4.1
Ca	12.2	11.4	4.2	24.8	4.9	7.2	4.3	1.5	18.9	6.7	22.3	21.3	14.4	35.1	6.7	4.4	4.6	3.2	5.3	0.7
Mg	1.5	1.5	0.5	3.4	0.6	1.3	1.4	0.2	2.6	0.8	1.8	1.7	0.7	3.0	0.7	1.5	1.2	0.5	2.6	0.8
K	1.5	1.4	0.2	3.4	0.8	1.3	1.4	0.1	2.6	0.8	2.0	1.6	0.7	4.2	1.1	1.1	0.8	0.2	3.5	1.0
Na	28.3	26.8	4.6	61.5	13.5	11.4	10.4	6.2	28.1	5.1	70.4	62.7	24.6	136.5	39.6	15.6	14.6	12.4	22.9	3.4
HCO_3	71.9	74.5	22.0	102.5	16.3	52.9	46.3	14.6	114.6	31.7	81.0	72.6	60.7	109.8	16.3	46.3	42.6	29.3	78.3	15.1
NO_3	3.0	2.1	0.5	10.2	2.5	2.7	2.7	0.2	5.4	1.3	2.2	2.4	0.1	4.3	1.5	3.2	3.3	0.3	6.2	1.6
SO_4	4.8	3.1	0.1	15.5	4.1	4.0	2.3	0.2	14.2	4.4	6.5	2.4	1.2	17.8	6.4	2.1	1.8	0.1	4.3	1.6
Cl	59.0	51.6	6.2	126.6	39.0	31.8	26.4	9.9	132.4	32.8	154.4	128.1	42.3	286.5	83.8	26.5	27.7	21.8	28.7	2.4
Fe	2.8	0.6	0.2	13.6	4.3	1.0	0.7	0.3	4.2	1.0	1.0	0.6	0.3	3.2	0.9	1.0	0.9	0.8	1.5	0.2
Mn	0.6	0.4	0.1	2.0	0.6	0.4	0.4	0.2	1.0	0.2	0.4	0.4	0.2	0.8	0.2	0.4	0.4	0.3	0.6	0.1
FC	0.3	0.0	0.0	4.0	1.0	0.5	0.0	0.0	4.0	1.2	1.9	2.0	0.0	5.0	1.9	2.6	3.0	0.0	6.0	2.4
E Coli	0.7	0.0	0.0	5.0	1.5	0.1	0.0	0.0	2.0	0.4	0.0	0.0	0.0	0.0	0.0	0.0	0.0	0.0	0.0	0.0

All units in mg/l except Temp (^{o}C), EC (μS/cm), FC-Faecal Coliform (per 100 ml), EC-E Coli (per 100 ml)

5.1 SURFACE WATER CHEMISTRY

The surface water temperatures in the basement area (BA) ranged from 23 to 30°C and from 26 to 30.5°C in the sedimentary area (SA). The electrical conductivity (EC) varies between 22.7 and 96.8 μS/cm and 40.1 and 72.1 μS/cm for the SA and BA respectively. The mean total dissolved solids (TDS) are 32.5 mg/l (BA) and 45.9 mg/l (SA) respectively. The EC and TDS are low in comparison to the WHO Standard of 1400 μS/cm for EC and 1000 mg/l for TDS, indicating that the waters are fresh. The Biological Oxygen Demand (BOD) varied from 0.46 to 3.80 mg/l and 0.40 to 0.80 mg/l for the BA and SA respectively. The Chemical Oxygen Demand (COD) vary between 0.18 and 0.46 mg/l and 0.16 and 0.80 mg/l for the BA and SA respectively. The mean pH for the BA and SA are 6.37 and 6.49 with ranges of 5.44 and 7.44 (BA) and 5.80 to 7.24 (SA). Considering the entire sample set, only 4 out of 21 samples had pH > 7. In addition, 35% and 45% respectively of the pH values in the BA and SA are not within the limits of World Health Organisation (WHO), of 6.5–8.5. Most of the samples are acidic in nature which is attributed to the fact that the surface waters flow through forest thereby leaching acidic water and organic matter in the process.

The dissolved oxygen (DO) varied from 5.0 to 8.1 mg/l for the BA and from 5.4 to 7.2 mg/l for the SA. The surface water samples show moderately high (standard deviation, SD = 18.3) variation in terms of hardness with values ranging from 4.8 to 55.5 mg/l as $CaCO_3$ for the BA and low (SD = 4.1) for the SA with values ranging from 11.4 to 22.3 mg/l as $CaCO_3$. However, all of the samples can be described as soft with hardness values of less than 60 mg/l (Freeze & Cherry 1979).

The mean concentration of cations in the surface water samples are Na (11.4 mg/l), Ca (4.30 mg/l), K (1.32 mg/l), and Mg (1.27 mg/l) and Na (15.6 mg/l), Ca (4.38 mg/l), K (1.08 mg/l), and Mg (1.49 mg/l) for the BA and for the SA respectively. The mean values of chloride concentration are 31.8 mg/l and 26.5 mg/l for the BA and SA. The maximum and minimum values of bicarbonate in the surface waters are 14.6 and 115 mg/l for the BA and 29.3 and 78.3 mg/l for the SA.

The amount of sulphate varies from 0.17 to 14.2 mg/l for the BA and 0.15 to 4.26 mg/l for the SA. The concentrations of the SO_4^{2-} are below the WHO (1993) limits of 250 mg/l. In the surface water samples, the nitrate concentrations varied from 0.24 to 5.44 mg/l for the BA and 0.32 and 6.20 mg/l for the SA. The concentrations are lower than that normally expected for unpolluted water, 0.124–9.92 mg/l (Custodio & Llamas 1983) and 11.3 mg/l (WHO, 1993) maximum recommended value for drinking purposes. The values of ion concentrations for the two Obudu plateau (BA) and Benue trough (SA) are not significantly different.

6 GROUNDWATER CHEMISTRY AND TYPES

Groundwater temperatures averaged 28.3°C and 28.7°C for the BA and SA respectively. The mean value of EC (TDS) for the two areas is 162 μS/cm (76.8 mg/l) (BA) and 412 μS/cm (204 mg/l) (SA) respectively. A plot of TDS in groundwater show increasing high values towards the SA (Figure 2). The probable explanation for this is that there is an increase in TDS along the flow path of groundwater and/or due to the reported brine deposits in parts of the SA (Ekwere & Ukpong, 1994; Tijani *et al.*, 1996; Tijani, 2004).

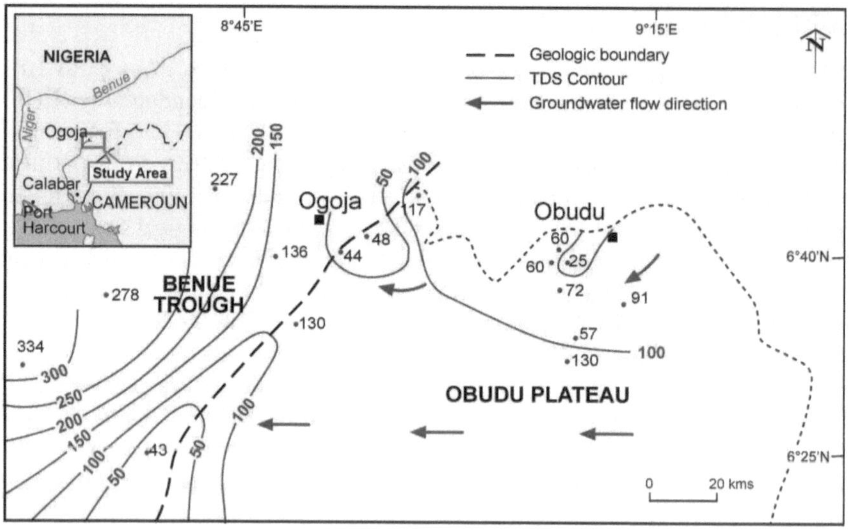

Figure 2. Contour map of TDS (mg/l) in groundwater.

The BOD varied from 0.02 to 12.4 mg/l and 0.09 to 2.50 mg/l for the BA and SA. The COD vary between 0.02 and 1.20 mg/l and 0.08 to 0.40 mg/l for the BA and SA respectively. The mean value of pH for the BA and SA are 6.21 and 6.23. In both the BA and SA, only 4 samples had pH values > 7. The acidic nature of the samples may be attributed to rainfall, and the lack of buffering material within the aquifers. The dissolved oxygen (DO) varied from 3.2 to 8.2 mg/l for the BA and 4.2 to 8.9 mg/l for the SA.

The hardness values (as $CaCO_3$) ranged from 13.4 to 100 mg/l for the BA and 40.3 to 94.3 mg/l for the SA. Data from Table 1 show that out of all the 33 samples considered for the BA, only 2 samples can be described as hard with hardness values of > 60 mg/l. For the SA, 9 water samples out of 15 are characterised by hardness values of greater than 60 mg/l as $CaCO_3$.

The data on chemical composition (Table 1) show that the mean concentrations of the cations in groundwater are: Na (28.3 mg/l), Ca (12.2 mg/l), K (1.48 mg/l), and Mg (1.49 mg/l) and Na (70.4 mg/l), Ca (22.3 mg/l), K (1.99 mg/l), and Mg (1.83 mg/l) for the BA and the SA respectively.

The mean value of chloride concentrations are 59.0 (BA) and 154 mg/l (SA). The concentration of bicarbonate in the groundwater varied from 22.0–102 mg/l (mean 71.9 mg/l) for BA and 60.7 to 110 (mean 81.0 mg/l) for the SA. Sulphate in the groundwater in the BA ranged from 0.12 to 15.5 mg/l with an average of 4.77 mg/l. The concentration of sulphate varied from 1.20 to 17.8 mg/l (mean 6.50 mg/l) for the SA. The mean values of sulphate concentrations in both areas are low ($\leq$ 10 mg/l). According to (Edmunds *et al.*, 1982) low sulphate ($\leq$ 10 mg/l) in groundwater is considered to be natural maximum concentrations derived from anthropogenic reactions.

Nitrate values ranged from 0.45–10.2 mg/l with mean of 3.05 mg/l (as N) for the groundwater samples from the BA and 0.11 to 4.27 mg/l (mean 2.16 mg/l) for samples from the SA. All the samples have lower concentrations reflecting little pollution (e.g. Custodio & Llamas, 1983) and less than 11.3 mg/l (WHO, 1993) maximum recommended value for drinking purposes.

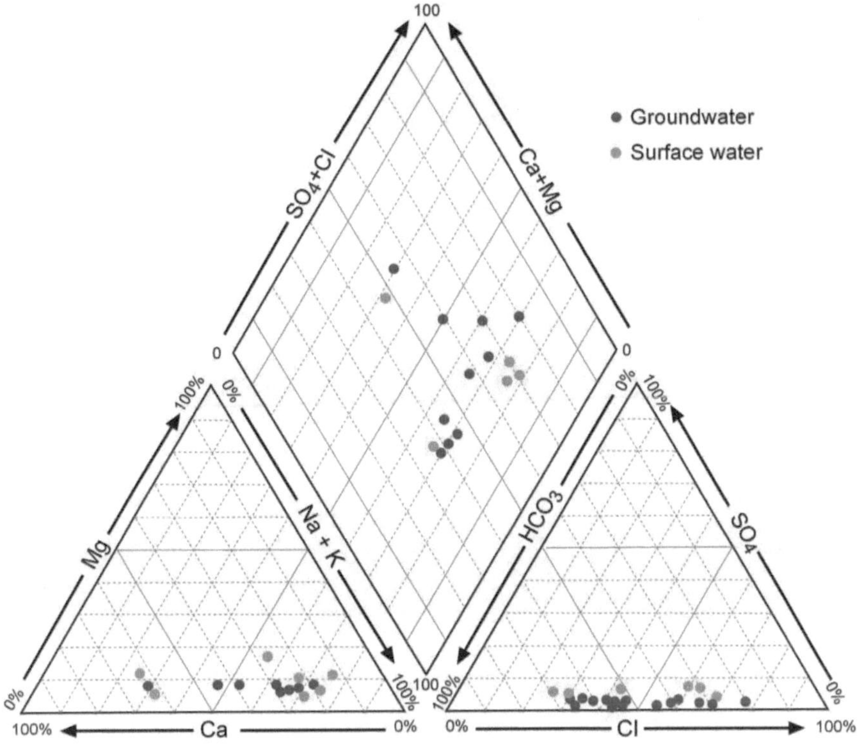

Figure 3a. Piper diagram for samples from Obudu Plateau based on average values.

7 HYDROGEOCHEMICAL FACIES AND CLASSIFICATION

The Piper diagram (1944) in Figure 3 shows the relative concentrations of the different ions from the individual samples based on average values for each location. Four types of hydrogeochemical facies based on the classification given by Deutsch (1997) are identified in the area. These include Calcium-Sodium-Bicarbonate-Chloride (Ca—Na—HCO_3—Cl), Calcium-Sodium-Bicarbonate (Ca—Na—HCO_3), Sodium-Bicarbonate-Chloride (Na—HCO_3—Cl), and Sodium-Chloride (Na—Cl) types. The water samples from the basement area are characterised mainly by HCO_3indicating active groundwater flushing and significant water rock interaction.

Classification of the water samples based on the proposed scheme by Chadha (1999) using a bivariate plot of $(CO_3^{2+} + HCO_3^-) - (Cl^- + SO_4^{2-})$ versus $(Ca^{2+} + Mg^{2+}) - (Na^+ + K^+)$ presented as Figure 4 show that for all the 78 samples, 26% are classified as Alkali metals (Na + K) > Alkaline metal (Ca + Mg) and weak acidic anions (HCO_3) > strong acidic anion (Cl + SO_4). This group represents contributions from water-rock interaction. Fifty eight (58%) percent are classified under the Alkali metals (Na + K) > Alkaline metal (Ca + Mg) and Strong acidic anion (Cl + SO_4) > weak acidic anions (HCO_3). This class represents mostly contributions from rainfall. Eight percent (8%) each belong to the Alkali metals (Na + K) < Alkaline metal (Ca + Mg) and weak acidic anions (HCO_3) > Strong acidic anion (Cl + SO_4) and Alkali metals

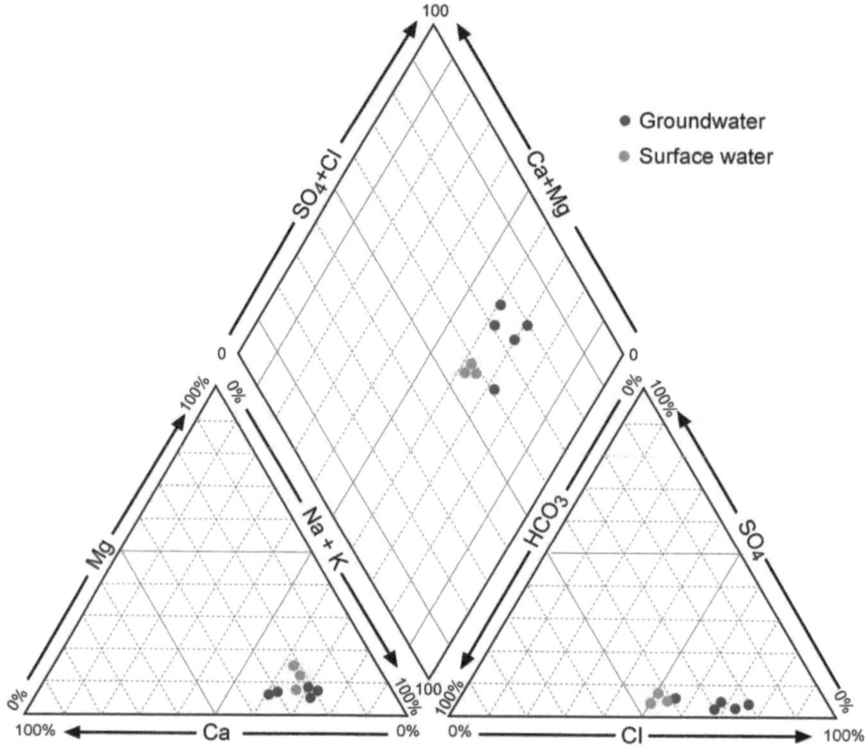

Figure 3b. Piper diagram for samples from the Benue Trough based on average values.

(Na + K) < Alkaline metal (Ca + Mg) and Strong acidic anion ($Cl + SO_4$) > weak acidic anions (HCO_3) respectively. This represents a transition between rainfall and silicate weathering.

8 MECHANISM CONTROLLING WATER CHEMISTRY

The ionic composition may be caused by several factors during the interaction. Hence, it is necessary to use the ionic ratios and plots to discriminate between them. The contribution of atmospheric sources to the dissolved salts has been discussed by many authors (e.g. Garrels & Mackenzie, 1971; Stallard & Edmund, 1983; Sarin *et al.*, 1989; Berner & Berner, 1996). Chloride is the most useful parameter for evaluating atmospheric input to water as it shows very little fractionation (Appelo & Postma, 1993).

Sodium and chloride inputs are likely to be mainly from rainfall and, therefore, will large reflect the ration observed in seawater. Cation exchange may account for a reduction in the Na concentration, and halite dissolution may account for high concentrations of Na aned Cl. The low concentrations of potassium in natural water are a consequence of its tendency to be fixed by clay minerals and participate in the formation of secondary minerals (Mathess, 1982).

In the basement areas, HCO_3, Ca^{2+}, Mg^{2+} may be derived from rock weathering. In the area a major proportion of these ions is derived from weathering, chiefly Ca—Mg silicates, mainly plagioclase and feldspar.

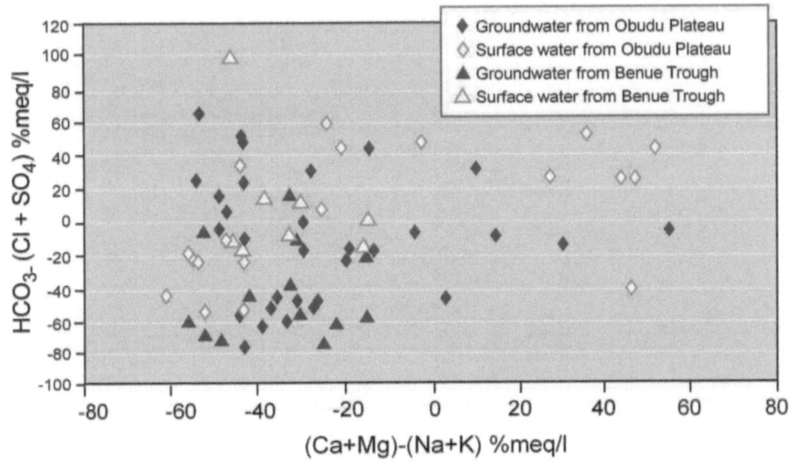

Figure 4. Hydrogeochemical classifications based on the bivariate plot of $(CO_3^{2+} + HCO_3^-) - (Cl^- + SO_4^{2-})$ versus $(Ca^{2+} + Mg^{2+}) - (Na^+ + K^+)$ of different water types.

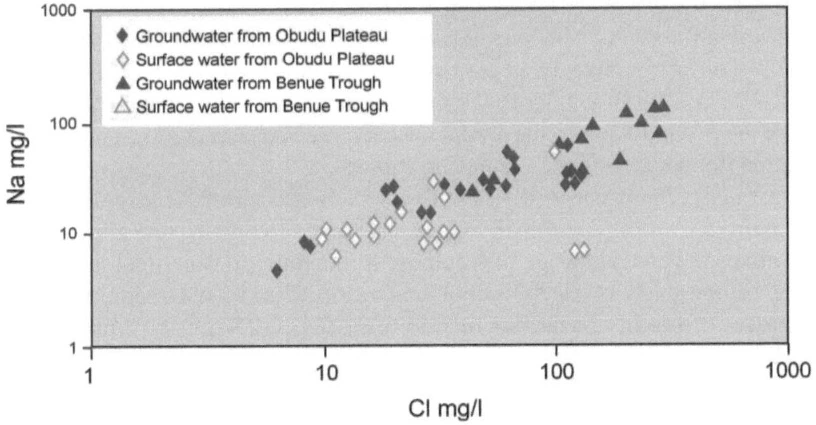

Figure 5. Plot for Na vs Cl the different samples.

The rock weathering in the area may be determined by two components mixing from dissolution of carbonates and silicates and the relative significance of these sources can be explained by ionic ratios. Figure 6 shows the plot of $Ca + Mg$ vs HCO_3 for the different source types. The water samples have an average $(Ca + Mg)/HCO_3$ of 0.19 (indicating contributions from carbonate weathering. In the basement areas, HCO_3, Ca^{2+}, Mg^{2+} may be derived from weathering of chiefly $Ca - Mg$ silicates, mainly plagioclase and feldspar:

More than 95% of the concentrations of iron and all of manganese in the water samples are higher than the maximum acceptable concentration of 0.3 and 0.1 mg/l respectively (Table 1). In this study, pH shows a negative correlation with Fe and Mn. This is attributed to the fact that low pH values result in higher solubility and mobility of elements in water. The waters in the area are also saturated with respect to Goethite (FeOOH), Hematite (Fe_2O_3), and near equilibrium with Siderite ($FeCO_3$) and Rhodochrosite ($MnCO_3$).

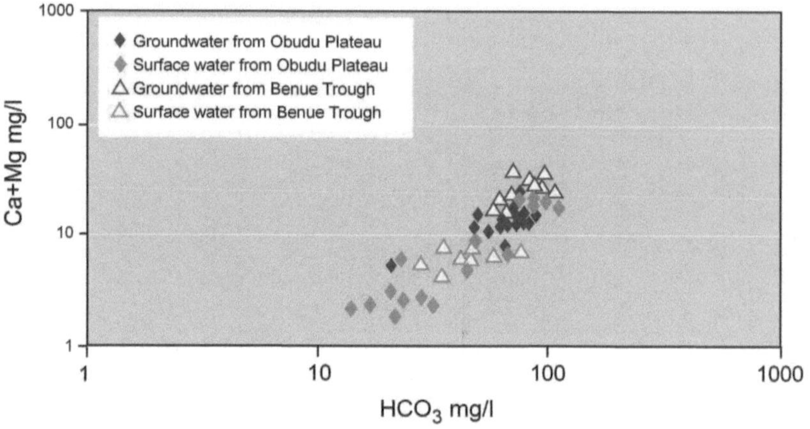

Figure 6. Plot for Ca+Mg vs HCO_3 for the different samples.

9 GROUNDWATER QUALITY

Groundwater quality assessment was made using pH, nitrate and biological parameters (BOD, COD, Coliform, Table 1) as contamination indicators. None of these parameters exceeds the WHO (1993) standards for drinking and domestic purposes indicating the potability of the groundwater in the area. However, the water may fail in other ways, such as the elevated iron and manganese concentrations.

In terms of agricultural purposes, parameters such as Sodium Absorption ratio (SAR), percent sodium (% Na) and residual sodium carbonate (RSC) were estimated. Richards (1954) classified the concentration of soluble salt in irrigation water (salinity hazard) into four classes on the basis of electrical conductivity, EC and SAR (sodium hazard). The different classes of salinity hazard include low, C1 (EC < 250 μS/cm); medium, C2 (EC 250–750 μS/cm); high, C3 (EC 750–2250 μS/cm); and very high, C4 (EC > 2250 μS/cm). The sodium hazard classes include: low, S1 (SAR < 10); medium, S2 (SAR 10–18); high, S3 (SAR 18–26); and very high, S4 (SAR > 26). According to Subramani *et al.* (2005) water with high EC leads to formation of saline soil a high Na leads to development of an alkaline soil. The Na or alkaline hazard in the use of water for irrigation is determined by absolute and relative concentration of cations and is expressed in terms of SAR and can be estimated by the formula: $SAR = Na/((Ca + Mg)/2)^{0.5}$.

If water used in irrigation is high in Na and low in Ca, the cation exchange complex may become saturated with Na. This can destroy the soil structure owing to dispersion of clay particles. The calculated SAR for the waters ranges from 0.18 to 2.83. The data show that the samples fall between C1 – S1 and C2 – S1 indicating low to medium salinity and low Na water for irrigation purposes for most soils and crops with little danger of development of exchange Na and salinity.

The sodium percentage (Na%) is calculated in meq/l from $(Na^+ + K^+)/(Ca^{2+} + Mg^{2+} + Na^+ + K^+) \times 100$. The Na % (22.41–80.15%) indicates that the waters are excellent to good for irrigation purposes based on the Wilcox diagram (1955) relating Na % and total concentration (Na < 80%; EC < 500 μS/cm; Total cation < 7 meq/l). When the concentration of sodium is high in irrigation water, sodium ions tend to be absorbed by clay particles, displacing magnesium and calcium ions. The exchange process of sodium in water for magnesium and calcium in soil reduces permeability and eventually results in soil with poor drainage. Hence air and water circulation is restricted during wet conditions and such soils are usually hard when dry (Collins & Jenkins, 1996; Saleh *et al.*, 1999).

In addition to the SAR and Na %, the excess sum of carbonate and bicarbonate in water over the sum of calcium and magnesium also affect the suitability of water for irrigation purposes. This is known as residual sodium carbonate (RSC), calculated according to Ragunath (1987) as $RSC = (HCO_3^- \; CO_3^{2-}) - (Ca^{2+} + Mg^{2+})$ where all concentrations are expressed in meq/l. The values of RSC for the present work varied between 0.7 and 1.01 meq/l. This indicates that the samples are suitable for all agricultural purposes since all the RSC values are less than 1.25 meq/l).

10 CONCLUSIONS

The water in the area is mainly acidic in nature. Calcium and sodium are the dominant cations and bicarbonate and chloride are the dominant anion. Four (4) types of hydrogeochemical facies are identified in the area. These include Ca—Na—HCO_3—Cl, Ca—Na—HCO_3, Na—HCO_3—Cl, and Na—Cl types. The water chemistry is largely controlled by rainfall and silicate weathering with minor contribution from carbonate weathering. The low ratio of Na/Cl supports contributions from rainfall and the low ratio of $(Ca^{2+} + Mg^{2+})/(HCO_3^- + SO_4^{2-})$, $(Ca^{2+} + Mg^{2+})/HCO_3^-$ indicate contributions of HCO_3 from biological activity. Ratios of Ca/Na, Mg/Na, and HCO_3/Na suggest combined silicate weathering and contribution of HCO_3^- from biological activity. The concentrations of iron in more than 95% of the water samples and all of manganese are higher than the maximum acceptable concentration of 0.3 and 0.1 mg/l respectively. The calculated shows parameters of SAR, %Na, and RSC show that the waters can be used for irrigation.

REFERENCES

APHA 1980. *Standard Methods for the examination of water and waste waters, 15th edition.* APHA, Washington.

Appelo, C. A. J. & Postma, D. 1993. *Geochemistry, groundwater and pollution.* Balkema. The Netherlands.

Banks, D., Reimann, C., Oddvar, R., Skarphagen, H. & Saether, O. M. 1995. *Natural concentrations of major and trace elements in some Norwegians bedrock groundwaters.* Applied Geochemistry, **10**, 1–16.

Barnes, S. & Worden, R. H. 1998. *Understanding groundwater sources and movement using water chemistry and tracers in a low matrix permeability terrain: the Cretaceous (Chalk) Ulster White Limestone Formation, Northern Ireland.* Applied Geochemistry, **13**, 143–153.

Berner, E. K. & Berner, R. A. 1987. *The global water cycle: geochemistry and environment.* Prentice-Hall, Englewood Cliffs, New Jersey, USA.

Chadha, D. K. 1999. *A proposed new diagram for geochemical classification of natural waters and interpretation of chemical data.* Hydrogeology Journal, **7**, 431–439.

Collins, R. & Jenkins, A. 1996. *The impact of agricultural land use on stream chemistry in the middle hills of the Himalayas, Napal.* Journal of Hydrology, **185**, 71–86.

Custodio, E. & Llamas, R. 1983. *Hydrologa Subterranea [Hydrogeology], 2nd Edn.* Omega, Barcelona, Spain.

Daessle, W. L., Sanchez, E. C., Camacho-Ibar, V. F., Mendoza-Espinosa, L. G., Carriquiry, J. D., Macias, V. A. & Castro, P. G. 2005. *Geochemical evolution of groundwater in the Maneadero coastal aquifer during a dry year in Baja California, Mexico.* Hydrogeology Journal, **13**, 584–595.

Deutsch, W. J. 1997. *Groundwater geochemistry. Fundamentals and applications to contamination.* Lewis Press, New York, USA.

Drever, J. I. 1982. *The geochemistry of natural waters.* Prentice Hall, Englewood Cliffs, New Jersey, USA.

Edet, A. E. 1993. *Hydrogeology of parts of Cross River State: Evidence from aero-geological and surface resistivity studies.* PhD, University of Calabar, Calabar, Nigeria.

Edet, A. E. & Okereke, C. S. 2005. *Hydrogeological and hydrochemical character of the regolith aquifer, northern Ogudu plateau, SE Nigeria.* Hydrogeology Journal, **13**, 391–415.

Edmunds, W. M., Bath, A. H. & Miles, D. L. 1982. *Hydrochemical evolution of the East Midlands Triassic sandstone aquifer, England.* Geochem et Cosmoch Acta, **46**, 2069–2081.

Ekwere, S. J. & Ukpong, E. E. 1994. *Geochemistry of saline groundwater in Ogoja, Cross River State.* Journal of Mining and Geology, **29**.

Ekwueme, B. N. 1991. *Geology of the area around Obudu Cattle ranch, southeastern Nigeria.* Journal of Mining and Geology, **29**, 277–282.

Ekwueme, B. N. 1998. *Geochemistry of Precambrian Gneisses of Obudu Plateau southeastern Nigeria.* Global Journal of Pure and Applied Sciences, **4**, 277–282.

Ekwueme, B. N. & Kroener, A. 2006. *Single zircon ages of migmatite gneisses and granulites in the Obudu plateau: timing of granulite-facies metamorphism in southeastern Nigeria.* Journal of African Earth Sciences, **44**, 459–469.

Freeze, R. A. & Cherry, J. A. 1979. *Groundwater.* Prentice Hall, Englewood Cliffs, New Jersey, USA.

Gaillardet, J., Dupre, B., Louvat, P. & Allegre, C. J. 1999. *Global silicate weathering and CO_2 consumption rates deduced from the chemistry of large rivers.* Chemical Geology, **59**, 3–30.

Garcia, M. G., Hidalgo, M. V. & Blesa, M. A. 2001. *Geochemistry of groundwater in alluvial plain of Tucuman province, Argentina.* Hydrogeology Journal, **9**, 597–610.

Garrels, R. M. & Mackenzie, F. T. 1971. *Gregor's denudation of the continents.* Nature, **231**, 382–383.

Gibbs, R. J. 1970. *Mechanism controlling world water chemistry.* Science, **17**, 1088–1090.

Idowu, J. O. & Ekweozor, C. M. 1993. *Petroleum Potential of Createaceous Shales in the Upper Benue Trough, Nigeria.* Journal of Petroleum Geology, **16**, 249–264.

Jalali, M. 2005. *Major ion chemistry of groundwaters in the Bahar area, Hamadan, western Iran.* Environmental Geology, **47**, 763–772.

Matheis, G. 1982. *The properties of groundwater.* Wiley, New York, USA.

Meybeck, M. 1987. *Global chemical weathering of surficial rocks estimated from river dissolved loads.* American Journal of Science, **287**, 401–428.

Negrel, P., Allegre, C. J., Dupre, B. & Lewin, E. 1999. *Erosion sources determined by inversion of major and trace element ratios and strontium isotopic ratios in river water: The Congo Basin case.* Earth Planet Science Letters, **120**, 59–76.

Kazemi, A. K. 2004. *Temporal changes in the physical properties and chemical composition of the municipal water supply of Shahrood, norrtheastern Iran.* Hydrogeology Journal, **12**, 723–734.

Ofoegbu, C. O. 1990. *The Benue Trough Structure and Evolution.* Friedr. Vieweg & Sohn, Braunschweig/Wiesbaden, Germany.

Pandey, K., Sarin, M. M., Trivedi, J. R., Krishnaswami, S. & Sharma, K. K. 1994. *The Indus river system (India-Pakistan): major ion chemistry, uranium and strontium isotopes.* Chemical Geology, **116**, 245–259.

Petters, S. W. & Ekweozor, C. M. 1982. *Petroleum Geology of Benue Trough and Southeastern Chad Basin, Nigeria.* AAPG, **66**, 1141–1149.

Piper, A. M. 1944. *A graphic procedure in geochemical interpretation of water analysis.* Transactions of the American Geophysical Union, **25** (6), 914–928.

Ragunath, H. M. 1987. *Groundwater.* Wiley Eastern Ltd, New Delhi, India, 563pp.

Saleh, A., Al-Ruwaih, F. & Shehata, M. 1999. *Hydrogeochemical processes operating in the main aquifers of Kuwait*. Journal of Arid Environments, **42**, 195–209.

Sarin, M. M., Krishnaswamy, S., Dilli, K., Somajajulu, B. L. K. & Moore, W. S. 1989. *Major ion chemistry of the Ganga-Brahmaputra and fluxes to the Bay of Bengal*. Geochem Cosmochim Acta, **53**, 997–1009.

Singh, A. K., Mondal, G. C., Singh, P. K., Singh, S., Singh, T. B. & Tewary, B. K. 2005. *Hydrochemistry of reservoirs of Damodar River Basin, India:weathering processes and water quality assessment*. Environmental Geology, **48**, 1014–1028.

Stallard 1980. *Major elements geochemistry of the Amazon River system*. PhD Thesis, WHOI-80-29, MIT,USA.

Stallard, R. F. & Edmond, J. M. 1983. *Geochemistry of the Amazon River. The influence of Geology and weathering environment on dissolved load*. Journal of Geophysical Research, **88**, 9671–9688.

Subramani, T., Elango, L. & Damodarasamy, S. R. 2005. *Groundwater quality and its suitability for drinking and agricultural use in Chithar River Basin, Tamil Nadu, India*. Environmental Geology, **47**, 1099–1110.

Tijani, M. N., Loehnert, E. P. & Uma, K. O. 1996. *Origin of saline groundwater in the Ogoja area, Lower Benue Trough, Nigeria*. Journal of African Earth Sciences, **23** (2), 237–252.

Tijani, M. N. 2004. *Evolution of saline water and brines in the Benue Trough*. Applied Geochemistry, **19** (9), 1355–1365.

Uma, K. O., Onuoha, K. M. & Egboka, B. C. E. 1990. *Hydrochemical facies, groundwater flow patterns and origin of saline waters in parts of western flank of the Cross River Basin, Nigeria*. In: Offeagbu, C. O. (ed.). The Benue Trough, 115–134.

Umeji, A. C. & Fitch, C. A. 1988. *The Precambrian of southeastern Nigeria: a magmatic and tectonic study*. Geological Survey of Nigeria Report, Kaduna, Nigeria, 12pp.

Wilcox, L. V. 1955. *Classification and use of irrigation waters*. US Department of Agriculture Circular 969, Washington DC, USA.

WHO 1993. *Guidelines for drinking water quality. Vol 1 Recommendations*. World Health Organisation, Geneva, Switzerland.

CHAPTER 24

Salinity problems in coastal aquifers: Case study from Port Harcourt city, southern Nigeria

A.E. Ofoma & O.S. Onwuka

Department of Geology, University of Nigeria, Nsukka, Enugu State, Nigeria

S.A. Ngah

Institute of Geosciences and Space Technology, Rivers State University of Science and Technology, Port Harcourt. Rivers State, Nigeria

ABSTRACT: Chemical analysis of groundwater in two separate areas of the coastal city of Port Harcourt, Rivers State of Nigeria, was carried out in order to determine the chemical distribution of the groundwater and evaluate the rate of saline water migration. Nineteen (19) samples were collected from the area close to the sea and creeks (Group A) and from an area inland (Group B). Samples of group 'A' are dominated by Na Cl and have average TDS of approximately 200 mg/l; samples from group 'B' are weakly mineralized (TDS approximately 50 mg/l), although still dominated by Na Cl. The study indicates that groundwater closer to the sea is affected by saline intrusion. Comparison of the groundwater chemistry of the group A samples in this study with samples from the same study area in 1990 show a deterioration in groundwater quality in the past 15 years.

1 INTRODUCTION

Port Harcourt, one of the coastal cities in the southern part of Nigeria, is located within the oil rich Niger Delta (Figures 1 & 2). The estimated number of people living in Port Harcourt in 1991 was about 289,459 (National Population Commission, 1991 census in Annual Abstract of Statistics, 1999 Edition). This has since increased enormously. As a result of oil exploration and production activities, there is a continuous increase in demand for fresh groundwater, the major source of urban water supply for domestic and industrial uses. The increased rate of groundwater abstraction may pose severe pressure on groundwater resources. In Port Harcourt, almost every home has a well (more than 65 boreholes exist within group A area and about 100 boreholes exist within group B area – see later). Also, most of the companies that generate chemical effluents discharge their wastes directly into the sea or creeks, without regard to the effects of these effluents on coastal aquifers and aquatic life.

Previous studies on groundwater quality and aquifer characteristics of the Niger Delta, discussed by Etu-Efeotor (1981), have indicated that two hydrogeochemical regimes exist in the study area, one inland and the other coastal, and that the iron content in the groundwater

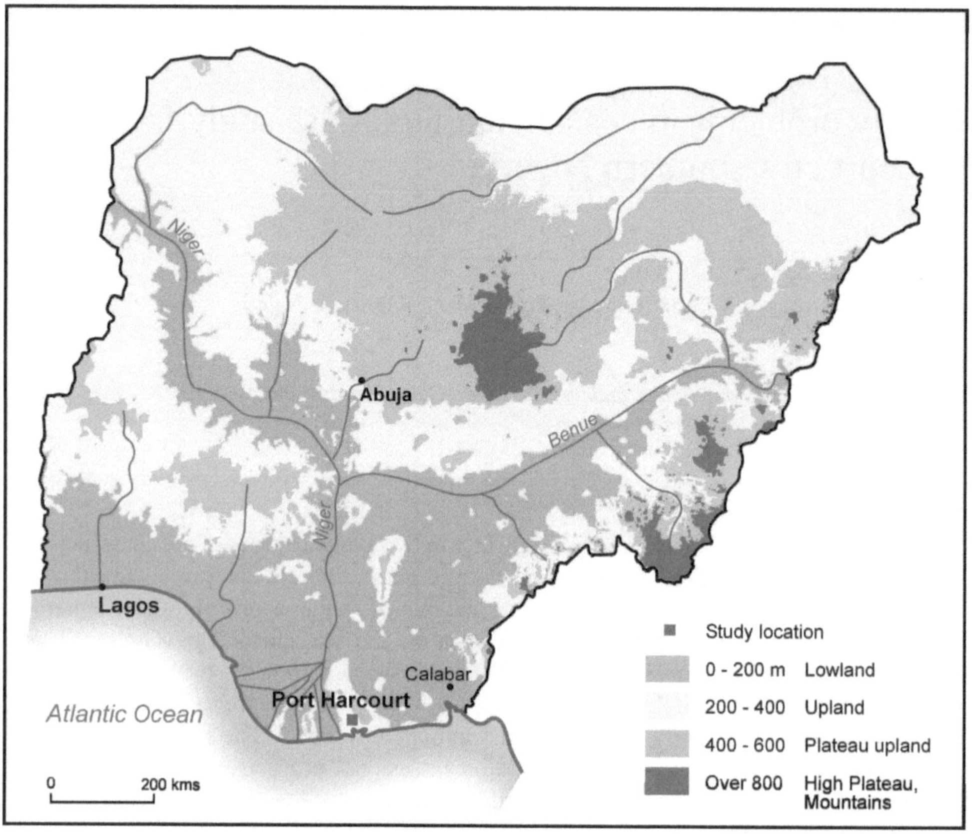

Figure 1. Map of Nigeria showing study location (Port Harcourt).

is higher than acceptable for drinking water. Etu-Efeotor & Odigi (1983) observed that water supply problems in the Eastern Niger Delta included salinity, bacteriological contamination and presence of undesirable ions. Amadi & Amadi (1990) outlined factors controlling saline water migration in coastal aquifers of southern Nigeria and observed that the chemistry of the natural waters in Port Harcourt and Degema areas changes with season. Etu-Efeotor & Akpokodje (1990) have identified one major and two sub-aquifer horizons within the geological/geomorphological units of the Niger Delta. Oteri (1990) delineated the extent of seawater intrusion in the coastal beach ridge of the Forcados, Niger Delta, mainly using geoelectrical survey data. Ngah (2002) observed that NO_3^-, SO_4^{2-}, and pH are higher in rainwater than in groundwater in the Port Harcourt area while Ofoma *et al.* (2005) observed low pH in the groundwater. Salinity problems in different parts of the world have been studied by many authors, including Gimenez & Morell (1997), Karro *et al.* (2004), Zhang *et al.* (2004) and Vives *et al.* (2005), who consider that saline water intrusion into aquifers results from over exploitation of the resources.

The main objective of this paper is to add to existing knowledge on the salinity problems in the coastal aquifers of Port Harcourt, southern Nigeria using, mainly, hydrochemical

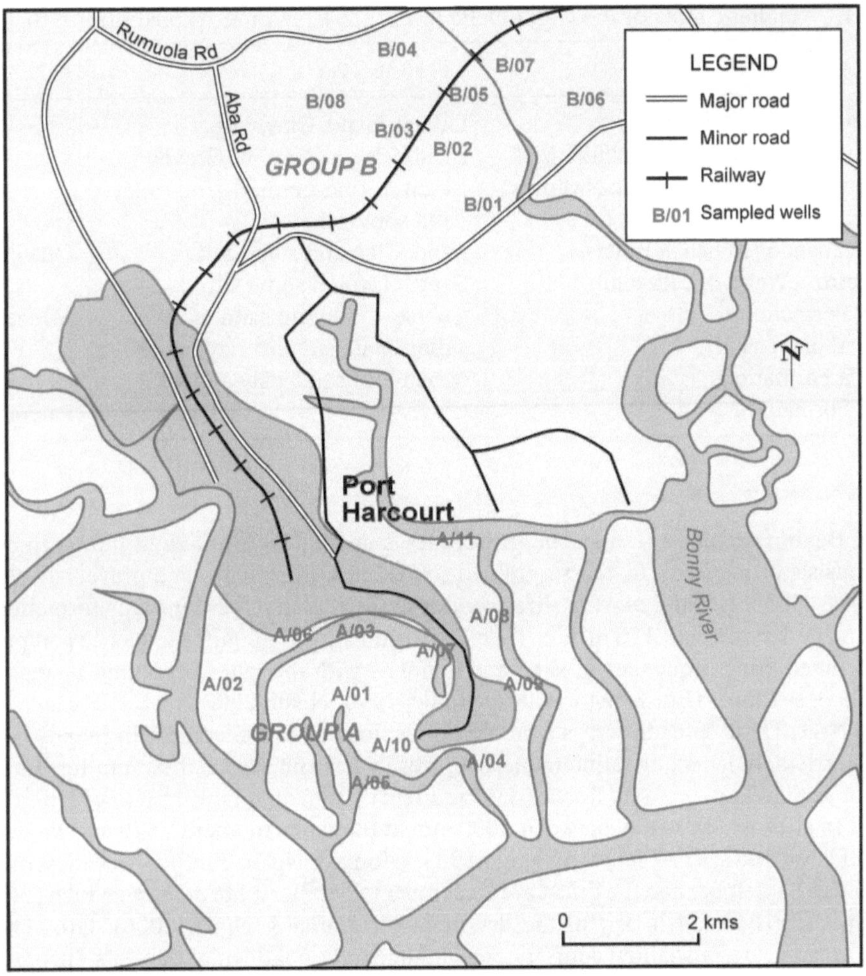

Figure 2. Map of Port Harcourt city showing sampled locations.

data. The study will also determine the hydrochemical facies distribution of groundwater in the study areas (groups A & B), with a view to predicting their water character.

2 GEOLOGY AND HYDROGEOLOGY

The development of the Niger Delta resulted from the formation of the Benue trough as a failed arm of a rift triple junction associated with the separation of Africa and South American continents and subsequent opening of the South Atlantic (Allen, 1965; Oomkens, 1974; Weber & Daukoru, 1975; Evamy, *et. al.* 1978; Whiteman 1982). The Niger Delta consists of three diachronous units, namely the Akata (oldest), Agbada and Benin (youngest) Formations.

Table 1. Geologic units of the Niger Delta (after Etu-Efeotor & Akpokodje 1990).

GEOLOGIC UNIT	LITHOLOGY	AGE
Alluvium.	Gravel, Sand, Clay, Silt.	
Freshwater backswamp, meander belt.	Sand, Clay, some Silt, Gravel.	
Mangrove and salt water/backswamps.	Medium fine-grained sands, clay and some silt.	
Active/abandoned beach ridges.	Sand, Clay and some Silt.	Quaternary
Sombeiro – Warri deltaic plain.	Sand, clay and some silt.	
Benin Formation. (Coastal plain sand)	Coarse – medium sand with subordinate silt and clay lenses.	Miocene
Agbada Formation.	Mixture of sand, clay and silt.	Eocene
Akata Formation.	Clay.	Paleocene

The Benin Formation (Oligocene to Recent) is about 2100 m thick at the basin centre and consists of medium- to coarse-grained sandstones, thin shales and gravels (Weber & Daukoru, 1975). It is the most prolific aquifer in the region. Overlying this formation are Quaternary deposits, 40–150 m thick (Table 1): an unconfined aquifer sequence comprising rapidly alternating sequences of sand and silt/clay, with the latter becoming increasingly prominent seawards (Etu-Efeotor & Akpokodje, 1990; Ngah, 2002).

The Niger Delta spreads across a number of ecological zones, comprising sandy coastal ridge barriers, brackish or saline mangrove, freshwater and seasonal swamp forests (Etu-Efeotor & Akpokodje, 1990). Recharge to aquifers is from direct infiltration of rainfall, the annual total of which varies between 5000 mm at the coast to about 2540 mm landwards (Etu-Efeotor, 1981; Etu-Efeotor & Odigi, 1983; Ofoma & Ngah, 2006). The wettest months are June to September and the driest are December to February, but at least an inch (2.5 cm) of rain is likely to fall in any of the dry months (Ofoma & Ngah, 2006). Groundwater in the area occurs in shallow aquifers of predominantly continental deposits encountered at depths of between 45 m and 60 m. The lithology comprises mixtures of sand in a fining upward sequence, gravel and clay (Etu-Efeotor, 1981; Etu-Efeotor & Odigi, 1983; Etu-Efeotor & Akpokodje, 1990; Ofoma & Ngah, 2006). Well yield is excellent, with production rates of 20,000 liters per hour common, and borehole success rate is usually high (Etu-Efeotor & Odigi, 1983). Across the area, measured transmissivity varies from 60 to 6000 m^2/d, hydraulic conductivity from 0.04 to 60 m/d, and storage coefficient from 10^{-6} to 0.13 (Amadi, 1986; Odigi, 1989; Amadi & Amadi, 1990). Surface water occurrence includes numerous networks of streams, creeks and rivers (see Figure 1).

3 METHODS OF STUDY

Sampling was carried out between July and September 2005, from shallow (<60 m deep) boreholes after well development. The samples were collected in small plastic containers that were first rinsed with the particular sample to be collected, and then analysed within 12 hours after collection. The conductivity and pH were measured on site, however, there is considerable uncertainty in the reliability of the pH measurements, so they have not been

Table 2. Physico-Chemical composition of Groundwater Samples in Group A.

Sample Nos.	Cond μs/cm	TDS (mg/l)	TSS (mg/l)	Ca (mg/l)	Mg (mg/l)	Na (mg/l)	Fe^{2+} (mg/l)	HCO_3 (mg/l)	Cl (mg/l)	SO_4 (mg/l)	NO_3 (mg/l)	Mn (mg/l)
SP/A/01	296	150	7	15	15	78	0.004	49	62	0.8	2.4	0.002
SP/A/02	573	295	8	47.5	106.5	64	0.03	81	86	2	4	0.03
SP/A/03	422	235	6	7.5	20.5	90	0.03	55	69	1.5	4.5	0.04
SP/A/04	521	270	5	27.5	78.5	91.6	0.04	58	78	2.5	5.5	0.03
SP/A/05	717	375	6	55	103	30	0.03	49	69	3.5	3	0.02
SP/A/06	429	230	7	47.5	30.5	68	0.02	26	46	1	0.9	0.003
SP/A/07	370	183	4	22.5	33.5	74	0.01	56	71	2	0.7	0.003
SP/A/08	268	142	3	5	21	63	0.01	12	32	0.7	2.5	0.005
SP/A/09	359	195	2	22.5	13.5	79	0.02	53	49	2.5	1.5	0.01
SP/A/10	236	140	7	17.5	4.5	80	0.01	51	56	1.5	1	0.004
SP/A/11	92	49	9	10	17.5	81	0.001	27.5	19	0.6	0.8	0.01
Average	389.36	205.82	5.82	25.23	40.36	72.6	0.019	47.05	57.91	1.69	2.44	0.014

reported here. Anal Concept Laboratories, Port Harcourt analyzed the cations using an Atomic Absorption Spectrophotometer (Perkin – Elemer AAS3110) and the anions using the colorimetric method with the UV-visible spectrophotometer WPAS110. Total dissolved solids (TDS) and total suspended solids (TSS) were analysed by gravimetric method and total hardness by a titrimetric method. Quality assurance measures consisted of analysis of multiple samples, which did not show significant variation. Also, a quick reliability check on some of the results, such as the TDS/Conductivity ratio, Ca verses $Ca + SO_4^{2-}$ and Na versus $Na + Cl^-$ show that the analysis is acceptable (Hounslow, 1995). A trilinear plot of the analyses was used to determine the hydrochemical facies distribution of the groundwater. In calculating values to be used in the trilinear plot, the values of cations may not exactly equal anions, due to analytical error and unreported minor constituents (Fetter, 1980).

4 RESULTS AND DISCUSSION

Hydrochemical results show that the range of parameters for the coastal group (A) samples are: EC 92–717 μs/cm, TDS 49–375 mg/l, Cl 19–86 mg/l, NO_3 0.7–5.5 mg/l, and HCO_3 12–81 mg/l (Table 2). In the upland group (B) samples, the waters are much less mineralised: EC 35–214 μs/cm, TDS 20–122 mg/l, Cl 8–36 mg/l, NO_3 0.4–0.9 mg/l and HCO_3 5.2–28 mg/l (Table 3).

Thus, the EC, TDS and Cl are much higher in water samples from the coastal group. Elevated chloride concentrations in coastal aquifers, with no corresponding elevation in anthropogenic contaminants such as nitrate, can indicate salt-water contamination (Etu-Efeotor, 1981; Amadi & Amadi, 1990). The results suggest that seawater is the major influence on the chemistry of the Port Harcourt coastal aquifers. In contrast, samples form the upland group are weakly mineralised with low Na and Cl concentrations. However, the Fe^{2+} concentrations of both groups of samples are low (Etu-Efeotor, 1981). Both groups of samples (A & B) are potable by WHO (1993) drinking-water standards for the constituents measured.

Table 3. Physico-Chemical composition of Groundwater Samples in Group B.

Sample Nos.	Cond µs/cm	TDS (mg/l)	TSS (mg/l)	Ca (mg/l)	Mg (mg/l)	Na (mg/l)	Fe^{2+} (mg/l)	Mn (mg/l)	Cl (mg/l)	SO_4 (mg/l)	HCO_3 (mg/l)	NO_3 (mg/l)
SP/B/01	47	26	8	4	3.5	11.8	0.004	0.003	8	0.7	8.5	0.9
SP/B/02	39	22	5	6	4	11.2	0.005	0.004	5	0.6	7.3	0.8
SP/B/03	36	21	4	5	5	7	0.002	0.005	4	0.6	21	0.8
SP/B/04	35	20	6	5	5	13	0.005	0.001	6	0.8	5.2	0.6
SP/B/05	59	34	7	7.5	4.5	12	0.005	0.001	7	0.5	11	0.4
SP/B/06	214	122	4	7.5	20.5	17	0.003	0.005	4	0.4	28	0.5
SP/B/07	125	71	3	7.5	6.5	10	0.002	0.005	3	0.3	15	0.6
SP/B/08	128	75	2	10	12	11	0.002	0.003	2	0.7	23	0.8
Average	85.38	48.88	4.88	6.56	7.63	11.63	0.0035	0.0034	4.88	0.58	14.88	0.68

Table 4. Comparison of water chemistry in Port Harcourt in 1990 and this study (2005).

Chemical parameters	Amadi & Amadi, 1990	This study (Group A)	% increase in average values 1990 & 2005
Total Hardness (mg/l)	6–26 (Ave. 14.1)	22–158 (Ave. 65.59)	365.18
Mg^{+} (mg/l)	2–11 (Ave. 5.4)	4.5–106.5 (Ave. 40.36)	647.41
Cl^{-} (mg/l)	18.5–45.4 (Ave. 26.8)	19–86 (Ave. 57.91)	116.08
HCO_3^{-} (mg/l)	7.3–26.84 (Ave. 14.6)	12–81 (Ave. 47.05)	222.26
CO_3^{-} (mg/l)	3.6–13.2 (Ave. 7.2)	41–90 (Ave. 54.36)	655
Total Dissolved Solids (mg/l)	31.2–162.5 (Ave. 97.3)	49–375 (Ave. 205.82)	111.53
Conductivity (µs/cm)	37.44–240 (Ave. 143.3)	92–717 (Ave. 389.36)	171.71

Ave. = Average

Amadi & Amadi (1990) had previously sampled groundwater from the group-A area of Port Harcourt, and comparison of the results of the present study with theirs, both carried out in the same rainy-season period of the year, revealed a large increase in the average values of some important chemical parameters (Table 4). These changes can be attributed to a possible reversal in the groundwater flow direction from land to sea, due to overexploitation of, or adverse withdrawal from, the aquifer, occasioned by the increase in population. For the group A samples, ranges of ion ratios were Cl^{-}/HCO_3^{-} 0.7–2.7, SO_4^{2-}/Cl^{-} 0.010–0.07, and Ca^{2+}/Mg^{2+} 0.14–2.4. Hem (1985) suggested that seawater intrusion into coastal aquifers may be indicated by sulphate ionic proportions similar to that in seawater, and by low Ca and Mg concentrations. The results for the SO_4^{2-}/Cl^{-} ratios and the Ca and Mg concentrations are consistent with intrusion of seawater into the shallow aquifers.

Variations in chemical composition of groundwater are described in terms of hydrochemical facies, represented in a trilinear diagram (Piper, 1944). The variations may be a function of lithology, solution kinetics or flow pattern (Fetter, 1980). The chemical character of the water from Group A is dominated by Na Cl type waters, for Group B, the waters are mainly Na Cl, but are also influenced by HCO_3 and Mg (see Figure 3).

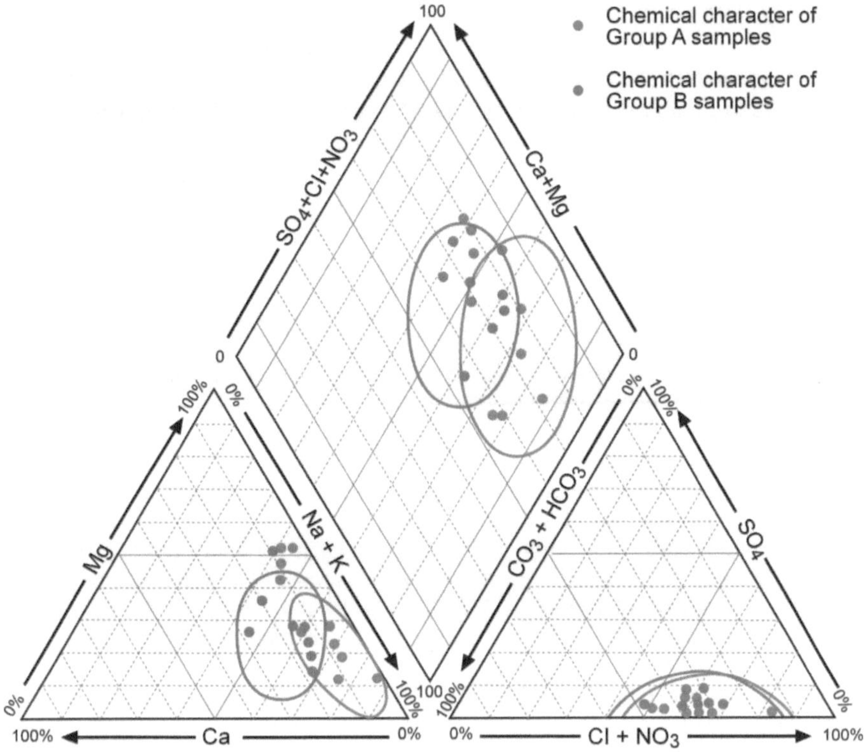

Figure 3. Piper diagram of Group A and B samples.

5 CONCLUSION

Information from geochemical data on groundwater in the Port Harcourt area suggests serious incursion of salt water into the coastal aquifers especially in the areas close to the sea (Group A). The saline intrusion is prompted by excessive withdrawal of water from the aquifers to meet the high demands of an ever-increasing population. Comparison of the results of the present analysis of groundwater samples from group 'A' with those obtained by Amadi & Amadi (1990), as well as the chloride/carbonate, sulphate/chloride and calcium/magnesium ratios for the present analysis indicate that quality is deteriorating in areas very close to the sea (group A). The quality of groundwater from areas a little farther upland (group B) is better.

Results of this study have also shown that the groundwater chemistry of the two study areas compares favourably with the WHO (1993) standards for drinking water for constituents measured. The water in both areas (groups A & B) is dominated by Na Cl.

The shallow sandy aquifers underlying Port Harcourt could be protected from the intrusion of seawater by controlled pumping of boreholes. However, little may be done at the present time to control effectively the excessive pumping of groundwater in Port Harcourt, due to the continuous influx of people in search of better and improved living conditions associated with the benefits of oil and gas related jobs.

ACKNOWLEDGEMENT

The Authors are grateful to Anal Concept Laboratories Port-Harcourt for the analysis of samples. We are also grateful to Chief D. O. Nwachukwu of Groundwater Engineering Port-Harcourt and Dr. C. Awalla of the Department of Geology and Mining, Enugu State University of Science and Technology for their criticisms and suggestions during the interpretation process. Finally, we thank Miss Chiamaka Nweke for typing the manuscript.

REFERENCES

Allen, J. R. L. 1965. *Late Quaternary Niger Delta and Adjacent areas*. American Association of Petroleum Geologists Bulletin, **49**, 547–600.

Amadi, P. A. 1986. *Characteristics of some natural waters from the Port Harcourt area of Rivers State*. Unpublished Master Thesis, University of Ibadan.

Amadi, U. M. P. & Amadi, P. A. 1990. Saltwater *migration in the coastal aquifers of southern Nigeria.* Journal of Mining Geology, **26**, 35–44.

Annual Abstract of Statistics (1999 Edition). 1999. Federal Office of Statistics Abuja. 498pp.

Etu-Efeotor, J. O. 1981. *Preliminary hydrogeochemical investigation of subsurface waters in parts of the Niger delta*. Journal of Mining Geology, **18**, 103–105.

Etu-Efeotor, J. O. & Odigi, M. I. 1983. *Water supply problem in the Eastern Niger delta.* Journal of Mining Geology, **20**, 83–195.

Etu-Efeotor, J. O. & Akpokodje, E. G. 1990. *Aquifer systems of the Niger delta.* Journal of Mining Geology, **26**, 279–284.

Evamy, B. D., Harembour, J., Kamerling, P., Knaap, W. A., Molloy, F. A. & Rowlands, P. H. 1978. *Hydrocarbon habitat of Tertiary Niger Delta*. American Association of Petroleum Geologists Bulletin, **62**, 1–39.

Fetter, C. W, Jr. 1980. *Applied Hydrogeology*. Bell and Howell Company Columbus, Ohio.

Gimenez, E. & Morell, I. 1997. *Hydrogeochemical analysis of salinization processes in the coastal aquifer of Oropesa, Spain.* Environmental Geology, **29**, 118–131.

Hem, T. D. 1985. *Study and interpretation of the chemical characteristics of natural water, 3rd edition*. USGS Water Supply Paper **2254**, 249pp.

Hounslow, A. W. 1995. *Water quality data analysis and interpretation*. Lewis Publishers, New York.

Karro, E., Marandi, A. & Vaikmae, R. 2004. *The origin of increased salinity in the Cambrian-Vendian aquifer system on the Kopli Peninsula, Northern Estonia*. Hydrogeology Journal, **12**, 424–435.

Ngah, S. A. 2002. *Patterns of groundwater chemistry in parts of the Niger delta.* 38th Annual International Conference of Nigerian Mining & Geoscience Society. [Abstracts], 39p.

Odigi, M. I. 1989. *Evaluating groundwater supply in Eastern Niger Delta, Nigeria*. Journal of Mining Geology, **25**, 159–164.

Ofoma, A. E. & Ngah, S. A. 2006. *Applicability of solute balance technique in estimating recharge: A case study of a paved and non-paved area of the Eastern Niger Delta, Nigeria.* Global Journal of Pure & Applied Science, **12**, 25–30.

Ofoma, A. E., Omologbe, D. & Aigberua, P. 2005. *Physico-chemical quality of groundwater in parts of Port Harcourt city, Eastern Niger Delta.* Journal of Water Resources, **16**, 18–24.

Oomkens, E. 1974. *Lithofacies relations in the Late Quaternary Niger Delta Complex*. Sedimentology, **21**, 115–122.

Oteri, A. U. 1990. *Delineation of saltwater intrusion in a coastal beach ridge of Forcados.* Journal of Mining Geology, **26**, 35–44.

Piper, A. M. 1944. A *graphical procedure in the geochemical interpretation of water analyses.* American Geophysical Union Transactions, **25**, 914–923.

Vives, L., Varni, M., & Usunoff, E. 2005. *Behavior of the fresh- and saline-water phases in an urban area in Western Buenos Aires province, Agentina.* Hydrogeology Journal, **13**, 426–435.

Weber, K.J. & Daukoru, E. M. 1975. *Petroleum Geological aspects of the Niger Delta.* Proc. 9th World Petrology Congress, **2**, 209–222.

Whiteman, A. J. 1982. *Nigeria: its petroleum geology, resources and potentials.* Graham and Trothman, London.

WHO 1993. *International Standards for drinking water and guidelines for water quality.* World Health Organization, Geneva.

Zhang, Q., Volker, R. E. & Lockington, D. A. 2004. *Numerical investigation of seawater intrusion at Gooburrum Bundaberg, Queensland Australia.* Hydrogeology Journal, **12**, 674–687.

Modelling Approaches to Groundwater Issues

CHAPTER 25

Groundwater flow and transport modeling in semi-arid Africa: Salt accumulation in river-fed aquifers

P. Bauer-Gottwein
Institute of Environment & Resources, Technical University of Denmark, Lyngby, Denmark

S. Zimmermann
Environmental Resources Management (ERM) Southern Africa, Johannesburg, South Africa

W. Kinzelbach
Institute of Environmental Engineering, ETH Zurich, Switzerland

ABSTRACT: In many river systems in semi-arid Africa, water is lost from shallow river-fed aquifers by infiltration and subsequent evapotranspiration. Since groundwater flow is generally directed from the river into the aquifer and evapotranspiration does not entirely remove the dissolved solids with the uptake water, salinity is continuously accumulated in the river-fed aquifer. This paper presents a simple conceptual and numerical model of the river-fed aquifer system. Salt accumulation by evapotranspiration and plant growth inhibition by elevated salinity levels are taken into account. The system reaches a steady state, in which the evapotranspiration is primarily controlled by groundwater quality and only to a minor degree by water availability (depth to groundwater table). A field dataset from the periphery of the Okavango Delta in Botswana (Shashe River Valley) is presented and the developed modelling concepts are applied to this system. Plant-salinity feedback mechanisms are shown to be essential for successful quantitative modeling of Shashe River Valley.

1 INTRODUCTION

Groundwater is an important water supply source for domestic and agricultural purposes in Africa (e.g. Giordano, 2006). The spatial distribution of groundwater over the continent is, however, highly heterogeneous. In several systems in the African continent, significant quantities of water are redistributed over larger spatial scales by river flow. River infiltration subsequently recharges shallow alluvial aquifer systems. Discharge from these systems occurs primarily by evapotranspiration and pumping. River-fed aquifers are thus important water resources for human use and sustain key ecosystems in semi-arid and arid Africa. Abstractions for human use directly reduce the "environmental flow" available to sustain riverine ecosystems. Water resources management needs to take into account both the hydrological and chemical dynamics of such systems. Since a major part of the groundwater in shallow river-fed alluvial aquifers is eventually lost to evapotranspiration, groundwater salinity increases. This affects the quality of the water resource and

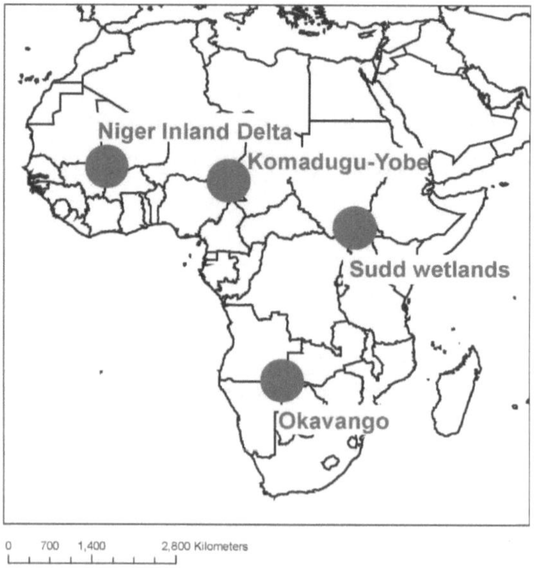

Figure 1. Four examples of infiltrating rivers and river-fed alluvial aquifers in Africa.

its suitability for human use. Moreover, important hydro-ecologic feedback mechanisms have to be taken into account when simulating river-fed alluvial aquifers: Transpiration is reduced by rising salinity levels, and consequently, groundwater flow is modified. Coupled simulations of groundwater flow and salinity transport are therefore required to properly analyse system dynamics.

Infiltrating rivers and shallow, river-fed aquifers abound in semi-arid and arid Africa. Due to the strong humidity gradients around the tropics, the presence of large sedimentary basins and changes in topographical relief, many rivers turn from gaining to losing (recharging) rivers. Some important and spectacular examples are the Okavango River (e.g. McCarthy *et al.*, 1998), the Komadugu-Yobe river system (e.g. Glenn *et al.*, 1995), the Niger River and Niger Inland Delta (e.g. Kuper *et al.*, 2003) and the Sudd wetland system on the River Nile (e.g. Sutcliffe & Parks, 1987). Figure 1 shows the location of these four systems in Africa. In all these cases, river discharge is generated in humid, relatively steep highlands underlain by bedrock. The rivers subsequently flow into extremely flat sedimentary basins, where the climate conditions change to semi-arid or arid.

Figure 2 shows maps of the drainage system, the land surface elevation and the surface geology in the four above-mentioned river systems. Data sources are the digital chart of the world (http://www.maproom.psu.edu/dcw/), the Shuttle Radar Topography Mission (SRTM, Slater *et al.*, 2006) and a digital geological map of Africa published by the USGS (Persits *et al.*, 2002). The diagram clearly shows that changes in surface geology correspond to changes in the appearance of the drainage network. The Okavango originates in the humid, crystalline Benguela Plateau in Southern Angola and flows into the Kalahari Basin. The Niger River originates in the mountains of Guinea near the border with Sierra Leone. It flows towards the Sahara Desert and once the river enters the Taoudeni sedimentary basin, it starts infiltrating and it forms the 30,000 km^2 Niger Inland Delta. The tributaries of the Komadugu-Yobe river system originate on the crystalline Jos plateau in Northern Nigeria.

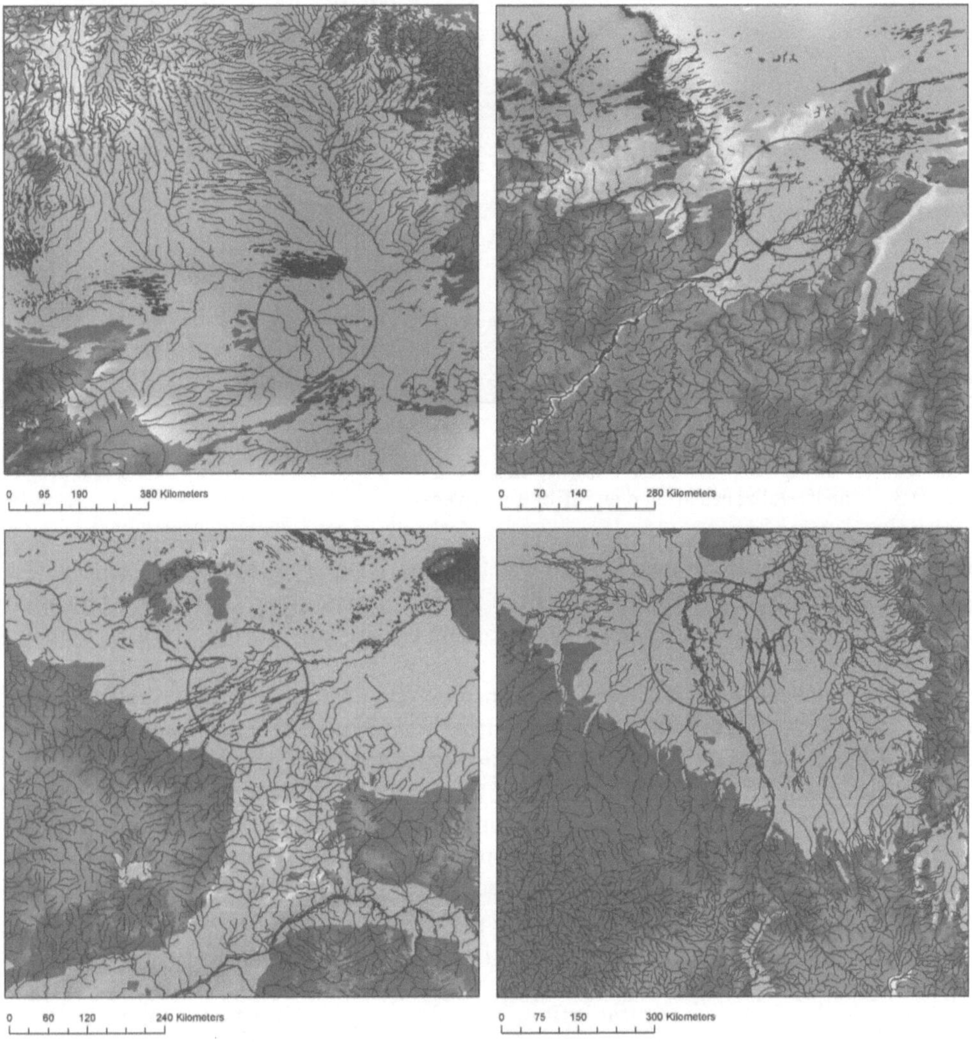

Figure 2. Okavango (upper left), Niger Inland Delta (upper right), Komadugu-Yobe (lower left) and Sudd wetlands (lower right). Topographic elevation is shown in colour, light shading represents sedimentary geology. The drainage network is indicated in blue. Note how the drainage network changes on entering the sedimentary basins.

On entering the sedimentary Chad basin, the rivers start bifurcating, infiltrating and losing water to evapotranspiration. Only a minor fraction of the water reaches the regional terminal drainage point of the region, Lake Chad. The Sudd wetlands (40,000 km^2) are formed by the White Nile on its transition from the East African Rift region into the Sudan basin.

Net water loss to the atmosphere from these rivers is substantial. In the case of the Okavango it amounts to almost 100% (e.g. McCarthy *et al.*, 1998), in the cases of the Komadugu-Yobe, the Sudd wetlands and the Niger Inland Delta, roughly 60–70% of the inflowing water is lost (Sellars, 1981, Kuper *et al.*, 2003, Sutcliffe and Parks, 1987). This water loss can be divided into four components: evaporation from open surface water

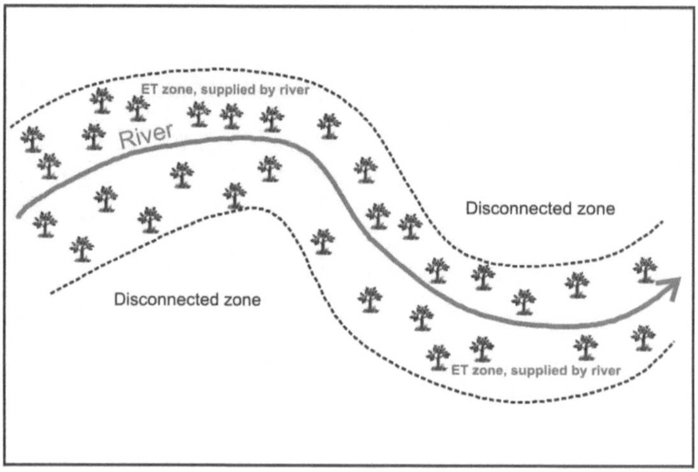

Figure 3. The lateral zones around an infiltrating river.

bodies, transpiration from open surface water bodies, capillary evaporation from shallow groundwater and phreatic transpiration from shallow groundwater. Capillary evaporation and phreatic transpiration from shallow aquifer are in the following referred to as phreatic evapotranspiration.

The ratio between evaporation (E) and transpiration (T) from open water bodies can be determined by plotting the successive downstream enrichment of stable isotopes $\delta^{18}O$ and $\delta^{2}H$, which reflects E only, versus the successive downstream enrichment of salinity, which reflects the sum of E and T. Dincer *et al.* (1978) presented this technique and applied it to the Okavango Delta. They derived E/T ratios ranging from 0.5 in summer to 1.0 in winter. The partition between the total water loss from open water bodies (E + T) and phreatic evapotranspiration from shallow groundwater recharged by the open water body can be determined from salt balance calculations. In the case of water loss from open water bodies, the salt accumulates in the surface water body, whereas infiltrating groundwater removes salts from the surface water body. Salt balance calculations for the Okavango suggest that, in this system, about one third of the losses occur as infiltration and subsequent phreatic evapotranspiration, whereas two thirds occur as direct evapotranspiration from wetlands (Bauer, 2004).

This paper focuses on phreatic transpiration as the most important loss mechanism from shallow river-fed alluvial aquifers. After introducing the conceptual model of river-fed alluvial aquifers, phreatic transpiration is discussed and the coupled governing flow and transport equations are developed. Shashe river valley is presented as an example of a shallow, river-fed alluvial aquifer in Africa.

2 CONCEPTUALIZATION OF THE SYSTEM

Typically the river water infiltrates into an alluvial aquifer, which is located around the river reach (Figure 3). Infiltration is driven by head gradients between the river and the surrounding shallow groundwater. These head gradients are maintained by evapotranspiration, which removes water from the alluvial aquifer. In the prototypical setting, the longitudinal water level slope in the river is very small compared to the transverse head

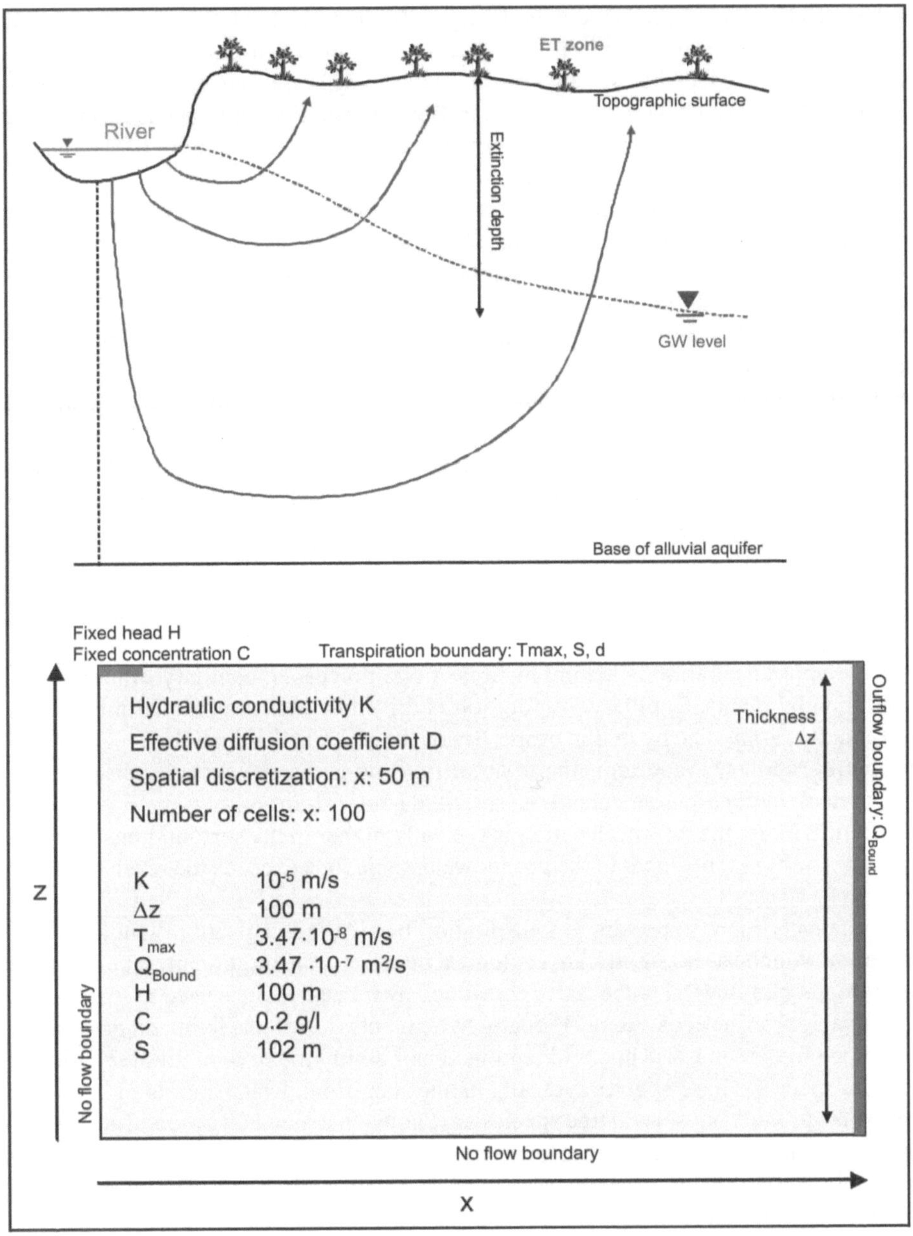

Figure 4. Upper panel: Vertical cross section of a river-fed alluvial aquifer including streamlines. Lower panel: Boundary conditions and parameters for the numerical implementation. The boundary flux was set to 10% of the maximum transpiration rate.

gradients in the alluvial aquifer induced by evapotranspiration. In this limiting case, the groundwater flow streamlines are perpendicular to the river and the flow geometry is approximately two-dimensional (Figure 4). Thus, the streamlines in the aquifer originate in the river and terminate at the evapotranspiration boundary located at the water table. Water particles travel only away from the river. Every water particle that has left the river is eventually consumed by evapotranspiration and no return flow from the aquifer back into the river takes place. The fact that water is entirely taken up by evapotranspiration has important implications for the salt balance of the aquifer. Both evaporation and transpiration only take up a minor fraction of the salinity contained in the water. The remainder of the dissolved solids is accumulated in the residual water body, i.e. here the aquifer.

In a further simplifying assumption, we neglect recharge to the river-fed alluvial aquifer from local precipitation. It is assumed that the recharge comes entirely from the river and that the local precipitation produces negligible recharge.

A minor flow component is assumed to pass from the alluvial aquifer out into the disconnected zone. This flow represents evaporative losses in the surrounding drylands, recharge to deeper aquifer units and other unspecified loss terms.

Under these simplifying assumptions, the river-fed alluvial aquifer can be understood and modelled as a rather simple system. Water recharges the aquifer laterally from the river and is taken up by evapotranspiration as depicted in Figure 4. The key process controlling the behaviour of the system is therefore evapotranspiration, which will be discussed in more detail in the subsequent section.

3 PHREATIC EVAPOTRANSPIRATION

Phreatic evapotranspiration is a combination of two processes, capillary evaporation and transpiration by plants. Capillary evaporation is driven by matrix potential differences in the unsaturated zone. The hydraulic properties of the soil and the depth to the groundwater table control capillary evaporation in the simplified "no-recharge" system considered here. In this system, hydraulic gradients in the river-fed aquifer transverse to the river are rather steep. Furthermore, the terrain elevation is typically higher in the surroundings of the river than in the riverbed. The depth to the groundwater table thus exceeds the reach of capillary evaporation.

Quantitatively more important is transpiration by riverine vegetation. Unlike capillary evaporation, which has no significant ecological benefits, transpiration must be regarded as an "environmental flow", i.e. the water consumed by transpiration is used to sustain potentially valuable riverine ecosystems. Phreatophytic plants (i.e. plants living on groundwater) are common in the semi-arid and arid regions. Their ability to tap groundwater varies from species to species. Grass species typically cannot tap groundwater that is more than one meter below ground, but several tree species are known to depend on groundwater, which is more than 10 m below the ground surface (Lamontagne *et al.*, 2005; Zencich *et al.*, 2002).

In humid systems, phreatic evapotranspiration is primarily controlled by the moisture content of the soil column. If the soil column is moist, groundwater uptake by evapotranspiration is negligible, since the processes water demand is satisfied by soil water. Here we consider semi-arid and arid systems, where water input by local precipitation is negligible and the only available water source is groundwater recharged by river infiltration.

In such systems, evaporation and transpiration thus depend on the depth to groundwater and the groundwater salinity. The standard transient one-dimensional groundwater flow equation including the evapotranspiration term for an unconfined aquifer reads

$$SY\frac{\partial h}{\partial t} = K\Delta z\frac{\partial^2 h}{\partial x^2} - ET(h, c), \qquad (1)$$

where h is the piezometric head (m), c the salinity of the groundwater (g/l), K the hydraulic conductivity (m/s), Δz the thickness of the aquifer (m), SY the specific yield of the aquifer (–) and ET the evapotranspiration rate (m/s).

Salinity transport is governed by the following equation

$$\frac{\partial c}{\partial t} = \frac{\partial}{\partial x}\left(D\frac{\partial c}{\partial x}\right) - \frac{\partial}{\partial x}(vc) + SY\frac{\partial h}{\partial t}c - \frac{ET(h,c)}{\Delta z}c_{ET}, \tag{2}$$

where D is the dispersion coefficient (m^2/s), v the pore velocity (m/s) and c_{ET} (g/l) the salinity of the water removed from the aquifer by evapotranspiration. If c_{ET} is smaller than the ambient salinity in the aquifer, salt accumulation occurs. In the following sections, it is assumed that the salinity and head dependence of evapotranspiration can be written in the following form:

$$ET(h,c) = ET_{\max} \times f(h) \times g(c), \tag{3}$$

where ET_{max} is the maximum evapotranspiration rate (m/s), f(h) ($0 \leq f(h) \leq 1$) is a reduction factor depending on the hydraulic head and g(c) ($0 \leq g(c) \leq 1$) is a reduction factor depending on the salinity.

4 GROUNDWATER LEVEL DEPENDENCE

Evapotranspiration rates generally decrease with falling groundwater levels. Evaporation and transpiration differ, however, in the dependence on depth to groundwater and the maximum depth from which water can be extracted.

4.1 *Evaporation*

Capillary evaporation from groundwater is significant, if the groundwater table is shallow, but rapidly decreases, once the groundwater table drops more than a couple of meters below the topographic surface (Figure 5). The curves in Figure 5 were calculated using the model of capillary evaporation presented by Milly, 1984. The model equations were solved numerically with MATLAB's partial differential equation solver (Skeel & Berzins, 1990). This general shape of the depth dependence of capillary evaporation has also been confirmed by experimental data (Coudrain-Ribstein *et al.*, 1998; Brunner, 2006).

In regional-scale hydrological models, where the solution of Milly's model or similar models for all model cells is computationally too expensive, depth dependence of capillary evaporation may be expressed with a simple exponential model:

$$f_E(h) = 1 \qquad \text{for } h \geq S, \text{ and}$$

$$f_E(h) = \exp\left(-\frac{S-h}{d}\right) \quad \text{for } h < S \tag{4}$$

The evaporation surface S (m) and the extinction depth d (m) depend on the local soil hydraulic parameters.

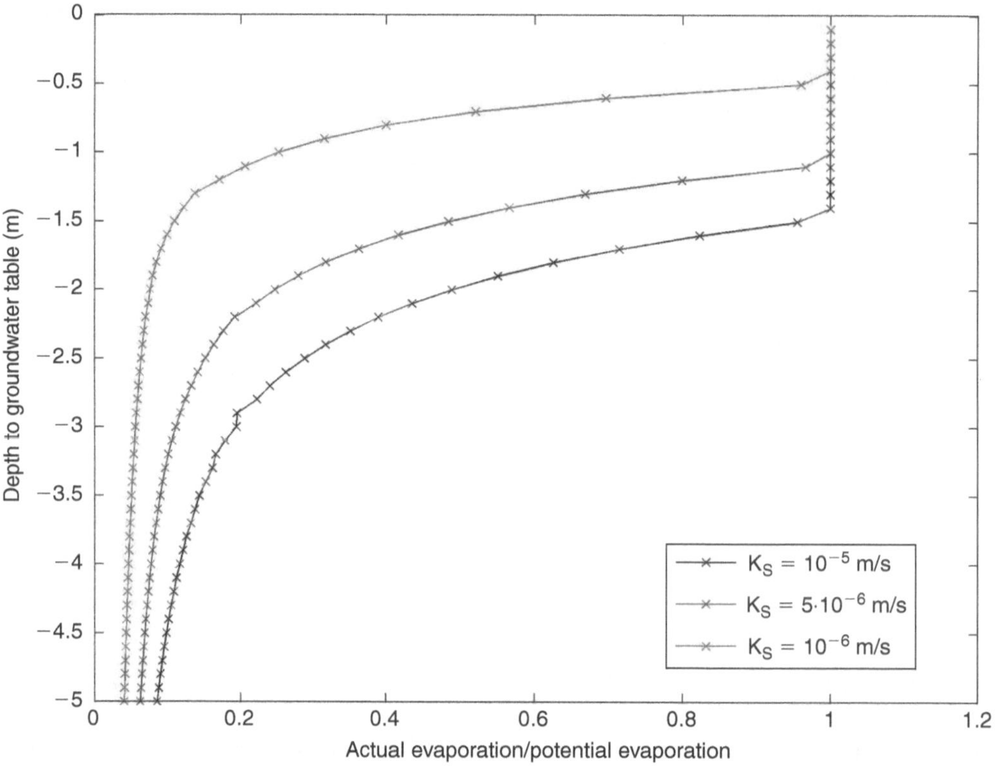

Figure 5. Relationship between capillary evaporation and depth to groundwater for different soil types. The three curves represent three different values of the saturated hydraulic conductivity. The remaining van Genuchten soil parameters are: Saturation water content $\theta_s = 0.4$, residual water content $\theta_s = 0.04$, $\alpha = 1.6$, $n = 1.3$.

4.2 *Transpiration*

The dependence of the phreatic transpiration rate on depth to groundwater is a function of the entire plant community's root distribution. In standard groundwater models, various relationships are used to express the dependence of phreatic transpiration on the depth to groundwater table. In general, the transpiration rate is assumed to decrease with increasing depth to the groundwater table. Two commonly used relationships are linear and exponential dependence of transpiration on depth to groundwater. The linear depth dependence can be formulated as:

$$
\begin{aligned}
f_T(h) &= 1 \quad \text{for } h \geq \text{S}; \\
f_T(h) &= 1 - \frac{S-h}{d} \quad \text{for S} - \text{d} \leq h < \text{S, and} \\
f_T(h) &= 0 \quad \text{for } h \leq \text{S} - \text{d}
\end{aligned}
\tag{5}
$$

Where S is the land surface elevation (m), h the piezometric head (m) and d the so-called extinction depth (m). This formula is implemented in the standard USGS MODFLOW

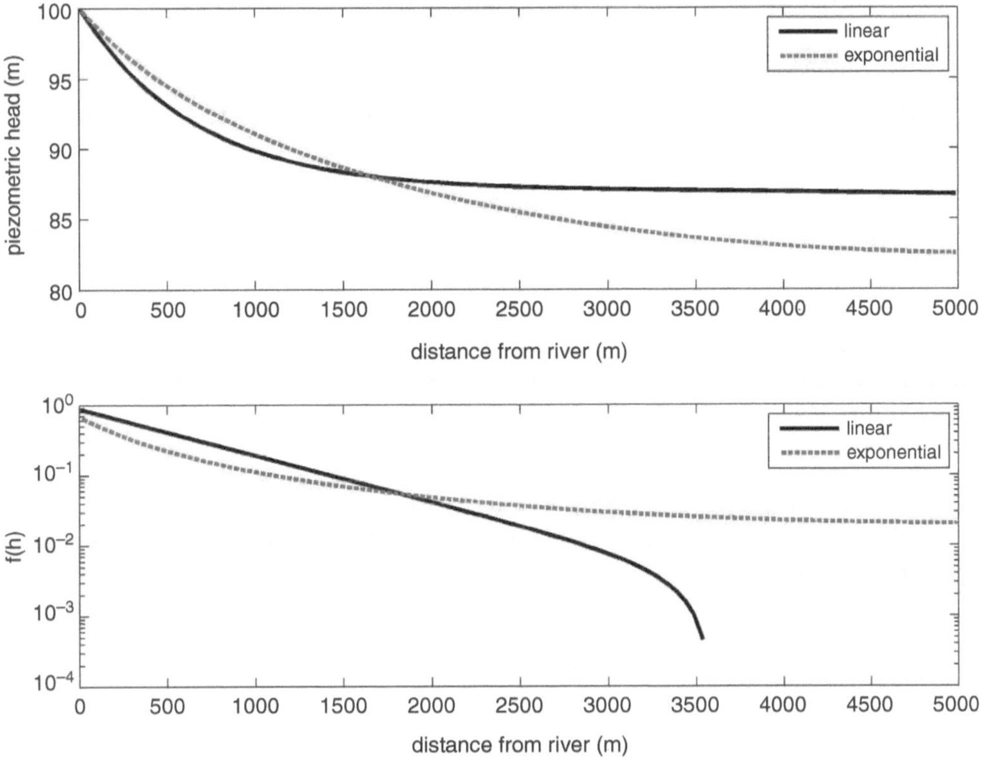

Figure 6. Profiles of hydraulic head and transpiration rates in the river-fed alluvial aquifer for linear and exponential dependence on depth to groundwater. Upper panel: Piezometric head; Lower panel: Transpiration rate.

groundwater modeling software (Harbaugh & McDonald, 1996a; Harbaugh & McDonald, 1996b). The exponential depth dependence reads (see equation 4):

$$f_T(h) = 1 \quad \text{for } h \geq \text{S, and}$$
$$f_T(h) = \exp\left(-\frac{S-h}{d}\right) \quad \text{for } h < \text{S}. \tag{6}$$

In Figure 6 the profiles of hydraulic head and transpiration in the river-fed aquifer are compared for linear and exponential dependence of the transpiration rate on depth and no dependence of the transpiration rate on salinity, i.e. g(c) = 1 (see equation 3). The extinction depth d was set to 15 m in the linear case and to 5 m in the exponential case, i.e. the exponential transpiration rate decreases to 5 percent of the maximum rate at 15 m depth. For model setup and parameters, refer to Figure 4. The linear depth dependence leads to a steeper hydraulic gradient in the vicinity of the river. Away from the river, the head quickly stabilizes and the hydraulic gradient converges to the residual value required to provide the boundary outflow. The head dependent reduction factor f(h) decreases from values close to one to zero in the linear case. In the exponential case, f(h) close to the river is much lower than in the linear case, but f(h) decreases much more slowly than in the linear case further away from the river.

5 GROUNDWATER SALINITY DEPENDENCE

5.1 *Evaporation*

In moderate salinity ranges, capillary evaporation is independent of the groundwater salinity. For the evaporation process, we generally assume

$$g_E(c) = 1 \tag{7}$$

5.2 *Transpiration*

Plant growth is inhibited by elevated salinity levels. If salinity levels exceed a species-dependent threshold, plants die off. This effect is well described in the literature for different plant species (e.g. Bernstein, 1975; Hoffman, 1980; Maas, 1986). In conclusion, growth inhibition by salinity is found to qualitatively proceed in three phases. Below a certain salinity level, plant growth is not inhibited at all. In the following transition zone, plant growth (and consequently transpiration rates) are successively reduced to zero. If the salinity level increases further, plant growth is no longer possible.

The dependence of transpiration rate on salinity can thus be expressed in the form of a smoothed step function:

$$g_T(c) = \frac{1}{2} erfc\left(\frac{c - c_S}{\sqrt{2}\sigma}\right) \tag{8}$$

where c_S is a threshold concentration (g/l) and σ (g/l) can't see in equation is a shape parameter describing the smoothness of the transition. Typical literature values for the threshold concentration and the shape parameter range from $c_S = 2$–8 g/l and $\sigma = 0.5$–4 g/l (Maas, 1986). Whereas salinity-growth relationships for individual crops are well studied, information for natural ecosystems is scarce.

6 SALINITY ACCUMULATION

Evapotranspiration accumulates salinity in the aquifer, if c_{ET} in equation 2 is lower than ambient groundwater salinity. River-fed alluvial aquifers are subject to continuous salt accumulation, since water is taken up by evapotranspiration and salinity is left behind. The accumulation term in equation 2 represents highly complex natural processes.

6.1 *Evaporation*

In the case of capillary evaporation, water first rises from the free water table to the so-called evaporative front. Below the evaporative front, the water transport is dominated by transport in the liquid phase, above the evaporative front, gas phase transport dominates. Minerals are precipitated and salinity is accumulated around the evaporative front. The more soluble minerals can re-dissolve, for instance during periods of high groundwater level or intensive rainfall. They are subsequently flushed back into the shallow aquifer. In groundwater applications (e.g. MT3DMS, Zheng & Wang, 1999), the accumulation process is typically parameterized as follows:

$$c_{ET} = c_{u,E} \quad \text{for } c \geq c_{u,E}, \text{ and}$$

$$c_{ET} = c \quad \text{for } c < c_{u,E} \tag{9}$$

Here, $c_{u,E}$ is a threshold uptake concentration. In the system considered here, accumulation by evaporation is minor, since the groundwater is fairly deep and evaporation consequently small.

6.2 *Transpiration*

Salt and water uptake by transpiration has been studied extensively (e.g. Garcia-Sanchez *et al.*, 2005; Grieve & Walker, 1983; Russell & Barber, 1960; Salim, 1989). However, there is no consensus on how to parameterize the process for groundwater applications. MT3DMS (Zheng & Wang, 1999) for instance suggests the following form:

$$c_{ET} = c_{u,T} \quad \text{for } c \geq c_{\text{u,T}}$$
$$c_{ET} = c \quad \text{for } c < c_{\text{u,T}} \tag{10}$$

where $c_{u,T}$ is a threshold concentration, which depends on the vegetation type. It is assumed that up to the threshold concentration $c_{u,T}$, plants take up all the solutes contained in the groundwater. Solute concentrations in excess of the threshold value lead to salinity accumulation in the aquifer. However, this parameterization of salt accumulation by transpiration must be used with great care since data from natural, mixed ecosystems are scarce and conclusions in the literature ambiguous. Further experimental research is needed.

7 SENSITIVITY OF THE COUPLED SYSTEM

If the salinity-transpiration feedback mechanisms are taken into account, a coupled water flow and salinity transport system must be solved numerically. The SEAWAT code (Guo & Langevin, 2002) was modified for this purpose. An upstream finite difference solver of the transport equation was used (Zheng & Wang, 1999). Figure 7 shows the resulting steady-state head, concentration and transpiration profiles in the river-fed alluvial aquifer for $c_{u,T} = 0$ g/l, $c_S = 2$ g/l and $\sigma = 1$ g/l. The extinction depth d was again set to 15 m in the linear case and 5 m in the exponential case. The other model parameters are as indicated in Figure 4. Figure 7 shows that the parameterization of the depth dependence becomes unimportant in this system. The transpiration rate is almost entirely controlled by groundwater quality. Significant transpiration occurs only close to the river, where salinity levels are low. Further away from the river, salinity quickly increases and the transpiration rate consequently breaks down. Salt accumulation in the uncoupled simulations proceeds to fairly high levels but is restricted by c_S in the coupled simulations. If the groundwater salinity becomes significantly higher than c_S, transpiration is switched off and salt accumulation stops. The transpiration pattern in the river-fed alluvial aquifer is thus primarily controlled by groundwater quality and to a much lesser degree by depth to groundwater or water availability.

One important result of the model is the steady-state width of the zone around the river, where plant life is possible. Here we analyze model results in terms of w_{10} which is the width of the zone where $f(t) \cdot g(c) >= 0.1$. To obtain a better understanding of the model behavior, the sensitivity of the width w_{10} with respect to the various model parameters is calculated. Exponential depth dependence is assumed and the base run is characterized by the set of parameters indicated in Figure 4 and d = 5 m, $c_{u,T} = 0$ g/l, $c_S = 2$ g/l and $\sigma = 1$ g/l.

Table 1 shows that the key parameters governing the behavior of this system are the outflow rate at the right-hand domain boundary and the parameters characterizing the

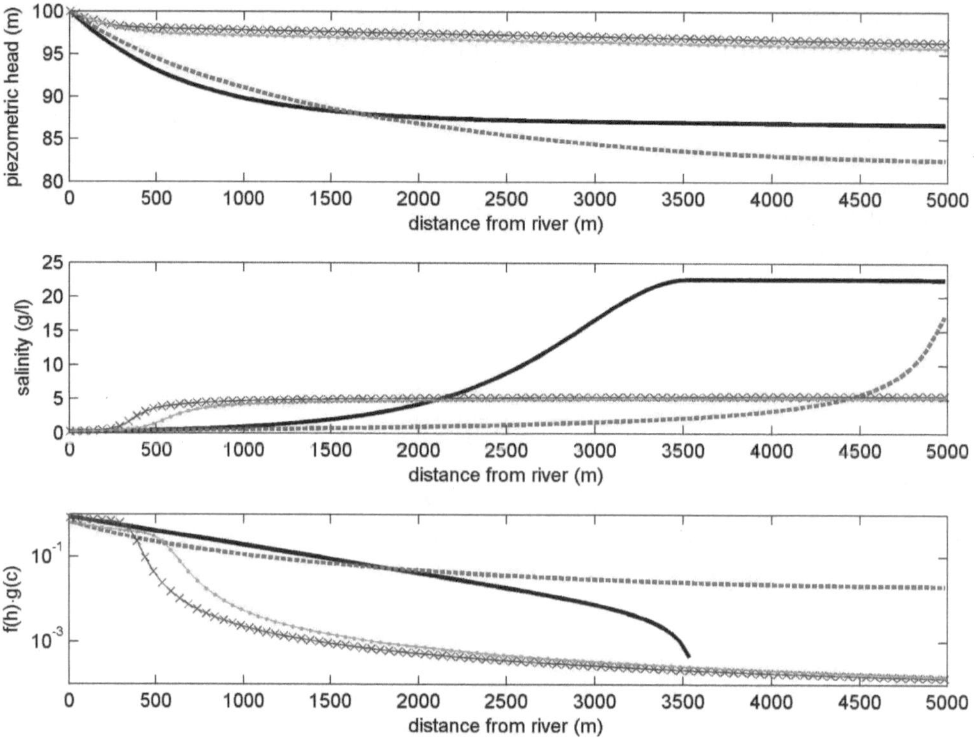

Figure 7. Head, concentration and transpiration profiles in the river-fed alluvial aquifer for linear and exponential depth dependence. Results from coupled (Eq. 4) and uncoupled simulations are shown. black solid line: linear; red broken line: exponential; blue crossed line: linear coupled; green dotted line: exponential coupled.

Table 1. The sensitivity of the width w_{10} with respect to the various model parameters.

Run	**w_{10} (m)**
Base run	585
$K = 2 \times K_{Base}$	535
$Q_{Bound} = 2 \times Q_{Bound,Base}$	1235
$c_S = 2 \times c_{S,Base}$	1085
$\sigma = 1.2 \times \sigma_{Base}$	735
$D = 2 \times D_{Base}$	585
$T_{max} = 2 \times T_{max,Base}$	285
$d = 1.1 \times d_{Base}$	535
$c_{u,T} = 0.1$ g/l	1185

reduction of the transpiration rate with increasing salinity. The outflow rate determines the flushing of the domain and the renewal of the water in the aquifer. Note, that no steady-state solution exists, if the boundary outflow is set to 0. On the other hand, salt accumulation and inhibition of transpiration by salinity become insignificant, if the outflow rate is set

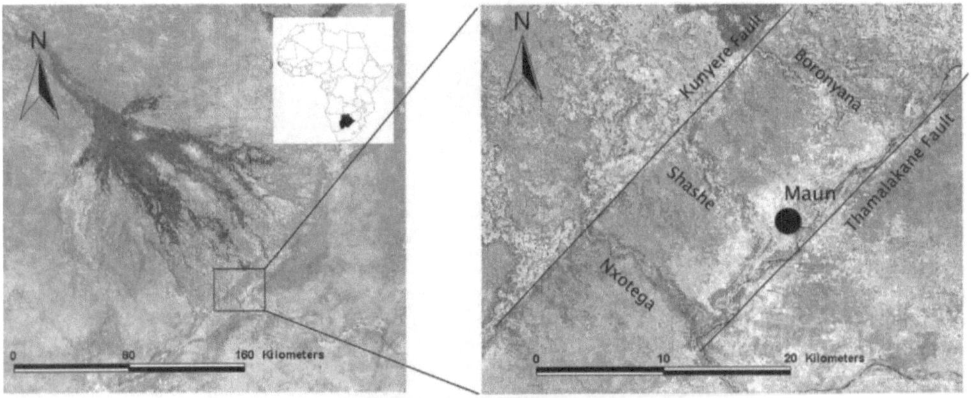

Figure 8. Location of Shashe River Valley (Reprinted from Bauer *et al.*, 2006 with permission from Elsevier).

to high values. In this simulation, flushing of the domain was assumed to occur through boundary outflow. Obviously, other mechanisms are potentially active in river-fed alluvial aquifers. For instance, low flow in the river might result in return flow from the aquifer and consequent salt removal. A strong rainfall event may flush the aquifer from above and periodic flow reversal over geologic time scales (dry and wet periods) may be of importance too. Increased salt tolerance of the plants (higher c_S and σ) also leads to increasing w_{10}, whereas changing the extinction depth d and the hydraulic conductivity K has only a minor influence. Increasing the maximum transpiration rate T_{max} leads to increased salinization of the aquifer and consequently to a smaller w_{10}. Increasing the uptake concentration $c_{u,T}$ leads to a higher w_{10}. The salinity is partly taken up by transpiration, which results in reduced salt accumulation in the system.

8 THE SHASHE RIVER VALLEY CASE EXAMPLE

The Shashe River Valley (SRV), one of the terminal rivers flowing away from the Okavango Delta, Botswana, is a good case example of a shallow river-fed alluvial aquifer. Both groundwater levels and groundwater salinities vary strongly in space and time. Successful quantitative modeling of this system required the inclusion of the transpiration-salinity feedback mechanisms discussed above. The following sections summarize the findings published in Bauer *et al.* (2006).

8.1 *Regional setting*

SRV is located at the periphery of the Okavango Delta in Botswana (Figure 8). As a consequence of tectonic activity (East African Rift Valley formation), Kunyere and Thamalakane normal faults cut the course of the palaeo Okavango River, as part of a graben structure on their north-western side, which has since been filled by wind- and waterborne sediments. In this graben, the Okavango Delta has formed (Gumbricht *et al.*, 2001; Hutchins *et al.*, 1976; McCarthy *et al.*, 1993; Modisi *et al.*, 2000; Scholz *et al.*, 1976; Thomas and Shaw, 1991). Most of the water flowing into the Delta is lost by evapotranspiration, however, small quantities of water cross the Kunyere fault and flow down to the Thamalakane River.

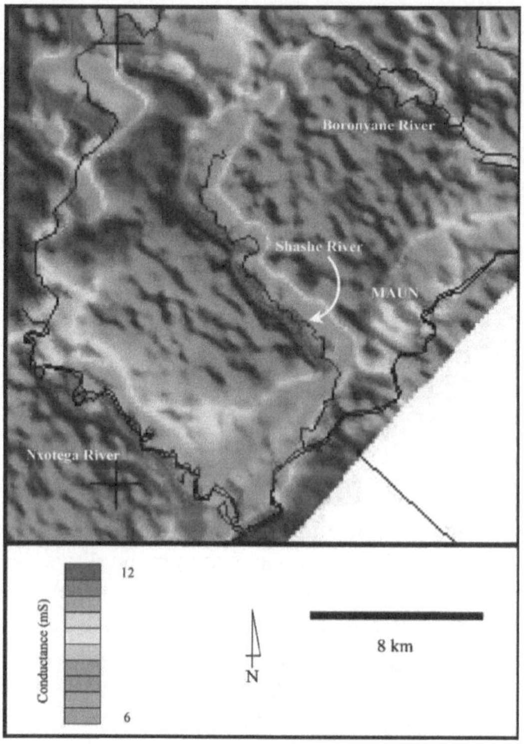

Figure 9. Freshwater (blue)/Saltwater (pink) distribution around SRV, seen from airborne EM (Reprinted from Bauer *et al.*, 2006 with permission from Elsevier).

Along those small outflowing channels, including SRV, fresh water has infiltrated in the shallow aquifers. These freshwater lenses can be clearly detected by airborne EM surveys (Figure 9). All channels act as recharge areas, while in the interfluve areas in between, highly saline water is almost stagnant. Hydrologic system boundaries are therefore given by the two fault systems perpendicular to a river and the parallel water divides between adjacent river valleys.

Where the groundwater has low salinity, and no seasonal flooding occurs, relatively vigorous riverine forest vegetation is present that satisfies its water demand at least partly by groundwater. Out in the interfluve areas, where the groundwater is highly saline, the vegetation cover is sparser and primarily rain fed (see Figure 8).

A vertical salinity distribution was mapped in a previous study by Water Resources Consultants, Botswana (Water Resources Consultants, 1997) using time domain electromagnetic sounding (TDEM). Results showed a superficial freshwater lens that extends down to a depth of approximately 60 m (Figure 10). Below that depth, the groundwater is highly saline, even in the centre of the river valley.

8.2 *Hydrogeology and water levels*

The main aquifer material of SRV consists of the wind blown Kalahari sands with a mean grain size of 250 micrometers. At a depth of about 200 m, the Kalahari sands are underlain by Ntane Sandstones, which are part of the late Permian Karoo formation. Locally, the

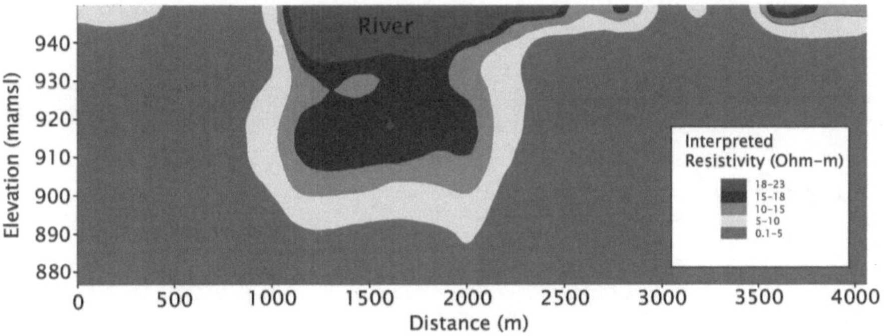

Figure 10. TDEM transect performed by WRC. (Reprinted from Bauer *et al.*, 2006 with permission from Elsevier).

Kalahari sands have been affected by precipitation of carbonates and silica, resulting in a highly cemented appearance and greatly reduced hydraulic conductivities. The main cemented horizons appear at two distinct depths, a first at about 20 m and a second at about 60 m depth.

Such hydraulic conductivity variations do not only appear vertically, but also horizontally. In the central part of the valley, the substrate consists of clean white sand, whereas in the interfluve areas the local substrate is presently affected by precipitation and is highly cemented. Therefore a significant contrast in hydraulic conductivity is expected between the fresh central part of the valley and its more saline surroundings. Measured horizontal hydraulic conductivities from pumping tests in the valley are around 1 m/day and in the interfluve areas usually less than 0.01 m/day.

Since SRV has been exploited for water supply over the last couple of decades, a number of boreholes were installed, both production wells and observation piezometers. The SRV study focused on a borehole transect consisting of 7 boreholes aligned between the centre of the river valley and the middle of the interfluve area. The general groundwater flow velocities in the longitudinal direction of the valley are small compared to the lateral flow velocities feeding the transpirative demand of the riverine forest. Historical water levels and water chemistry data, as well as recently acquired isotopic tracer data for the borehole transect, were used.

Figure 11 shows the water levels in the boreholes measured at different times. SRV used to receive regular flooding prior to 1991. After 1991, the river valley has remained dry. The combined action of pumping and transpiration in the absence of recharge lead to a general decline in water levels in SRV. Water level declines have been constant over time over the entire observation period (approx. 1 m/year). At the end of the last infiltration period in 1991 (i.e. when the Shashe River last carried water), the water level gradient was directed from the centre of the river valley out into the interfluve areas. That gradient has now been reversed due to transpiration and pumping in the absence of recharge from the river.

8.3 *Salinity and tracer data*

Figure 12 depicts superficial groundwater salinities along the research transect for two points in time in the years 1997 and 2003. In the upper aquifer layers, salinity is continuously increasing from the centre of the river valley towards the interfluve area. The boreholes

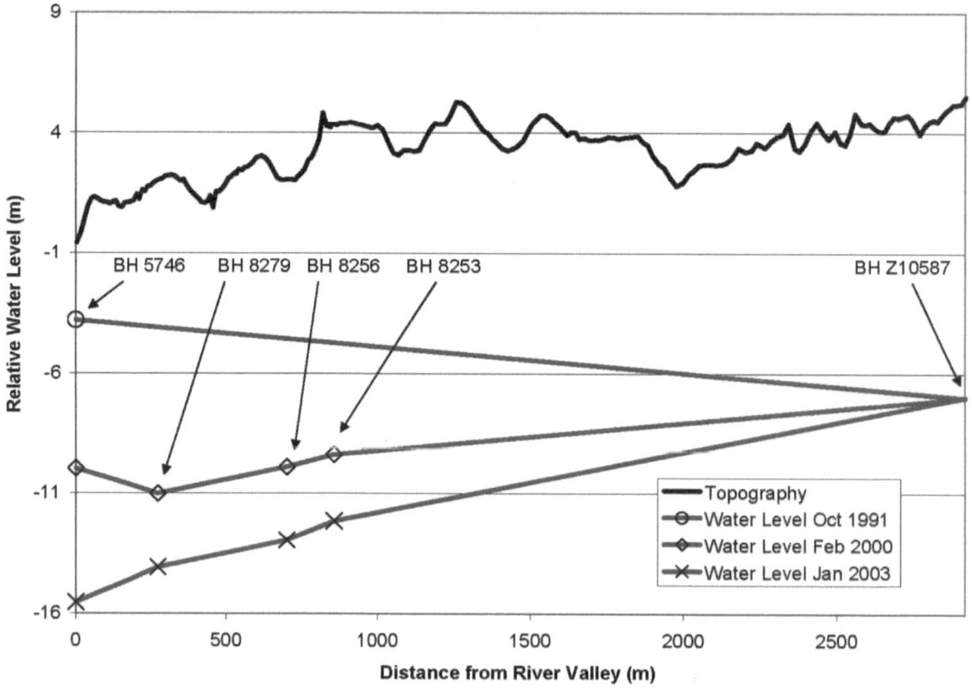

Figure 11. Water levels along the research transect for different points in time. (Reprinted from Bauer *et al.*, 2006 with permission from Elsevier).

located within SRV display salinity values close to the typical values of Okavango surface water. In the interfluve areas, the groundwater is highly saline and the total dissolved solid concentration is typically around 20 g/l. The transition zone between fresh and saline water is located near boreholes 8253 and 8256. The temporal sequence reveals that the saline front has advanced into the freshwater lens. This is due to the reversal of the hydraulic gradient, as indicated in Figure 11.

Stable isotopes (Deuterium and O-18) are enriched, if a certain water mass has been subject to evaporation. Since the Okavango waters experience evaporation during their slow passage through the Delta, surface water stable isotope signatures are elevated at the periphery of the Delta (Dincer *et al.*, 1978). In SRV, stable isotopes can thus be used to delineate the extent of swamp water infiltration Enriched water originating from the surface was found in borehole 5746. Boreholes 8256 and 8253 display values in transition to the background isotope signature, which is present in the interfluve borehole Z10587. Tritium is used as an age tracer for short time scales (half-life 12.3 years, Schlosser *et al.*, 1989; Solomon & Cook, 2000). The Tritium contents indicated recent waters in boreholes 5746 and 8256 and old waters in the other boreholes. C-14 has a half-life of 5730 years and can be used to date much older waters. Here the dissolved inorganic carbon was used for dating (Kalin, 2000; Mook, 1980; Plummer *et al.*, 1983). While Tritium just indicated pre-1970 waters in boreholes 8253 and Z10587, radiocarbon data gave a more detailed age distribution. BH 8253 is located close to the infiltration zone, the age difference between BH 8253 and 8256 being about 100 years. Borehole Z10587 had a relative age of several

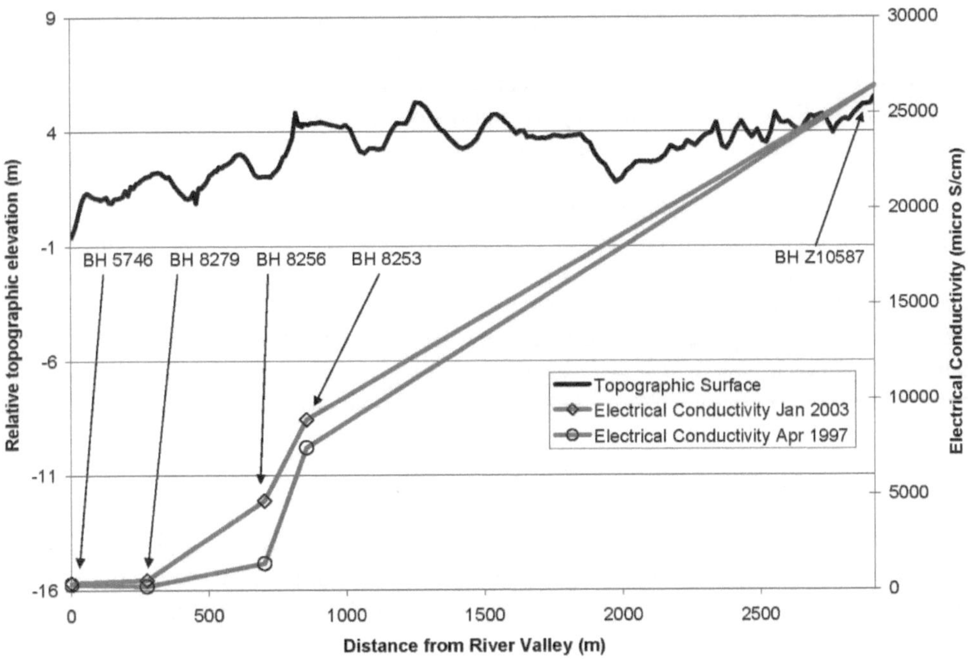

Figure 12. Groundwater EC along the research transect for 2 points in time (Reprinted from Bauer *et al.*, 2006 with permission from Elsevier).

thousand years, which indicated its isolation from the infiltration lens. In conclusion, all tracer data suggested a marked contrast in age and origin between the groundwater located within SRV and the groundwater located in the interfluve areas.

8.4 *Interpretation of the shashe river valley dataset*

The combined water level and water quality dataset from SRV was interpreted using a numerical groundwater flow and transport model of the river valley and its surroundings. The most important findings from the Shashe River Valley data, that the model was expected to reproduce, were the following:

Uniform drawdown in space in the entire river valley indicated that the water losses by transpiration were much more important than the water losses by pumping.

Uniform drawdown over time indicated that the transpiration rate was practically independent of the depth to groundwater. The rate of drawdown was the same in 2003 and in 1991, although the water table had fallen about 15 m in the meantime. Depth to groundwater table was thus never limiting the transpiration.

Although the depths to groundwater were almost the same in the river valley and in the interfluve areas, no transpirative groundwater losses occurred in the interfluve areas, as confirmed by the old ages of the interfluve groundwater. This indicated that transpiration in SRV was primarily limited by water quality.

Quantitative system modelling was performed and it was demonstrated that the hydrology of Shashe River Valley could only be understood, if the feedback mechanisms between transpiration and salinity were included in the model. Model results turned out to be most sensitive to the parameters governing the salinity dependence of transpiration (c_T and σ) and to the uptake concentration $c_{u,T}$. (see Bauer *et al.*, 2006 for details). A realistic and accurate representation of the plants' salt and water dynamics is therefore essential for successful predictive modeling of shallow river-fed aquifer systems.

The width w_{10} is about 500–600 m in the case of SRV. It can be estimated from the width of the phreatophytic vegetation belt around the river, as seen for instance from LANDSAT ETM satellite imagery Figure 8). If the system is assumed to be in steady state, w_{10} reflects the relative magnitudes of T_{max} and Q_{bound} and the salt tolerance of the local plant community (c_s, σ). The parameter set that yielded the best fit to the available observations in SRV was $T_{max} = 4 \cdot 10^{-4}$ m/d, $Q_{bound} = 5 \cdot 10^{-3}$ m^2/d, $c_s = 2.34$ g/l and $\sigma = 1$ g/l.

9 CONCLUSIONS

River-fed alluvial aquifers are key water resources in Africa and elsewhere in the world. They are controlled by phreatic transpiration. Phreatic transpiration is a function of depth to the groundwater and water quality. In the case of the Shashe River Valley, water quality is the main constraint for phreatic transpiration. In general, management of shallow river-fed aquifers involves trade-offs between water allocation to human use and water allocation to natural ecosystems. High transpiration losses lead to increased river water infiltration and consequently to water shortage downstream. Water supply wells placed in river-fed aquifer systems are generally vulnerable due to the presence of saline groundwater. The SRV wellfield for instance had to be replaced due to deteriorating water quality in the wells. Understanding the relationship between the salt and water uptake by plants and the state variables of the aquifer (hydraulic head and salinity) is a key issue for the sustainable management of water resources in semi-arid and arid regions. Further research should target the relationship between groundwater quality and transpiration for individual plants and for typical natural plant communities. More plant physiological and ecological research is also needed to achieve a realistic representation of the plant's and plant communities' salinity uptake mechanisms in standard groundwater models.

REFERENCES

Bauer, P. 2004. *Flooding and salt transport in the Okavango Delta: Key issues for sustainable wetland management*, Diss. No. 15436, ETH Zürich, Zürich.

Bauer, P., Held, R., Zimmermann, S., Linn, F. & Kinzelbach, W. 2006. *Coupled flow and salinity transport modelling in semi-arid environments: The Shashe River Valley, Botswana*. Journal of Hydrology, **316**, 163–183.

Bernstein, L. 1975. *Effects of Salinity and Sodicity on Plant-Growth*. Annual Review of Phytopathology, **13**, 295–312.

Brunner, P. 2006. *Water and salt management in the Yanqi Basin, China*. ETH Zurich, Zurich, Switzerland, 169 pp.

Coudrain-Ribstein, A., Pratx, B., Talbi, A. & Jusserand, C. 1998. *Is the evaporation from phreatic aquifers in arid zones independent of the soil characteristics*. Comptes rendus de l'Academie des Sciences, Paris, Sciences de la terre et des planetes, **326**, 159–165.

Dincer, T., Hutton, L. G. and Khupe, B. 1978. *The study of flow distribution, surface-groundwater and evaporation-transpiration relations in the Okavango Swamp, Botswana with stable isotopes* IAEA Symposium on Isotope Hydrology. IAEA., Neuherberg, Germany, 3–26.

Garcia-Sanchez, F., Botia, P., Fernandez-Ballester, G., Cerda, A. & Lopez, V.M. 2005. *Uptake, transport, and concentration of chloride and sodium in three citrus rootstock seedlings.* Journal of Plant Nutrition, **28**, 1933–1945.

Giordano, M. 2006. *Agricultural groundwater use and rural livelihoods in Sub-Saharan Africa: A first-cut assessment.* Hydrogeology Journal, **14**, 310–318.

Glenn, E., Thompson, T. L., Frye, R., Riley, J. & Baumgartner, D. 1995. *Effect of salinity on growth and evapotranspiration of Typha domingensis.* Pers. Aquatic botany, **52**, 75–91.

Grieve, A.M. & Walker, R.R. 1983. *Uptake and Distribution of Chloride, Sodium and Potassium-Ions in Salt-Treated Citrus Plants.* Australian Journal of Agricultural Research, **34,** 133–143.

Gumbricht, T., McCarthy, T. S. & Merry, C. L. 2001. *The topography of the Okavango Delta, Botswana, and its tectonic and sedimentological implications.* South African Journal of Geology, **104**, 243–264.

Guo, W. & Langevin, C. D. 2002. *User's guide to SEAWAT: A computer program for simulation of three-dimensional variable-density ground-water flow.* USGS, Techniques of Water-Resources Investigations 6-A7, Tallahassee, Florida.

Harbaugh, A. W. & McDonald, M., G. 1996a. *User's Documentation for MODFLOW-96, an update to the U.S. Geological Survey Modular Finite-Difference Ground-Water Flow Model.* USGS, Reston, Virginia.

Harbaugh, A. W. & McDonald, M. G. 1996b. *Programmer's Documentation for MODFLOW-96, an update to the U.S. Geological survey Modular Finite-Difference Ground-Water Flow Model.* USGS, Reston, Virginia.

Hoffman, G. J. 1980. *Salinity in irrigated agriculture.* In: M. E. Jensen (Editor), Design and operation of farm irrigation systems, ASAE, Beltsville, 145–185.

Hutchins, D. G., Hutton, L. G., Hutton, S. M., Jones, C. R. & Loenhert, E. P. 1976. *A summary of the geology, seismicity, geomorphology and hydrogeology of the Okavango Delta.* Geological Survey Botswana, Gaborone.

Kalin, R. M. 2000. *Radiocarbon dating of groundwater systems.* In: P. G. Cook & A. L. Herczeg (Editors), Environmental tracers in subsurface hydrology. Chapter 4. Kluwer Academic Press, Dordrecht, 111–144.

Kuper, A., Mullon, C., Poncet, Y. & Benga E., 2003. *Integrated modelling of the ecosystem of the Niger river inland delta in Mali.* Ecological Modelling, **164**, 83–102.

Lamontagne, S., Cook, P. G., O'Grady, A. & Eamus, D. 2005. *Groundwater use by vegetation in a tropical savanna riparian zone (Daly River, Australia).* Journal of Hydrology, **310**, 280–293.

Maas, E. V. 1986. *Salt tolerance of plants.* Applied Agricultural Research, **1**, 12–26.

McCarthy, T. S., Bloem, A. & Larkin, P. A. 1998. *Observations on the Hydrology and Geohydrology of the Okavango Delta.* South African Journal of Geology, **101**, 101–117.

McCarthy, T. S., Green, R. W. & Franey, N. J. 1993. *The influence of neo-tectonics on water dispersal in the northeastern regions of the Okavango swamps, Botswana.* Journal of African earth Sciences, **17**, 23–32.

Milly, P .C. D. 1984. *A Simulation Analysis of Thermal Effects on Evaporation from Soil.* Water Resources Research, **20**, 1087–1098.

Modisi, M. P., Atekwana, E. A., Kampunzu, A. B. & Ngwisanyi, T. H. 2000. *Rift kinematics during the incipient stages of continental extension: Evidence from the nascent Okavango rift basin, northwest Botswana.* Geology, **28**, 939–942.

Mook, W. G. 1980. *Carbon-14 in hydrogeological studies.* In: P. Fritz & J.C. Fontes (Editors), Handbook of Environmental Isotope Geochemistry. Elsevier Sc. Publ. Co., 49–74.

Persits, F. *et al.* 2002. *Map showing geology, oil and gas fields and geologic provinces of Africa, ver. 2.0.* Open File Report 97-470A, version 2.0, USGS.

Plummer, L. N., Parkhurst, D. L. & Thorstenson, D. C. 1983. *Development of reaction models for ground-water systems.* Geochimica et Cosmochimica Acta, **47**, 665–686.

Russell, R. S. & Barber, D. A. 1960. *The Relationship between Salt Uptake and the Absorption of Water by Intact Plants.* Annual Review of Plant Physiology and Plant Molecular Biology, **11**, 127–140.

Salim, M. 1989. *Effects of Salinity and Relative-Humidity on Growth and Ionic Relations of Plants.* New Phytologist, **113**, 13–20.

Schlosser, P., Stute, M., Sonntag, C. & Münnich, K. O. 1989. *Tritiogenic* ^{3}*He in shallow groundwater.* Earth and Planetary Science Letters, **94**, 245–256.

Scholz, C. H., Koczynski, T. A. & Hutchins, D.G., 1976. *Evidence of incipient rifting in Southern Africa.* Geophysical Journal of the Royal Astronomical Society, **44**, 135–144.

Sellars, C. D. 1981. *A Floodplain Storage Model Used to Determine Evaporation Losses in the Upper Yobe River, Northern Nigeria.* Journal of Hydrology, **52**, 257–268.

Skeel, R. D. & Berzins, M. 1990. *A Method for the Spatial Discretization of Parabolic Equations in One Space Variable.* Siam Journal on Scientific and Statistical Computing, **11**, 1–32.

Slater, J. A. *et al.* 2006. *The SRTM data "finishing" process and products.* Photogrammetric Engineering and Remote Sensing, **72**, 237–247.

Solomon, D. K. & Cook, P. G. 2000. ^{3}*H and* ^{3}*He.* In: P.G. Cook and A.L. Herczeg (Editors), Environmental Tracers in Subsurface Hydrology. Kluwer Academic Press, Boston.

Sutcliffe, J. V. & Parks, Y. P. 1987. *Hydrological Modeling of the Sudd and Jonglei Canal.* Hydrological Sciences Journal, **32**, 143–159.

Thomas, D. S. G. & Shaw, P. A. 1991. *The Kalahari Environment.* Cambridge University Press, Cambridge.

Water Resources Consultants 1997. *Maun Groundwater Development Project, Phase 1: Exploration and Resource Assessment.* Department of Water Affairs, Botswana, Gaborone.

Zencich, S. J., Froend, R. H., Turner, J. V. & Gailitis, V. 2002. *Influence of groundwater depth on the seasonal sources of water accessed by Banksia tree species on a shallow, sandy coastal aquifer.* Oecologia, **131**, 8–19.

Zheng, C. & Wang, P.P. 1999. *MT3DMS: A modular three-dimensional multispecies transport model for simulation of advection, dispersion, and chemical reactions of contaminants in groundwater systems.* Documentation and user's guide, US Army Corps of Engineers, Washington, DC.

CHAPTER 26

Three-dimensional modelling of a coastal sedimentary basin of southern Benin (West Africa)

M. Boukari
Faculté des Sciences et Techniques, Université d'Abomey-Calavi, Bénin Cotonou, Rep. of Bénin

P. Viaene
DHI Water & Environment, Denmark

F. Azonsi
Direction-Générale de l'Hydraulique, Bénin

ABSTRACT: Three-dimensional modelling of groundwater flow was carried out for the coastal aquifer system of Godomey (Benin, Western Africa), in order to determine the probable sources of salt-water intrusion in an area of intensive pumping. The model, under the assumption of steady-state with mean abstraction rates recorded over the period 1991–2000, was used to investigate the impact of current levels of abstraction (as of 2000) as well as the likely impact of a predicted increase in pumping to occur by the year 2011. The simulation results indicate that salt-water intrusion under current conditions comes only from Nokoué Lake, a shallow, salt-water lake. When a higher production rate was modelled (for the year 2011, again under the assumption of steady-state), it was observed that the potential for salt-water intrusion increases significantly, now including potential for intrusion not only from the lake, but also from recharge related to salt-water lagoons along the coast and possible salt-water upflow derived from deeper aquifers within the system. By taking into account the results of the model, an initial management strategy was developed to limit salt-water intrusion. These results are considered only as preliminary because of the assumptions of steady-flow and neutral density. Therefore management strategies are an interim measure during the period required to develop a more sophisticated management model that will include transient flow and density effects.

1 INTRODUCTION

Mathematical modelling is an excellent tool to study the hydraulics of groundwater systems subject to natural or anthropogenic stress (Gangopadhyay & Das Gupta, 1995). However, modelling of complex aquifer systems is a difficult task which can require enormous time and effort. Difficulties lie mainly in the design, calibration and validation of a model, starting from a limited quantity of data describing the characteristics and behaviour of the aquifer as well as the stresses on the groundwater system. The model must be designed in such a manner that it represents sufficiently the dynamics of the system of flow, such that it can guide both monitoring of the behaviour of the aquifer and the management of the groundwater resource.

The problem of salt-water intrusion in coastal aquifers is well-known throughout the world as challenging to model (e.g. Newport, 1977; Cheng & Chen, 2001; Zhang *et al.*, 2004). It is recognized that salt-water intrusion can cause significant economic and environmental problems. Moreover, corrective actions required to suppress this problem are often extremely expensive (Cheng & Chen, 2001). Unfortunately, the complex interplay among density effects, limited data, and complex geological formations makes precise modelling of salt-water intrusion challenging. As a result, initial modelling, neglecting density effects, may provide a tool with which to outline short-term management strategies for aquifer protection during periods in which more complete characterization and development of a density-dependent model is being pursued.

The Bénin population is evaluated to 6,752,569 inhabitants by the third General Census of the Population and the Habitat in year 2002 (INSAE, 2003). Life expectancy at the birth is currently 54 years and the distribution by sex indicates 51.4% women. Bénin

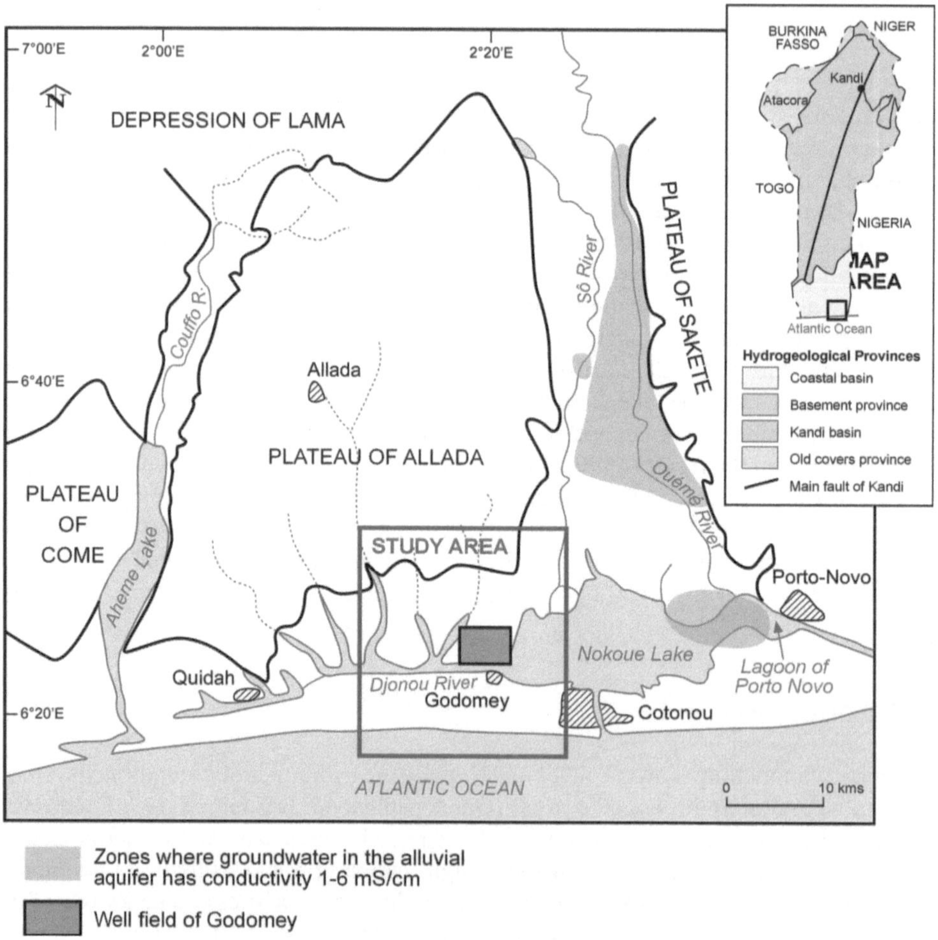

Figure 1. Bénin hydrogeological provinces and study area location.

(Figure 1) is a relatively well watered country, whose average annual rainfall is between 900 and 1300 mm. A great part of the rainfall is taken away by evapotranspiration (potential evapotranspiration is 1350 mm in the south and 1750 mm in the north). The remaining water is partitioned between runoff to a relatively dense river network and recharge to groundwater. The principal rivers are tributary of the great Niger river in the North, and of the Atlantic Ocean in the South (Ouémé and Couffo in particular, via Nokoué lake and Ahémél Lake respectively, Figure 1).

According to the Document of National Policy of Water (MMEH, 2005), the internal and external contributions to Bénin water resources are estimated to approximately 25 billions cubic metres per annum. The same document notes that the increase in the population would reduce, in a drastic way, the quantity of water available per capita, by supposing that these resources remain constant. In 1955 for example, the population of Benin amounted to 2,111,000 inhabitants and the availability in water resources per capita and per annum was 12,622 m^3. In 1990, this population was 4,622,000 inhabitants, and the availability in water resources per capita and per annum was not any more 5825 m^3. According to conservative estimates (MECCAG, 2000a, 2000b), the population of Bénin could reach 11,337,000 inhabitants in 2025, which would reduce the quantity of water available per capita and per annum to 2293 m^3; less conservative estimate of population increase would reduce the water available per capita to less than 2000 m^3. These estimates assume that water availability in Benin remain the same, and do not decrease as a result of climatic variability.

These estimates of future water availability would place the Bénin in the category of the countries with grave, even catastrophic shortage of water. Such a situation is likely to generate serious problems between the various users, if preventive measures are not taken in time, for controlling withdrawals and well managing the wastes. This control will have to be done as well with regard to the internal resources suitable for Bénin, but also to the resources shared with the close countries. Within this last framework, a project entitled "Joint Management of Coastal Aquifers in the Gulf of Guinea" was initiated between five countries (Benin, Côte d'Ivoire, Ghana, Nigeria and Togo), for the coastal transboundary aquifer systems located along the Gulf of Guinea shared by these five West Africa countries, under the aegis of UNESCO and UNEP. This project is currently submitting to GEF for financing. Moreover, on the strategic level, Bénin opted for an integrated approach to management of its water resources. The action plans were worked out and are currently implemented to achieve the Millennium Development Goals, as regards drinking water and sanitation (Direction Générale de l'Hydraulique, 2000; MMEH, 2006a, b).

With the present stage of knowledge of the various hydrogeological units and respective local climatic conditions, the total mean annual recharge of the aquifers of Bénin is estimated to approximately 1870 Mm^3 (Direction de l'Hydraulique, 2000). According to the same source, in 2003, the quantities exploited annually for drinking water reach only approximately 30 Mm^3 in urban environment and 45 Mm^3 (10,000 functional production wells approximately, at a rate of 12 m^3/day per well) in rural areas, for which it is necessary to add approximately 45 Mm^3 of informal withdrawal rate in urban environment as in rural areas, so be it, on the whole, approximately 120 Mm^3. These exploited water quantities come, for more than 97%, from the groundwater resources. The water quantities used for irrigation and industry are not known, but remain relatively low.

Four hydrogeological provinces can be identified within Bénin, according to the type of porosity and the geographical location of the formations (Figure 1): (i) a vast central province, with dominant migmatito-gneissic and granitic formations, corresponding to the "Structural unit of Bénin plain" (Basement SS hydrogeological province), (ii) a north-western province, with dominant quartzito-sandstone and schisto-pelitic formations, corresponding to the "Tectono-structural unit of Atacora and its (tabular) foreland" (Old covers hydrogeological province) and two hydrogeological provinces with dominant sandy, conglomeratic and argillaceous formations, corresponding, on the one hand, (iii) to the "Sedimentary basin of Kandi" (hydrogeological basin of Kandi) in the North-East, and on the other hand, (iv) to the "Coastal sedimentary basin" (coastal hydrogeological basin) in the South. The first two provinces consist of discontinuous aquifers where fracture flow dominates; the two last are broadly continuous aquifers with primary porosity and intergranular flow.

On the whole, the guarantee of the availability of water at lower cost for the present and future generations in Bénin, poses many current and emerging problems, among which the crucial problem of the hydrogeological knowledge of the country. In addition, some of the aquifers are threatened by pollution, because of anthropogenic activities. Groundwaters begin to be polluted by domestic, agricultural, even industrial waste (Sagbo, 2000), in addition to the problem of salt-water intrusion (Boukari, 1998).

In Benin, salt-water intrusion has affected aquifers in all coastal regions of the country. The present discussion is focused on one such area, the Godomey zone of south-central Benin (in the region of the population centre of Cotonou). This area is indicated in Figures 1 and 2. The drilling of the first production well in the study area dates from 1956 (Slansky, 1962; SGI, 1981). Pre-production values of the hydraulic heads in this area are unknown. Since 1956, the number of production wells has gradually increased. For example, 8 additional wells were drilled during the period of 2001–2002, thus placing at 24 the number of production wells in the region, each pumping 22–24 hours per day. The pumped discharge has continuously increased, from 5000 m^3/d in 1970, to 40,000 m^3/d in 2001 and to 50,000 m^3/d in 2004. The impact on the resource of such intensive pumping in a region of ten square kilometers is observed in the form of a progressive decline of the heads in observation piezometers (Boukari, 1998) and, since the end of 1980, by salt-water intrusion in select production wells (Boukari, 1998; Gnaha & Adjadji, 2001). The salt-water intrusion continues to progress, successively impacting the four production wells closest to Lake Nokoue. In spite of these observations, the company responsible for managing the wells has hesitated to limit production based on economic arguments. This is in part due to the fact that the study area is the closest to end users of the water supply (e.g. ten kilometers from Cotonou) and provides very good discharge (mean discharge of 100 to 150 m^3/h per well).

Prior studies have been undertaken in this region of Benin in order to understand the hydraulic and hydrochemical conditions present in this region (Pallas, 1988; Turkpak International-SCET-Tunisie, 1991; SOGREAH/SCET-Tunisie, 1998). However, these models were integrated over the entire southern region (large scale) and generally modelled this complex aquifer system as a single aquifer.

This paper presents the results of the use of MODFLOW (MacDonald & Harbaugh, 1988) and MODPATH (Pollock, 1989) to analyze this multilayered aquifer system under the initial assumptions of constant density and steady-flow. The main objective of modelling this system under these assumptions was to obtain an initial conceptual understanding of flow within this system with specific attention paid to zones of recharge and discharge. Such a conceptual model will provide an initial estimate of the impact of intensive pumping on system behavior (e.g. drawdown, water balance) and identification of the probable source(s) and direction(s) of movement of salt-water intrusion. While it is recognized that the assumption of steady state in the absence of density effects does not adequately describe this system, this relatively simple initial modelling effort helps to identify the need for immediate management, thus allowing time for development, calibration and verification of models that will include transient behaviour, detailed recharge and complex geologic characterization, as well as the impact of variable density. Hence, the current effort represents a transition tool between the current condition of zero assessment and the future development of a fully three-dimensional, transient, density-dependent effort.

2 HYDROGEOLOGICAL SETTING

This study was undertaken in the region of intensive groundwater extraction near Godomey, located near the south-eastern edge of the plateau of Allada, in the coastal sedimentary basin of Benin (Figure 1). This region is exploited to supply drinking water for the city of Cotonou, located 10 km to the east. Cotonou and the surrounding urban development (with approximately 1 million inhabitants) represent the largest centre of population and industry in Benin.

The plateau of Allada covers approximately 60 km in the north-south direction by 40 km in the east-west direction. Lithologically, as it shown in Figure 2, it is comprised of three layers (a sandy-clay layer on the top overlying a clayey-sand which, in turn, overlies a sand layer) of Pleistocene-Pliocene-Miocene age, resting in angular unconformity on a clay/marl Eocene substratum (Slansky, 1962; Istituto Ricerche Breda, 1987; Boukari, 1998). Of ochre red colour and having a mean thickness of 15 m, this sandy-clay layer is locally called 'Terre de Barre' (TB). Taken together, the layers comprising the plateau have a combined thickness of approximately 50 m to the north grading to 150 m to the south with a south-eastern dip. Geologists in western Africa refer to these layers as the 'Continental Terminal' (CT).

The littoral plain (LP), of Quaternary age, commences on the southern edge of the CT and provides a geological connection to the ocean (Figures 1 & 2). Within the LP and at the southern edge of the CT the lithological structure becomes more complex, marked by the presence of clay layers of variable thickness and continuity (Slansky, 1962; Istituto Ricerche Breda, 1987; Oyédé, 1971; Boukari *et al.*, 1995). As observed through existing well logs, the upper 150–200 meters of the LP consists of a homogeneous, sandy layer (20 to 30 m thick) resting on a clay layer which varies in thickness from north to south between 15 and 30 m. Below the clay exists a relatively thick and heterogeneous layer, alternating between sand or clayey-sand layers separated by thin layers of clay (5 to 10 m) which are more or less continuous.

The eastern and western edges of the plateau are marked by two large depressions (Figure 1): that of the river system Ouémé-Sô-Nokoué Lake in the east and that of the river system Couffo-Ahémél Lake in the West. These lakes and rivers contain salt or brackish water. Indeed, the rivers flow into the Nokoué lake and the Ahémél Lake respectively (in the north), which, in their turn, communicate with the Atlantic Ocean in the South. During the rainy season, fresh water from the rivers pours into the lakes, while during the dry season, it is the oceanic water which runs towards these lakes. The increase of brackish waters within the lakes then causes backward flow towards the north in the rivers (on 25 to 30 km from the Sô and Ouémé mouth in particular; Texier and Colleuil, 1980). The geology of these depressions is comprised of alluvial deposits that show a heterogeneous lithological structure with significant clay layers alternating with sand.

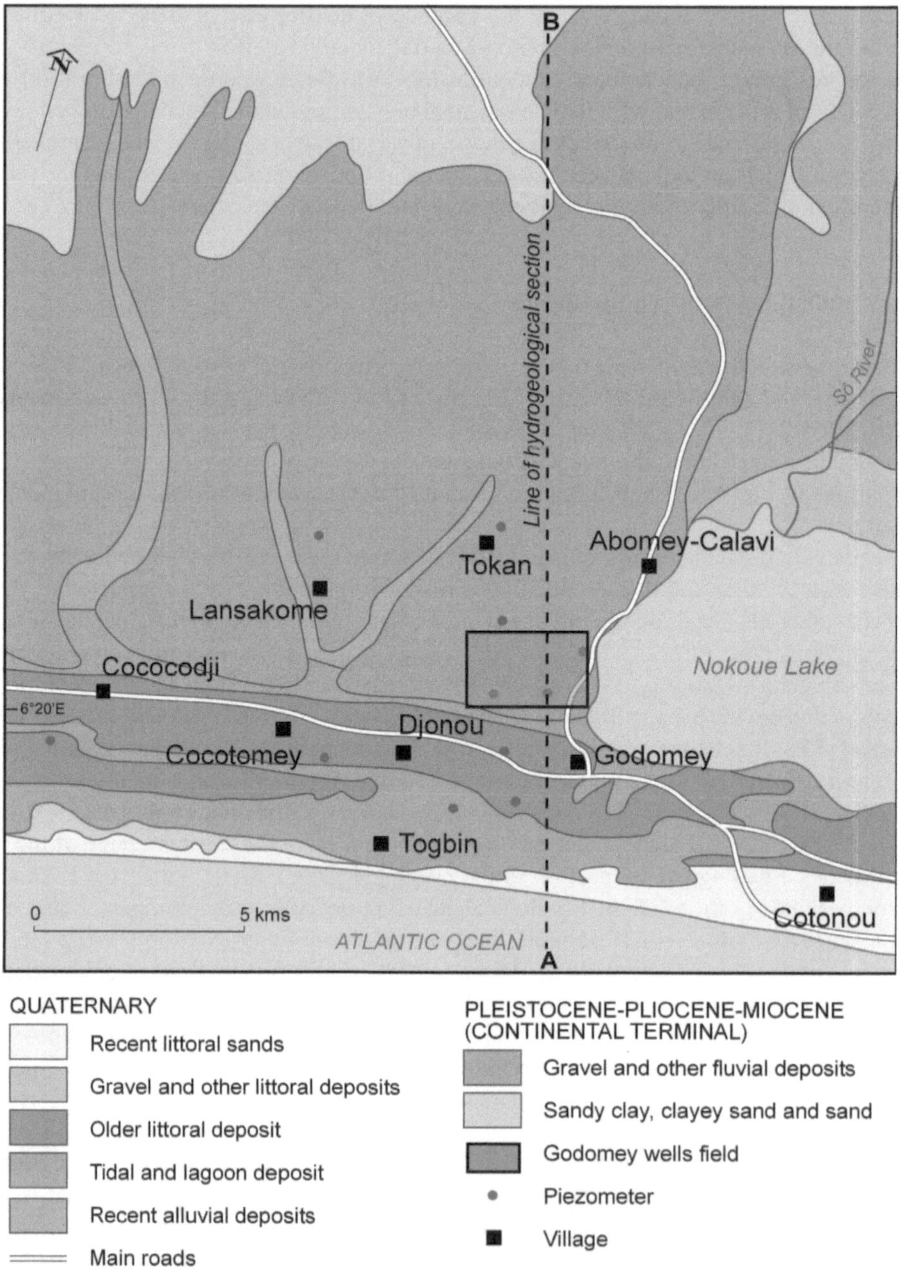

Figure 2. Geological map of the Godomey area with observation wells.

The CT is an unconfined aquifer on the major part of the plateau but, in certain locations, is confined by local clay layers. In the southern portion of the CT and within the LP, however, the system becomes more complex with the unconfined aquifer gradually transitioning in the southerly direction to form a layered system which can be subdivided into three distinct aquifers. Hydraulic continuity between the aquifer of the northern CT and the layered aquifer system of the southern CT and the LP system has been documented by several investigators (SGI, 1981; Maliki, 1993, SOGREAH/SCET-Tunisie, 1998; Boukari, 1998).

The zone of production (Godomey wells field) is within the layered aquifer system in the southeast portion of the CT and covers a surface of area slightly less than 9 square kilometers (Figure 2). As of 2001, 19 production wells remained in production; 3 wells in the eastern portion of the field, had been abandoned due to salt-water intrusion (due to concentrations of NaCl exceeding 250 mg/l). A fourth well in this same region has demonstrated increasing salt concentration, but had not yet been abandoned as of 2001.

Daily outputs from each of the wells are known. Response of the aquifer system to pumping has been monitored at 32 piezometers (with a weekly frequency in the rainy seasons, and semi-monthly in the dry seasons) over the period 1991 through 2001. For the upper water-table aquifer, relatively few piezometers are available. As a result, piezometer data were supplemented with data from large-diameter wells which were monitored between 1991 and 1993. This use of limited (time) data was justified by the observation that the mean heads in the available piezometers in the upper aquifer for the period 1991–1993 were not significantly different than the heads in these same piezometers averaged over the period 1991–2000.

3 THE GROUNDWATER MODEL

3.1 *Conceptual model*

The numerical model, as delimited on the basis of geological, hydrological and piezometric criteria, covers a land-surface area of approximately 750 km^2 (Figure 2). It extends approximately 27 km to the north of the coastline to a boundary arbitrarily defined by the mean piezometric contour +15 m amsl. In the south, the model extends beyond the coastline to the mean bathymetric contour of −10 m, which coincides with approximately 4 km at the greatest distance into the ocean. It is bounded to the east by a combination of Nokoué Lake, the channel connecting Nokoué Lake to the ocean within the town of Cotonou, and the Sô river. It extends to the west to slightly beyond the Dati valley. The ArcView/GeoEditor software (GeoEditor, DHI Water & Environment unpublished data, 2002) was used to construct a conceptual geological model of the site (through interpolation of features observed on local lithological exposures).

The system was modelled as a homogeneous, horizontally-isotropic ($K_x = K_y \neq K_z$), three-dimensional aquifer system with seven layers of which four are aquifers and three are confining layers. Utilizing Visual MODFLOW, a variably spaced grid was constructed (Andersen *et al.*, 1992; Ebraheem *et al.*, 2004). The highest level of discretization was used in the regions surrounding production wells with elements of size approximately 230 m (N–S) by 250 m (E–W). Element dimensions gradually increased with distance from the wells to a maximum dimension on any element of 1850 m along the edges of the grid. It consists of 55 rows and 50 columns, for a total of 2750 elements per layer. Thickness of the layers varies from cell to cell. Absence of a layer was modelled by introducing a very small thickness of the layer in those cells where the layer is absent.

Under the natural conditions, the flow general direction of groundwater is directed SSW. The whole of the system was thus turned of an angle of approximately 15 degrees so that the flow lines are parallel to the grid.

Drilling data from village and urban water supply wells as well as geological cross sections (IGIP-GKW-GRAS, 1989; Maliki, 1993; Boukari *et al.*, 1995; Boukari, 1998) were digitised and entered into a GIS database. Hydrogeological data such as surface elevation and measured groundwater levels were interpolated with GIS tools (Gossel *et al.*, 2004). The resulting database was used to calculate the model layers bottoms, tops and thickness.

3.2 *Boundary conditions and parameters of the model*

To the north, the first two layers (unsaturated and/or non-existent in this area), as well as the lowest 2 layers, are inactive cells. The third to fifth layers are set as constant heads to coincide with a mean piezometric contour of 15 m amsl as identified from analysis of historical field data. To the south, those elements of the first layer which underlie the ocean are assigned hydraulic heads of 0 m. No flow boundaries are used on the southern edge of the grid for all other layers. It is recognized that this boundary condition imposes vertical flow that is unlikely to match reality. This boundary condition will be further studied in extensions of this work that include density and transient effects. A constant head of 0.5 m is assigned for all elements in the top layer underlying Nokoué Lake, as well as at all elements along the channel connecting Nokoué Lake to the ocean and the Sô river. Layers 2–7 along the eastern boundary are set as no flow conditions based on the assumption that the center of Lake Nokoue, combined with the channel and the So river, represents a groundwater divide in the East-West direction. Further study would be required to assess this assumption. No flow conditions are applied along the West and North-West edges of the model. These conditions are justified based on observations of the distribution of the piezometric surface in these regions.

The eastern part of the river Djonou, which is the nearest river to the well field, is incorporated into the top layer using the river option in MODFLOW. The Bakamé and Dati rivers and other depressions in the LP and in the region along the eastern border of the plateau were modeled as simple drains using the drain option in MODFLOW. The top layer of the model was simulated as a free surface such that the water-table was able to fluctuate in response to recharge.

Estimates of total porosity for the sand and clayey-sand layers of the CT, lie between 39 and 42%, with a mean value of 40% (SGI, 1981). The porosity of the surface layer of the TB is between 34 and 36% (GIGG, 1983). Estimates of the coefficient of storage of the lower aquifers lie between 10^{-3} and 10^{-5} (IGIP, 1989; TurkPak/SCET-Tunisia, 1991). For the LP, the total porosity of the littoral and dune sands (for which the thickness reaches 6 m) is likely higher than 40% with a 20% specific yield. The total porosity is estimated to be 35% (7 to 10% specific yield) in the underlying fine sands.

Transmissivity data are available for several zones within the model, particularly the lower aquifers which are subjected to intensive pumping (Slansky, 1962; Pallas, 1988; Boukari, 1998). Specifically, there have been pumping tests run on each of the wells drilled in the Godomey well field. The values of transmissivity in general lie between 10^{-3} and 10^{-2} m^2/s.

Mean precipitation varies spatially, increasing from approximately 1000 mm/a in the northern region of the study site to 1200 mm/a along the coastline. There is also substantial variability in annual precipitation. Based on the variation in precipitation, several zones are commonly identified in estimating recharge with rates varying between 20 and

110 mm/a (Pallas, 1988; Maliki, 1993), with one study identifying rates as high as 300 mm/a (Hohenheim, *unpublished data*).

For the steady-state simulations, the respective means of the monthly yield for each well during the period 1991–2000 were calculated and applied to the model. These range for the production wells from 500 to 1800 m^3/d. This period coincides with the period for which measured hydraulic heads are available. In addition to withdrawals from the production wells, a diffuse withdrawal occurs in this region due to construction and use of a multitude of private wells (in general with large diameter). As the location and yield of these private wells is generally unknown, this form of pumping is not included as a sink. It is implicitly assumed here that the impact of these wells is negligible in comparison to uncertainty in the magnitude and distribution of recharge.

3.3 *Calibration of the groundwater flow model*

Calibration of the model at steady state was performed through a trial-and-error iterative process, using the mean well yields measured in the Godomey aquifer system from 1991 to 2000 as known production rates. Mean measured piezometer heads for the period 1991–2000 were used as the comparison data set. Initial estimates for the horizontal hydraulic conductivities (K_x and K_y), vertical hydraulic conductivity (K_z), conductance for the surface water features (lagoons, rivers and shallows) and recharge were based on analysis of the existing data.

As indicated in Figure 3, four lithological types were defined: two sands and two clays. Estimates of the conductance for the surface-water features were determined within MODFLOW based on estimates of thickness and vertical hydraulic conductivity of the sediments

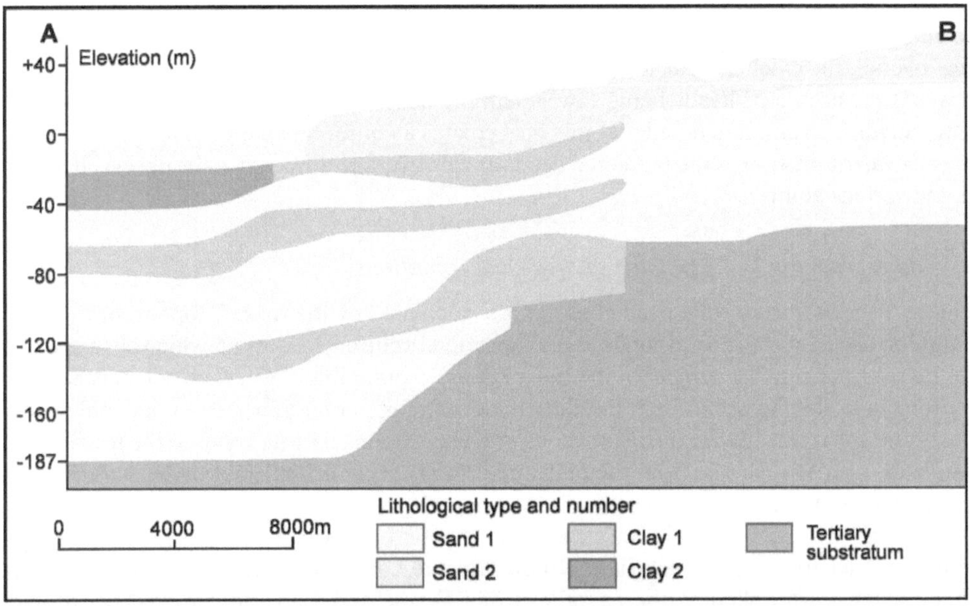

Figure 3. Hydraulic conductivities along cross section A-B of the Godomey system aquifer as they are implemented in the numerical groundwater model.

Table 1. Hydraulic conductivities along cross section A-A′ of the Godomey system aquifer as they are implemented in the numerical groundwater model.

Lithological types	Horizontal conductivities (m/s)	Vertical conductivities (m/s)	Storage coefficients	
			S_s	S_y
Sand 1	2×10^{-4}	2×10^{-5}	10^{-6}	0.15
Sand 2	10^{-5}	10^{-6}	10^{-5}	0.1
Clay 1	10^{-6}	10^{-7}	–	–
Clay 2	10^{-7}	10^{-7}	–	–

underlying these features. By taking account of the values of infiltration measured or estimated by various authors, records of precipitation, and the sediments and thickness of the unsaturated zone, recharge (assumed to be uniform in space within three zones on the surface of the aquifer not overlain by the ocean) was estimated.

The hydraulic conductivities and rates of recharge were adjusted so as to reproduce in a reasonable fashion the mean observed heads in the piezometers. This calibration effort led to estimates of recharge of 25 mm/a for the northern part of the plateau, 100 mm/a for the intermediate zone and 200 mm/a for the littoral low zone. Further, the hydraulic conductivities of the four lithological units were estimated to have values shown in Figure 3 and Table 1.

It is recognized that the steady-state model will not represent transient behavior, particularly in the region immediately around the production wells. The predicted heads near the wells will tend to be lower than those observed in the field due to the steady-state assumption. Despite this, the calibration of the steady-state model appears satisfactory (Boukari, 2004, 2006) as supported by the results shown in Figure 4. This figure shows that overall; the calculated heads correlate well with the observed heads, with a slight bias towards the predicted heads being lower than the observed heads and a standard error of estimate of 13.8 cm. It is noted that this bias creates a conservative prediction with respect to salt-water intrusion as the predicted heads are, on average, lower than the observed heads in the various aquifers.

3.4 *Interpretation and extension of historical conditions*

Figure 5 shows steady state head contours, as calculated by the model, for two of the four aquifers: layer 1 which corresponds to the superficial aquifer, and layer 5 which corresponds to the confined aquifer subject to the most intensive pumping. Several prominent features are shown in the Figure 5. First, the flow lines in the upper aquifer lead to the conclusion that there are at least two significant zones of recharge contributing water to the production wells: (a) diffuse recharge from the plateau north of the well field, and (b) recharge in the region south of the well field north of the Djonou lagoon, thus suggesting a groundwater divide between the well field and the ocean. Second, there appears to be four primary zones of discharge: (a) the well field, (b) the Nokoué Lake-Sô-Ouémé river system, (c) the topographic lows linked to the as valleys of Bakamé and Dati, and (d) the ocean and depressions linked to the coastal lagoons (south of the apparent groundwater divide). Third, the centre of the cone of depression resulting from the well field is predicted to have a steady

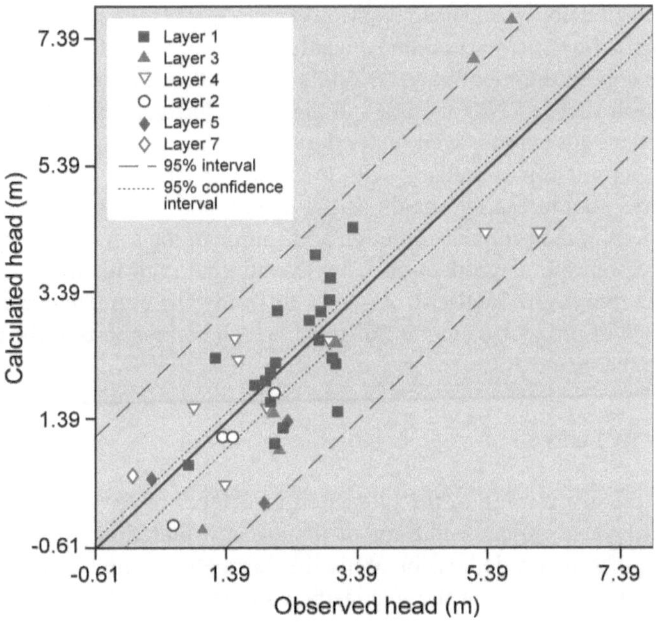

Figure 4. Calculated vs observed heads for the calibration effort.

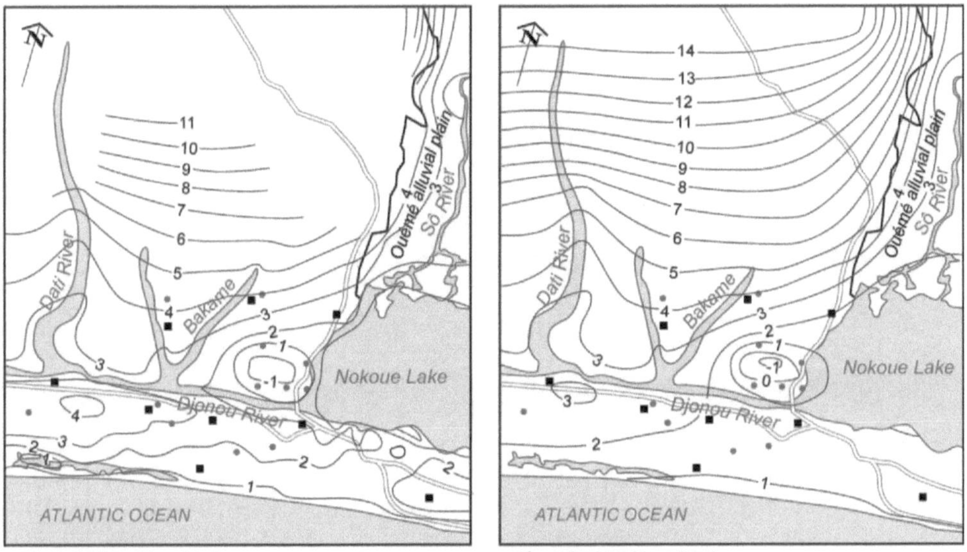

Figure 5. Contours of the calculated heads at the water table aquifer or layer 1 (left) and the second lower confined aquifer or layer 5 (right) for the steady state, corresponding to the period of 1991–2000 mean yields of the pumping wells.

state hydraulic head below mean sea level (lower than −1 m in the superficial aquifer and lower than −2 m in the confined aquifer). Fourth, heads are predicted to be below sea level at some locations in all the model layers. While the predicted cone of depression extends only to the western shore of Nokoué Lake in the unconfined aquifer (due to the specified boundary conditions), it extends well under the lake in the confined aquifer.

Based on this simulation, a scenario with increased production from the well field was simulated in order to address two goals of the overall study of this aquifer system. The first goal is to identify the probable direction and source of the salt-water intrusion, as well as its dependence on rate of production. The second goal is to identify the likely impact of an increase in rate of production from the well field. The simulated scenario involved increasing the rate of production to 56, 262 m^3/d, which corresponds to the predicted level of production in the year 2011.

4 SCENARIO RESULTS

Figure 6 shows the distribution of the piezometric surfaces of the main aquifers, the first and fifth model layers, under conditions of increased production. Results illustrate that the steady state piezometric depression along the perimeter of the region of pumping will increase, reaching −6 m amsl in the unconfined aquifer and −8 m amsl in the confined aquifer (layer 5 in the model). Clearly the increase in rate of production will have dramatic effect on the distribution of hydraulic head (e.g. Gaus, 2000).

MODPATH was used to provide further insight into both the original simulation and the simulation with the increased rate of production. In particular, MODPATH is used in this

Figure 6. Contours of the calculated heads in the water table aquifer or layer 1 (a) and the second lower confined aquifer or layer 5 (b) for the steady state corresponding to the higher rate of production (year 2011).

context to define potential flow routes to the production wells and likely zones of recharge impacting the production wells. Specifically, reverse particle tracking was used to follow particles from the production wells backwards in time to their recharge locations.

The furthest extension of these lines away from the well field in the unconfined aquifer is interpreted as identifying the region of primary recharge impacting the well field. This interpretation is supported through examination of vertical cross-sections which indicate where flow paths intersect the surface (discussed below). The furthest extent of the lines away from the well field in the confined aquifer is an indication of the limit of vertical leakage into this aquifer from the overlying confining layer. It is noted that, in general, the pathlines for the unconfined aquifer indicate local recharge for the unconfined aquifer (contributing to the production wells). The pathlines for the confined aquifer indicate an unbalanced source of water (larger zone of contribution from the north) even at depth within the aquifer system.

Figure 7 shows north-south vertical cross sections through the zone of production. These results indicate that the majority of water produced at the wells under both current rates of production and increased rates of production is derived from surface recharge, including both local (shallow aquifer), intermediate (mid-level aquifers), and regional contributions (deep aquifer). In comparing Figures 7a and 7b, it is noted that the zone of recharge captured by the production wells is appreciably wider under increased production. However, even under increased pumping, the model predicts the presence of the groundwater divide in the LP north of the coastline. Although this latter observation implies that there may not be significantly increased risk of salt-water intrusion directly from the ocean, it does indicate that water in the region of the lagoons, which is brackish, will represent a larger contribution to the recharge impacting the wells, thus increasing the threat of salt-water intrusion from this source.

Figure 8 presents vertical cross-sections trending east-west through the production zone. These cross-sections provide a more dramatic portrayal of the extension of the capture area of the production wells under the condition of increased pumping. Under the conditions of historical mean production (Figure 8a), contributions to the production wells from the region bounding the lake are limited primarily to surface recharge along the border of the lake and from upflow from the lower aquifers. When conditions are extended to projected production for the year 2011 (Figure 8b), the model predicts significant contribution of water directly from Lake Nokoue, thus representing a major, long-term threat of salt-water intrusion. This threat takes on two forms. First, relatively rapid influx of salt-water is possible in the upper aquifers through direct recharge of water from Lake Nokoue. Second, the model indicates that, if Lake Nokoue represents a groundwater divide, recharge from the lake may impact the quality of water in the lower aquifers over longer time periods, thus representing a long-term threat to water produced at the well field.

Water balances were calculated for five zones as defined in Figure 9: the zone corresponding with the surface area of the cone of depression in the superficial aquifer (original simulation), and four zones bounding the exterior perimeter of the first zone and covering the remainder of the model grid. Table 2 presents the results of these calculations for both simulations. Under the assumption of historical mean pumpage, the model predicts that a total of 35,589 m^3/d of water moves through the aquifer system. Interestingly, the predicted recharge within the calculated zone of depression for the upper aquifer is 23,577 m^3/d. The implication of this result is that the majority of the water withdrawn from the deeper aquifers under historical pumping conditions is of local origin, with minimal volumetric

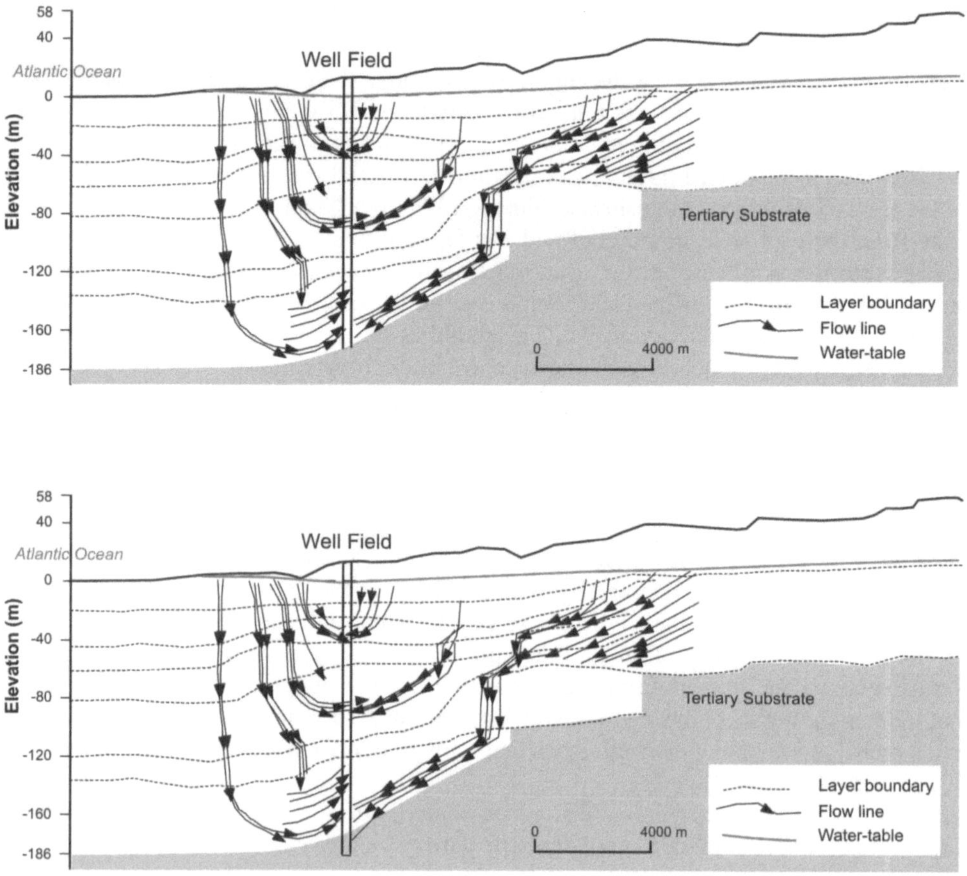

Figure 7. Steady state pathlines in North-South trending, vertical cross-sections calculated using the historical mean well yields (a) and predicted for the projected increase in well fields representing water use in 2011(b).

contribution from the more distant sources of recharge. Further, the contribution of the region of Lake Nokoue to the total influx to the aquifer system, under historical pumping, is a very small percentage of total recharge to the aquifer.

The situation changes dramatically when under conditions of increased pumping. Under higher production, the rivers and drains contribute substantial recharge to the aquifer. Further, the contribution from the region of Lake Nokoue increases dramatically (nearly an order of magnitude) to become comparable to the contributions of the northern and southern zones. Although less significant, the contribution of the southern zone also increases (accounted for primarily by contributions along the eastern portion of this zone). Hence, the model predicts that the increase in the production at the well field will be balanced primarily by increased inflows from the region of Lake Nokoue and the LP in the south, i.e. the two zones which are of greatest concern with respect to salt-water intrusion. These zones represent threats not only from the point of view of the potential direct intrusion of salt-water from surface water with high salt content, but also from the increased rate of development of the surface lands which therefore accelerates reduction in local recharge

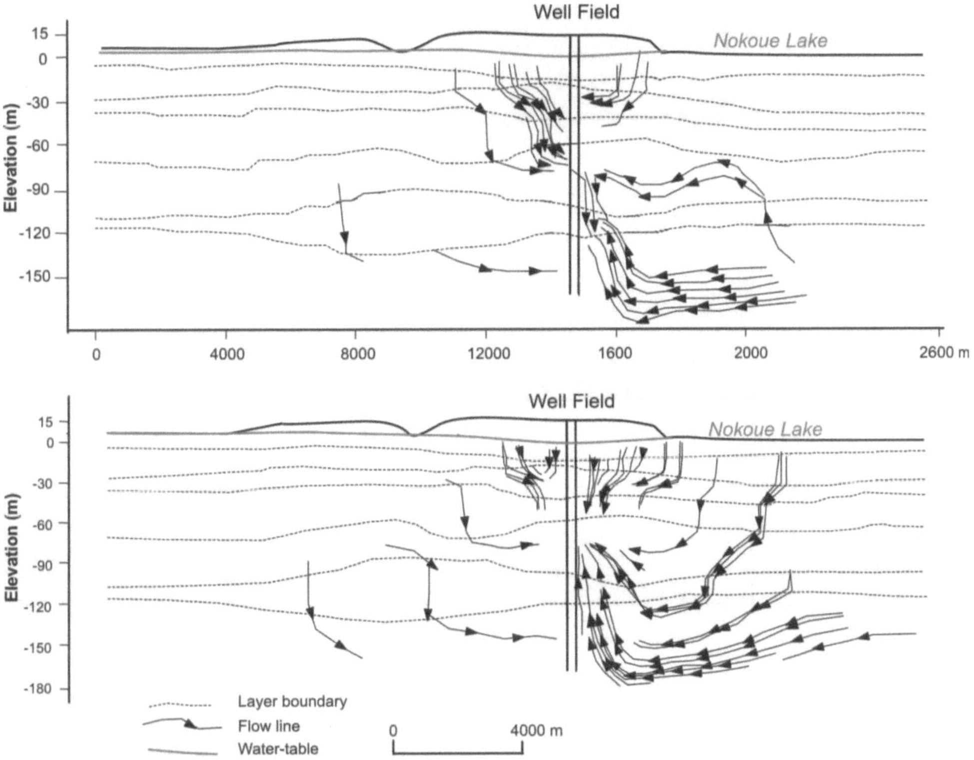

Figure 8. Steady state pathlines in vertical cross-section oriented along the east-west axis through Godomey, calculated (top) for the mean rate of production during the period of 1991–2000 and (bottom) for the projected well production in 2011.

and introduces new sources of contamination. Specifically, the zones of concern coincide with the most desirable lands surrounding the town of Cotonou. Finally, these results may underestimate the impact of increased production due to an observed pattern of reduced precipitation, and therefore recharge, observed for this region and apparently related to climatic change (Vodounon-Totin, 2003).

5 IMPLICATIONS FOR WATER RESOURCE MANAGEMENT

As noted, these initial modelling efforts demonstrate several general aspects of problems associated with continued production of groundwater in the study region. These include infiltration of waters from Lake Nokoue and the coastal lagoons and increased contribution of water, with increasing production, from these regions which contain salt-water and/or brackish water. These observations lead to several observations regarding management of this aquifer system:

1. The rate of production within this current zone should be held constant or decreased during the time in which a more sophisticated model can be developed, calibrated and verified. The result of this modelling effort provides substantial support for the possibility

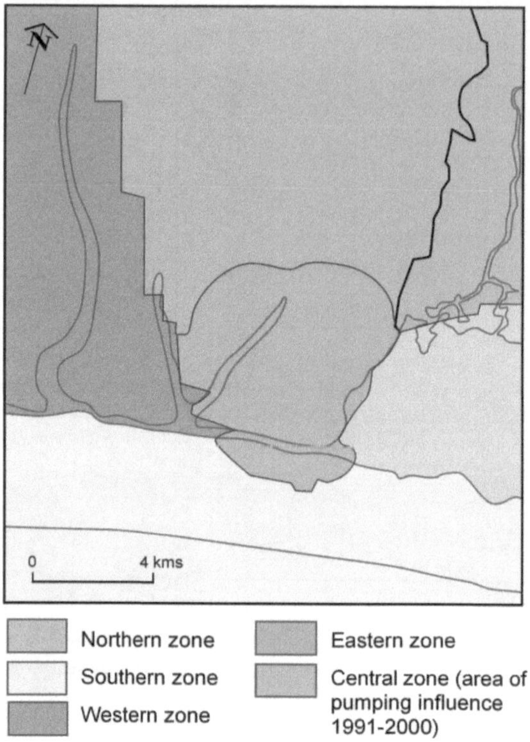

Figure 9. Water balance zones (zone 2 corresponds to cone of depression at the water table for the 1991–2000 mean well yields) of the model.

Table 2. Comparative water balance of the model for steady state of the period of 1991–2000 mean yields and 2011's projected yields.

	Zone influenced by pumping for 1991–2000 period in steady state (88 km^2)			
	Input (m^3/day)		Output (m^3/day)	
Source or target	2000	2011	2000	2011
Recharge	23,577	23,577	0.00	0
Pumping wells	0	0	22,613	56,262
Lake Zone	1512	10,643	3,713	3,424
Northern zone	7,770	10,527	1,226	554
Southern zone	2,281	10,196	679	557
Western zone	450	2,291	1,045	432
Other inputs (rivers, drains, etc.)	0	4,700	6,310	709
Total input/output	35,589	61,937	35,589	61,937

that increased pumping may lead to increased recharge from Lake Nokoue and, possibly, the lagoons along the coastal region.

2. It is apparent that the current zone of production will not be able to support increased production into the future and it is therefore necessary to seek locations for development of new well fields. Based on this initial modelling, development should be focused substantially to the west of the current region to take advantage of pathlines currently not contributing to the water supply of Cotonou.
3. Changes are required with respect both to land use (to avoid introducing additional impermeable surfaces in the southern region) and water use (to reduce demand) in order to extend the useful life of the groundwater resource. Only through such changes is there a possibility of approaching a sustainable rate of pumpage.

6 LIMITATIONS OF THE MODELLING

It is recognized that the modelling effort presented herein has several limitations. Among these are the assumptions of constant density within the fluid phase and steady-flow conditions. Further, a number of assumptions have been made with respect to the boundary conditions. Nevertheless, this modelling effort is considered to be of substantial utility for the short-term management of groundwater resources in southern Benin. Such utility is based on the argument that this effort provides support of short-term management strategies allowing time for development of more sophisticated modelling based on longer-term observation of transient behavior (both hydraulic and water quality) and field characterization.

Justification of this simplified model is based on assessment of the impact of the various assumptions and limitations of this model. First, the assumption of constant density leads to prediction of reduced vertical migration from the shallow aquifers to the deeper, production aquifers. It also leads to prediction of greater vertical upflow from the lowest aquifer into the zone of production. As, in the short term, it appears that the primary route of migration of salt-water is vertically downward from Lake Nokoue, the assumption of constant density will therefore underestimate the potential for vertical migration from Lake Nokoue. As migration from Lake Nokoue is predicted, using the present model, to be a significant source of salt-water inflow to the well field, this result is considered conservative in leading to the conclusion that production must be managed in the short term to limit the impact of salt-water intrusion from the lake.

More difficult is the impact of the assumption of steady flow. However, by simulating the distribution of heads at steady-state subject to the production rates predicted in 2011, the water levels at the production wells should be minimized (as compared to transient drawdowns), thus maximizing the impact of the production wells on the regional distribution of groundwater flow. Hence, once again, the simulation is expected to be conservative with respect to prediction of the potential for salt-water intrusion.

With respect to the assumed boundary conditions, there are two concerns. First, the vertical boundary condition for layers 2–7 along the southern edge of the numerical grid forces vertical upflow in this region of the model. The primary region of interest is, in the present study, the region bounding the production wells near Godomey. As noted in the results, the channel region (north of the coastline) is predicted to be a groundwater divide under current assumptions. This is consistent with recent measures of hydraulic head in this region. Hence, there is confidence that, for the region bounding the production wells, the assumption regarding the southern grid boundary is likely to have minimal impact on these modelling results, at least within the region of the production wells.

Second, no-flow conditions are assumed for layers 2–7 along the eastern boundary of the numerical grid where it underlies the So river and Lake Nokoue. Difficulty in specifying this boundary results from a lack of piezometric data in the vicinity of this boundary. However, the large hydrologic impact of Lake Nokoue combined with regional analysis of the groundwater flux in the vicinity of the Sô river provides support for this region containing a north-south groundwater divide. Hence, the assumption of no flow along this boundary is considered as a reasonable first approximation that will be the subject of further study during extensions of this modelling effort.

7 CONCLUSIONS

Within this manuscript, initial modelling efforts are discussed that provide short-term insight into the threat of salt-water intrusion into the production wells used to supply water for the city of Cotonou, Benin. While the model is based on a number of assumptions, it provides substantial insight into the likely flow pathways for salt-water intrusion to the production wells. Observations from this initial modelling study indicate that continuing salt-water intrusion into the well field is likely with the primary source being infiltration from Lake Nokoue. As such, this preliminary model provides for definition of short-term management strategies.

Extension of this model to develop long-term management strategies must include addition of density-dependent flows under transient hydraulic conditions. Further, the hydraulic boundary conditions along the eastern and southern boundary of the numerical grid must be studied further to ensure either that assumptions made regarding these boundaries have minimal impact on the numerical solution and/or data requirements to further quantify these boundaries are identified.

ACKNOWLEDGMENTS

This research was partially funded by the Danish International Development Agency (Danida, Denmark), International Research Centre for the Development (IRCD, Canada), Société Béninoise pour l'Eau et l'Electricité (Bénin), and the U.S. National Science Foundation. This paper has benefited from helpful review of Stephen Silliman, Civil Engineering and Geological Sciences, University of Notre Dame Indiana (USA) and two other reviewers made valued comments and suggestions on the text. The authors thank the editors of this special issue and extend their thanks to Ouorou Moussa, Afouda Eric and Falola Emilola who provided many contributions, in particular for organizing the data used in this effort.

REFERENCES

Andersen, M. P. & Woessner, W. W. 1992. Applied groundwater modelling: simulation of flow and advective transport. Academic Press, New York.

Boukari, M., Alidou, S., Oyédé, L. M., Gaye, C. B. & Maliki, R. 1995. Identification des aquifères de la zone littorale du Bénin (Afrique de l'Ouest): hydrodynamique, hydrochimie et problèmes d'alimentation en eau de la ville de Cotonou. African Geoscience Review, **12**, 1, 139–157.

Boukari, M. 1998. *Fonctionnement du système aquifère exploité pour l'approvisionnement en eau de la ville de Cotonou sur le littoral béninois. Impact du développement urbain sur la qualité des ressources.* Thèse de doctorat ès-Science. Université C. A. Diop de Dakar, Sénégal.

Boukari, M., Moussa, O., Azonsi, F. & Viaene, P. 2004. *Model for the groundwater flow in the aquifers of the "Continental Terminal" and the littoral Quaternary of the coastal sedimentary basin of Benin (West Africa).* IMPETUS Integrated water resource management of tropical river basin Conference proceedings. Cotonou, Bénin. 65pp.

Boukari, M., Viaene, P. & Azonsi, F. 2006. *Modélisation tridimensionnelle du système aquifère du bassin sédimentaire côtier du Sud-Bénin (Afrique de l'Ouest).* UNESCO (PHI) International Conference on water security and hydrological extremes: forwards a sustainable development in Africa. Abuja, Nigeria.

Cheng, J. M. & Chen, C. X. 2001. *Three-dimensional modeling of density-dependent salt water intrusion in multilayered coastal aquifers in Jahe River Basin, Shandong Province, China.* Ground Water, **39**, 128–136.

Direction de l'Hydraulique. 2000. Vision Eau 2025 Bénin. Rapport Direction de l'Hydraulique, Cotonou, Bénin. 28pp.

Ebraheem, A. M., Riad, S., Wyciskm P. & Sefelnasr, A. M. 2004. *A local-scale groundwater flow model for groundwater resources management in Dakhla Oasis, SW Egypt.* Hydrogeology Journal, **12**, 714–722.

Gangopadhyay, S. & Das Gupta, A. 1995. *Simulation of salt-water encroachment in a multi-layer groundwater system.* Hydrogeology Journal, **3**, 74–88.

Gaus, I. 2000. *Effects of water extraction in a vulnerable phreatic aquifer: consequences for groundwater contamination by pepticides, Sint-Jansteen area, The Netherlands.* Hydrogeology Journal, 2000, **8**, 218–229.

Gnaha, F. C. & Adjadji, C. A. 2001. *Evolution quantitative et qualitative des ressources en eaux souterraines captées dans le périmètre de pompage intensif de Godomey: impact sur l'approvisionnement en eau potable de l'agglomération de Cotonou Mémoire de Maîtrise.* Université d'Abomey-Calavi, Bénin.

Gossel, W., Ebraheem, A. M. & Wicisk, P. 2004. *A very large scale GIS-based groundwater flow model for the Nubian sandstone aquifer in eastern Sahara (Egypt, northern Sudan and eastern Libya).* Hydrogeology Journal, **12**, 698–713.

IGIP-GKW-GRAS 1989. *Plans directeurs et études d'ingénierie pour l'alimentation en eau potable et l'évacuation des eaux pluviales, des eaux usées et des déchets solides. Ville de Cotonou. Rapport final.* Ministère de l'Industrie des Mines et de l'Energie, Cotonou, Bénin.

INSAE, 2003. *Troisième recensement général de la population et de l'habitation, février 2002: synthèse des résultats.* Rapport Direction des Etudes Démographiques MECCAG. Cotonou, Bénin. 27pp.

Istituto Ricerche Breda 1987. *Etude de cartographie géologique et prospection minière de reconnaissance au Sud du 9ème parallèle. Rapport final.* Ministère des Mines de l'Energie et de l'Hydraulique, Cotonou, Bénin.

MacDonald, M. G. & Harbaugh, A. W. 1988. *A modular three-dimensional finite-difference ground water flow.* USGS Open-File Report **83–875**.

Maliki, R. 1993. *Etude hydrogéologique du littoral béninois dans la région de Cotonou (AO)* Thèse de 3ème cycle; Université Cheikh Anta Diop de Dakar, Sénégal.

MECCAG 2000a. *La population au Bénin: évolution et impact sur le Développement.* Rapport Policy Project/USAID. MECCAG, Cotonou, Bénin.

MECCAG 2000b. *Etudes nationales de perspectives à long terme (NLTPS-Bénin 2025):dynamique démographique, question agraire et urbanisation au Bénin.* Rapport MECCAG, Cotonou, Bénin.

MMEH 2005. *Document de politique Nationale de l'Eau: la gouvernance de l'eau au service du développement du Bénin.* Rapport MMEH, Cotonou, Bénin, 19pp.

MMEH 2006a. *Stratégie nationale de l'approvisionnement en eau potable en milieu rural 2005–2015.* Rapport Direction Générale de l'Hydraulique. Cotonou. Bénin 68pp.

MMEH 2006b. *Stratégie nationale de l'approvisionnement en eau potable en milieu urbain 2006–2015.* Rapport Comité de pilotage de la stratégie de l'AEP en milieu urbain (à valider). Cotonou. Bénin 28pp.

Newport, B. D. 1977. *Salt water intrusion in the United States.* Report 600-8-77-01, Environmental Protection Agency. Washington DC. United States.

Oyédé, L. M. 1991. *Dynamique sédimentaire actuelle et messages enregistrés dans les séquences quaternaires et néogènes du domaine margino-littoral du Bénin (Afrique de l'Ouest).* Thèse de Doctorat Univ Bourgogne et Univ Nat du Bénin, 302pp.

Pallas, P. 1988. *Contribution à l'étude des ressources en eau souterraines du Bassin côtier du Bénin, Confrontation ressources-besoins.* Rapport final, Ministère de l'Equipement et des Transports, Cotonou, Bénin.

Pollock, D. W. 1989. *Documentation of computer programs to compute and display pathlines using results from the US Geological Survey Modular three-dimensional finite difference groundwater flow model.* Scientific Software Group, Washington, DC. United States.

Sagbo, O. M. 2000. *Pollution des nappes alimentant la ville de Cotonou en eau de consommation: vulnérabilité et urgence des mesures de protection des aquifères.* Mémoire de DESS, Université Nationale du Bénin, Cotonou, Bénin, 69pp.

Slansky, M. 1962. *Contribution à l'étude géologique du Bassin sédimentaire côtier du Dahomey et du Togo.* Mém BRGM, **11**, Orléans, France.

SGI 1981. *Etude de la nappe aquifère de Godomey. Rapport final.* Société Béninoise d'Electricité et d'Eau, Cotonou, Bénin.

SOGREAH/SCET-Tunisie 1998. *Etude de la stratégie nationale de gestion des ressources en eau du Bénin: assistance à la définition de la stratégie nationale de gestion des ressources en eau du Bénin.* Rapport final, Ministère des Mines, de l'Energie et de l'Hydraulique, Cotonou, Bénin.

Texier, H. & Colleuil, B. 1980. *Le lac Nokoué, environnement margino-littoral béninois: bathymétrie, lithofaciès, salinités, mollusques et peuplements végétaux.* Bull. Inst. Géol. Bassin d'Aquitaine, **28**, 115–136.

Turkpak International-SCET-Tunisie 1991. *Inventaire des ressources en eaux souterraines au Bénin.* Rapport final, Ministère des Mines, de l'Energie et de l'Hydraulique, Cotonou, Bénin.

Vodounon-Totin, H. S. 2003. *Changements climatiques et vulnérabilité des ressources en eau sur le plateau d'Allada : approche prospective.* Mém. Maîtrise, Université d'Abomey-Calavi, Bénin, 101pp.

Zhang, Q., Volker, R. E. & Lockington, D. A. 2004. *Numerical investigation of sea water intrusion at Gooburrum, Bundabeg, Queensland, Australia.* Hydrogeology Journal, **12**, 674–687.

CHAPTER 27

Groundwater modelling and implication for groundwater protection: Case study of the Abidjan aquifer, Côte d'Ivoire

K.J. Kouame, J.P. Jourda, & J. Biemi

Laboratoire des Sciences et Techniques de l'eau et de l'Environnement (LSTEE) et Centre Universitaire de Recherche et d'Application en Télédétection (CURAT), UFR des Sciences de la Terre et des Ressources Minières, Côte d'Ivoire

Y. Leblanc

Richelieu Hydrogéologie Inc., Richelieu (Québec), Canada

ABSTRACT: The domestic water supply of Abidjan depends on groundwater abstracted from local sandy aquifer formations. The quality and quantity of this resource are threatened by surface activities and population growth, caused partly by war, which resulted in a rapid and massive displacement of populations towards Abidjan. This displacement of the population exerts an enormous pressure on groundwater resources. This study uses a numerical groundwater flow model to quantify the available groundwater resources of the Abidjan aquifer and to assess the risks associated with the presence of potential polluting activities. The conceptual hydrogeological model comprises two sand layers overlying a presumed impervious rock formation and the lateral limits of the aquifer are physical water bodies such as rivers and the Atlantic Ocean. The hydrogeological parameters of the aquifer were determined from previous studies. The aquifer recharge is provided by the river leakage and rainfall which has been computed by a water balance calculation. The numerical model was built using MODFLOW and calibrated in steady-state flow with initial groundwater head measurements from 1978 and then validated in transient flow with piezometric data from 1992 and 2005. Predictive simulations in transient flow with projected groundwater withdrawal from 2005 to 2030 were carried out. In order to define protection areas of this aquifer, the wells capture zones were calculated using MODPATH. The results indicate that the Abidjan aquifer can be used for future groundwater supply and the methods employed can help to manage and protect the aquifer.

1 INTRODUCTION

Abidjan district (Figure 1) is located in the south of Côte d'Ivoire and consists of ten communities and three sub-prefectures (Bingerville, Songon and Anyama). With an area of 2119 km^2, it extends to 40–50 km around the Abidjan city. The population in the district

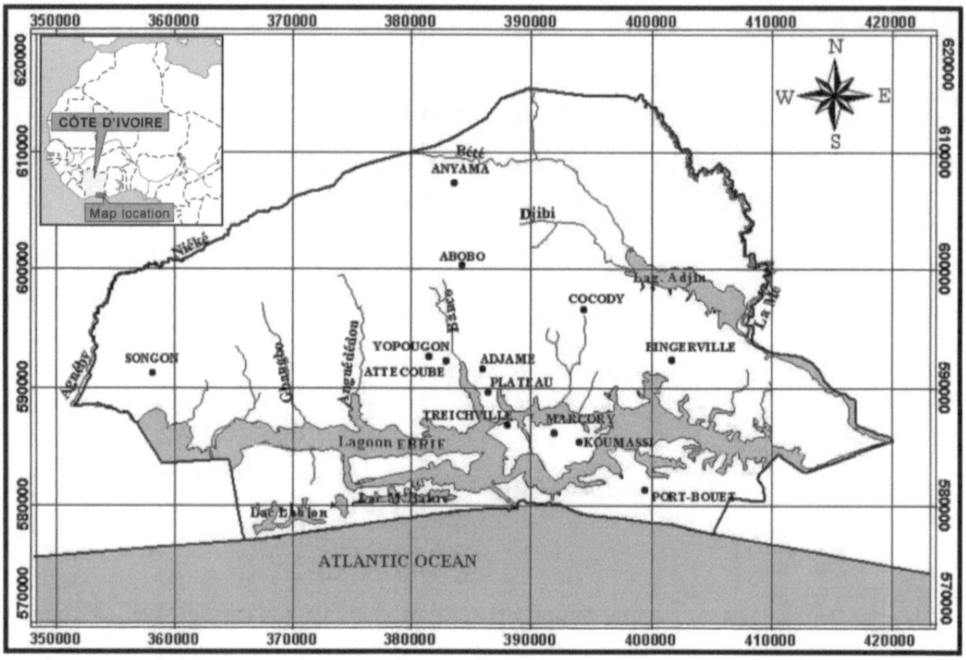

Figure 1. Location of study area.

increased from 3,125,890 in 1998 to 3,660,682 in 2003 due to the socio-political crisis in the country since September 2002.

The coastal sedimentary basin underlying the area comprises sediments of Cretaceous to Quaternary age. It presents enormous groundwater potential. These groundwater resources are contained in three aquifer layers: only the Continental Terminal aquifer (locally called the Abidjan aquifer) is exploited for the drinking water supply (Aghui & Biémi, 1984; Jourda, 1987).

Abidjan is the economic capital of Côte d'Ivoire. It is an example of a metropolis in full geographical and demographic growth. This growth generates problems for water supply to the population. Annual production of the Water Distributive of Côte d'Ivoire Society (SODECI) increased from 7,131,564 m^3 in 1993 to 12,418,617 m^3 in 2002. Even with this increase in supply, it is inadequate to satisfy the increasing demand.

To meet the demand, SODECI wishes to increase its production capacity according to the demographic growth for the next 25 years (until 2030). This necessitates developing a mathematical model of Abidjan aquifer to predict the effect of such exploitation. It is also necessary to determine the capture zone of the wells and the aquifer vulnerability in order to keep a good groundwater quality for future generations.

2 METHODOLOGY

The modelling task requires several stages such as the data acquisition and interpretation, design of conceptual and numerical models, model calibration and verification, the realization of predictive simulations and finally the critical analysis of the results obtained. The following flow chart describe the methodology used (Figure 2).

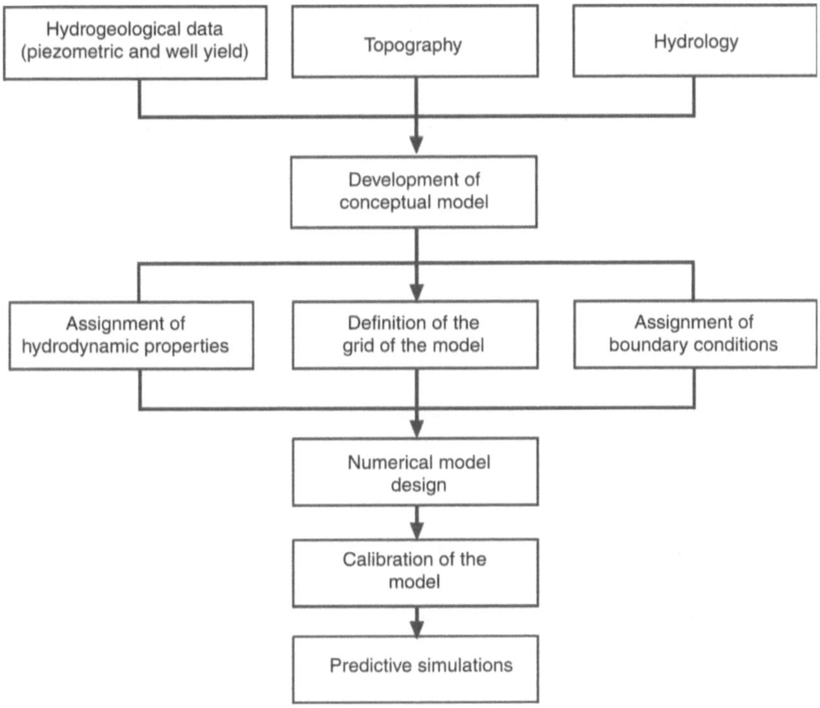

Figure 2. Flow chart for the design of the numerical model of Abidjan aquifer.

The conceptual model development was based on data from geological and hydrogeological studies carried out on Abidjan aquifer by several authors (Adou, 1971; Guérin, 1962; Loroux, 1978; Tastet, 1979; SOGREAH, 1996; Jourda, 1987). From a synthesis of these works, hydraulic parameters such as hydraulic conductivity (permeability), storage coefficient and porosity of fine and coarse sands layers, were obtained. Piezometric data for 1978, 1992 and 2001 were obtained from SOGREAH (1978), Oga (1998) and SODECI (2001). The abstraction for various stations was collected by HYDROEXPERT (2001). Topographic map and bedrock topography were also used for the design of this model.

The numerical model was developed using MODFLOW, a widely used code, developed by the United States Geological Survey (USGS) through the commercial software *Visual MODFLOW* version 3. provided by *Waterloo Hydrogeologic* (MacDonald & Harbaugh, 1988).

3 CONCEPTUAL MODEL

The hydrostratigraphic unit of Abidjan which forms the aquifer constitutes a closed hydrogeological basin whose boundaries are as follows:

- in extreme north, the brooks which recharge the aquifer by imposing high piezometric heads;
- southwards, the lagoon of Ebrié, which discharges the aquifer by imposing heads close to 0 m;
- in extreme west and east boundaries, rivers (Agnéby and Lamé) also discharge the aquifer; by imposing piezometric constant heads corresponding to their rise;

Table 1. Yield of pumping stations from 1960 to 2005.

Pumping stations names	Withdrawals (m^3/d)		
	1960–1978	1978–1992	1992–2005
Anonkoua Kouté	0	21,960	31,440
Zone nord	24,000	25,824	32,592
Adjamé nord	37,680	13,848	20,880
Niangon	0	33,000	40,608
Zone ouest	49,200	37,488	37,920
Zone est	29,040	28,776	32,304
Nord Riviéra	0	32,496	33,062
Riviera centre	0	16,704	19,992
Plateau	1 800	1 800	0

- rivers of the aquifer (Banco, Gbangbo, Anguédédou) drain this one and maintain piezometric constant heads of water or little variables in the zones no disturbed by the exploitation. Along these small rivers, one admits the existence of conditions to imposed potential (Kouadio, 1997).

The top of Abidjan aquifer comprises the topography which varies from 0 to 100 m in elevation (SOGREAH, 1996). The bottom of the aquifer is located at the contact of fine sands with the top of the crystalline basement (Birimian aged schists) with elevations between +30 and −150 m according to bedrock map established by General Company of Geophysics (CGG) in 1977 and 1981–1982 (Aghui & Biémi, 1984). The thickness of the aquifer varies between 20 m in north and 160 m in the south (Jourda, 1987).

The conceptual hydrogeological model has the following two hydrogeological units:

- an aquifer made of sands which covers the entire territory. This unit can be subdivided in two sub-units represented by fine sands at the base and coarse sands at the top of the sequence;
- an aquitard formed by the Birimian schists and which constitutes the impervious base of the aquifer.

Finally, the conceptual model includes abstraction from the well fields (Table 1).

4 NUMERICAL MODEL DESIGN

4.1 *Grid design*

The structure of the numerical model used is identical to SOGREAH (1996). This model is double-layered with the higher layer representing coarse sand and the lower layer fine sand (SOGREAH, 1996). The bottom of map being used as reference for the model is obtained by digitalization of the topographic map 1974 (IGCI, 1974).

For the space discretization (grid) of the model, a square cell of 1 km × 1 km size was used on the zone of study. The size of the grid was then refined to 250 m × 250 m size around

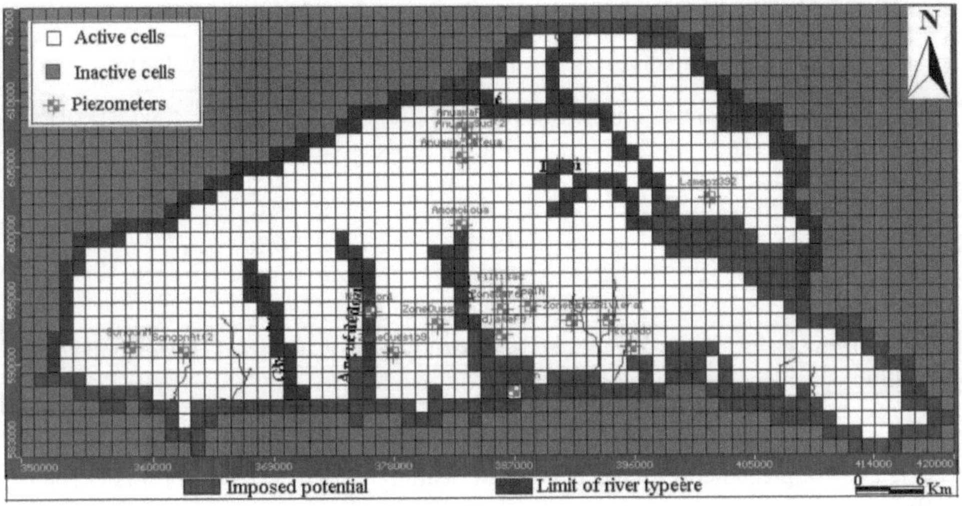

Figure 3. Boundary conditions of the model.

well fields. The part of the grid which did not belong to the aquifer, were inactivated. Thus the model consists of 20,564 cells distributed in 194 columns and 106 lines.

4.2 *Boundary conditions*

The boundary conditions of the model are represented on Figure 3. They consist of boundary conditions of "constant head", "river" and "recharge" type. The head assigned in boundary conditions of constant head and river corresponds to the rise in the rivers represented, while the conductance of the river boundary type was determined by trial and error. The assigned dimensions are as follows:

- constant head in north: 60 m;
- constant head in south: 0.2 to 0 m.

Recharge is assigned uniformly on the whole of the first layer of the model and corresponds to 274 mm/a, where data have been obtained by establishing a water budget from monthly precipitation data.

4.3 *Hydrogeological properties*

The documented hydrogeological properties have been assigned into the numerical model and then were adjusted during the calibration process. Table 2 shows the initial and calibrated hydrogeological properties that were assigned in the numerical model.

5 MODEL CALIBRATION AND VERIFICATION

The calibration of this model aims at reproducing the measured heads in the whole of Abidjan aquifer by adjusting various parameters in the range of realistic values (Leduc, 2005). The calibration of the model was carried out in steady-state mode, assuming that

Table 2. Adjusted hydrodynamic parameters of Abidjan aquifer.

Hydraulic parameters	Initial values	Adjusted values
Hydraulic conductivity		
Coarse sand	4×10^{-4} m/s	5×10^{-4}m/s
Fine sand	10^{-4} m/s	2×10^{-4}m/s
Effective porosity	1.5%	2 %
Specific yield	15 %	20 %

the 1978 piezometric data set were reflecting a steady-state established under small aquifer abstraction which occurred since 1960.

Adjustments of parameters were always close to documented values (Table 2). The calibration was carried out in a "manual" way; a calibration by "grouping" because of insufficiency of piezometries data. This type of calibration proceeds as follows (Leblanc, 1999).

- adjustment of the parameters;
- simulation of the model;
- comparison of the error values of calibration (if these errors are large then one takes again these operations).

These operations are carried out until low error values of calibration are attained so that representing a good calibration between heads "observed" on the ground and those "calculated" by the model (Gurwin & Lubezynski, 2004). When the model was sufficiently calibrated, parameter verification was done in transient state by comparing time vs head calculated graph with the piezometric data set of 1992.

5.1 *Model calibration*

For 1978 piezometric and well yield data, calibration provided a good agreement between observed and calculated heads (Figure 4). In steady-state flow mode, the value of the Normalised RMS is 4.63% (<10%) indicating that the model is well calibrated. The distribution of the differences between observed and calculated heads illustrates the good calibration obtained in permanent mode starting from piezometric data of 1978.

5.2 *Model validation*

The validation of the model in transient mode was carried out by using initial heads (1978) from piezometric data provided in steady-state mode with increasing well yield data. The groundwater levels measured in 1992 were then compared with those calculated by the model. Nine piezometers existing in 1978 were still available in 1992 for the validation of model in transient mode.

Figure 4 indicates a very good agreement between model and observation values (Normalised RMS is 2.7%). The proximity of the points representing piezometers of 1992 around the first bisectrix reveals well that the difference between calculated heads and those observed is not significant. Indeed, seven of the nine piezometers considered, are within a confidence interval of 95%. The model can thus be regarded as calibrated and

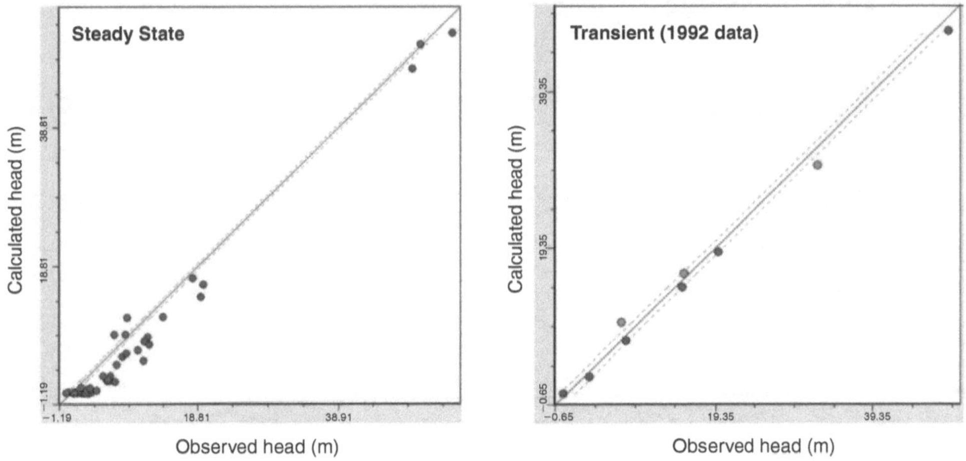

Figure 4. Calculated heads vs. observed heads for the steady state and transient modes.

validated with these piezometers of 1992. Nevertheless, significant differences between observed heads in piezometers at Anonkoua kouté 2 (2.06 m) and Niangon 1 (2.01 m) are to be noted. These differences are caused by the exploitation of boreholes during the piezometric program and also due to piezometers which are located not far from the pumping stations.

6 RESULTS OF MODELLING

The calibrated model was used for the following predictive simulations:

- To reconstruct the initial head distribution of Abidjan aquifer before pumping occurred (1960);
- To reconstitute the 1978 head distribution of Abidjan aquifer (first documented pumping data);
- To predict the evolution of the piezometry and drawdown from 2005 to 2030 according to increasing groundwater retrieval;
- Finally, to determine capture zones for well fields whose determination can lead to delimit the protection zones of the Abidjan aquifer.

6.1 *Initial head distribution of Abidjan aquifer before pumping occurred (1960)*

In order to reconstruct the initial level of Abidjan aquifer before any exploitation, the model was run in steady-state mode without any active well. Figure 5 shows the resulting piezometric map. This map shows an initial groundwater flow direction from north (55 m) to the south (1 m). Groundwater flow directions are oriented towards lagoons of Agnéby and Lamé. Head contour lines vary between 55 m (north) to 1 m (south).

6.2 *1978 Head distribution of Abidjan aquifer (first documented pumping data)*

The piezometric map of 1978 was obtained by assigning the first documented pumping data for the 1960–1978 periods. A steady-state flow was assumed for that year because the

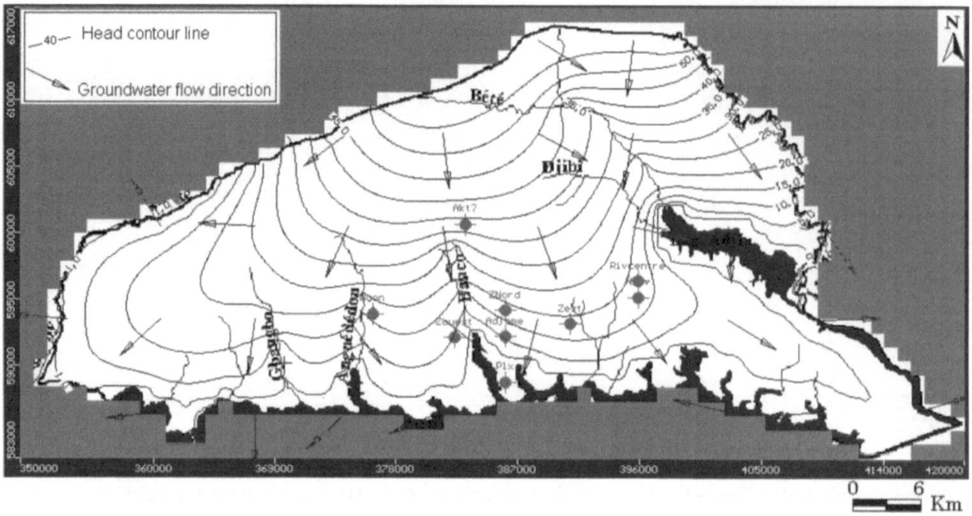

Figure 5. Initial contour head reconstructed in steady state mode (year 1960).

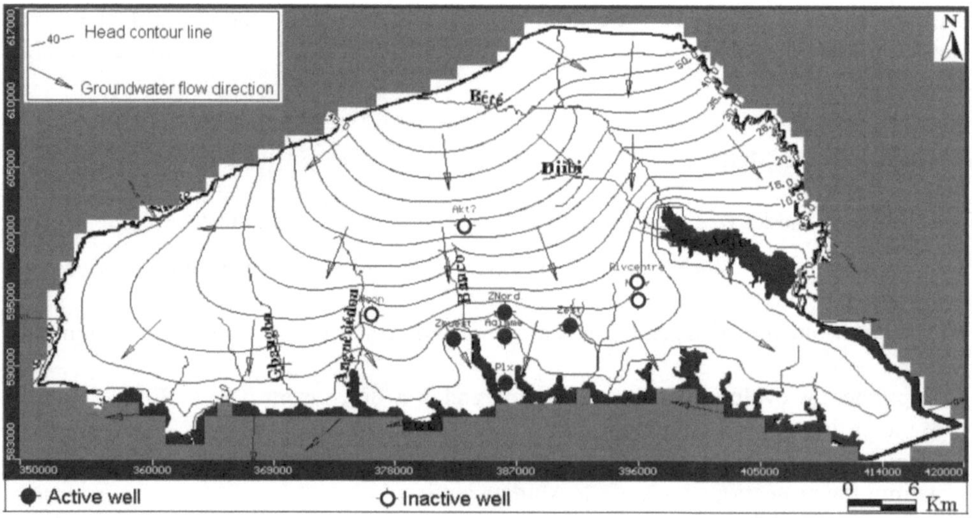

Figure 6. Piezometry calculated in 1978 in steady state mode.

pumping rate was relatively small at that time. Figure 6 shows the resulting piezometric map. Disturbances in piezometers are only located around active pumping stations (Plateau, Adjamé Nord, Zone Nord, Zone Ouest and Zone Est). Northern, eastern and western parts of the model are not affected by these well fields.

The water budget for this simulation is summarized in Table 3. This table shows that the principal input of the system is recharge due to the infiltration of precipitations (Faye, 1998), which is 908,840 m^3/d. In 1978, discharge by pumping reaches approximately 169,920 m^3/d and accounts for approximately 19% of the contributions by net recharge. These values show that pumping did not have a great influence on the aquifer – there were only five pumping stations including 36 exploited boreholes (SOGREAH, 1996) in 1978.

Table 3. Water budget calculated after steady state calibration in 1978.

Input	m^3/day	Output	m^3/day
Constant head	290 390	Constant head	589 280
Well	0	Well	169 920
Recharge	908 840	Recharge	0
River leakage	34 870	River leakage	475 000
Total	1 234 100	Total	1 234 200

Table 4. Projected groundwater abstraction from well fields

Pumping stations names	Withdrawals (m^3/d)	
	2005–2015	2015–2030
Anonkoua Kouté	38 000	43 000
Zone nord	34 992	37 000
Adjamé nord	9 300	27 480
Niangon	37 000	48 000
Zone ouest	48 000	50 000
Zone est	33 000	43 992
Nord Riviéra	38 000	43 992
Riviera centre	20 000	21 288
Plateau	0	0

6.3 *Piezometry and drawdown evolution from 2005 to 2030 according to increasing groundwater retrieval*

This simulation was carried out in transient state flow by gradually updating the pumping rates of existing well fields and adding new well fields such as Anonkoua kouté, Niangon and Riviera (Centre and North). The simulated pumping well abstraction for the predictive simulations are shown in Table 4.

Figure 7 shows the modelled groundwater levels for 2005, and Figure 8 the drawdown. These show important differences relative to the 1978 piezometric map. Except around the well fields, all head contour lines remained unchanged on the entire model compared to 1978. The highest value of drawdown was observed in Anonkoua kouté 2 (14.32 m) while piezometers of Akakro (1.57 m), Anyama-Adjamé (2.66 m) and Akouédo (3.31 m) recorded relatively small drawdown values.

Calculated drawdown from piezometers located in well fields is function of the increase in pumping rate. The water shortages observed in the last few years in Abidjan can be explained by technical breakdowns but also by drawdown of water level. This drawdown can involve the pump draining of borehole causing its stop. These differences are caused by the increase of the pumping rate and the introduction of new pumping stations in the interval.

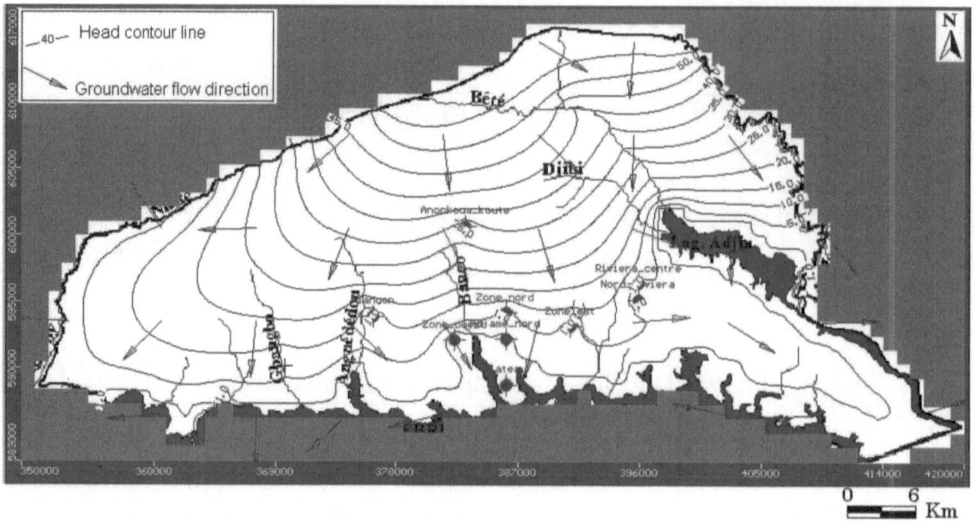

Figure 7. Piezometry calculated in transient state flow mode for 2005.

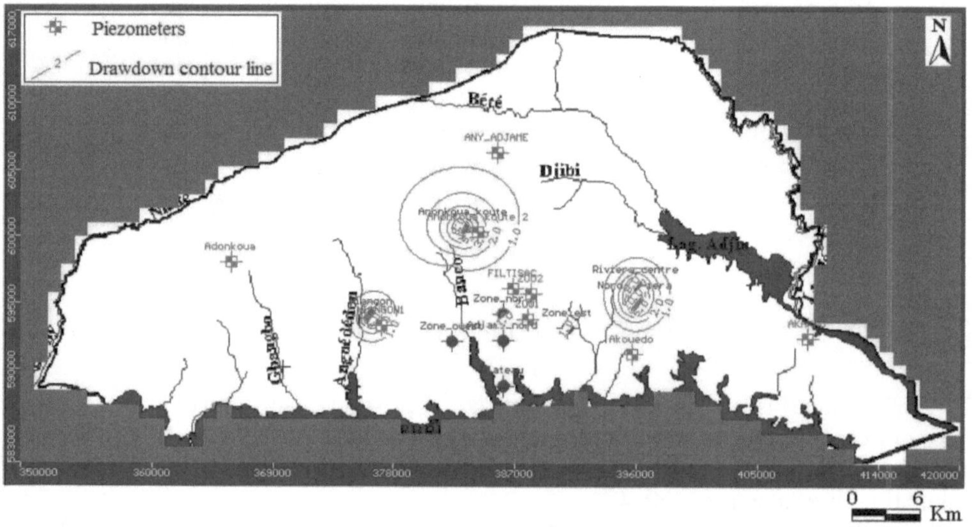

Figure 8. Drawdown calculated in transient state flow mode for 2005.

The transient state flow simulation from 2005 to 2030 presents modification of head contour lines and groundwater flow directions. Examples of the model output for 2015, and 2030 are shown in Figure 9. The following observations are interpreted from the results.

In 2015, modifications result in: (1) projection of head contour line 5 m above Riviera (Centre and Nord), Zone Est and Zone Nord stations; (2) a piezometric cone of 1 m around Riviera (Centre and Nord) stations; and (3) the progression of head contour line 25 m up to Anonkoua Kouté station.

In 2020 and 2030, modifications result in: (1) increase in drawdown across most of the southern part of the model, reducing water-levels to around 1 m; (2) the movement of head

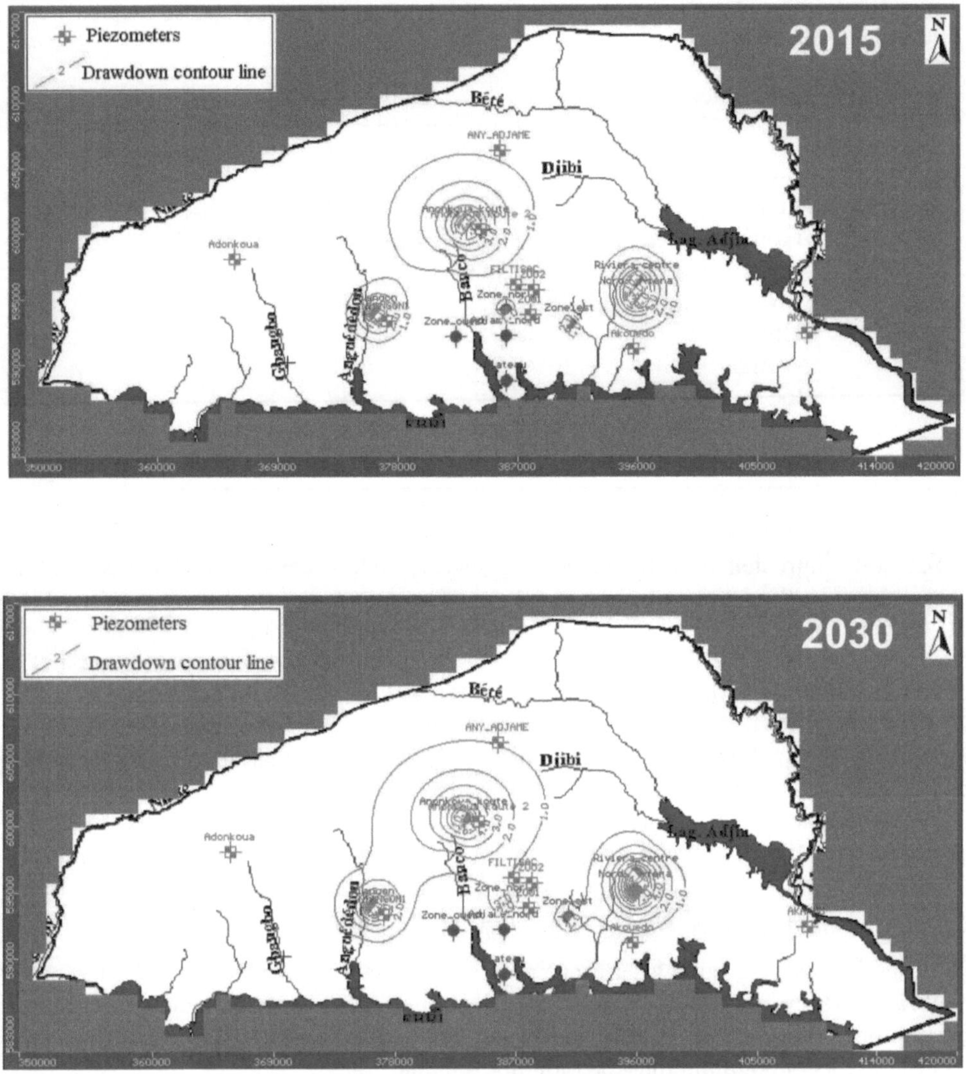

Figure 9. Maps showing the drawdown evolution in the aquifer from 2015 to 2030.

contour line 25 m to north of the Anonkoua Kouté station; and (3) progression of the 1 m piezometric line above Zone Nord station.

The north, eastern and western parts of Abidjan aquifer are little disturbed by the increase of the discharge on pumping stations for the period 2005–2030. Only the model zone occupied by the majority stations is most disturbed.

Table 5 shows the calculated drawdown for 2015, 2020 and 2030 modelling scenarios. All calculated values are at higher than 1 m. It is clear that groundwater abstraction since 1960 has greatly influenced the water-levels of Abidjan aquifer. The drawdown varies between 1 and 15.71 m. The highest drawdown was recorded at Anonkoua kouté 2 and the lowest at Akakro (1.57 m).

Table 5. Drawdown calculated into piezometers compared to 1978 piezometric data.

Piezometers	Drawdown (m)		
	2015	2020	2030
Adonkoua	6.22	6.22	6.24
Akouédo	3.42	3.5	3.64
Anonkoua kouté 2	15.29	15.49	15.71
Anyama-Adjamé	2.92	2.99	3.08
Filtisac	4.04	4.19	4.44
Niangon 1	5.99	6.38	6.86
Zoo 1	4.35	4.47	4.7
Zoo 2	5.12	5.23	5.44
Akakro	1.57	1.57	1.57

Salt-water intrusion from the lagoons is also possible because drawdown cone (1 m) projection towards lagoons of Ebrié and Adjin. The drawdown contour of the water level of Abidjan aquifer delimits the zone of influence of pumping stations. The limit of these zones corresponds to the distance where the drawdown caused by pumping is negligible. The circumscription by the drawdown contour (1 m) of well fields; Anonkoua Kouté, Niangon, Zone Nord and Riviera Centre, Nord-Riviera, Zone Est; shows that those well fields are interfering with each other by 2030 (Figure 9). Any other installation of pumping station or boreholes inside this surface would be disadvantageous because it could increase the drawdown and cause the draining of certain wells. Therefore, future groundwater abstraction projected by SODECI will certainly have an impact on water level of Abidjan aquifer.

6.4 *Delineation of protection zone for well fields*

An essential goal in groundwater protection is to estimate the zone of contribution of the well fields (Rasmussen & Rouleau, 2003). The capture zone or zone of contribution is the portion of territory in which the groundwater contributes to abstraction from a pumping well. The risks of pollution must be minimized in that zone.

To delineate capture zones, the inverse pathways of virtual water particles were simulated in periphery of the cells to which pumping wells were assigned. MODPATH computes the pathlines of water particles, from their infiltration point to the well. The delineation of these zones was carried out using the numerical model in steady state flow with abstraction for 2005. A steady-state flow is adequate because isochrones are not influenced by varying heads (Rasmussen & Rouleau, 2003; Xu & Tonder, 2002).

Time markers were used to delineate the 200 and 550 days traveling time (isochrones). These time markers are delineating respectively bacterial and viral protection zones inside the capture zone. The 200 and 550 days are similar to legislation in Canada (Rasmussen & Rouleau, 2003) for delineating groundwater protection zones for wells.

Figure 10 shows the pathlines representing capture zones for each one of the well fields for the projected groundwater retrieval of 2030. The 200 and 550 days isochrones are also represented in order to show their bacterial and viral protection zones. Virtual particles of

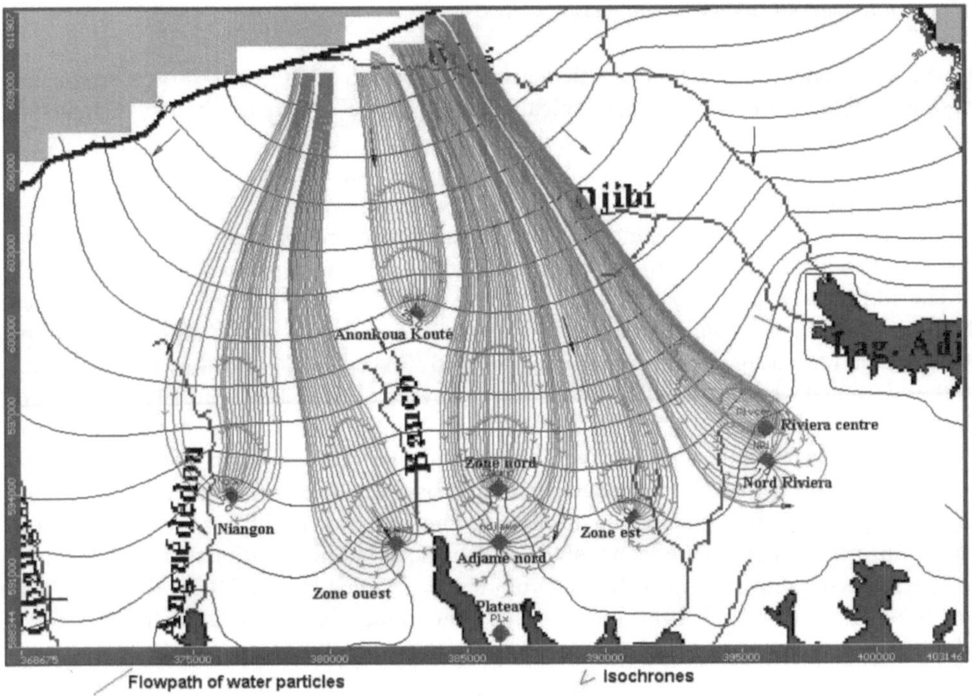

Figure 10. Flowpaths of water particles delimiting zones of contribution of capture fields of Abidjan aquifer.

Table 6. Maximum distance which water particles are collected.

Capture fields	200 days distance (km)	550 days distance (km)	Maximum distance (km)
Anonkoua Kouté	4.89	7.36	9.79
Zone nord	11.52	13.99	18.89
Adjamé nord	13.86	17.51	16.76
Niangon	10.9	13.7	16.10
Zone ouest	13.13	15.93	17.87
Zone est	15.55	18	20.23
Nord Riviera	17.17	18.67	20.01
Riviera centre	14.62	17.03	18.90

water collected by the well fields come all from north at different distances. These zones are to be protected from contamination sources, particularly those which are persistent in an underground environment. Inside these zones, activities or any deposit likely to harm directly or indirectly the groundwater quality of Abidjan aquifer must be prohibited or regulated. Distances to which water particles are collected by capture fields are presented in Table 6.

7 DISCUSSIONS AND CONCLUSIONS

Numerical modelling is a tool of calculation and prediction. It is based on a simplified design of highly complex hydrogeological settings. The reliability of a model depends on the data which is used to define and calibrate it.

The purpose of the numerical model developed in this study was to simulate the behaviour of the Abidjan aquifer for projected groundwater abstraction from 2005 to 2030, based on its reaction from the past. The model was calibrated from initial data in steady-state flow mode and validated in transient flow from 1978 to 1992 with the piezometric data of 1992.

The additional exploitation envisaged by the SODECI from 2005 to 2030 is considered feasible regarding the obtained results, but it will have an impact on the water-levels of the Abidjan aquifer. The projected piezometry for 2030 allowed the description of the influence of the additional well field exploitation envisaged by the SODECI on Abidjan aquifer (from 258,490 m^3/d in 2005 to 314,750 m^3/d in 2030, which is an increase of 18%). By 2030, the drawdown caused by this additional withdrawal will vary between 1 and 16 m compared to initial head values. The highest drawdown cone was calculated at piezometers, Anonkoua kouté 2 with 15.71 m, Niangon 1 with 6.86 m and Adonkoua with 6.24 m. Western and south-eastern parts of the model are slightly influenced by this additional exploitation.

The projected drawdown is different from those obtained by HYDROEXPERT (2001) and SOGREAH (1996). This difference can be due to the code of modelling used, the size of grid, the number of well fields considered, data used and especially to the distribution of the annual recharge on the model. The assumption on constancy and uniformity of the recharge (2005–2030) seems to be an optimistic condition while referring in climatic conditions and especially to the land use of the study area. The land use of the study area especially that of the well fields by constructions can make the ground impermeable. This situation in certain places can lead to a reduction in the net recharge of the aquifer. Thus, the assumption according to which the recharge of the aquifer would remain constant during all simulation and would uniformly be distributed does not completely correspond to the field reality into any place. These values of drawdown can thus be regarded as being those which would be obtained in the zones where this assumption would be true.

A decrease in the annual recharge of 20% (i.e. from 247 to 200 mm/year), would drain the model cells adjacent to pumping wells, thus causing overpumping of the aquifer. It is therefore necessary to establish some weather stations on all the districts of Abidjan in order to determine the real recharge of the aquifer because the pumping rate of wells also depends on the recharge.

The drawdown contour obtained in 2030 progresses towards the lagoons (Adjin, Ebrié, Lamé). This behaviour of the drawdown may result in salt-water intrusion. This salt-water intrusion will occur when the withdrawals projected by the SODECI reverse the groundwater flow direction from south towards north. The foreseeable risk of saltwater intrusion is certain, but a good distribution of future production wells could decrease that risk. A network of observation wells for the monitoring of salinity would be necessary in the south of the aquifer while observation wells would be necessary into the protection area to identify potential contaminants before they reach the well fields.

From the pathlines of virtual water particles computed by the model, capture and protection zones of well fields have been delimited. The defined areas can be used by Côte d'Ivoire's legislation to protect the SODECI well fields. Indeed, most of these areas are already occupied by constructions without adequate sanitation system, thus endangering groundwater quality.

It is advisable to specify that the reliability of values produced by the model is a function of the reliability of data used as input data of the model. For the continuation of this project, a simulation of groundwater chemistry (e.g. nitrate, hydrocarbons) is recommended in order to evaluate the fate and transport of pollutants into the Abidjan aquifer.

ACKNOWLEDGEMENTS

The authors thank UNESCO, Nairobi office for providing necessary support for the research on this subject. They would like to thank also Y. Leblanc for his constructive comments and contribution to the improvement of this paper.

REFERENCES

Adou, A. 1971. *Etat actuel de l'étude hydrogéologique de la région d'Abidjan.* Rapport N°269, SODEMI, Ministère des mines, Abidjan.

Aghui, N. & Biémi, J. 1984. *Géologie et hydrogéologie des nappes de la région d'Abidjan et risques de contamination.* Annales de l'Université Nationale de Côte d'Ivoire, Série C, **20**, 331–347.

Faye, S., Gaye, C. B. & Faye,. A. 1998. *Modélisation du fonctionnement hydrodynamique du système aquifère du littoral nord du Sénégal. Simulation de prélèvements supplémentaires pour réduire le déficit de distribution d'eau potable de la région de Dakar.* Hydrogéologie, **1**, 13–22.

Guérin, G. V. 1962. *Hydrogéologie en Côte d'Ivoire.* Bulletin de la Direction de la Géologie et de la Prospection Minière (DGPM), **2,** 40 pp.

Gurwin, J. & Lubezynski, 2004. *Modelling of complex multi-aquifer systems for groundwater resources evaluation. Swidnica study case (Poland).* Hydrogeology Journal, **12**, 627–639.

HYDROEXPERT 2001. *Transposition du modèle de la nappe d'Abidjan sous TALISMAN.* SODECI, Abidjan, 28 pp.

IGCI 1974. *Cartes topographiques au 1/50 000 de la région d'Abidjan.* Institut Géographique de Côte d'Ivoire, Abidjan.

Jourda, J. P. 1987. *Contribution à l'étude géologique et hydrogéologique de la région du Grand Abidjan (Côte d'Ivoire).* Thèse de doctorat de 3ème cycle, Université scientifique, technique et médicale de Grenoble, 319 pp.

Kouadio, B. H. 1997. *Quelques aspects de la schématisation hydrogéologique: cas de la nappe d'Abidjan.* Mémoire de DEA des Sciences de la Terre option hydrogéologie, Université de Cocody, 66 p.

Leblanc, Y. 1999. *Prédiction de l'effet du décapage d'une mine à ciel ouvert sur l'hydrogéologie locale à l'aide de la modélisation numérique.* Systèmes Geost. International, Laval, Québec, 23 pp.

Leduc, C. 2005. *Modélisation hydrogéologique Orsay 2004 – 2005.* Notes de cours de modélisation numérique en hydrogéologique, Université de Paris-Sud, Paris.16 pp.

Loroux, B. F. 1978. *Contribution à l'étude hydrogéologique du bassin sédimentaire côtier de Côte d'Ivoire.* Thèse de doctorat, Université de Bordeaux I, No.1429.

MacDonald, M. G. & Harbaugh, A. W. 1988. *A modular three dimentional finite-difference ground water flow model.* USGS, book 6 modeling techniques, Washington, USA.

Oga, M. S. 1998. *Ressources en eaux souterraines dans la région du Grand-Abidjan (Côte d'Ivoire): Approches hydrochimique et isotopique.* Thèse de Doctorat de l'Université de Paris XI Orsay, 211 p.

Rasmussen, H. & Rouleau, A. 2003. *Guide de détermination d'aires d'alimentation et de protection de captage d'eaux souterraines.* Centre d'étude sur les ressources minérales, Université de Québec à Chicoutimi; contrat du ministère de l'environnement du Québec, 182 pp.

SODECI 2001. Résultats de la campagne piézométrique de 2001. SODECI, Abidjan.

SOGREAH 1996. *Etude de la gestion et de la protection de la nappe assurant l'alimentation en eau potable d'Abidjan. Etude sur modèle mathématique. Rapport de phase 2: Présentation du modèle mathématique et des résultats du calage.* Ministère des Infrastructures Economiques, Direction et Contrôle des Grands Travaux (DCGTX), Abidjan, 30 pp.

Tastet, J. P. 1979. *Environnements sédimentaires et structuraux quaternaires du littoral du Golfe de guinée (Côte d'Ivoire, Togo, Bénin).* Thèse de Doctorat d'Etat ès sciences, Université de Bordeaux 1, 181 pp.

Xu, Y. & Tonder, G. J. V. 2002. *Capture zone simulation for boreholes located in fractured dykes using the linesink concept.* Water South Africa, **28**, pp. 165–169.

CHAPTER 28

Geostatistical assessment of the transmissivity of crystalline fissured aquifer in the Bondoukou region, north-eastern Cote d'Ivoire

T. Lasm

UFR STRM, Laboratoire des Sciences et Techniques de l'Eau et de l'Environnement, Université de Cocody, Abidjan, Côte d'Ivoire
Laboratoire d'Hydrogéologie, UMR 6532 HydrASA CNRS, Université de Poitiers, Poitiers, France

M. Razack

Laboratoire d'Hydrogéologie, UMR 6532 HydrASA CNRS, Université de Poitiers, Poitiers, France

M. Youan Ta

CURAT: Centre Universitaire de Recherche et d'Application en Télédétection, Université de Cocody, Abidjan, Côte d'Ivoire

ABSTRACT: This study is focused on the geostatistical assessment of the transmissivity of crystalline fissured aquifers in the Bondoukou region of north-eastern Cote d'Ivoire. In this region, few transmissivity data are available due to the lack of boreholes. Therefore, geostatistical methods were used to perform an estimation of this important hydraulic parameter over the whole region (2400 km^2). The variographic analysis of this parameter shows that 'raw' transmissivity does not display any spatial structure. The logarithmic transform of the transmissivity (logT) is however a spatially structured variable. The use of the kriging method led to the estimation of logT over the whole study area. The geostatistics indicated a large nugget effect which indicates that the spatial correlation between individual boreholes is weak, even at short distances. This emphasises the importance of fractures in supplying these boreholes. Despite the lack of strong spatial correlation, the geostatistical procedure developed in this paper proved useful to understand variations in transmissivity. By including a large nugget effect, kriging has reliably reproduced the measured data and provided a tentative estimate of the transmissivity of the Bondoukou fractured aquifers.

1 INTRODUCTION

Groundwater resources management and optimal exploitation require an accurate knowledge of hydraulic properties of aquifers. Transmissivity (T, L^2/T) is a major hydraulic parameter of aquifers and is generally evaluated using pumping tests (e.g. Kruseman & deRidder, 1973; Castany, 1982). However, available pumping test data are often few, which do not allow analysis and assessment of the distribution of transmissivity on large areas using classical methods. This problem is compounded by the fact that aquifers are often heterogeneous and accordingly their transmissivity can be strongly variable in space. This problem can be overcome with the use of geostatistics.

Many authors (Delhomme, 1976; Aboufirassi & Marino, 1984; Ahmed & deMarsily, 1987, 1993; Roth *et al.*, 1996; Roth & Chiles, 1997; Taborton *et al.*, 1995; Fabbri, 1997; Lasm, 2000; Sinan & Razack, 2005) have performed geostatistical analysis of aquifer hydraulic parameters such as transmissivity or permeability. They have shown that a geostatistical approach can lead to a reliable estimation of the transmissivity field. Most of these geostatistical approaches are however related to porous aquifers. There are few works concerning geostatistical estimation of crystalline fissured aquifers parameters.

In Cote d'Ivoire, several studies of the transmissivity of discontinuous crystalline aquifers have been undertaken (Soro, 1987; Biemi, 1992; Savane, 1997; Savane *et al.*, 1997; Lasm, 2000; Lasm *et al.*, 2004; Razack & Lasm, 2006). As a general rule, schist appears more transmissive than granite. These observations corroborate the previous findings by Engalenc (1978). In the Bondoukou region, transmissivity studies are only fragmentary and cover isolated small areas. Assessing the spatial distribution of this parameter over the whole region proves to be important in view of future modelling tasks. The objective of this paper is thus to assess the applicability of geostatistics to the estimation of the transmissivity of the Bondoukou crystalline fissured aquifers in Cote d'Ivoire.

A univariate geostatistical approach is a twofold analysis. The first step is to describe the spatial correlation between sample points. This is achieved by calculating the variogram (Isaak & Srivastava, 1989; Kitanidis, 2000). The second step is to provide the best estimation of the variable at unsampled points. Contrary to other estimation techniques, geostatistics take into account the observable spatial correlation between sample points to predict the variable at unsampled points. The geostatistical estimator (called 'kriging' in the literature) is a BLUE (Best Linear Unbiased Estimator) and is characterised by the two following properties: 1) the average of the errors of estimates is zero, and 2) the variance of the errors of estimates is minimum. Kriging is a linear geostatistical estimation method, which enables the estimation of a regionalized variable (Z) at any point in space, based on its measured values at other locations. One should note that kriging takes into account: i) the spatial positions of the point to be estimated and the known points; and ii) the spatial variability of the variables through the variogram. The variance of the errors of estimate is an indicator of the confidence to be granted to the kriging estimates. The lower the variance, the higher the confidence.

The work presented here, concerns an area located at the north-east of Cote d'Ivoire (Bondoukou) where boreholes available for the supply drinking water are few. The aim is the assessment of the transmissivity of these fractured aquifers based on a geostatistical approach, and a better hydrogeological knowledge of these aquifers in order to implement optimal exploitation of their water resources.

2 GEOLOGY

The study area is located in the north-eastern part of Cote d'Ivoire between the latitudes West 7°55′ and 8°30′ and longitudes North 2°40′ and 3°20′, to the East of the major fault of Sassandra (Figure 1). It forms part of the Baoule-Mossi domain of Cote d'Ivoire. This domain comprises a set of Precambrian gneissose rocks, constituting the basement

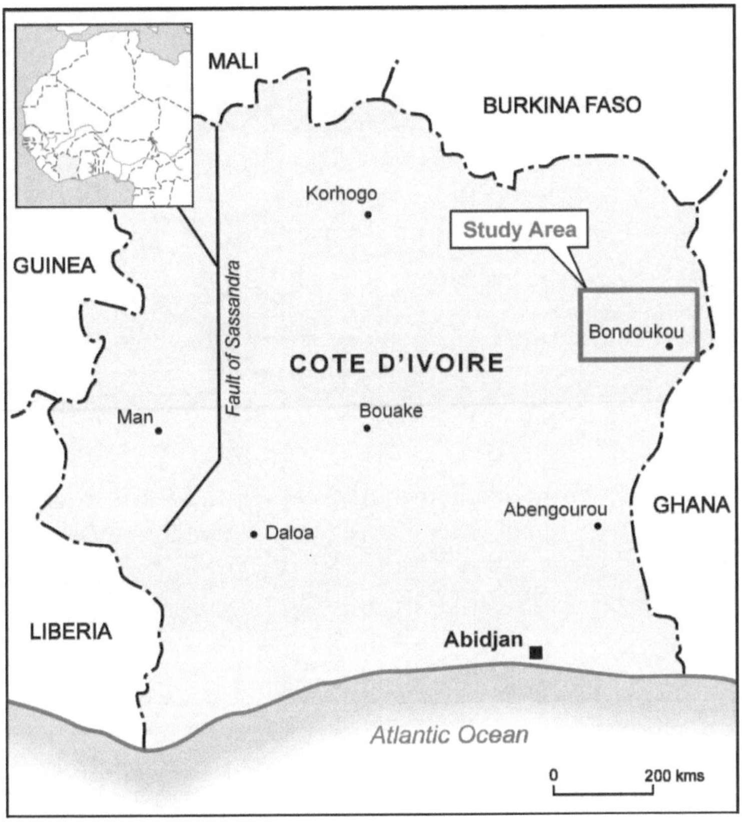

Figure 1. Location map of the study area.

of supracrustal formations which have volcanic, sub-volcanic and sedimentary origin, deposited in furrows or intracratonic basins (Kouamelan, 1996). These formations were differentiated from the mantle between 2200 Ma and 2300 Ma ago and characterize the Birimian, which groups the West Africa formations with age between 2400 Ma and 1600 Ma. The Eburnean orogenesis is the major tectono-metamorphic event which has most affected these formations to the east of the fault of Sassandra. From a geological viewpoint, three lithological fields can be distinguished (Figure 2): i) a volcanic and sedimentary unit, constituted of several different petrographic sub-units, outcroping in the southern and central areas of the sector; ii) an intrusive unit constituted primarily of metamorphized granodiorite and secondarily of granites and tonalites; iii) a tarkwaîen unit constituted of post-tectonical detritical formations represented by conglomerates, sandstones and arkoses.

Jourda (2005) has shown that the area is intensely fractured. He performed a critical and synthetic study of the fracturing of this region using airphotos, remote sensing (satellite images) and geological field surveys. He concluded that the tectonics of this region is polyphased and complex.

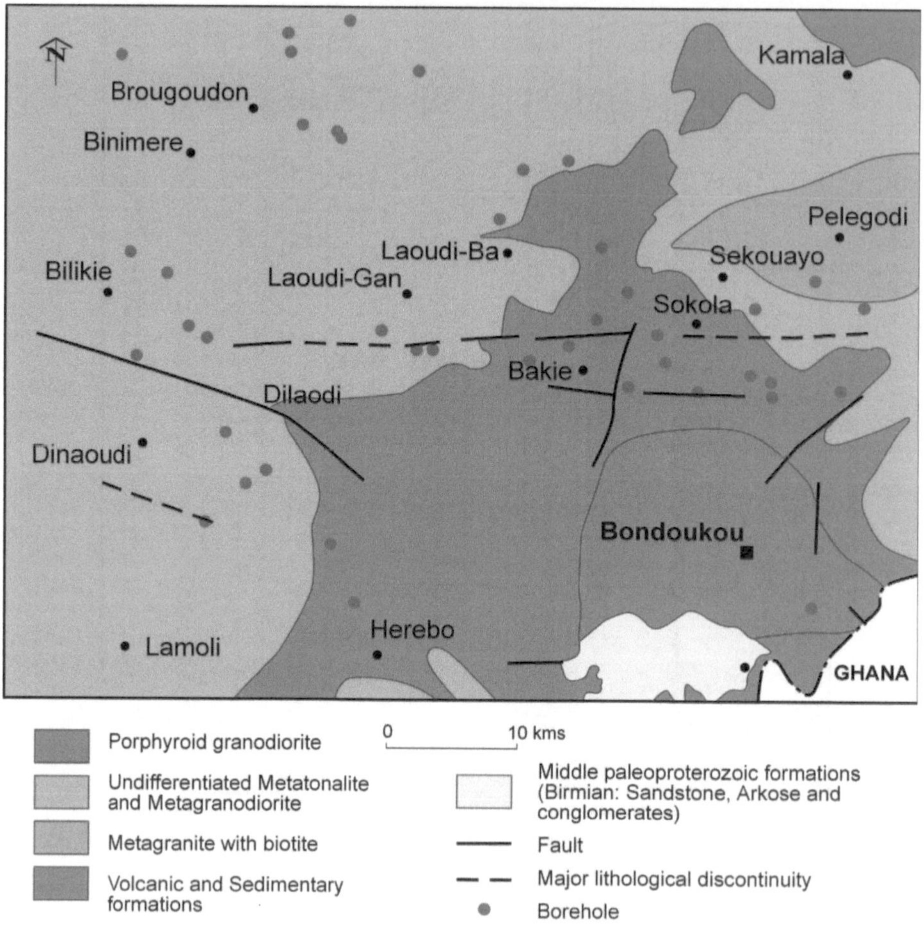

Figure 2. Geological map of the study area with distribution of the boreholes.

3 DATA AND METHODS

3.1 *Data*

Data available for this study are borehole records used to elaborate the transmissivity distribution over the study area (Figure 2). Boreholes were drilled in several types of rocks among which are included granites and schists. Indeed, the latter rocks have different hydrogeological behaviour. A critical and synthetic study of the hydrogeology of the area was recently performed by Lasm (2000). Boreholes have mainly been drilled in granitic rocks (89%). Very few boreholes (only 8) are drilled in the schist aquifers. This very small number of boreholes drilled in schists prevents from performing any quantitative analysis of these aquifers and draw relevant conclusions. Thus the few transmissivity values determined in the schists aquifers are not taken into account in the following geostatistical analysis.

Transmissivity values were determined from the interpretation of pumping tests data (Kruseman and deRidder, 1973; Castany, 1982). The north-eastern area of Cote d'Ivoire is equipped with large diameter wells and boreholes. Most boreholes are isolated and exploit water resources contained in fractures. Pumping tests were carried out at each borehole. It was thus possible to determine the aquifer transmissivity at 65 locations.

3.2 *Methods*

Transmissivity values of the granite aquifers were calculated using pumping recovery data with the use of Cooper-Jacob method, in order to reduce the influence of the turbulent head loss on the drawdown in the borehole. The distribution of boreholes over the area is not homogeneous. Indeed, boreholes do not cover the totality of the study area and the measurements are unequally distributed. The absence of sampled values close to an estimated point obviously influences the quality of the estimate. It is possible with the use of geostatistics to determine the estimation errors corresponding to various sampling networks (Taupin *et al.*, 1998). The variogram allows describing the spatial structure of the variable at hand and points out the deterioration of the correlation between the measured points when the distance increases. The computed variogram identifies some specific characteristics of the studied variable: continuity, anisotropy, size of regionalization, presence of imbricated structures, etc. Estimating the variable after its spatial (or variographic) analysis, requires adjusting an analytical function to the experimental variogram.

The use of a variogram allows the passage from local knowledge to regional knowledge. It points out spatial heterogeneities which would not be displayed through traditional studies. It is defined by the equations (1) and (2):

$$\gamma(h) = \frac{1}{2}\,Var\,[Z(x+h) - Z(x)] \tag{1}$$

$$\gamma(h) = \frac{1}{2}E\,[(Z(x+h) - Z(x))^2] \tag{2}$$

where: *Var [Z(x)]* is the variance at point x; h is the vector of module $(x - x')$; E is the mathematical expectation.

Note that originally the term 'variogram' was used to indicate the function $2\gamma(h)$ and therefore $\gamma(h)$ is strictly known as the 'semivariagram'. However, since geostatisticians use only γ(h), they took the practice to call the function γ(h) by the term 'variogram'. The assumptions underlying the definition of the variogram are detailed in basic books on Geostatistics (e.g. Isaak & Srivastava, 1989; Kitanidis, 2000; Journel & Huijbregts, 1978). The variogram expresses the half average square difference between two points separated by h. Given a set of i measurements $z(x_1), z(x_2)\ldots z(x_i)$ of a spatial variable z (e.g. transmissivity), the experimental variogram is computed using the following equation (3):

$$\gamma(h) = \frac{1}{2N}\sum_{i=1}^{N}[z(x+h) - z(x)]^2 \tag{3}$$

where N is the number of pairs $[z(x+h), z(x)]$ separated by the distance h.

The plot of the experimental variogram is analysed near the origin and at distances h comparable to the size of the study domain. The behaviour of the variogram near the origin gives information on the presence of variability at the scale of the sampling span. In general a discontinuity at the origin in the experimental variogram is indicative of fluctuations at a scale smaller than the sampling interval (i.e. microvariability). Or it may indicate random observation errors. The behaviour of the variogram at large distances (comparable to the domain scale), determines whether the spatial variable is stationary. For such stationary variable, the experimental variogram should stabilise around a value called the *sill*. The value of h at which the sill is obtained is indicative of the scale at which two measurements of the spatial variable become uncorrelated. This distance is called the *range* (or the correlation length).

The kriging techniques can thereafter be used to perform an estimation of the spatial variable over the whole study domain. Kriging takes into account the spatial structure of the variable, characterized through the experimental variogram, which should previously be adjusted by a theoretical analytical model. Interested readers can find more detailed presentations in the following references (Isaak & Srivastava, 1989; Razack, 1984; Journel & Huijbregts, 1978).

Data processing was performed using software VARIOWIN (Panatier, 1996) and GEO-EAS (Geostatistical Environmental Assessment Software) (Englund & Sparks, 1991).

4 RESULTS AND DISCUSSION

The approach developed in this paper to perform the estimation of the transmissivity of the Bondoukou granite fissured aquifers consists of the following steps: i) logarithmic transform of the available raw transmissivity sample data, T to logT; ii) evaluation of the experimental variogram of logT data; iii) modelling of the logT experimental variogram; iv) estimation of logT using kriging; v) backtransform of logT to T to assess the transmissivity field. Before proceeding to the geostatistical estimation, statistical characteristics of the raw transmissivity data are discussed first.

4.1 *Statistical analysis of raw transmissivity data*

The available raw transmissivity data vary between 6.57×10^{-7} m^2/s and 1.27×10^{-4} m^2/s, spanning several orders of magnitude. Some transmissivity values are isolated compared to the whole data. They were determined from pumping tests at three boreholes located at Koumekara (6.57×10^{-7} m^2/s), Nafenbeni (7.17×10^{-7} m^2/s) and Neguere (1.27×10^{-4} m^2/s). Apart these outliers, the remaining transmissivity data (i.e. 96%) range between 1.08×10^{-6} m^2/s and 5.37×10^{-5} m^2/s. These values are consistent with those met in West Africa and in particular in Cote d'Ivoire in the Baoule-Mossi region at East of the Sassandra fault. In the remainder of this study, the isolated transmissivity values were removed in order to preserve a certain continuity of the data.

As stated above, raw transmissivity values (T) are first transformed into logarithmic values (logT) before undertaking the geostatistical analysis. A summary statistical analysis of logT data is given. Figure 3 shows the frequency distribution of logT. The diagram presents a symmetrical form and can be described suitably by a normal law. A Chi-square test shows that the adjustment of a normal distribution to logT is acceptable at the significance threshold of 10% (Table 1). The fitting models to logT experimental variogram are shown in Table 2. The lognormality of the transmissivity is well recognised in the literature, as well for porous aquifers (Delhomme, 1976; Aboufarissi & Marino, 1984; Ahmed & de Marsily, 1987; Razack & Huntley, 1991) as for fractured aquifers (Huntley *et al.*, 1992; Bracq & Delay, 1997; Fabbri, 1997; Jalludin & Razack, 2004).

4.2 *Assessment of the transmissivity of the Bondoukou aquifers*

It is recommended in the geostatistical literature to perform kriging using normal data rather than data which have skewed distributions. A normal distribution improves the geostatistical estimation (Ahmed *et al.,* 1988). A lognormal transformation is often used to this end (Lasm, 2000; McGrath & Zhang, 2003). In the present case (transmissivity of the granite fissured reservoirs of the Bondoukou Region), the results of the statistical analysis of the raw transmissivity data has clearly shown that the use of a logarithmic transform achieves normality in the transmissivity data. Accordingly the variographic analysis (i.e. calculation of the experimental variogram; adjustment of an analytical function to the experimental variogram) is performed using the logarithmic transform of the data.

4.2.1 *Variogram of logT*

The variogram for the data is shown in Figure 4 and clearly displays, near the origin, a significant *nugget effect* (more than 70% of the total dispersion). At large distances, the experimental variogram fluctuates around the *sill*, indicating that logT is stationary. These characteristic indicate that the data have only a poor spatial relationship. The fitted model indicates a *range* (distance at which the sill is obtained) of several kilometres (10 km). These parameters (nugget effect, sill, range) characterize the spatial structure or *regionalization* of the logarithm of the transmissivity.

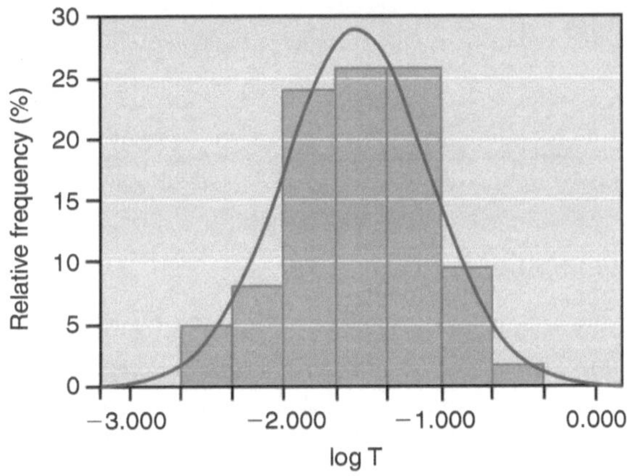

Figure 3. Log T distribution and fitting of a normal law.

Table 1. Chi-square fitting test of a normal law to logT distribution.

Distribution law	Calculated $\chi 2$	Theoretical $\chi 2$ ($\alpha = 10\%$)	Degree of freedom
Normal	1.26	7.78	4

Table 2. Fitting models to logT experimental variogram.

Model	Range a (km)	Sill C (km^2)	Sill-Nugget $C-C_0$ (km^2)	Nugget effect C_0 $(m^2/h)^2$	Average deviation
Spherical	8.96	0.213	0.057	0.156	3.58 E−3
Exponential	10.08	0.213	0.063	0.150	3.36 E−3

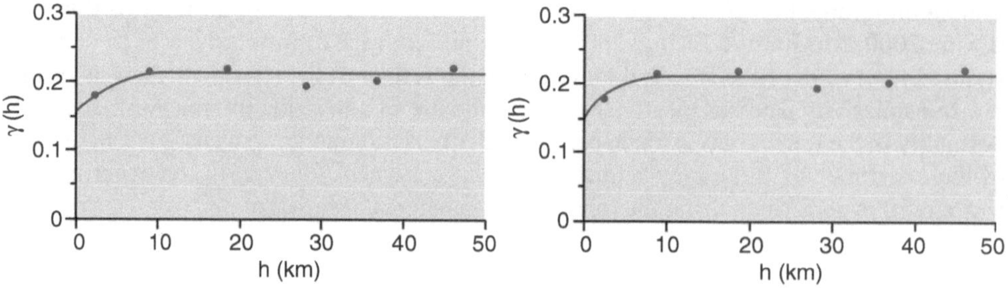

Figure 4. Variogram of logarithmic values of transmissivity. Fitting of a spherical model (left) and exponential model (right).

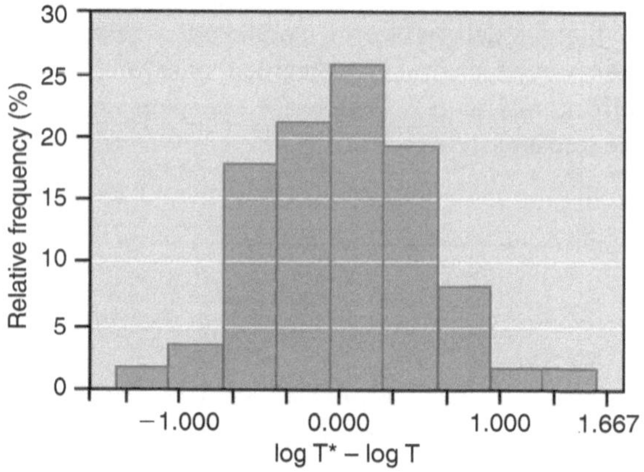

Figure 5. Frequency distribution of logT estimate errors.

4.2.2 *Estimation of the transmissivity of the Bondoukou reservoirs*

The variogram model shown in Figure 4 and described in Table 2, with the available logT sample data (62 values) were used to perform a kriged estimation of logT over the whole domain on a grid with a mesh size of 3 km × 3 km. A cross-validation of the selected variogram model can be carried out in order to check its validity. This procedure is described in details in (Clark, 1986; Gascuel-Odoux *et al.*, 1994; Razack & Lasm, 2006). Its purpose is to analyse the estimation errors. In fact, the procedure eliminates a single value (Z_i) from the data set and perform a kriged estimation (Z_i^*) at this location. This is repeated for all the data set. If the variogram model is valid, then the following results should be verified: the average of the actual errors (Me) should be zero and their variance (σ_e^2) should be a minimum; the ratio of the variance of the actual errors to the average kriging variance ($\sigma_e^2/\sigma_K{}^2$) should be one. The values found in this study are: Me $= -0.03$ and $\sigma_e^2/\sigma_K{}^2 = 0.92$. The average of the actual errors (Me) and the ratio $\sigma_e^2/\sigma_K{}^2$ are respectively close to 0 and 1 and are in conformity with the theoretical values. It can be concluded that the exponential variogram model is valid. The frequency distribution of the estimates errors is shown on Figure 5. The frequency distribution is symmetrical and can be described by a normal law, which again is an indication of the model validity.

The kriged logT map is used afterwards to make an estimation of the transmissivity over the study area through a backtransform of logT to T. Some noteworthy points should be

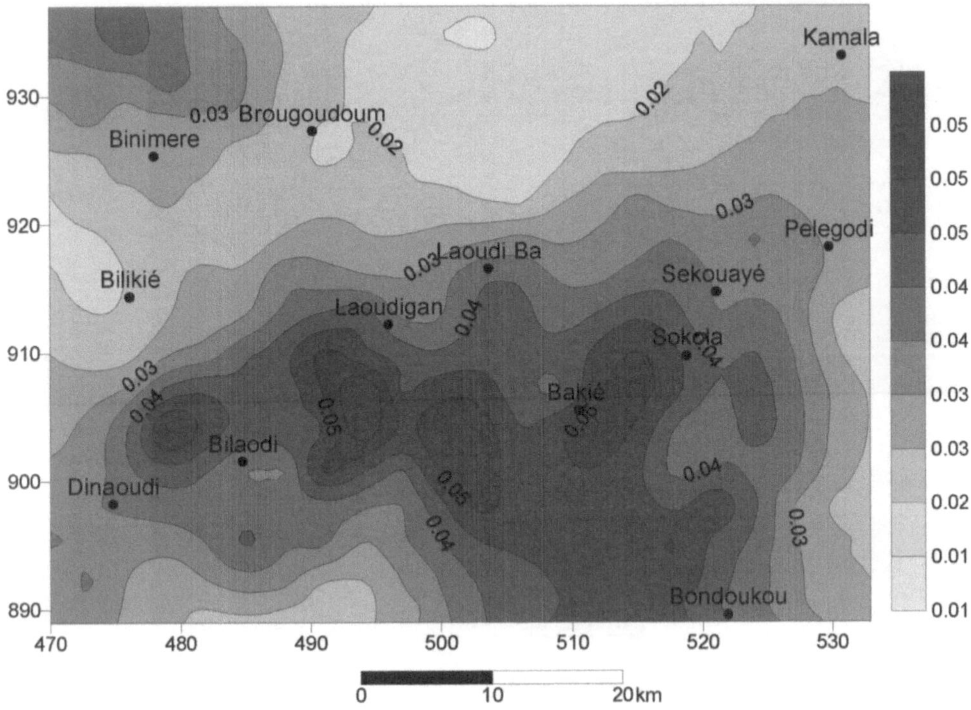

Figure 6. Transmissivity estimation map obtained by a backtransform logT to T.

emphasized at this stage. The logarithmic transform T to logT is a non-linear transform. In this case, when the kriged unbiased estimates (logT) are backtransformed, then their unbiasedness property is lost. The transmissivity values estimated using this procedure is no longer unbiased. The estimated transmissivity values are mapped on Figure 6. This geostatistical procedure makes it possible to obtain an assessment of the transmissivity over the whole study domain.

The kriging procedure also calculates the estimation variance (or its square root the estimation standard deviation). The estimation standard deviation map presents zones of weak and strong values Figure 7). This map is an indication about the quality of the estimates. The smaller the standard deviation value, the higher the accuracy of estimates. Highest values of the standard deviations of estimate are found towards the limits of the study area. These high values can be explained by the skin effect and by the presence of zones where data points are lacking.

5 DISCUSSION

The range of the experimental variogram of logT is equal to 10 km. This distance, which is an indication of the scale of the spatial structure of the variable at hand (logT) is significant. It means that two points in the field separated by a distance less than the range remain correlated with each other. This large correlation length may be explained by patterns of weathering in the study domain. Similar studies undertaken in various areas of Cote d'Ivoire

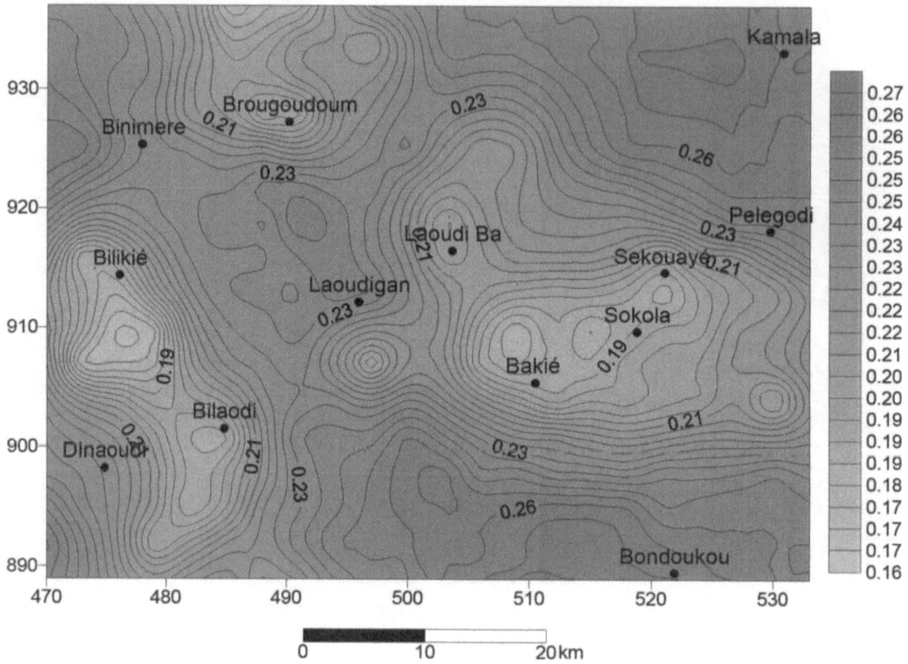

Figure 7. logT kriging standard deviations map.

(Lasm, 2000; Soro *et al.*, 2001, Jourda, 2005; Razack & Lasm, 2006) gave comparable results. The following values for the variogram ranges were found: a = 6.6 km in western Cote d'Ivoire; a = 8.4 km in the south.

Of significance in this study is the large nugget effect, which indicates that the spatial correlation between individual boreholes is weak, even at short distances.

These statistics indicate that on a statistical viewpoint, both observed and estimated transmissivity data sets are similar. This result is quite interesting and confirms the robustness of the estimation. In the study domain, the spatial distribution of observed and estimated values of transmissivities is also similar. The geostatistical procedure produces satisfactory distribution of high and low values of transmissivity in relation with the available knowledge. On this basis, the estimates of the transmissivity of the Bondoukou region proved coherent and are acceptable.

6 CONCLUSIONS

This study aimed to estimate the transmissivity of the highly fractured aquifer of Bondoukou in North-Eastern Cote d'Ivoire. The main findings of this study:

1. Transmissivity values are spread over several orders of magnitude, revealing the strong heterogeneity of the aquifer. Measurements of transmissivity are log-normally distributed.
2. The variographic analysis indicates that logT is a stationary variable, characterised by a significant nugget effect (discontinuity at the origin of the variogram due to

micro-regionalization and/or measurement errors). The data have been modelled with a range equal to 10 km.

The range, which is an indication of the scale of the spatial structure of the variable at hand (logT) is significant. It means that two points in the field separated by a distance less than the range remain correlated (albeit weakly) with each other. This large correlation length may be explained by patterns of weathering or fracturing in the study domain. Similar studies undertaken in various areas of Cote d'Ivoire (Lasm, 2000; Soro *et al.*, 2001, Jourda, 2005; Razack & Lasm, 2006) gave comparable results. The following values for the variogram ranges were found: a = 6.6 km in western Cote d'Ivoire; a = 8.4 km in the south.

Of significance in this study is the large nugget effect, which indicates that the spatial correlation between individual boreholes is weak, even at short distances. This emphasises the importance of fractures in supplying these boreholes. Two boreholes drilled 30 m away from each other may intercept different fractures, and therefore the aquifer properties may be unrelated.

Despite the lack of strong spatial correlation, the geostatistical procedure developed in this paper proved useful to understand variations in transmissivity. By including a large nugget effect, kriging has reliably reproduced the measured data and provided a tentative estimate of the transmissivity of the Bondoukou fractured aquifers. This significant improvement of the knowledge about these aquifers is a basic stage that will permit to undertake further tasks regarding the modelling of these aquifers. The overall goal is to elaborate a representative numerical tool for a sustainable management of these groundwater resources.

ACKNOWLEDGEMENTS

The authors would like to thank two anonymous reviewers for their valuable and helpful comments on this paper.

REFERENCES

Aboufirassi, A. & Marino, M. A. 1984. Cokriging of aquifer transmissivity from field measurements of transmissivity and specific capacity. Mathematical Geology, **16** (1), 19–35.

Ahmed, S. & de Marsily, G. 1993. *Cokriged estimation of aquifer transmissivity as an indirect solution of the inverse problem: A practical approach.* Water Resources Research, **29** (2), 521–530.

Ahmed, S. & de Marsily, G. 1987. *Comparison of geostatistical methods for estimating transmissivity using data on transmissivity and specific capacity.* Water Resources Research, **23** (9), 1717–1737.

Ahmed, S., de Marsily, G. & Talbot, A. 1988. *Combined use of hydraulic and electrical properties of an aquifer in a geostatistical estimation of transmissivity.* Ground Water, **26** (1), 78–86.

Biémi, J. 1992. *Contribution à l'étude géologique, hydrogéologique et par télédétection des bassins versants sub-saheliens du socle précambrien d'Afrique de l'Ouest: Hydrostructurale, hydrodynamique, hydrochimie et isotopie des aquifères discontinus de sillons et aires granitiques de la Haute Marahoué (Côte d'Ivoire).* Thèse de doctorat ès Sciences Naturelles, Univ. Abidjan, Côte d'Ivoire, 493pp.

Bracq, P. & Delay, F. 1997. *Transmissivity and morphological features in chalk aquifer: a geostatistical approach of their relationship.* Journal of Hydrology, 191, 139–160.

Castany, G. 1982. *Principles et méthodes de l'hydrogéologie.* Dunod Université, Paris.

Chauvet, P. 1994. *Aide-mémoire de géostatistique linéaire*. Cahiers de Géostatistique, Ecole des Mines de Paris, Fontainebleau, France, 210pp.

Clark, I. 1986. *The art of cross validation in geostatistical applications*. Proc. 19th International APCOM Symposium, Penn. State Univ., USA, April 1986, 211–220.

Delhomme, J. P. 1976. *Application de la théorie des variables régionalisées dans les sciences de l'eau*. Thèse de Docteur-Ingénieur, Ecole des Mines de Paris, Fontainebleau-Université de Pierre et Marie-Curie, France, 160pp.

Delhomme, J. P. 1978. *Kriging in hydrosciences*. Advances in Water Resources, **1** (5), 251–266.

Engalenc, M. 1981. *L'eau souterraine dans les roches cristallines de l'Afrique de l'Ouest*. Géohydraulique, **3**.

Englund, E. & Sparks, A. 1991. *GEO-EAS, Geostatistical Environmental Assessment Software*. US EPA Report 600/8-91/008, 186.

Fabbri, P. 1997. *Transmissivity in the geothermal Euganean Basin: A geostatistical analysis*. Ground Water, **35** (5), 881–887.

Fahy, J. C. 1981. *Hydraulique villageoise en Côte d'Ivoire. Situation au 30 septembre 1979*. Bulletin of BRGM, Sér. II, Sect. III, **4**, 327–333.

Gascuel-Odoux, C., Bovin, P. & Walter, C. 1994. *Eléments de géostatistiques*. In: Des processus pédologiques. Ed. Actes, 217–247.

Huntley, D., Nommensen, R. & Steffey, D. 1992. *The use of specific capacity to assess transmissivity in fractured rock aquifers*. Ground Water, **30** (3), 396–402.

Isaaks, E. H. & Srivastava, M. R. 1989. *An introduction to applied geostatistics*. Oxford University Press, Oxford, UK, 561pp.

Jalludin, M. & Razack, M. 2004. *Assessment of hydraulic properties of sedimentary and volcanic aquifer systems under arid conditions in the Republic of Djibouti (Horn of Africa)*. Hydrogeology Journal, **12**, 159–170.

Jourda, J. P. 2005. *Méthodologie d'application des techniques de télédétection et des systèmes d'information géographique à l'étude des aquifères fissurés d'Afrique de l'ouest. Concept de l'Hydrotechniquespatiale: cas des zones tests de la Côte d'Ivoire*. Thèse de Doctorat ès Sciences Naturelles, Université d' Abidjan, Côte d'Ivoire, 429pp.

Journel, A. G. & Huijbregts, C. J. 1978. *Mining geostatistics*. Academic Press, New York, 600pp.

Kitanidis, P. K. 2000. *Introduction to Geostatistics. Application in Hydrogeology*. Cambridge University Press, Cambridge, UK.

Kruseman, G. & de Ridder, N. A. 1973. *Analysis and evaluation of pumping tests data*. ILRI Press Wageningen, The Netherlands, 230pp.

Kouamelan, A. N. 1996. *Géochronologie et géochimie des formations archéennes et protérozoïques de la dorsale de Man en Côte D'Ivoire. Implications pour la transition Archéen-Protérozoïque*. Thèse Université d'Rennes (France), 284pp.

Lasm, T. 2000. *Hydrogéologie des réservoirs fracturés de socle: Analyse statistique, et géostatistique de la fracturation et des propriétés hydrauliques*. Application à la région des Montagne de Côte d'Ivoire (Domaine archéen). Thèse Université d'Poitiers (France), 274pp.

Lasm, T., Kouamé, F., Oga, M. S., Jourda, J. P., Soro, N. & Kouadio, B. E. 2004. *Etude de la productivité des réservoirs fracturés des zones de socle. Cas du noyau archéen de Man-Danané (Ouest de la Côte d'Ivoire)*. Revue Ivoirienne des Sciences et Technologie, **5**, 97–115.

McGrath, D. & Zhang, C. 2003. *Spatial distribution of soil organic carbon concentrations in grassland of Ireland*. Applied Geochemistry, **18** (10), 1629–1639.

Pannatier, Y. 1996. *VARIOWIN: Software for Spatial Data Analysis in 2D*. Springer- Verlag, New York, USA.

Razack, M. 1984. *Application des méthodes numériques à l'identification des réservoirs fissurés carbonatés en hydrogéologie*. Thèse Doct ès Sci Univ Montpellier, France, 384pp.

Razack, M. & Huntley, D. 1991. *Assessing transmissivity from specific capacity in a large and heterogeneous alluvial aquifer*. Ground Water, **29** (6), 856–861.

Razack, M. & Lasm, T. 2006. *Geostistical estimation of the transmissivity in a highly fractured metamorphic and crystalline aquifer (Man-Danane Region, western Ivory Coast).* Journal of Hydrology, **325** (1-4), 164–178.

Roth, C., Chilès, J. P. & Fouquet, C. 1996. *Adapting geostatistical transmissivity simulations to finite difference flow simulators.* Water Resources Research, **32** (10), 3237–3242.

Roth, C. & Chilès, J. P. 1997. *Modélisation géostatistique des écoulements souterrains: comment prendre en compte les lois physiques.* Bull BRGM Hydrogéol. **1**, 23–32.

Savané, I. 1997. *Contribution à l'étude géologique et hydrogéologique des aquifères discontinus du socle cristallin d'Odienné (Nord-Ouest de la Côte d'Ivoire). Apports de la télédétection et d'un Système d'Information Hydrogéologique à Référence Spatiale (S.I.H.R.S.).* Thèse Doctorat ès Sciences Naturelles, Université d'Abidjan, Côte d'Ivoire, 386pp.

Savané, I., Doumouya, I. & Doumbia, L. 1997. *Une approche à partir de modèles statistiques pour la détermination de la productivité des puits en contexte de socle cristallin dans la région d'Odienné (Côte d'Ivoire).* Hydrogéologié., **4**, 19–26.

Sinan, M. & Razack, M. 2005. *Management of regional aquifers using a combined procedure based on geostatistical and GIS tools (Haouz Groundwater of Marrakech, Morocco).* Proc. European Water Resources Assoc. Conf., Menton (France), Sept 2005.

Soro, N. 1987. *Contribution à l'étude géologique et hydrogéologique du Sud-Est de la Côte d'Ivoire.* Bassin versant de la Mé. Thèse 3ème cycle, Univ. Grenoble 1, Inst. Dolom., France, 218pp.

Soro, N., Savané, I., Ouattara, A. & Fofana, S. 2001. *Approche géostatistique de la variabilité spatiale des écoulements souterrains dans les aquifères du Sud-Ouest de la Côte d'Ivoire.* Revue Bioterre, **2** (1), 85–100.

Tarboton, K. C., Wallender, W. W., Fogg, G. E. & Belitz, K. 1995. *Kriging of regional hydrologic properties in the western San Joaquin Valley, California.* Hydrogeology Journal, **3** (1), 5–23.

Taupin, J. D., Amani, A., Lebel, T. 1998. *Variabilité spatiale des pluies au Sahel: une question d'échelles. 1. Approche expérimentale. Water Resources in Africa during the 20th Century.* IAHS Publications, **252**, 143–155.

CHAPTER 29

Technical note: Hydrogeological mapping in Africa

W. Struckmeier
BGR, Geozentrum Hannover, Hannover, Germany
Chairman of the IAH Commission on Hydrogeological Maps (COHYM)

ABSTRACT: This short note describes the context of regional and continent wide hydrogeological mapping in Africa. In particular it introduces the new hydrogeological map for Africa derived from the Worldwide Hydrogeological Mapping and Assessment Programme (WHYMAP, 2008).

Hydrogeological mapping in Africa has got a rather long history. The first hydrogeological maps in the modern style were produced in Northern Africa (Morocco) together with a general legend which provided a methodological guideline for creating hydrogeological maps (Ambroggi & Margat, 1960). This led to a new type of map legend that paved the way for the International Legend for Hydrogeological Maps issued by UNESCO, the International Association of Hydrogeologists (IAH) and the Commission for the Geological Map of the World (CGMW), (Struckmeier & Margat, 1995).

The African continent has undergone various phases of hydrogeological mapping in different scales and continental groundwater related maps in colour were produced in various formats, such as:

- Two Africa maps (scale 1:20 million) annexed to the UNTCD publication "Groundwater in North and West Africa, 1988 (Major Hydrogeological Formations and Groundwater Resources of Africa);
- International Hydrogeological Map of Africa, five sheets at the scale of 1:5 million plus one legend sheet, (OACT, 1986–1992).

A draft for a new digital hydrogeological map of Africa, scale 1:5 million, has been prepared recently by the French Geological Survey (BRGM) and a number of African partners, as a contribution of the African Geo-Information System (SIG Afrique).

In addition, a number of regional hydrogeological maps have been prepared or are under preparation:

- Hydrogeological Map of the Arab Region and adjacent areas, scale 1:5 million (ACSAD 1988);
- Carte de potentialité des ressources en eau souterraine de l'Afrique Occidentale, scale 1:5 million (BRGM 1986);
- Hydrogeological Map of the SADC Region (in progress).

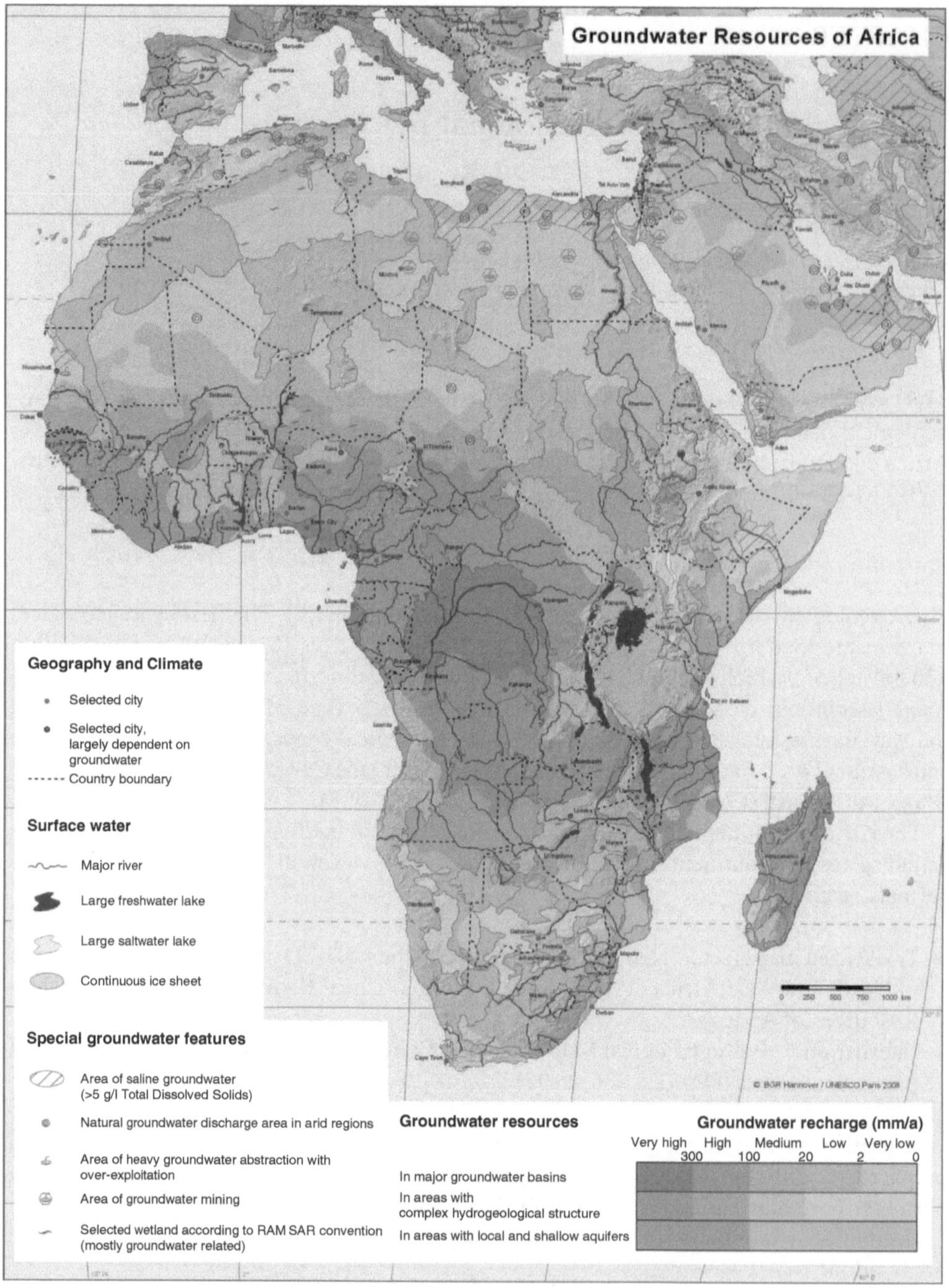

Figure 1. A scan showing a newly compiled hydrogeological map of Africa, developed under the WHYMAP programme (WHYMAP, 2008).

The groundwater conditions of the African continent vary greatly, as shown on Figure 1, derived from the Worldwide Hydrogeological Mapping and Assessment Programme (WHYMAP, 2008). The geology and thus the hydrogeological typology of Africa is principally characterised by old basement outcrops (most of the areas shown in brown colour), sedimentary aquifer basins (blue or green, depending on the complexity of their structure), young volcanic effusive rocks (green, mainly in East Africa), as well as huge desert sand areas and important alluvial aquifers (blue).

This great variety of aquifer types is subjected to a typical climate pattern decisive for the rainfall and recharge conditions within the African continent. The light colours of blue, green and brown highlight those climatic zones in which the rainfall is very unreliable and the average annual recharge from rainfall is below 20 litres per square metre (<20 mm/a). Hence, the aquifers in these zones are prone to groundwater mining. They should only be used under socio-economic conditions of drought and poverty, and their exploitation must be based on a solid operation plan and an exit strategy for reducing groundwater availability.

Most of the natural groundwater basins and complex aquifer structures in Africa are segmented by political boundaries, which make them trans-boundary units. A first appraisal executed by UNESCO yielded about 40 of such trans-boundary aquifers, but new, more detailed inventories of the trans-boundary aquifers have been started recently in Western and Southern Africa by the African groundwater surveys under the umbrella of UNESCO and IAH (Internationally Shared Aquifer Resources Management/ISARM). This has already raised the number of trans-boundary aquifer systems to more than sixty.

REFERENCES

Ambroggi, R. & Margat, J. 1960. *Légende générale des cartes hydrogéologiques du Maroc.* IAHS, **50**, Gentbrugge, Belgium.

ACSAD 1988. *Hydrogeological Map of the Arab Region and adjacent areas 1:5,000,000.* Arab Centre for the Studies of Arid Zones and Dry Lands, Damascus, Syria.

BRGM & Geodynamique 1986. Carte de potentialité des ressources en eau souterraine de l'Afrique Occidentale 1:5,000,000, with explanation. BRGM, Orléans, France.

OACT 1986–1992. *International Hydrogeological Map of Africa 1.5,000,000.* (5 map sheets and explanatory notes; 1 legend sheet). Organisation Africaine de Cartographie et de Télédétection, Alger, Algeria.

Struckmeier, W. F & Margat, J. 1995. *Hydrogeological Maps. A Guide and a Standard Legend.* IAH International Contributions to Hydrogeology, **17**, Hannover, Germany.

WHYMAP 2008. Groundwater resources of the world. BGR/UNESCO. accessed at: http://www.whymap.org .

United Nations 1988. Ground Water in North and West Africa. Natural Resources/Water Series **18**, New York, US.

Author index

Subject index

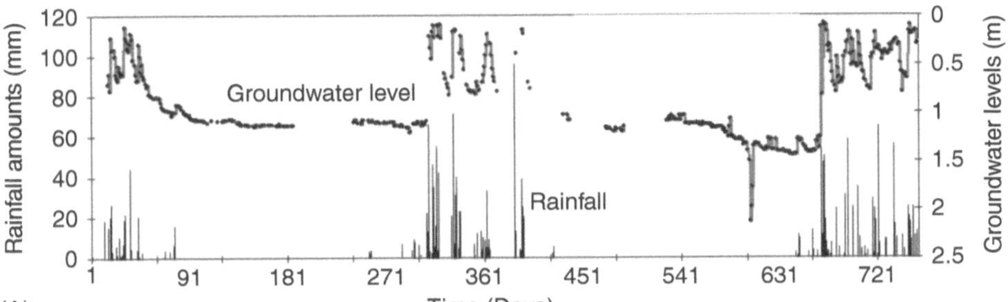

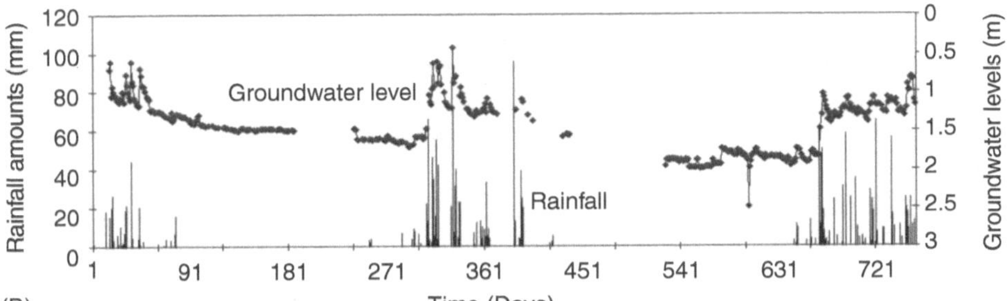

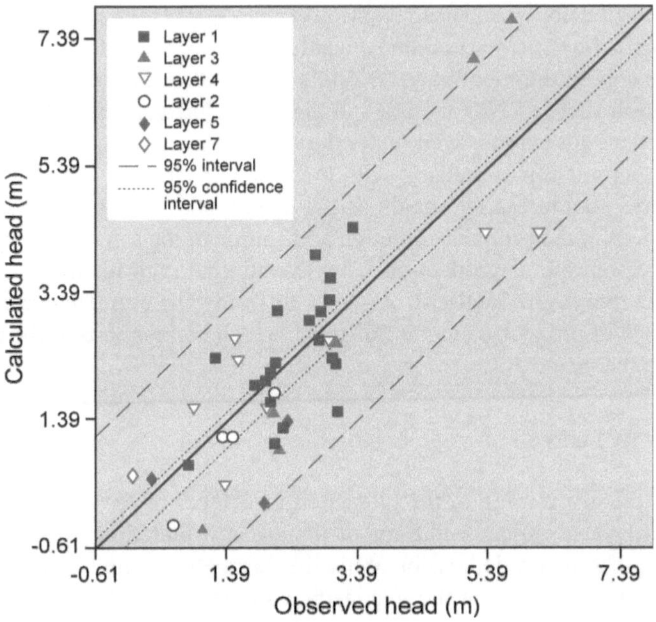

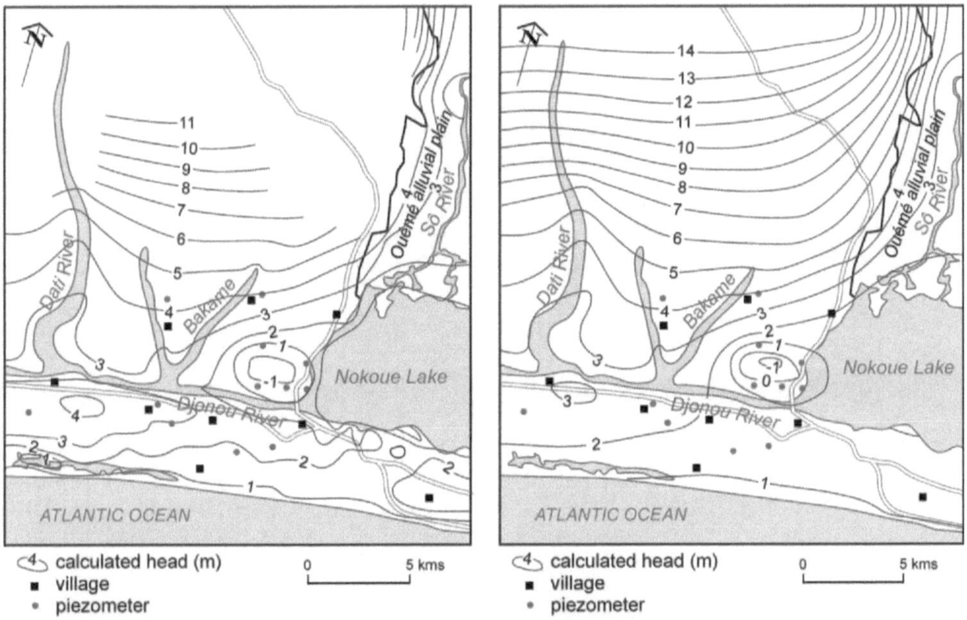

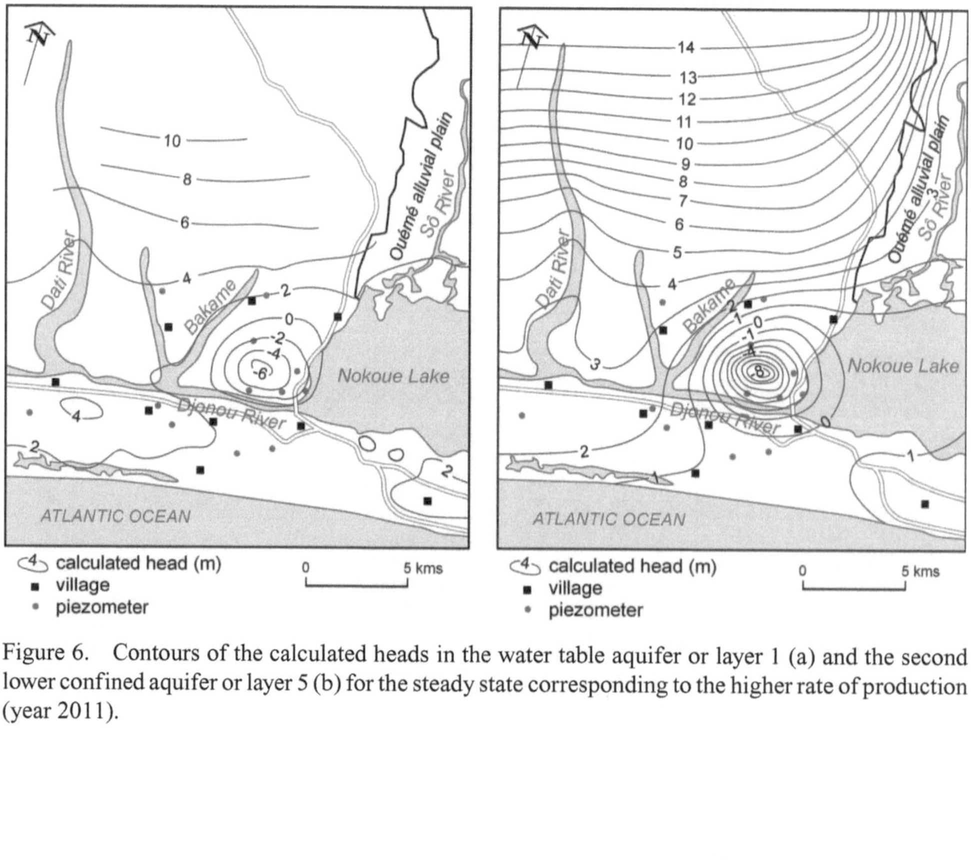

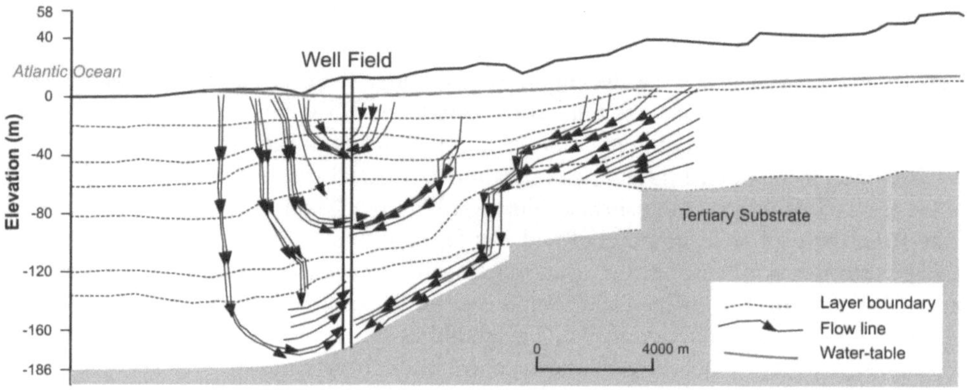

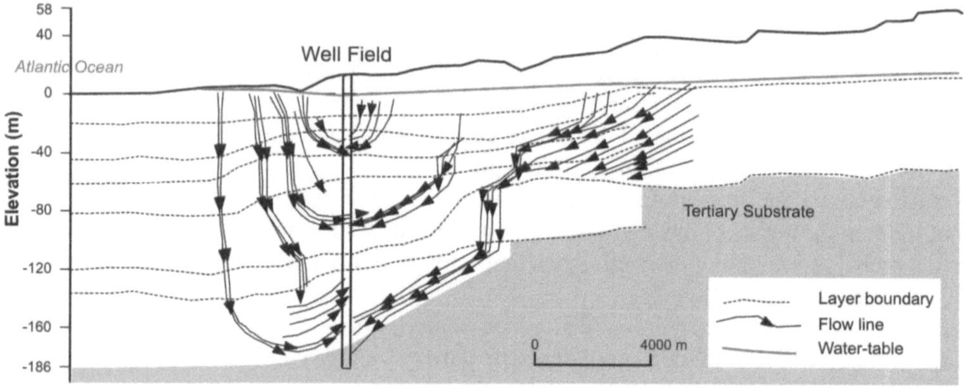

